The Global and the Local

The Global and the Local

An Environmental Ethics Casebook

By

Dale Murray

BRILL

LEIDEN | BOSTON

Cover illustration: Luisita Certain Murray.

Cataloging-in-Publication Data is available online at http://catalog.loc.gov
LC record available at http://lccn.loc.gov/2016057395

Typeface for the Latin, Greek, and Cyrillic scripts: "Brill". See and download: brill.com/brill-typeface.

ISBN 978-90-04-33998-9 (paperback)
ISBN 978-90-04-33999-6 (e-book)

This paperback is also published in hardback under ISBN 978-90-04-33997-2.

This book is printed on acid-free paper and produced in a sustainable manner.

To SD, for all of your love, charm, and dedication.
To LCM, just for being you.

Contents

Acknowledgements

From the first inklings that I wanted to write a casebook to its completion now, the work found here has taken about ten years, albeit with various fits and starts. As can be imagined of any work with such a long gestation period, there are countless people to thank. I shall do my best to give credit where credit is due, hopefully without any serious omissions.

First of all, the publishing team at Brill deserves tremendous credit for providing their patient labor in bringing this book to press. I wish to wholeheartedly thank Jennifer Pavelko, the philosophy acquisitions' editor at Brill, for beginning the process, securing reviewers to vet an earlier version, and essentially "greenlighting" the project. I would also like to thank her assistant, Meghan Connolly, and the production editor, Michael J. Mozina, for shepherding the work the remainder of the way to its completion. I am very grateful to an anonymous reviewer for his/her careful reading of the manuscript and for offering several excellent suggestions for improvement.

For research support, I am in deep debt to the library staffs at the University of Wisconsin-Baraboo/Sauk County, the University of Wisconsin-Richland, and the University of Wisconsin-Madison. In particular, Ann Vogl and Cate Booth at UW-BSC were extremely helpful in locating source materials, as were too many reference librarians to name at UW-Madison. Their cheerful and efficient assistance made research much easier than it ought to have been and is much appreciated. I also must thank Kana Ewing for his diligently assistance with the index.

I am extremely fortunate to have so many fine colleagues throughout the UW Colleges Philosophy Department. I am amazed at their enthusiastic support and dedication despite difficult times for our institution. Mark Peterson deserves special recognition as he carefully read and astutely commented on segments of the work. Evan Kreider gave me the sage idea of including both study questions at the beginning of each case study and review questions at the end. Roger Rigterink alerted me to take a serious look at de-extinction, and Dan Putman encouraged me to tackle the issue of which species should be saved based on their ecological worth and in the face of scarce resources. Dean Kowalski offered me useful tips on how to solicit potential publishers and pitch the casebook. I thank them all for their greatly helpful suggestions. My colleague in sociology at UW-BSC, Annette Kuhlmann, not only shared pots of coffee, but also read through the "Walleye War" case study and gently steered me in some places where I went off track. I am very grateful for the encouragement all have shown for this project and am so lucky to have colleagues and friends whose great minds are only exceeded by extraordinarily generous hearts. Any mistakes unearthed in this book are mine alone.

As this is meant to be a teaching text, I should thank the students of my two most recent environmental ethics classes at UW-Baraboo (Fall 2015) and UW-Richland (Spring 2016). They were my willing guinea pigs in this endeavor, reading through the case studies, offering insight, constructive criticism, and deeply appreciated encouragement. Derek Carden, Marc Longo, Malachi Persche, Kana Ewing, and Wayne Gilmour went the extra mile to give me specific suggestions for improvement and even some intriguing possible cases for future editions. The suggestions from all of the students in these classes have made this book much better than it otherwise would have been.

Since perhaps the greatest commodity for academics is time, I would be greatly remiss if I failed to thank the UW System, UW Colleges, and my department for granting me a year's sabbatical leave over the 2014–2015 academic year. This period of uninterrupted time to write and reflect allowed me the opportunity to complete the vast majority of this casebook. Also, the Wisconsin Institute for Public Policy and Service at the University of Wisconsin-Marathon awarded me a grant in 2012, providing me support to investigate the ethical implications of frac sand mining in Wisconsin. The WIPPS director, Eric Giordano, deserves special thanks for facilitating a pair of talks on the topic at the WIPPS Center on the UW-Marathon campus. Three departmental colleagues there, Mark Brown, Doug Hosler, and Jason Zinzer all raised excellent questions at the lectures and gave warm support for the project. The fruits of that work can be found in some of the case studies in this book.

I thank Cindy McVenes and the UWBSC Campus Culture and Community Committee for inviting me to discuss frac sand mining as part of their Distinguished Lecture Series. Jessica Laeseke, the Regional Director of Continuing Education (and my former student!) also graciously asked me to give a similar talk for the "Love of Learning" lecture series at the UW-Richland campus. My former Dean at UW-BSC, Thomas Pleger, even put me in touch with family members living in areas with heavy frac sand mining activity, which allowed me the opportunity to learn more about the topic "on the ground."

Along with the aid of many colleagues, tangible and intangible contributions of family lay down the foundation for completing long projects such as this. I am deeply appreciative and consider myself very fortunate to have an extremely nurturing support group. My mother, Frances Murray and father, Raymond Murray (both of whom recently passed), always provided loving personal support even if they didn't fully understand my work. My father-in-law, Phil Certain, unfailingly sent along articles he gleaned from newspapers and magazines relating to environmental topics. He and my mother-in-law, Melinda Certain, provided encouragement, insight, and childcare—all ingredients that help greatly both with writing books and living life. Finally, my greatest loves and inspirations; my wife, Heather, and my daughter, Luisita, deserve the most gratitude. They both remind me daily of the most important things in life. Without their loving support none of this work would be possible, nor would it have any meaning even if it were.

Introduction

Perhaps one of the most troubling and controversial issues facing us today is how we ought to interact with our environment. Many scientists warn that the Earth is warming at an alarming rate and most likely this is due to human activities. A good deal of the world's economy is based upon agricultural, transportation, mining, and drilling practices whose by-products (and sometimes end products) are highly toxic. Ultimately, the technologies upon which the current economy relies are not particularly efficient—in the extraction of fuel, our overall input of energy is much greater than the actual output. For any product we manufacture, a great deal of time and energy must be expended in its creation. Once more, much of the energy used in this unclean, inefficient process is nonrenewable. Consider as well the number of species that have gone extinct and continue to vanish at astonishing rates, again correlated with our business and recreational practices. This approach to how we work and play on Earth, claim an increasing number of natural and social scientists, philosophers, politicians, and common citizens, is unsustainable. Without radical alteration of our lifestyles and dramatic changes in our economic policies, many believe we head to our own destruction.

Notice that so far all of these are *descriptive* claims (simply concerning the way things *are*). They are subject to critical scrutiny, but will ultimately be decided by scientific investigation (or more worryingly, political rhetoric). Some believe the sky is falling and others charge they are merely being Chicken Littles. In the midst of this debate, come a bevy of *normative* questions (that is, questions about what we *ought* to do). Human societies have a strong tendency toward harboring conflicting interests. While clean water and air are certainly social goods, so are money and readily accessible products. Legitimate governments, usually seen as stewards of rights and interests, are charged with supporting these often-competing claims. What should be done when there are only so many resources (such as time, effort, and money), yet so many genuine interests to nurture?

Of course, there is also the possibility that there are no obligations to other human beings. But this is a skeptical position that doesn't hold up well to close examination. With that said, even if we assume there are obligations to other persons, only half the battle has been fought. There could be quite justifiable needs that cannot be fully met because other pressing ones are being attended to at the moment. Giving priority to some needs over others in a society with many vying interests is perhaps the most daunting task faced by human institutions.

 | DOI 10.1163/9789004339996_002

Even assuming there are perils in continuing our current way of living, we have to consider why we ought to change our behaviors. Ought we to curtail environmental devastation because of other people? Assuming we have obligations to other people (which could be quite minimal), does this mean we have a duty not to pollute so as to offer a high quality or even barely adequate environment to them? But to what extent do we owe such obligations? Are there not already legal mechanisms available to stop some from harming others? If you can show I have harmed your health by depositing toxic wastes that spoil your land and cause you cancer, you have grounds for suing me. But surely, until I have done you some demonstrable harm, I should not be stopped from engaging in activities allowing me to make money or enhance my quality of life.

Perhaps plausible replies to these questions and concerns already exist. Maybe we should care about the effects of polluting on other people (especially if we know it will have some adverse effects on them), simply because there is something morally problematic about consciously creating dissatisfaction to other people. One might also appeal to what is known as the *precautionary principle* as a reason for not posing risks to others. The precautionary principle relies upon the idea that in the face of uncertainty of whether engaging in some activity might cause harms to others (especially if they are severe), the burden of proof is upon the actor to show that said activity will *not* pose a threat. While this might be a promising strategy, advocates of the precautionary principle need to give justifications for its adoption and also show how it applies in each controversial case.

Once more, of what concern is it to other currently living humans if I engage in activities that greatly reduce or have adverse effects on *non-human* animal or *plant* species? Now one possibility is that some humans enjoy having animals and/or plants around. They desire the experience of observing other species—indeed a good deal of ecotourism is based on highlighting entertaining, majestic, and/or exotic animals and plants. These appear to be significant *aesthetic* reasons for not destroying flora or fauna (that is, reasons which are matters of taste or etiquette). But on the other hand, are these compelling enough *moral* reasons, if they are moral reasons at all? Why should *I constrain* my polluting activities (perhaps restricting my ability to make a livelihood) because of *your preferences* for beauty or elegance? After all, we don't normally think that aesthetic judgments, such as whether chocolate tastes better than vanilla can be shown to be correct or incorrect. And even if we *could* do so, it is not at all clear we mean that chocolate would be *morally* better or worse than vanilla, or could even know what that would mean.

But the challenge may be greater than this. Some argue the reason why I am supposed to refrain from activities that may benefit me, but are environmentally harmful, is due to their *potential* for harming other humans. However, is it plausible to say that this is an either/or proposition? Is there a certain limit up to which I can pollute and still not cross the line of harming or even potentially causing you harm? If we use the criterion of *potentially causing you harm*, then we need to be prepared to restrict many activities of humans. Are we prepared to accept the perhaps worse consequences of restricting the *liberty* of some (perhaps many) for the benefit of others? Furthermore, is it *ever* justifiable to restrict the liberty of anyone (especially if their activities benefit them) to avoid the *risk* of harming others?

Another possibility is that I shouldn't destroy the environment because we should worry about those who come after us. One might think it is selfish and shortsighted to degrade the environment now at the expense of other humans who will live here on Earth later. Correspondingly, in the environmental ethics literature, a great deal is made of *future generations*—those individuals who come after our current generation of humans. With respect to future generations, we can think of individuals who already exist (children, grandchildren, great grandchildren), but in the context of environmental ethics we usually think of those who are not yet born. Now it is easily conceivable that current adults have obligations to their own children. But it may become increasingly hard to think we have obligations to *other people's children* or at least to know what the content of those obligations might be. More to the point, what duties do I have to *other people's future offspring*?

Additionally, there is a problem with respect to obligations we may have to members of what are called future generations, for they do not yet exist. Normally we think we owe obligations to those who have *rights*. For instance, if you are an innocent person who has the right to life, I at least think I have the minimal obligation not to kill you. But, we are assuming in this case that you are a living, breathing being. I may respect your right to life by virtue of any number of reasons: you are a creature capable of rational thought, or capable of feeling pleasure and pain, or are simply human, or simply alive. Notice though that for members of future generations, there *no existing beings* whose rights are to be respected or whose interests can be known precisely. This is problematic. It is not clear non-existent entities can have rights, and consequently, it is not clear why I have any obligations to future generations. It is already controversial that human embryos have rights, let alone fertilized human ova. It would seem to be even more debatable that "potential" human beings could possibly have such rights. *Prima facie* (i.e., on the surface), it would appear to be very difficult to maintain that those not yet born are owed greater obligations than fertilized human eggs, oocytes, fetuses, and embryos. Consequently, those who think we should protect the environment and not deplete its resources for the sake of future generations are going to need to develop a robust case for their position.

But perhaps we have duties that go beyond merely those to humans or even to those potential humans who are not yet born. Maybe I should cease and desist from harming the environment because non-human animals are endangered by my activities. There is little question that human activities have adversely affected fauna. The number of extinct animal species has increased dramatically: almost in all cases due to overhunting or overfishing them, or perhaps more often due to destruction of their habitats. But is there anything wrong with this? As noted above, it is possible that the destruction of species of non-human fauna and flora could someday harm humans, but is there anything in and of itself wrongful about exterminating all penguins, for example? We clearly do something to the penguins, but is it unjust? Do we violate the rights of penguins? Can penguins have rights? What does it mean for something that can't make claims (if penguins have a language, we surely can't understand it) to have rights?[1] This would even seem to go further in the case of plants. It may be plausible to defend the idea that animals should be taken into moral consideration by appealing to their sentience—that is, to argue they feel pleasure and pain, and thus have interests in not being harmed.[2] Yet, it is not at all reasonable to contend that plants should be saved for the same reason. Hence, on what grounds do we think we have obligations to flora? Certainly, it wouldn't be on some basis of rights.

All I have said here should not be construed as hostile to the idea that we ought to take seriously a number of environmental issues, nor do I suggest there is no obligation at all of humans to non-human animals, plants, or the environment generally. But it *does* point to the moral challenge that faces those who advocate for environmental reform. If we are to expect people to actively engage the environmental issues of our time and in the future, a convincing case must be made to do so. Only through presenting strong arguments in favor of treating the environment in a way quite different than humans have historically is change likely to come.

---

1 See William Baxter, *People or Penguins: The Case for Optimal Pollution* (New York: Columbia University Press, 1974).

2 See Peter Singer, *Animal Liberation* (New York: New York Review/Random House, 1975). Also see Joel Feinberg, "The Rights of Animals and Unborn Generations," in his *Rights, Justice and the Bounds of Liberty: Essays in Social Philosophy* (Princeton, NJ: Princeton University Press, 1980).

PART 1

Theory and Applied Theory

Before we can delve into cases in environmental ethics, we need to say something about ethical theories more generally and how to apply them properly. Broadly speaking, *ethical theory* is the study of right and wrong.[1] As mentioned earlier, normative ethical theory concerns what should or ought to be the case. Some of the questions it addresses are: How should I conduct myself? What actions should I commit or not commit? What moral rules ought I to follow? What sort of person should I be? There is certainly overlap in the ways we can answer these questions. For instance, we may find that I should be the type of person who adheres to morally permissible rules and principles, whatever they may be. Moreover, we might think we can't possibly know whether someone is a morally good person without investigation of what acts he/she has committed. Even further than that, we might think there is no way we can tell whether someone exhibits a morally upstanding character without assessing *both* his/her ethical actions and his/her intentions for following the morally correct rules.

Nevertheless, while we note the ways in which these questions may beg for interconnected answers, each query serves as the background to a different type of normative moral theory. A *normative moral theory* is a set of systematic explanations for a particular view of the nature and justifications of good and evil. If the question emphasized is "What acts should we commit?" then this points toward a consequentialist theory. That is, whether something is morally good or bad depends upon the results of actions. One common consequentialist theory is *utilitarianism*. On this moral theory, acts are good depending upon how much happiness or pleasure they generate, and are bad based on how much pain or unhappiness they cause for a moral community.

With that said, some consequentialist theories (including utilitarianism) could also be thought of as "teleological." The term *"teleology"* comes from the ancient Greek word "telos" which means "purpose" or "end" and "logos" which, in turn, means roughly "study of" or "science." Put together, teleology could be thought of as the "study of purposiveness." So, a *teleological theory* is goal-oriented and one way to track whether we are meeting goals properly is to evaluate the consequences of actions.[2]

---

1 There is one step back from this as well. *Metaethics* involves questions concerning what it means to say that something is good or evil, but such questions are a bit too remote for our purposes here.

2 Note that utilitarianism is not the only teleological theory and that not all teleological theories need be consequentialist. For example, as we shall see, Aristotle's virtue theory is teleological in the sense that he thought things are morally right insofar as they perform their proper function.

Alternatively, if we think the main question of morality is "What moral rules should I follow?" then we are likely drawn to deontological theories. *Deontology* derives from the ancient Greek word "deon" which means "duty" or "obligation." Deontological theories are rule-oriented. So, according to deontologists, the morally right thing involves doing one's duty and adhering to the correct moral rules. While there are certainly other deontological views, perhaps the paradigmatic example in philosophy is the moral theory of Immanuel Kant (1724–1804).

And finally, if we take the central question of morality to be "What sort of person should I be?" then the approach that fits most comfortably is virtue theory. *Virtue theory* is a "character-based" ethic—the emphasis is placed not so much on which action the morally good person should take or which rules he/she should follow, but instead on what kind of traits should the morally good person display. Character in this case has more to do with the attitudes and dispositions we have, and the intention we show in living a morally good life. With that said, we can now look more closely at these three major normative moral theories.

CHAPTER 1

General Moral Theories

Utilitarianism

Utilitarian ideas have been espoused for a long time. Indeed, we find the notions of calculation and happiness in both Socrates' (469–399 BCE) and Aristotle's (384–322 BCE) moral views. However, it was Jeremy Bentham (1748–1832) and John Stuart Mill (1806–1873) who first systematized and fully developed the principles of utilitarianism. The motto of this theory is "the greatest good for the greatest number." Utilitarianism is a teleological theory in the sense that it has a clearly stated goal (to accrue the greatest amount of happiness for the greatest number of people). As mentioned earlier, it is also a consequentialist theory in that we assess how well we are meeting the goal by tracking the results of actions. For example, we would need to determine if we should support a green lifestyle in a community by seeing if its aggregate members are indeed happier or more satisfied by adopting green policies versus not doing so.

The basic moral principle of utilitarianism is often called the "principle of utility" or "Greatest Happiness Principle." While there are a number of versions of utilitarianism that may formulate this principle in different ways, there is a basic unified formulation that can be culled from them. The basic idea is that all policies or actions ought to be accepted on the basis that they *maximize* net utility (understood as either satisfaction or happiness) of everyone in the moral community.[1] So, actions are right insofar as they promote happiness and wrong insofar as they promote suffering. Note also that what determines whether an act is morally right or wrong is not dependent upon the nature of the act or the intention one has while performing it. As Mill says, "He who saves a fellow creature from drowning does what is morally right, whether his

1 Traditionally, utilitarianism has been understood as an anthropocentric theory (see the section called "Anthropocentric Environmental Ethics" in Chapter 2). This means it is human-centered and this bears out in including only humans in the moral community. However, this doesn't have to be the case. For instance, if we think those who can feel pleasure and pain should be considered members of the moral community, then it seems non-human animals ought to be included as well.

 | DOI 10.1163/9789004339996_003

motive be duty or the hope of being paid for his trouble."[2] All that matters (morally speaking) to the utilitarian are the consequences of actions.

There are a few other basic characteristics of utilitarianism beyond the greatest happiness principle and the understanding that it is a teleological and consequentialist theory. First of all, happiness is equated with the good according to this theory. That is, all that the good *is* simply happiness.

Second, happiness is taken to be not merely the only intrinsic good, but also the highest good. For something to be intrinsically good means it is good in and of itself, and is not dependent on anything else for its moral worth. To say that happiness is the highest good means there are no other competing values that are more important.

Third, utilitarianism depends on there being a way to measure pleasure and pain, and has a calculability requirement. The calculability requirement just means we need some way to add and subtract happiness. The ability to measure happiness is significant for the simple reason that we can't add and subtract happiness without having some unit of measurement.[3] The calculability requirement is vital to utilitarianism because the only way we can choose between different courses of action we face is to be able to calculate and try to predict which act or rule will provide us with the greatest happiness or pleasure.

The fourth characteristic of utilitarianism is that each member of the moral community affected by an action or policy is to be counted equally. The idea here is that each member of the moral community is being considered equally in his/her interest in being capable of feeling pleasure and pain. Also, no one member's pleasure and pain is worth more or less by virtue of who he/she is. For example, my ability to feel happiness "counts" as just much as the president's (his/her happiness does not count twice more than mine because of the politically influential position he/she holds). If some action causes the president 33 hedons of pleasure, then that is exactly what it is worth. If some action causes me 71 hedons, then that is exactly what it is worth.

The fifth characteristic of utilitarianism is that it appeals to empirical evidence to project which course of action to take. Empirical evidence is merely information that is collected via observation. For utilitarians, we can only know what is in fact good by appeal to our experiences of the world. It should be said that Mill and Bentham were both rather optimistic that we could use the power of science (which subscribes to the notion that truth is revealed through careful observation) to solve our societal problems. As Mill noted,

2 John Stuart Mill, *Utilitarianism*. Oskar Priest, ed. (Indianapolis, IN: Hackett, 1957), p. 24.

3 Bentham designated the term "hedon" as the unit of measurement for pleasure.

"All the grand sources of human suffering are in a great degree, many of them almost entirely, conquerable by human care and effort."[4]

Perhaps more relevant to environmental ethics is how utilitarian elements are employed by those embroiled in recent debates concerning the use of natural resources. Cost-benefit analysis involves the idea that one policy is better than another if it produces the most net benefit (or the least net cost). Utility in this case is usually equated with money and cost-benefit analysis is really considered to be a measure of efficiency. Like the standard form of utilitarianism, in making a decision one examines the likely costs of her action and weighs them against the possible benefits. If the benefits outweigh the costs, then one should proceed with the project, if not, then she shouldn't.

While cost-benefit analysis will be important to understanding how businesses and governmental agencies make decisions, we should recognize that such choices are not usually moral ones. They are important insofar as they may take the interests of other people into account. For example, there may be costs of plowing under farm fields to develop a shopping mall (say the developer needs to pay for environmental impact studies done on the property) whereby the company must at least acknowledge that humans, non-human animals, and plants could be overly burdened by the development. Or we might consider whether new mines for energy resources ought to be developed (especially in pristine, environmentally sensitive areas). Agencies may have to take account of the economic benefits (in terms of job creation and domestic energy independence) to a community from having more mines for coal versus the loss of aesthetic beauty, the cost of lawsuits, and the potential loss of other economic opportunities (such as tourism). But again, notice that such costs are put in terms of money, not necessarily in units of happiness.

Utilitarianism has proven to be an influential theory in environmental ethics, especially as it extends into issues of animal welfare. We shall see Peter Singer's use of it in this context below.

Kantian Deontology

While utilitarians care primarily about the consequences of actions, the philosopher, Immanuel Kant, was suspicious that one could ever use consequences, feelings, or the will of God as the basis for morality. Instead, Kant thought the moral law has its own intrinsic value, and thus no appeal needs to be made outside of it to assess its moral value. This means the value of a moral

4 Mill, *Utilitarianism*, p. 20.

law comes independently of the consequences of following it, the feelings we have about it, or whether it agrees with the will of a supreme being.

Kant has interesting and intuitively plausible reasons for this stance. First, there is the worry that one does not freely choose what he/she wishes when under the influence of sentiments. Michael Sandel has brought this out lucidly when he has us consider the Sprite soda advertisement imploring you as the consumer to "Obey your thirst!" Think of what this is really saying. The suggestion of the expression is (as Kant would have put it had he encountered the ad!) to concede control of yourself to *heteronomous* impulses (i.e., your desire for something thirst-quenching and sweet).[5] But this is not acting *autonomously* (i.e., acting according to a law you give to yourself), at least if you are merely giving in to a feeling whether it be for sex, food, violence, etc. Autonomous activity then, according to Kant, requires acting for reasons rather than emotions. Also, feelings can shift over time, sometimes quite dramatically. At times, even similar events draw quite different reactions from us. Kant's point is that we are subject to what is the equivalent of rule by whim, with inconsistent reactions of moral blame, moral indifference, or moral condemnation based on the emotions we are having at the time. This can lead to explicit (or even implicit) bias leaving moral judgment unstable at best, and unfair at worst.

As mentioned, deontology is heavily invested in the idea that what is most important, morally speaking, is to do one's duty. Kant is no exception, but in some ways his brand of deontology is even more demanding. Complete moral understanding cannot be achieved simply by doing one's duty, but also requires knowledge of what one's duty is. The idea is that one is only truly free (or in Kant's terms, autonomous) when she understands *why* she is doing what she is doing. This includes our moral lives as well; the autonomous moral agent comprehends what the moral law requires, and also knows that the motivation for doing the right thing comes from the moral law itself. The idea of knowing what the right reasons are for committing a moral act is in Kant's terms "acting out of duty." In Kant's view, to be a fully moral agent, one must have knowledge of what must be done, why it must be done, with the realization that one should do the right thing for the right reasons.

Kant thought the only thing that is unconditionally good is a good will. He expresses this poetically—even if the good will were unable to accomplish its task by some misfortune, it "like a jewel, would still shine by its own light as

5 Michael Sandel, *Justice: What's the Right Thing to Do* (New York: Farrar, Straus and Giroux, 2010), p. 108.

something which has full value in itself."[6] The will is the capacity to choose between different courses of action, or as Kant would put it, a faculty for acting according to a conception of law.

Another way of understanding a good will is that it acts in accordance with the categorical imperative. An *imperative* is a command. For something to be *categorical* means that it must be *universal* (i.e., apply to all—in this case, to all rational beings) and *absolute* (i.e., apply in all cases). The categorical imperative is the one universal, absolute, and supreme (highest) moral law that must always be followed. Yet, there are different formulations of the categorical imperative. Kant actually thought there were five formulations.[7] He expressed the first formulation in the following way: "Act only according to that maxim whereby you can at the same time will that it should become a universal law."[8] In this formulation, we can think of the categorical imperative as being like the criterion that all other moral laws must meet.

How do we know when a proposed moral law (what Kant calls a "maxim") meets the requirements of the categorical imperative? Luckily, we can derive a test from the categorical imperative to be used to see if any maxim passes muster with it (what has been called by some Kant scholars the universalizability test). The first formulation of the categorical imperative basically means that whatever rule I wish to follow, I have to make sure I can will that all others should be morally allowed to do it too. This basically ensures I am not biasing decisions in my favor. Laws I should follow ought to be universalizable. So, according to Kant, any rational person has the ability to see if the rules they wish to follow are morally acceptable. One merely needs to propose a possible moral law (a maxim) and put it to the test. To pass the universalizability test, the maxim cannot be contradictory. This makes sense in Kant's ethical theory, considering that he puts such a premium on ensuring that morality is a rational system. In fact, when looking closely at the categorical imperative and the universalizability test, we can see that they serve in part as a logic test for moral claims.

6 *Grounding for the Metaphysics of Morals*. Trans. James Ellington. (Indianapolis: Hackett, 1981), p. 8.

7 Despite Kant's claim, a number of scholars contend there are only four formulations. No matter what, there are only two key formulations that really concern us here.

8 *Grounding for the Metaphysics of Morals*. Trans. James Ellington. (Indianapolis: Hackett, 1981), p. 30.

Virtue Ethics

As already mentioned, virtue ethics (also known as virtue theory) takes seriously the moral character of individuals. Perhaps the best way to describe virtue ethics is to begin by reflecting on people we look up to morally speaking, whether they are family members, friends, scientists, physicists, religious leaders, etc. Why do we admire them? More than likely, we will list a series of traits such as tenacity, courage, forgiveness, generosity, kindness, and the like. These positive characteristics we pick out are what are also known as virtues. Of course, people exhibit bad moral characteristics as well; we may find them to be boorish, vindictive, cowardly, stingy, rash, etc. Typically, these traits are called vices.

Virtue ethics has its roots not only in ancient Greek philosophy (in the work of Plato and particularly Aristotle) but also in early Chinese thought. While the theory has enjoyed periods of great prominence and waned at other times, it was most recently resurrected in the 1950's via an article by G.E.M. Anscombe, who thought that utilitarian and Kantian approaches to moral philosophy were sometimes inadequate to handle moral problems.[9] Many have criticized both utilitarianism and deontology for not properly addressing several issues that virtue ethics takes very seriously, such as the development of moral character and its relationship with motives, the nature of virtues, practical moral wisdom, friendship, the family, the role of emotions in morality, and moral education. In the light of this, it is not surprising that feminists who are interested in studying the marginalization of women have been attracted to virtue ethics. The theory acknowledges the central place of the moral relationships of family and friends, as well as the strong role sentiments can play in the construction of our moral lives.

In some respects, virtue theory seems to be more personal than other moral theories. Perhaps the attention virtue ethicists pay to familial and friendship ties applies more readily to our everyday lives. Discerning how one ought to be as a moral character forces reflection on how we relate to our parents, siblings, children, and co-workers. How should we treat those individuals closest to us, and what sort of treatment is fair? Is fairness even the most important moral value in guiding us through our interactions in close personal relationships? Moreover, how do we navigate a tendency to being partial to one's family and friends (perhaps thought of as the virtue of loyalty) versus general notions of justice and fairness to other members of our society in general? To what degree should obligations to my local community and my country override obligations

9 See her "Modern Moral Philosophy," *Philosophy* (1958) 33: 1–19.

to people in other societies? How much does the particular situation I find myself impact the way I behave towards others? Does loving your children equally mean treating them all in the same way? Virtue ethics would also seem to directly relate to responsibilities one has to children, to whom you may be offering guidance. What lessons should we teach them, and do we act as we speak? Reflection on moral education feeds back to our moral attitudes and may remind us of our own hypocritical stances and double standards.

For virtue ethics, a virtue such as honesty is not merely a tendency to tell the truth, but is a more deeply ingrained moral disposition. To truly possess a virtue is to be a certain sort of person who exhibits the virtue as a habit. In fact, Aristotle thought the truly virtuous man progressed to the point where he would just do the virtuous thing without even thinking about it. To bring this point home, think of someone you admire who just seems to have a knack for knowing the right thing to say to someone when she is distraught, or has the right kind of calming demeanor to defuse a tense or potentially dangerous situation. You might notice these individuals seem to conduct themselves well effortlessly. However, what may now really be easy has likely come from a long history in this person's life of practicing doing things in a calm and collected way. That practice has deeply instilled the appropriate virtues into his/her moral character.

Yet, we shouldn't be misled to think of virtue ethics as if it is in a vacuum, with good character completely cut off from good actions. Aristotle's famous "function argument" makes plain that the goodness of human moral character is dependent upon excellent, distinctly human *activity*.[10] Virtue ethicists also care deeply about the moral reasons people have for acting as they do. The motivations and justifications for choosing one course of action over others cannot be severed from one's character. Unlike both deontology and utilitarianism, the central moral focus of virtue ethics is on how one might live a flourishing human life, instead of simply fixating on the goodness or badness of the state of the world or one's actions. This is mainly because, as Aristotle noted, living a virtuous life is a key element for living a flourishing life.

Value

Whether we take the slogan on bumper stickers to "recycle, reuse, restore" seriously depends on our notions of value. We need to take a closer look at the types of values environmentalists and others appeal to when they discuss the

10 See his *Nicomachean Ethics*. Trans. W.D. Ross. (New York: Oxford University Press, 1998).

importance of protecting the natural world. This means beginning with some very general distinctions between types of values.

We can value something for itself or for its own sake, which is what is usually called its *intrinsic* (or *inherent*) value. When we express that our love is unconditional (say, for our family members) this means we love them not for what they can do for us, but for who they are. We would love them regardless of whether they offered us any benefit. Even though this example gives us some idea of what intrinsic value is like, it doesn't tell the entire story. This is mainly because whatever has intrinsic value isn't dependent on other things for its value. A test to see if something has intrinsic value is to ask whether, even if nothing else existed, if that entity would have value. Consider why a sunset might be valuable. If we thought it would be valuable even if for some reason no one could perceive it, then the sunset would possess intrinsic value.

Contrast this to *extrinsic* (or *instrumental*) value. This is the sort of value something has because someone values it. If we value something extrinsically, we don't value it for its own sake, but for the sake of something else. Simply put, if when we reflect upon why non-human animals are valued and if all of the reasons go back to how amused we humans are by them, through the affection they supply, or because we enjoy the products they generate, then such creatures are merely extrinsically or instrumentally valuable.

Some environmentalists believe the environment is valuable instrumentally. Other environmentalists think individual flora or fauna are valuable in and of themselves, while still others insist that whole ecosystems possess intrinsic value. It is important to note the relationship between intrinsic and extrinsic value. These are not necessarily mutually exclusive categories. Certainly, one might argue flora and fauna are only intrinsically valuable (though this is rare), or that they are only extrinsically valuable. But, it is also possible to think the environment is valuable both intrinsically *and* extrinsically. For example, one could say the environment is intrinsically valuable—its value doesn't derive from someone else valuing it, but also could think that *on top of that* the environment is also valuable because of its usefulness.

Another type of value sometimes discussed in environmental ethics is *prima facie* value. The term "prima facie" literally means "on the surface." In our context, if something has prima facie value, this means its value can sometimes be overcome by some stronger competing value. Another way of understanding prima facie value is to think about rules of thumb. Rules of thumb admit of exceptions. As a rule of thumb, I shouldn't kill other human beings. Yet, there might be exceptions to this rule—say, someone is intent on killing me with a machete and my only way of protecting myself is by using lethal force. So, certain circumstances may allow for us to make exceptions to the rule. In this case, the prima facie rule of 'thou shalt not kill' has an exemption for self-defense. But, some might think that even under these extreme circumstances, I shouldn't kill and this stance reveals a different type of value and corresponding type of rule. These individuals believe "thou shalt not kill" is a rule with *absolute* moral worth. If something has absolute moral value, then that value should be respected no matter what. Absolute moral principles are thus rules that should be followed without exception.

CHAPTER 2

Applications and Approaches to Environmental Ethics

Applied Ethics

So far we have explored ethical theory generally and some types of normative moral theories we need to know, since they are referred to from time to time in environmental ethics. The term "applied ethics" came into currency mainly in the 1970's. It was during this time that philosophers began addressing moral problems in society in a way they had not really done before. Out of this new attention to issues and difficulties outside of the "ivory tower" (especially as they applied to the professions) emerged new philosophical sub-disciplines such as *medical ethics* and *business ethics*. Pressing issues in medical ethics include debates over the moral permissibility of abortion, euthanasia, the use of human subjects in medical research, and the allocation of resources for health care. More recently, other issues have arisen, due in large part to advances in medical technology, such as human cloning, gene transfer, gene patenting, and the privacy of patient medical information. *Business ethics* explores issues such as manipulative or false advertising, affirmative action, corporate social responsibility, sustainability, whistleblowing, intellectual property, and product safety.

Around this time, with the institution of the Clean Water and Air Acts in the United States in 1972, another area of applied ethics arose. Examples of environmental devastation were becoming increasingly evident and more people began working to draw attention to these problems. This doesn't mean there were not already some earlier touchstones imploring us to take on a new moral relationship to nature. Aldo Leopold's *A Sand County Almanac*, published posthumously in 1949,[1] and Rachel Carson's *Silent Spring* in 1962,[2] are both now seen as landmark works in the environmental conservation literature. However, despite these earlier works, the academy as a whole only began to shed wider critical light on the problems of environmental ethics in the 1970's.

For our purposes here, we can define *environmental ethics* simply as the area of philosophy that studies values in the natural world. I leave this

1 (Oxford, U.K.: Oxford University Press, 1949).

2 See her *Silent Spring* (Boston: Houghton Mifflin, 1962).

 | DOI 10.1163/9789004339996_004

definition extremely wide on purpose. There has been the temptation to define environmental ethics as an area of philosophy dealing with the morality of human interactions with nature. But as we will see this initially straightforward definition has been extremely controversial among those working in the field.

Most of what was seen as environmental crises during the earlier stages of environmental awareness remain with us today—pollution of air, ground, and water, vanishing species of flora and fauna, and overuse of resources (especially timber, oil, coal, and water). Such issues alerted federal and state environmental regulatory agencies that more needed to be done to protect species and their habitats. Since that time, another more global issue has come to the forefront—climate change. The role of environmental ethics has been to discuss what types of moral obligations we have (if any at all) to the environment itself, its members, and/or other humans with respect to our relations in the natural world. With the growing number of issues and their increased complexity, a variety of perspectives and approaches to environmental ethics have been developed.

Anthropocentric Environmental Ethics

Considering the difficulties of convincing the large number of people necessary for a sustained environmental movement to accept the premise that the environment itself or some its members (e.g., plants, animals, ecosystems) are worthy of protection because of its/their intrinsic worth, many environmentalists take a more straightforward approach. They instead use an *anthropocentric* (human-centered) perspective,[3] which can either mean that humans alone have intrinsic moral worth,[4] or humans have significantly greater inherent worth than any non-human things.[5] Aristotle took a position relevantly similar to this when he said, "Nature has made all things for the sake of man."[6]

3 See for example, Alan Gewirth, "Human Rights and Future Generations," in Michael Boylan, ed. *Environmental Ethics* (Upper Saddle River, NJ: Prentice Hall, 2001), pp. 207–211, Onora O'Neill, "Environmental Values, Anthropocentrism, and Speciesism," *Environmental Values* (1997) 6(2): 127–142, John Passmore, *Man's Responsibility for Nature* (New York: Scribner, 1974), and William T. Blackstone, "Ethics and Ecology," in his *Philosophy and Environmental Crisis* (Athens, GA: University of Georgia Press, 1972), pp. 16–42.

4 This could be thought of as anthropocentrism in the strong sense.

5 This could be considered to be a weaker sense of anthropocentrism.

6 Aristotle, *Politics*, Trans. Ernest Barker (Oxford, U.K.: Oxford University Press, 1948), bk. 1, ch. 8.

He went on to assert that the moral value of non-human entities in nature is merely extrinsic or instrumental.

Environmentalists from an anthropocentric perspective attempt to show that the reasons why we should support a sustained, healthy, and beautiful environment is because it is essential to human well-being. They argue the value of nature is thus derived from how it affects humans. From this perspective, one might claim national parks are valuable because they provide solitude to humans who are sometimes overburdened by societal pressures and job expectations. Or they might reason that forest areas are important due to the wood they provide for human projects. Some may even go so far as to say that preservation of ecosystems is important because of their *potential* value to humans in the future. Given that other important discoveries of useful substances for the production of drugs (including Taxol from the bark of the Pacific yew tree for the treatment of breast and ovarian cancer), there is promise that other natural resources could become immensely valuable to pharmaceutical companies and patients in the future in ways currently unknown. For this reason, we must take great care in our evaluation of which resources are "useful" and not hastily degrade or exploit flora and fauna.

Despite this nod to the instrumental value of the environment, anthropocentric positions in ethics have a very hard time showing why we shouldn't treat non-human animals as we wish. Consider Kant's ethics, which as we already observed considers the good human will to be the only intrinsically good thing. Kant's only appeal to why we should not torture animals is because it could lead to the morally bad consequence that such actions might become habitual, making us more inclined to torture or otherwise harm other human beings. More specifically, Kant suggested that cruelty to a dog might promote the building of a character desensitized to cruelty to humans.[7] But without further evidence to show this would be the case, there seems to be no independent reason why humans couldn't simply treat animals any way they wish.

Anthropocentrism is also charged with not taking non-human interests into account when developing policies focused on human interests. However, defenders of anthropocentric views admit that while some from this perspective adopt practices that (largely) disregard the environment, most make decisions that can align well with stable and healthy ecosystems. Anthropocentric views are sometimes classified in two ways, as *strong anthropocentrism* and *weak anthropocentrism*. Strong anthropocentrism means that, "All value is explained

7 See Immanuel Kant, "Duties to Animals and Spirits," Trans. Louis Infield, *Lectures on Ethics* (New York: Harper and Row, 1963).

by reference to satisfactions of felt preferences of human individuals."[8] In contrast, *weak anthropocentrism* means that "some value is explained by reference to satisfaction of felt preferences of a human individual or by reference to its bearing upon the ideals which exist as elements in a worldview essential to determinations of considered preferences."[9]

There is a good deal of jargon here and the meanings of these terms require unpacking. First, a "felt preference" is one's desire or need that can be (at least temporarily) met by some experience. A "considered preference" is the desire or need one has after careful reflection and deliberation, and is consistent with a rational perspective based on a coherent scientific basis and moral theory. So, in effect, strong anthropocentrism endorses an action or policy that might arise from merely a passing and casual desire we have, and would thus presumably justify instant gratification. However, weak anthropocentrism leaves the door open for adopting policies that while satisfying human interests, focus on our long-term interests or human ideals, such as community or aesthetic appreciation of nature.

While weak anthropocentrism might take the edge off from the widespread belief that anthropocentrism will often mean that when there are conflicts between humans and nature, the former will win out, this has done little to convince a number of ethicists that anthropocentrism is an inadequate environmental perspective.

Biocentrism vs. Ecocentrism

Many early environmental ethicists insisted that if the field was to be considered distinct from others, it had to reject anthropocentrism. Earlier, we considered a broad definition of environmental ethics. That was in response to charges by biocentric and ecocentric critics of anthropocentrism that defining the field narrowly with respect to human interactions with nature was both inaccurate and irresponsible. The main contention was that if we merely try to extend traditional ethical theories to environmental matters, non-human entities receive short shrift by definition. One interpretation of anthropocentrism seems to suggest this conclusion. As Tim Hayward puts it, anthropocentrism is the view that privileges values, attitudes, and practices which give "exclusive

8 Bryan Norton, "Environmental Ethics and Weak Anthropocentrism," *Environmental Ethics* (1984) 6(2): 131–148 at 134.

9 Bryan Norton, "Environmental Ethics and Weak Anthropocentrism," *Environmental Ethics* (1984) 6(2): 131–148 at 134.

or arbitrarily preferential treatment to human interests as opposed to other beings [or the environment]."[10]

Peter Singer notes the possible arbitrariness of the anthropocentric view—that its proponents give no strong reasons for endorsing—when he charges the human-centered approach is "speciesist."[11] *Speciesism* is the arbitrary assumption that only the interests of human beings matter morally at the expense of non-human animals, without providing any justification beyond the fact that they are different species. Other pioneers in environmental ethics, even when they didn't agree on which exact type of non-human entity should be thought to have intrinsic value, did at least target difficulties with anthropocentrism.[12]

Those from a non-anthropocentric perspective at least share an insistence that not only humans have intrinsic or inherent moral worth. But there is at least one significant divide between those who argue conventional ethical theories are ill-equipped to deal with environmental issues. The approach of one camp, believing individual plants and animals to be intrinsically valuable, has come to be called *biocentrism*. Others, however, find this position to be theoretically problematic and instead claim that whole ecosystems have intrinsic moral value. This position is now known as *ecocentrism*. No matter the perspective, the idea is that morality has as much to do with what will be in the best interests of the environment as it does with us, and human activities should accommodate these interests.

Deep Ecology[13]

If we define radical as anything that questions conventional wisdom or usual practices, then deep ecology is a straightforwardly radical position. It endorses

10 Hayward, "Anthropocentrism: A Misunderstood Problem," *Environmental Values* (1997) 6: 51. Note that Hayward himself does not necessarily raise this challenge, but instead, simply reports it.

11 See Singer, *Animal Liberation*.

12 See Arne Naess, "The Shallow and the Deep: Long-Range Ecology Movements," *Inquiry* (1973) 16: 95–100, Holmes Rolston, "Is There an Ecological Ethic?" *Ethics* (1975) 85: 93–105, Richard Routley (later Sylvan), "Do We Need a New, an Environmental Ethics?" in *Philosophy and Science, Morality and Culture, Technology and Man*, Proceedings of the XVth World Congress of Philosophy (Sofia, Bulgaria: Varna, 1973), pp. 205–210, Peter Singer, "All Animals are Equal," *Philosophical Exchange* (1974) 1: 103–116.

13 I wish to thank Mark Peterson for his helpful comments on this presentation on deep ecology and the following one on social ecology.

a unique metaphysical view[14] with a radical perspective of the self. According to deep ecologists, we are *not apart* from nature (or separated) from it, but instead humans are just *a part* of nature. This is not merely a social perspective, but also a metaphysical position. The idea of being at one with nature means the human individual is linked to nature *by identity*. According to deep ecologists, I can extend the boundaries of what I consider to be myself through my identification with nature.

In support of this position, consider what happens in our everyday lives. The coffee I drink in the morning enters me and combines with me through biochemical processes that may even alter my psychological state. The food I eat literally nourishes my body and is incorporated into me for my survival. Even the air I breathe contains molecules that for better or worse become part of me. Of course, these inputs can interact with me in constructive or destructive ways. No matter what, realizing we are simply a part of nature is a crucial first step to seeing its health really is my health. My true self-realization depends on my understanding that I should think of myself in this larger ontological sense of self.[15] Deep identification with nature and members of other species provides a truly enriched life. This is quite different from how humans have historically related to nature, seeing it as something different and inferior. Deep ecologists add that humans have mainly regarded this detachment from nature as an overture to controlling it, using it merely "as a resource" to be exploited for our benefit.

The deep ecologist view of the self as being identified with nature has important implications. Notice already that by changing the metaphysical view of the human relationship with the natural world changes *how* we value things. If we are truly a part of nature, we are implicated in the health of the environment and our ethical perspective has to change. It also reveals why the conventional metaphysical view of humans vis-à-vis the natural world can be dangerous. In the eyes of deep ecologists, most humans live under the illusion that they are simply atomized individuals detached from nature. This makes it easier for our species to create ecological crises. We delude ourselves into thinking that when we pour toxins into rivers, or bury plastics in the ground, that we have "taken care of them." If they are "out of sight" then they are "out of mind" and hence no longer our problem because they are "out there."

Deep ecology gets its name for its depth when compared to what is sometimes called "shallow" or mainstream environmentalism. Deep ecologists argue that "shallow" environmentalists accept the dubitable metaphysical

14 *Metaphysics* is the study of the first principles of reality.

15 "Ontological" means roughly "related to being."

assumptions noted earlier that tend to be individualist and anthropocentric. For example, Arne Naess observed that the chief objective of "shallow environmentalists" is to "fight against pollution and resource depletion" to assure "the health and affluence of people in the developed countries."[16] This, says the deep ecologist, only allows for cosmetic policy changes distracting us from genuine environmental reform. Instead, deep ecologists stress a different worldview with respect to nature—a perspective they believe necessary to confront the basic assumptions about our usual relationship with it. As a consequence, deep ecology challenges not only what normally "count" as environmental problems, but also conventional approaches to arguing for and against ecological measures.

Deep ecology has been criticized by a number of commentators for being too radical. Some of the principles of deep ecology (called by Arne Naess and George Sessions "The Deep Ecology Platform") point out general environmental degradation originating from the existence of too many humans. Perhaps the most controversial principle for which deep ecologists argue is that the survival of ecosystems depends on, and the flourishing of human life is consistent with, the reduction in the number of humans existing.[17] To be fair, deep ecologists certainly don't advocate direct reductions of humans through killing off members of our species. However, taking this principle to its logical conclusion is still highly controversial, as it at least appears to be consistent with slowing the human birth rate considerably to reduce humanity's ecological footprint. It also seems to reveal that deep ecology is a straightforwardly ecocentric position.

Likewise, deep ecologists would likely endorse political measures such as the "one-child" policy in China (only recently revoked). This brings to light another contentious issue—that non-human life is not only intrinsically valuable, but is at least on an equal footing with human life. Some argue that the implication of the first principle of the deep ecologist platform is even stronger. That is, it suggests that *future* increases in human lives should be sacrificed for the good of non-human lives or for the biosphere generally. Critics of deep ecology believe this implies that if all life forms are inherently valuable, not only are non-human lives *as* important as human lives, but also in some cases are even *more* important to deep ecologists.

However, deep ecologists can answer these charges. They could make a general point that their critics are too reliant on an outmoded Newtonian

16 Naess, "The Shallow and the Deep."

17 Bill Devall and George Sessions, *Deep Ecology: Living as if Nature Mattered* (Layton, UT: Gibbs & Sons, 1985).

worldview. The deep ecologist could add that since humans are not separated from the world, this also affects how things are valued. It simply isn't the case that ground squirrels and marine estuaries would have to have the *same* or *greater* value than humans just because they are inherently valuable.

Social Ecology

The pioneer of the social ecology movement was Murray Bookchin, a Marxist social commentator who agrees with deep ecologists that a radical transformation needs to occur to end environmental devastation.[18] With that said, social ecology, like deep ecology, is a radical position. However, Bookchin didn't appeal to a faulty conventional metaphysical view as the main cause of environmental problems as deep ecologists do. Instead, he notes that patterns of domination are carried over from social oppression to domination of the environment. The patterns of domination we witness in social structures become ingrained in the culture, accepted as appropriate social norms, and internalized as the beliefs and actions of ordinary citizens. In sum, environmental problems ultimately have a *social* cause and as a response, environmentalism should be understood as a social movement.

While traditionally our role in nature has been a negative one, Bookchin still saw great promise in human intervention to curtail and remediate environmental damage. By putting our powers of intelligence and abilities of social organization to use, we can in a sense "speak for" the environment by working in its interest instead of exploiting it. Bookchin believed that instead of changing our metaphysical view of human relations to nature, environmental reform hinges on revising social structures to end both class and environmental domination.

Bookchin observed that most human societies are structured hierarchically, with those at the higher levels of the structure exerting immense control over those in lower status levels. We find this pattern expressed through our social, political, and religious institutions, with top-down authority, power, and control holding everything into place. Bureaucrats claim to speak for higher "social interests" and then proceed to aid one class, ethnic, or racial group in mastering others.

Additionally, according to Bookchin, the ideology of the free market has caused the gap in wealth and power between socioeconomic classes and this

18 See both his *The Philosophy of Social Ecology* (Montreal: Black Rose Books, 1990) and *The Ecology of Freedom* (Palo Alto, CA: Cheshire Books, 1982).

has transformed both human beings and the natural world into mere commodities. In lieu of these oppressive arrangements, Bookchin advocated a just society with absolute equality of all people in a way maximizing human freedom. The goal is to break individuals of the urge to dominate and he thinks that can only be achieved though establishing a society of equals. Individuals who truly possess political freedom, unlocked from social conditioning and institutional control, will not feel the need to react by oppressing others and dominating nature.

For members of lower social classes to break the bonds of oppression, Bookchin prescribes grassroots organization as a way of expanding political voice to people who have been historically left out of democratic decision-making. Although the details are a bit fuzzy, it appears Bookchin's political ideal is to have a confederation of small communities with a greater share of local control. This extends into what Bookchin sees as a viable way to attack environmental problems, based largely on the premise that local communities are able to gather the most reliable information for combating their own issues. Such political organization infuses a direct participatory democratic element with respect to the environment—decision-making comes mainly from the local level and promotes the political voice of more citizens, as they see they can have important effects on their own communities.

In many ways, Bookchin thought the plight of oppressed humans and the environment are interconnected. In fact, he saw the relations between interdependent parts of nature as a possible blueprint for how human societies could be reshaped into more harmonious and equitable arrangements. Bookchin observed that in an ecosystem different species form symbiotic relationships, with no one species being any more significant than any other.[19] He saw this model of equality and mutually beneficial arrangements as one that humans could utilize in organizing their own societies.

Ultimately, one might question how far social ecology really is from deep ecology. For instance, recall this approach to environmental ethics is supposed to be different in that deep ecologists believe the causes of environmental problems stem from a faulty metaphysical view of the human/nature relationship, whereas social ecologists blame class and racial division. But then again, given Bookchin's Marxist position, he surely has metaphysical assumptions of his own and these would seem to be *non-individualistic.* This at least raises the question of whether Bookchin's view that broad class, racial, and gender

---

19 "What is Social Ecology?" in Michael Boylan, ed. *Environmental Ethics* (Upper Saddle River, NJ: Prentice Hall, 2001), pp. 62–76.

relations constitute the "real world" is so different from the deep ecologist's holistic metaphysical view after all.

Ecofeminism

Ecofeminism (or ecological feminism) is a relatively new type of environmental ethic. The term "ecofeminism" was first used by Francoise d'Eaubonne in 1974.[20] Some of the most prominent ecofeminists include Karen Warren and Val Plumwood.

Ecofeminism (like social ecology) distinguishes itself from deep ecology by finding the causes of environmental practices not in particular metaphysical views of the world, but instead in social practices. Consequently, some environmental ethicists see ecofeminism as being part of social ecology, as they both take seriously the role that social institutions and patterns play in environmental degradation. However, ecofeminists generally agree that one pervasive social pattern particularly harmful to the environment arises from a patriarchal or male-centered view of the world. As Karen Warren explains, ecofeminism is "the position that there are important connections ... between the domination of women and the domination of nature."[21] She adds that, "The goal of ecofeminism is to develop theories and practices concerning humans and the natural environment that are not male-biased and that provide a guide to action in the prefeminist present."[22]

Domination and ways of fighting it are key themes in ecofeminism. Ecofeminists charge that those who have perpetuated male domination of women have tried to defend their position by stressing the biological differences between the sexes. Those arguing for why males should be the leaders in society point to what they see as different and inferior natural characteristics of females (i.e., they are weaker physically and too emotional to make rational decisions, etc.).

Since males have historically been at the helm of political, economic, and religious institutions, it is not surprising they have made major decisions concerning human relations with the environment. Predictably, according to ecofeminists, men have structured the human/nature relationship as one of

20 See her *Le Feminisme ou la Mort* (Paris: Pierre Horay, 1974).

21 See her "The Power and Promise of Ecological Feminism," *Environmental Ethics* (1987) 9: 3–20.

22 Warren, *Ecofeminist Philosophy: A Western Perspective on What It is and Why It Matters* (Oxford, U.K.: Rowman & Littlefield, 2000).

difference—of nature being "out there," irrational, and needing to be brought under the rational control of humans. The structure of patriarchy is "dualist" and one of its operational principles is the "logic of domination." The origins of this domination are substance dualism (mind versus body) and gender dualism (men versus women). After identifying biological differences between the sexes, it was a short step for the patriarchal order to accept one as more desirable than the other and to control the other it sees as inferior.

As alluded to earlier, ecofeminists claim that male-controlled institutions have established a link between women and nature. Women have been associated with nature because they are seen as being more intuitive and more attentive to "bodily features" and not to the world of the mind. Moreover, nature has historically been thought of as feminine. This is reflected in common expressions such as "Mother Earth." Even the terminology that is conventionally used to describe environmental destruction is telling. For example, land developers and strip miners are said to "rape" the earth to exploit its resources. On this line of interpretation, just as men see women as objects (property to subject to their wills), they see nature as a repository of resources to be utilized.

Despite the problematic dimensions of comparing nature to femininity, some ecofeminists think the promise of ending oppression may come from this close association of women with nature. Women and nature have both been thought to embody birth and renewal. Thus, this shared value is reason for ecofeminists to embrace environmental causes and to demonstrate the importance of the special features of both women and nature.

It should be said that while there are common threads running through ecofeminism, this doesn't mean there are not different brands of it.[23] While we've noted some ecofeminists stress there are essential features of women and men making them different, others strongly deny this claim. The latter argue that nurturing, for example, is a role that has been ascribed to women socially and the patriarchal hierarchy has devalued women in this role. But given the correct social conditions, men can also be nurturers.

Yet, since ecofeminists see the patriarchal oppression of women as being of chief importance, only its elimination will root out other sorts of oppression. Thus, not only do ecofeminists think environmental problems are linked with the oppression of women. Additionally, they think such issues will only be solved by first fixing the problems of inequalities between men and women and the domination from which they emerge.

---

23 For a discussion of this point, see Val Plumwood, "Ecofeminism: An Overview and Discussion of Positions," *Australian Journal of Philosophy* (1986) 64: 120–38.

Leopold's Land Ethic

The work of Aldo Leopold, while it doesn't really constitute a separate, full-fledged, systematic environmental theory on its own, does provide guidelines around which philosophers have tried to build such a theory. In *A Sand County Almanac*, Leopold not only demonstrates ways in which governmental policies have caused environmental problems directly, but also how well-intentioned measures at conservation can create perverse incentives leading to further degradation of ecosystems. Beyond this, he sketched a different approach to how humans should view the environment, adding some general normative guidelines to govern our actions toward it.

Leopold's land ethic was in part a reaction to the crudely economic approach of cost-benefit analysis most humans took to the environment during his lifetime. He carefully noted the environment's aesthetic value and emphasized that we have an ethical obligation to conserve nature. Leopold wrote, "That land is a community is the basic concept of ecology, but that land is to be loved and respected is an extension of ethics."[24] His position is really a prescription to extend our conception of what exactly the moral community is—to entertain the notion of a land ethic that "simply enlarges the boundaries of community to include soils, waters, plants, and animals, or collectively, the land."[25] His position can be classified as ecocentric, with ecosystems serving as the primary moral objects. Defining *ecosystems* can be difficult, but the general idea is that it is a natural system—a whole with interdependent and interactive parts, yet within a geographical area. So, a bay or marine estuary is an ecosystem, since it has a regular circulation of water (a flushing time), rocks and ledges that support flora and fauna, with predators and prey that live in the estuary (some of which may return at a certain time each year to breed). But again, there will always be a degree of vagueness with defining what exactly an ecosystem is.

Leopold thought this unity has inherent value. Instead of first imagining nature as a bounty of resources to be exploited, we should consider it to be "a fountain of energy flowing through a circuit of soils, plants and animals."[26] The land forms what Leopold calls a "biotic pyramid" with many smaller organisms at the bottom and gradually fewer as we get the top. The complexity of the organisms also increases the higher you look in the pyramid. The biotic pyramid reflects the interdependent nature of creatures within it. Larger animals depend on insects for sustenance, which in turn rely on plants, which in

24 Leopold, *A Sand County Almanac* (Oxford, U.K.: Oxford University Press, 1949), pp. vii–ix.

25 Leopold, *A Sand County Almanac*, p. 251.

26 Leopold, *A Sand County Almanac*, p. 251.

turn depend on the soil. Perhaps the most important idea to take away from Leopold's understanding of the biotic pyramid is that those species at the top are not of primary moral interest. The chief value of the pyramid lies at its bottom as soils and other materials serve as the foundation essential for the existence of the rest of the pyramid.

In order to speak of the moral value of the land, Leopold employs a medical model. Ecosystems are good insofar as they are healthy, bad insofar as they are unhealthy. If there is erosion and not enough soil remaining to sustain abundant life, the ecosystem is degraded, sick, and bad. But where there is a diverse array of flourishing flora and fauna, with soil and rocks intact, the ecosystem is healthy and good.

Again, while Leopold's approach is not developed into a full-fledged environmental ethic, he supplies some normative principles with regard to the environment. The general principle he promotes is that a thing "is right when it tends to preserve the integrity, stability, and beauty of the biotic community. It is wrong when it tends to do otherwise."[27] By the *integrity* of a system, Leopold means it contains a unity of interdependent parts that constitute a unique character. Even when individual members perish, the core of the ecosystem will remain. Interestingly, this is a way in which Leopold's land ethic can be distinguished from biocentric views. Recall that from the biocentric perspective, individual members of species are seen as units of intrinsic moral value. However, from Leopold's ecocentric perspective, individual members of species may need to be sacrificed to leave the rest of the ecosystem functional. This is consistent with controls on invasive species, for example, that threaten whole ecosystems. Or it may mean culling some members of a population to prevent spread of disease or to keep overpopulation from imperiling the ecosystem. Leopold himself was an avid hunter, but a careful one, tracking how many deer and other animals he killed.

Second, a system possesses the virtue of *stability* when it changes only gradually. Leopold didn't think an ecosystem should never change. Even without human intervention ecosystems have individual members that come and go, its boundaries may change from time to time, and there is a sense in which some alteration is necessary for a healthy environment. The worry is that human interventions can and have induced rapid changes causing stresses ecosystems can't handle. The call for stability is a reminder to take care not to push ecosystems past their functional limits.

Third, the good ecosystem possesses *beauty*. Again, it is important to caution that while Leopold is thinking of beauty in the usual way as a purely aesthetic

27 Leopold, *A Sand County Almanac*, pp. 224–225.

concept, this is not all he has in mind. In its application to the environment, his focus is on beauty thought of as the harmonious interactions between interdependent parts with diverse entities functioning smoothly in a greater unity. Our moral charge then is to preserve these systems, acting according to the three stated guidelines. Of course, the idea is not just to preserve *an* ecosystem, but that its health contributes to health of the Earth as a whole biotic community.

Extensions of Utilitarianism

Perhaps the most prominent contemporary thinker who has applied utilitarian thought to environmental ethics is Peter Singer. Singer's work has focused mainly on the moral standing of non-human animals. His 1975 book, *Animal Liberation*, has been lauded by animal rights advocates as a landmark step in the promotion of the ethical treatment of animals.[28]

According to Singer, the prevalent moral feature for both human and non-human animals is *sentience*. That is, moral value should be attached to those beings with the capacity to feel at least some modicum of pleasure and pain. This, of course, fits well with Singer's acceptance of utilitarianism, which defines good and evil as the experience of pleasure and pain, respectively. Normally, discomfort is an undesirable state to be in and most organisms indicate this by attempts to remove themselves from adverse conditions. One function pain serves is to alert organisms to something potentially deleterious to their health. For both of these reasons (that discomfort is not desirable and that pain can be indicative of threats to self-preservation) organisms that can feel pleasure and pain have interests in not being harmed. Singer observes a link between sentience and interests, with pleasure being the product of having one's interests satisfied. While Singer does not argue for animal *rights*, he thinks their *interests* should be taken into account just like the interests of humans. Other environmental thinkers have, however, tried to claim Singer's work serves as the basis for animal rights by extension. They think that since having an interest is a proper basis for having rights, any sentient creature would, consequently, possess rights.

But without using the language of animal rights, Singer recognizes an interesting implication of linking sentience with the interests of fauna. If it were true that any creature capable of sentience has interests in its own protection (even if they are not completely conscious of those interests), then many

28 See Singer, *Animal Liberation*.

non-human animals would also qualify as entities with moral status. He grants that non-human animals may well have *different* interests than we do, but just because they are different doesn't mean they are any less significant. Recall the characteristic I mentioned above in the section on utilitarianism—that each member of a moral community is to count equally with respect to its ability to feel pleasure and pain. As my ability to feel pleasure and pain is just as morally significant as the President's, so is a dog's with the same capacity.

Some will balk at this suggestion from the outset, claiming it is specious to think human and animal interests are equal. But Singer would retort by asking on what basis should they be thought unequal? Throughout history most humans have given little to no moral weight to animal interests. This, Singer charges, is a case of speciesism (see "Biocentrism vs. Ecocentrism"). Recall that speciesism is not taking animal interests into account merely because they are non-human species and giving preferential treatment to members of one's own species simply because they are of one's own species. Speciesism is, to Singer, just as morally indefensible as racism or sexism. Just as a human being should not be discriminated against merely because of their sex, race, or ethnicity, no being should be discriminated against merely on the grounds of being a member of a certain species. Once more, since animals have some vital interests similar to ours, they ought to be given equal weight.

Singer does not mean here that animals should have all of the rights and privileges that human beings do. For instance, he is not endorsing the absurd consequence that animals should be allowed to vote, drive, etc.—after all, animals have no interests in voting. Singer is merely acknowledging that non-human animals have interests that should be taken into account, such as not being caused unnecessary suffering just for the minor benefit of humans, or due to a certain human callousness to their interests.

Two additional items should be noted here. First of all, some philosophers consider questions of animal rights as lying outside of the bounds of environmental ethics. Some philosophy textbooks reflect this by having separate sections for animal rights and environmental issues. Second, I also mentioned earlier that many early environmental ethicists insisted an adequate environmental ethic could not depend upon traditional ethical theories because they were inherently anthropocentric.[29] This means the birth of environmental ethics as a field came with skepticism toward the use of normative theories such as utilitarianism with respect to environmental issues.

Part of the proposed separation between animal rights and environmental ethics may also come from the tenor of works like Singer's, which still justifies

29 See Robert Routley on this point.

the existence of animal interests based on comparisons with human interests. But note that even in showing the possibility that many non-human animals have interests in not being abused or exploited, this doesn't extend to the rest of nature. While both human and non-human animals may have interests and/or rights based on the shared feature of sentience, plants and rocks do not. While plants might have things *in* their interest, *they do not have* interests. Hence, it is not clear how Singer's argument could possibly be extended to other things in the natural world, or even to the environment itself. Thus, it is hard to see how Singer's view is really an *environmental* ethic. Yet, even environmental ethicists who oppose the use of anthropocentric positions with respect to environmental matters would have to admit that animals have long endured the same kind of disregard as nature has generally. It is from this perspective that the major discussions of animal rights and interests should be included in any sustained engagement with environmental ethics.

Extensions of Deontology

Kantian moral theory has been expanded to a certain extent to include non-human animals and ecosystems. Two theorists who have spearheaded such work include Tom Regan[30] and Paul Taylor.[31] Regan's work is inspired by Kant's notion of universal duties, a deep respect for human ends, and rights. Where Regan diverges from Kant is in determining what should qualify as an autonomous being. Recall that Kantian deontology appeared to be hostile to taking the environment seriously as a moral entity. Kant thought only entities with the requisite level of rational capacity could qualify as moral agents. He certainly did not think non-human animals possess this quality. Consequently, it is doubtful at least prima facie that non-human entities would have much moral value. Also on Kant's view, if animals have any moral value at all, it is always only in the instrumental sense. In the example I gave earlier (see "Anthropocentric Environmental Ethics"), it is doubtful that a dog has any value in and of itself on Kant's view. Any prohibition on torturing it is not to *promote* the value of equal respect for dogs, but merely for *curtailing* a possible habit of causing harm to humans. In other words, the moral concern of torturing a dog has nothing to do with the dog's interests from a Kantian perspective,

30 See his *The Case for Animal Rights* (Berkeley, CA: University of California Press, 1983).

31 See both his *Respect for Nature* (Princeton, NJ: Princeton University Press, 1986) and "The Ethics of Respect for Nature," *Environmental Ethics* (1981) 3: 197–213.

but only as a means to avoiding a possible future violation of respect for persons.

How then can Regan possibly apply Kant's theory? He simply broadens the definition of what a moral subject ought to be and fits it upon the rest of Kant's deontological moral framework. He proposes that any being with complex perceptual powers and emotional responses should be considered to be a morally valued subject. Like Singer, Regan suggests sentience is certainly evidence of the appropriate level of complexity for moral consideration. And as Singer does, he sees the notion of interests to be significant—the capacity to pursue actions and goals with some degree of independent initiative should qualify a being for moral status. Regan regards the many animals that carry these capacities to be "*subjects-of-a-life*" which consequently have intrinsic moral value. Since no being with intrinsic moral value should ever be treated merely as a means to another's end, these subjects-of-a-life should not merely be exploited for human benefit. As individual members of species can hold intrinsic value, Regan's position would likely be categorized as biocentric.

In contrast to Regan's view, Taylor's is intriguing because while it qualifies as an extension of Kantian ethical insights to entities beyond rational humans, it also relies on Aristotelian principles. Taylor also takes a position unlike Singer's and Regan's in that his view is not dependent on the attribute of sentience to extend moral value to non-human animals. In lieu of sentience, he appeals to what he calls a "teleological center of a life."[32] Recall that proper functioning is central to Aristotle's ethical theory. All living things have a "telos" or function, which is inborn or natural to them, is to be sustained, and is to serve as an end for which they strive. This striving is not isolated to humans. Taylor thinks non-human animals and plants have central functions unique to them. Animals strive to preserve themselves when threatened; plants "move" toward light (known as heliotropism). In Aristotle's terminology, the acorn "strives" to become a tree and fulfill its function to be a flourishing member of its species.

But how do we know plants and non-human animals are good in and of themselves just because they possess some function? Can non-human animals be valued objectively even if they have a purpose? Doesn't this teleological argument beg for some self-awareness for an entity to have purposive action? If so, in what way do less complex non-human animals or plants (which do not appear to have this level of self-reflection) possibly have a purpose?

---

32 Paul Taylor, *Respect for Nature* (Princeton, NJ: Princeton University Press, 1986) and "The Ethics of Respect for Nature," *Environmental Ethics* (1981) 3: 197–213.

There are a couple of factors we need to consider in answering these questions. Non-human animals may not be aware of their "telos" and plants definitely are not. Yet, Taylor would suggest they need not be, or at least plants and animals have functions not dependent upon them being conscious of such purposes to possess them. We can know whether an organism is meeting its potential simply by observing it. A healthy animal doesn't reveal distressed behaviors and has few limitations on its normal range of capacities.

The upshot of Taylor's view is that we can know any being with a purpose has inherent value due to its having some *observable* purpose (even if the being itself isn't cognizant of it). Also, for us to be able to observe some operational goal of an organism (such as those of respiratory or circulatory systems) means that goal must already be "there" objectively. This is because we don't subjectively and randomly "come up with" some systematic explanation for how an organism works. Taylor concludes that if we extend concern for the well-being of individual, goal-oriented beings to the environment more generally, we could argue for the intrinsic value of any organism that can achieve its own good.

Extensions of Virtue Ethics

There have been some recent developments in the extension of virtue ethics into environmental ethics. To understand how, we need to comprehend what modern virtue ethicists think is a major advantage of their approach to moral theorizing over deontological and consequentialist ones. Most deontologists and consequentialists, as we have seen, see "goodness" or "rightness" as central moral concepts, even if achieving them comes through different methods. However, they usually don't go beyond that to engage other potentially normative concepts such as loyalty, bravery, charity, prudence, and the like. Virtue ethicists note that "goodness" and "rightness" are "thin" concepts in our moral world. These theorists open us to the possibility that our ethical lives are more complex, adding the importance of "thick" moral concepts such as kindness, justice, and honesty. Paying attention to these richer and fuller concepts of morality, we get a different understanding of the moral world beyond merely knowing which actions or intentions are right or wrong, or what kinds of entities are good or bad.

By employing the robust variety of moral concepts from virtue ethics, and by giving careful consideration to the full development of moral character, we also may be able to adopt a promising, alternative approach capable of handling the rapidly changing and complicated issues of environmental sustainability.[33]

On the face of things, it wouldn't seem that virtue ethics is a particularly good candidate theory for extension into environmental issues. As with Kant, Aristotle's version of virtue theory seems predicated on the idea that humans are the genuine moral focus. Aristotle exalts human flourishing, which as I mentioned earlier calls for practicing the virtues. A flourishing human life is, in large part, a virtuous life. Yet, this emphasis on *human* flourishing seems to have no application to the non-human world. After all, if flourishing comes with the *practice* of virtues, then it is unclear how flourishing could possibly relate to non-human animals, plants, soils, marine estuaries, rock formations, and the like. Such entities do not make moral decisions and some of them don't act at all.

Furthermore, as some virtue ethicists note, Aristotle thinks part of what it means to develop an appropriate moral character comes by our interactions with other people. In his political theory, in fact, the only way a just society can be achieved is through deliberations engaged in by humans. Humans are, according to Aristotle, essentially social animals. The great necessity of social interaction isn't merely a concern for us as citizens, but also as individual humans. Aristotle is one of only a few philosophers who discussed the moral value of friendship extensively. Friendship allows us the opportunity to practice the virtues of loyalty, cheerfulness, kindness, honesty, and many others (notice how many of these virtues require more than one person to be practiced). These relationships are vital to us living a full and meaningful existence. Through our genuine friendships, which Aristotle took to mean that one values, loves, respects, and cares for a friend for his/her sake, we become genuinely human.

But one might argue there doesn't seem to be any reason in principle why we couldn't extend our relationships to non-human creatures. Indeed pet owners may well have very strong affections for any number of non-human animals. Even nature enthusiasts who enjoy seeing (and to a certain extent perhaps interacting with) animals in the wild, could be said to have important relations with non-humans. Perhaps more importantly, one's moral development comes from his/her practicing the virtues with respect to other people. It is through showing love, respect, benevolence, courage, and the like toward others that we show our capacities to be proper friends. But why must these relationships

33 Ronald Sandler, *Character and Environment: A Virtue-Oriented Approach to Environmental Ethics* (New York: Columbia University Press, 2007).

only be with humans? Surely, we can exhibit love and care for non-human animals as demonstrated by kindly acts toward them. Some philosophers have thought it reasonable to think that perhaps to properly construct our characters requires we show love, respect, and care for the non-human natural world as well.[34]

But nonetheless, there are reasons to pause at how "Aristotelian" the proposed extensions of virtue ethics could plausibly be, at least if the concept of friendship is relied upon to do it. This extension of concern beyond humans would seem to require that one's showing care for others need not be reciprocated; while it could be argued that nonhuman animals might show signs of affection for humans, certainly plants and other inanimate natural objects do not. There is the additional issue of what level of commitment humans should give to things that do not have the ability to reciprocate. Finally, it is not clear that human/non-human animal relations would qualify as "friendships" in an Aristotelian context, for while humans could perhaps act "for the sake of non-human entities," it is unclear non-human animals could do so, and completely implausible to think marine estuaries, rock formations, sunflower seeds, or forests could.

---

34 See John Barry, *Rethinking Green Politics* (London: Sage, 1999). Also see John O'Neill's "The Varieties of Intrinsic Value," *Monist* (1992) 75: 119–137 and *Ecology, Policy and Politics* (London: Routledge, 1993).

PART 2

The Cases

CHAPTER 3

What is Natural? Does It Matter?

Case 1 – Sometimes You Might Be Able to Bring Them Back, but Should You? –

Study Questions

1) What is meant by "de-extinction"?
2) Which species of extinct animals might be candidates for de-extinction and what might make some more desirable to bring back than others?
3) What are some possible ecological benefits of de-extinction? What are some possible pitfalls of it?
4) What are some moral arguments in favor of de-extinction? What are arguments against de-extinction?

Perhaps you are a fan of *Jurassic Park*, Michael Crichton's best-selling book that spawned a blockbuster movie with the same title from Stephen Spielberg. While the special effects in the film were fabulous, many people claim (including many scientists) that *Jurassic Park's* effect on them went well beyond the stunning visual display. The underlying allure of the story came from the mosquito preserved in amber and the notion that blood it had extracted from a dinosaur carried the DNA possible, through advanced biotechnology, to re-create the Tyrannosaurus Rex, triceratops, and of course, the velociraptor. The excitement and horror of the story is generated by thoughts of the terrifying prospects of humans having to survive the onslaught of such creatures. But also, the possibility that long-extinct species might be able to return to our contemporary era left a deep impression. Of course, the idea of reviving any extinct species was pure fantasy, or so at least many of us thought just a decade ago.

All of this may be changing as our ever-growing understanding of the genomes of several animals, combined with germ-line and somatic cell engineering, sets in motion the possibilities of bringing back a number of extinct species to our world. These developments have brought a new word to our vocabulary—*de-extinction*. It has also provided a significant challenge to what has thus far been a seemingly ironclad contingent fact—extinction is forever.

Technological advances have arrived at a point where scientists from a number of different fields thought it prudent to come together recently to

 | DOI 10.1163/9789004339996_005

begin the first widely available public forum on the subject. There was a special independently organized daylong "De-extinction" series of TEDx talks (Technology, Entertainment, and Design) held in Washington, D.C. in March of 2013.[1] Of course, no one at this meeting was suggesting we should bring back a stegosaurus or like kind of animal. But other species, some recently extinct and others that have been gone much longer are strong possibilities for revival.

Some of the methods of de-extinction are actually quite conventional. Humans have fostered the back breeding of various animals for thousands of years. The practice is currently taking place in Western Europe where scientists are trying to recreate the aurochs, which went extinct in the 1600s. Current cows are derived from aurochs. Breeders can selectively breed for certain traits the aurochs had that are not manifest in current cows, such as longer horns, bigger bulk, or aurochs-specific coloration. This is a low-tech and viable way towards de-extinction of a species. The investigators in these trials have had some success, as the resulting offspring indeed do look much like aurochs.

While de-extinction *will not* mean rebirth of *identical* members of a species, it can be accomplished by approximating close reconstructions. That is, the extinct species' genomes would be reconstructed through information gleaned from preserved DNA (which would come from fossils and museum specimens). This could be combined with genetic information from closely related living species using interspecies somatic nuclear transfer (cloning).[2]

No matter how strange ancient species from prehistoric times look to us now, they share something with every other species. We share the same genetic language and are all constructed from the same DNA, the instructions that can

---

1 TED as an organization sponsors talks on several topics accessible to a worldwide audience via the Internet. TED now allows for free licenses for independently organized talks under its banner (known as TEDx events) but with one caveat, they are to be organized around a location as opposed to a topic (e.g., TEDx Boston, TEDx Rio de Janeiro, etc.). The TEDx event on de-extinction was rare in that it was organized around a topic as opposed to a location. The TED organization permitted the event because the subject matter was seen as unique, pressing, and potentially paradigm shifting. See John Fahey, "A New Century of Exploration," *TEDx De-extinction/National Geographic* (online video), March 15, 2013. Available at http://longnow.org/revive/tedxdeextinction/ (accessed December 20, 2014).

2 See Robert Lanza, "The Use of Cloning and Stem Cells to Resurrect Life," *TEDx De-extinction/National Geographic* (online video), 2013. Available at http://longnow.org/revive/tedxde-extinction/ (accessed December 2014), George Church, "Hybridizing with Extinct Species," *TEDx De-extinction/National Geographic* (online video), 2013. Available at http://longnow.org/revive/tedxdeextinction/ (accessed December 2014), and Ben Novak, "How to Bring Passenger Pigeons All the Way Back," TEDx De-extinction/National Geographic (online video), 2013. Available at http://longnow.org/revive/tedxdeextinction/ (accessed December 2014).

be used to program a cell to develop into first an embryo and eventually a mature organism. This raises many interesting possibilities. Carl Zimmer (one of the speakers at the TEDx De-extinction conference) has us imagine what it would be like if we could get a thumb drive with the DNA sequence of some extinct species. Furthermore, imagine you could obtain a DNA molecule with the matching sequence and transfer that DNA into a cell, consequently coaxing it to divide. Would you be able to bring such a species back from extinction, even if it had been gone for millions of years?

It likely wouldn't work for dinosaurs, as Zimmer explains, for several reasons. Recall the premise of *Jurassic Park* is that after taking blood from a dinosaur, a mosquito then became preserved in amber. In the story, the blood of the dinosaur was then in turn extracted from the mosquito and from these blood cells a dinosaur was cloned. This is where the science portion of this science fiction story falls apart. DNA disintegrates over time after an animal dies. The preservation of DNA relies on many factors, most of which scientists do not completely understand. But there are some things we do know. How quickly an animal is buried after death, the depth at which it is buried, and most importantly, the consistency of the temperature at which that tissue is kept, will all dictate how long that animal's DNA is able to survive. Immediately upon death, enzymes start slicing apart the DNA and bacteria feed on it to the point where there is only an array of fragments remaining thousands and millions of years later. Scientists have even been able to determine how quickly this degradation process can occur—sort of like in the case of uranium, DNA has a half-life. We have lost too much DNA of dinosaurs over millions of years for viable revival as the deterioration has occurred continuously.[3]

Yet, recently in the scientific community there have been quite serious discussions about the possibilities of resurrecting a host of recently extinct species such as the thylacine, the ivory-billed woodpecker, and the passenger pigeon. Not only have recently extinct animals been considered for de-extinction. Some Pleistocene megafauna are thought to be plausible candidates as well, including the wooly mammoth, the wooly rhinoceros, and the Irish elk.

Cloning has been a possible pathway for resurrecting extinct creatures as well as conserving currently imperiled species populations.[4] The first cloned

---

3 Carl Zimmer, "(Some) Extinction is (Not Necessarily) Forever," *TEDx De-extinction/National Geographic* (online video), 2013. Available from http://longnow.org/revive/tedxdeextinction/ (accessed December 2014).

4 See Robert Lanza et al., "Cloning of an Endangered Species (*Bos gaurus*) Using Interspecies Nuclear Transfer," *Cloning* (2000) 2:79–84, Pasqualino Loi et al., "Genetic Rescue of an Endangered Mammal by Cross-species Nuclear Transfer Using Post-mortem Somatic Cells,"

mammal, Dolly the sheep, was born just three years after the theatrical release of *Jurassic Park*. The cloning of Dolly was a very difficult endeavor for Dr. Ian Wilmot's team, involving many failed attempts, with Dolly herself dying after a very short life due to malformed lungs. Nonetheless, the cloning of Dolly represented a significant breakthrough in genetics and scientists have made huge developments in cloning techniques since that time. Hundreds of animals are now cloned every year.

Cloning can be accomplished with genetic material frozen for a considerable amount of time. Zimmer notes that a mouse frozen for 16 years supplied DNA extracted by Japanese scientists. Nuclei (the sacks that contain DNA) pulled out from the frozen mouse were inserted into the mouse eggs of a female who produced healthy pups. As mentioned earlier, the key factor in cloning a dead animal is not so much the time that its DNA has been around, but instead the consistency of the temperature of the immediate environment of the donated genetic material.[5]

This experiment leads to the possibility of finding frozen material from extinct species that might be preserved well enough to result in a viable clone. Wooly mammoths became extinct 3,400 years ago with frozen samples of its tissue existing in Siberia, the Arctic, and Canada. Perhaps somewhere in those samples, geneticists could find an intact and viable nucleus.

But even if such a nucleus can't be found, there are other possibilities for reviving extinct creatures. While DNA falls apart as part of the decomposition process, it is possible to find missing parts of genomes. These can then be pieced together to figure out what the original genes looked like. Researchers have developed novel techniques to extract and isolate desired DNA from other materials (including bacteria and fungi) that are drawn out in the extraction process. The missing material can be found via a number of different conduits. DNA can be extracted from the fossils of saber-toothed tigers, now extinct for 10,000 years. From this, scientists have been able to build out the sequence of the saber-toothed tiger genome. But material needn't only come from fossils of extinct animals themselves. Scientists have found bits of giant ground sloth DNA in its fossilized dung and can build its genome by looking at the DNA left behind. The elephant bird of Madagascar went extinct 1,800 years ago, which

---

Nature Biotechnology (2001) 10: 962–964, and J. Folch et al., "First Birth of an Animal from an Extinct Sub-species (*Capra Pyrenaica Pyrenaica*) by Cloning," *Theriogenology* (2009) 71: 1026–1034.

5 Hendrik Poinar, "Not All Mammoths Were Wooly," *TEDx De-extinction/National Geographic* (online video), 2013. Available at http://longnow.org/revive/tedxdeextinction/ (accessed December 2014).

was an animal that grew to be 10 feet tall and weighed 8,000 lbs. However, its DNA can be pulled out of the fossilized remains of its eggshells. DNA can also be pulled out of the dirt. In Alaska, scientists can dig through the permafrost and pull out DNA fragments to isolate mammoth DNA. Between the DNA obtained from dirt and frozen mammoth tissue (a mammoth preserved in the permafrost will have around 50% of its DNA being mammoth),[6] scientists now have much of the information they need to construct the entire genome of the species. Scientists can now figure out what gives mammoths hair, what allows them to survive in extremely harsh, cold environments, and perhaps even what made them go extinct in the first place. Research of this type may even allow us, in a more general way, to learn why some species are more vulnerable to extinction than others. This information could contribute to a growing corpus of work allowing us to avert the extinction of currently imperiled species.

George Church (a geneticist and molecular engineer at Harvard University) and others are working on ways to use this genetic information to bring back species from extinction, with the help of other species. While scientists cannot yet construct an entire DNA molecule from a sequence, Church and other synthetic biologists have been able to synthesize large chunks of it. If we can identify what characteristics "make a mammoth a mammoth" and differentiate them from other species, one may be able to insert them into the genome of an existing species, such as an Asian elephant. This would result in a cell with Asian elephant DNA with mammoth genes in it. Since most of the wooly mammoth genome has been reconstructed,[7] the goal would then be to implant this into an Asian elephant and see if this results in a mammoth.[8]

With that said, there are several steps to bringing an extinct species back to life and they are not all easy. First, as mentioned, scientists need to sequence the genome of the extinct species. They need a blueprint of the genes that code for the proteins that in turn, make the body cells of the organism. Then these genes need to be inserted in the chromosomes, because this is the way cells carry genetic material. The next step is to first insert the chromosomes into the nucleus, then into the cells, and coax the cells to divide.

Finally, researchers need to place the embryo into a surrogate mother and the surrogate has to take the fetus to term. This can be problematic since there might be a significant size difference between the fetus of an extinct animal and the normal fetus of the surrogate. In the case of the wooly mammoth, the Asian elephant surrogate is much smaller than the mammoth and it is unclear

6 Poinar, "Not All Mammoths."

7 Poinar, "Not All Mammoths."

8 Carl Zimmer, "(Some) Extinction."

if the surrogate will be able to successfully carry the fetus to term without endangering both the surrogate and the fetus. Even upon a successful birth, a baby mammoth will need direction in how to survive and behave like a mammoth. Also, they will need a hospitable environment to come back to.[9]

Despite these difficulties, some may argue the techniques used for de-extinction of some species or being proposed for others have already been utilized for what is called *genetic rescue*. Endangered species such as the Florida panther and the Scandinavian wolf have been involved in this procedure. Basically, new individuals from other closely related species are introduced into a breeding program to provide the genetic diversity, and eventually the numbers, for the endangered species to recover. In the case of the Florida panther, eight female cougars from Texas were brought in to crossbreed with the panthers, as they are a closely related subspecies. When this integration began in 1995, there were about two-dozen Florida panthers remaining. Today, the population has swelled to 100–180 panthers and is climbing. These cats have shown fewer genetic abnormalities than what was afflicting recent previous generations that resulted from inbreeding.

IVF procedures, sometimes using banked semen from members of an imperiled species, have been used to increase their populations, such as in the case with the black-footed ferret. At the San Diego Zoo, the most extensive collection of the cells of extinct animals in the world is part of what is called "The Frozen Zoo." These cells have been used extensively in biomedical research, as well as in conservation efforts. Having such a repository with samples readily available allows for the enhancement of work done on the sequencing of the genomes of several extinct species. Scientists may be able to reprogram adult cells to take on the functions of pluripotent stem cells. Induced pluripotent stems cells could be an additional conduit for the rescue of species due to increased genetic variation, as in some species they have produced sperm and eggs. For some "doomed" species, rapid development and integration of this technology may be the only way to save them, such as with the Northern white rhinoceros, of which there are only seven left, and only four of those have reproductive capability.

As you can probably guess by now, much of the philosophical debate with respect to de-extinction is moral in nature. Beyond the question of whether extinct creatures *can* be brought back to life is the puzzle of whether they *should* be. We can begin with what some see as the promise of de-extinction

---

9 Beth Shapiro, "Ancient DNA: What it is and What it Could Be," *TEDx De-extinction/National Geographic*, (online video), 2013. Available from http://longnow.org/revive/tedxdeextinction/ (accessed December 2014).

and the ethical reasons for pursuing it. Supporters of de-extinction base their arguments not only on the value of wonder (seeing once extinct animals revived would likely inspire awe of nature) or on the values of resurrecting lost species as a conservation tool. They also think considerations of justice ground the argument for de-extinction. One argument for de-extinction comes from *restorative justice*. Since human activity has caused a plethora of species to go extinct and we may now have the financial and technological capabilities to bring them back, we ought to do so. Hunting and destruction of habitat on a large scale eradicated species such as the passenger pigeon, the thylacine, and the dodo bird. These are anthropogenic causes of species elimination that leave humans with a specific obligation to revive them. This argument involves an element of "making up for" or restoring a species from a great wrong that was done to it. In sum, since the activities of humans contributed greatly to the loss of these animals, but we have since developed (or could develop) means to bring these species back, there is a moral requirement that we do so.[10]

Another argument for de-extinction is that perhaps it can reestablish lost value in the world. Obviously, there are several different senses in which a species might have value. It could be ecologically valuable and perform functions valuable to ecosystems. Or, a species could be valuable to humans who desire having it around. A species could provide other *instrumental* value to humans (perhaps by cleaning water through filter feeding). Some even think each species has *intrinsic* value (i.e., is valuable in and of itself). Given these many types of value a species may have, it is not surprising that some argue the problem with extinction is that it inherently involves loss of value. We can have serious debates about what type of value it makes sense to say a species has, but it seems unlikely that any species possesses *no* value. The wonder of de-extinction is that it might allow the opportunity to reclaim some of that lost value.

Holmes Rolston has argued the *natural-historical value* of species is quite great. Others believe, especially with keystone species, there is a deep dependence of other species on them. *Keystone species* create and maintain ecosystems vital to a bevy of other species. We already have evidence that the reintroduction of certain species shows what immense ecological value they

---

10 See Michael Archer, "Second Chance for Tasmanian Tigers and Fantastic Frogs," *TEDx De-extinction/National Geographic* (online video), 2013. Available from http://longnow.org/revive/tedxdeextinction/ (accessed December 2014), and Stuart Brand, "The Dawn of De-extinction: Are You Ready?" *TED Long Beach*, 2013. Available from http://www.ted.com/talks/ stewart_brand_the_dawn_of_de_extinction_are_you_ready.html (accessed December 2014).

can have. Since the reintroduction of wolves to Yellowstone National Park, conservation biologists and ecologists have acknowledged not only their positive ecological effects, but also how they even enhance the physical geography of the area. George Monbiot points up some of these advantages.[11] Ecologically, wolves have helped to increase the number of rabbits by curbing the number of their predators, such as coyotes. Monbiot adds that for humans to regain a sense of adventure, *rewilding* of once wild places can reinvigorate us—at least those of us who are "bored" with the status quo of a relatively "tame" level of biodiversity.

Advocates of de-extinction insist it is not merely regaining what was lost in the extinction of various animal species, but that de-extinction offers a promise[12] of gaining value that perhaps we've never had. This is likely true if a species has been extinct for some period of time (more than one generation). Beyond just the tremendous scientific and ecological achievements of successful de-extinction, scientific knowledge would likely be gained from examining the behaviors, phenotypes, and genetic structures of animals and their descendants. In these ways, advocates of de-extinction claim there is clear scientific value in pursuing it.

Cultural capital may also be created by de-extinction. In this line of thought, we could imagine those concerned with the environment might be given new hope in conservation by seeing formerly extinct animals in zoos, wildlife parks, and other locations. This may inspire further conservation efforts. Cultures sometimes take pride in animals that once were prevalent or held a high historical place in their regions and this value should not be undersold. Scientists have argued that some species such as the passenger pigeon in the US and the thylacine in Tasmania serve as examples of this kind of valuing. As the philosopher Ronald Sandler has noted, "When revived species are valued by people in these and other ways, and so desired by them, then those de-extinctions have value for that reason."[13] Ecotourism may also arise to provide economic value as people are drawn to witness revived species. As an addition to the conservation biology toolkit, more public support could be marshaled as de-extinction may provide a "shot in the arm" to what some may see as a tired, well-worn

11 George Monbiot's *Feral: Rewilding of the Land, the Sea, and Human Life* (Chicago: University of Chicago Press, 2014), pp. 84–85.

12 See Jacob Sherkow and Henry T. Greely, "What if Extinction is Not Forever?" *Science* (2013) 340: 32–33.

13 Sandler, "The Ethics of Reviving Long Extinct Species," *Conservation Biology* (2013) 28(2): 354–360. See also his *The Ethics of Species* (Cambridge: Cambridge University Press, 2012).

conservation strategy of merely trying to manage and lessen loss of habitat and still existing species.

As we have already seen with genetic rescue, some believe cloning techniques refined via de-extinction could even serve imperiled species. DNA could be taken from such species and preserved so that, should it go extinct, it could be revived later, when habitat is re-established enough to give the species a second chance. Or perhaps cloning techniques could be utilized to fortify a greatly imperiled species before it goes extinct. Some believe a morally responsible scientific community would take such measures to try to keep more species from going extinct due to human activities.

For all of these well-intentioned arguments for pursuing de-extinction, it is unclear if they really hold up under close critical scrutiny. Even if some plausible arguments for de-extinction obtain, they may not be strong enough to show that the practice should be a top-tier priority for conservationists. As we shall see, Sandler brings forth a number of doubts about the relative strength of the case for de-extinction. In the background, some worry that while geneticists and others doing primary work in the biological sciences (including synthetic and cell biologists) and paleontology are proceeding "full steam ahead," ecologists have not really been involved in the conversation about de-extinction. De-extinction faces the same kinds of challenges as the reintroduction of currently existing species. That is, revived, extinct species may be returned to a world where most ecosystems have passed them by. After all, one challenge to the reintroduction of species is the concern they will become invasive. Couldn't revived species cause unintended spillover adverse effects on other species in the ecosystems to which they are added?

With that said, Sandler raises several counterpoints to the arguments in favor of de-extinction. With respect to the justice claim made by some of its advocates, it is very difficult to get a handle on what it means to say "we owe something to species" whose existence was taken away by anthropogenic means. There are a host of concerns here. If we mean by that statement that humans *as a species* caused harm to another species, we have to look much more closely at what it means for something to be a species and to what kind of entity duties can be owed.

Normally, we owe obligations to those with decision-making capacities. Part of the make-up of these individuals is that they have rational capacities (even if they don't always put them to use). They engage in goal-directed behavior—they have agency. Do these characteristics map on to a species? It doesn't appear so. It is hard to see what the goal-directed behavior of Florida panthers as a species might be. Without this, as Sandler says, "It is not possible to impair their capacity to achieve their ends."[14] Species do not have conscious plans for action. They don't have minds. Thus they lack attitudes, desires, and intentions. You cannot harm or wrong a species because it doesn't have aims and interests. And you can't redress something to some entity you have not wronged.

When we say we owe something to, say, the aurochs, perhaps we mean something else. One might argue it isn't that we owe some sort of restorative justice to the species *Bos Primigenius*. Instead, since human activities eliminated them, we owe something to (the memories? legacy?) of each *individual* aurochs wiped out. But this doesn't follow either. The aurochs which once roamed Europe are now gone. How does bringing them back serve as a measure of justice toward non-living aurochs? Moreover, the generations that caused the extinction of the aurochs no longer exist. Why should present generations of humans have the obligation (and thus have the burdens of expending time, effort, and treasure) on de-extinction of this species? Critics of de-extinction thus claim the restorative justice argument seems to fall flat in light of these counter arguments.

Another possibility might be that some sort of wrong has been done to the natural world in eliminating certain species. Correspondingly, the argument could be that humanity has an obligation to "make right" for having eliminating species by reintroducing or resurrecting them. But then again, this sort of argument will not work for the same reasons that they will not work for restoring justice to an extinct species. Ecosystems and nature have no more goal-directed capacities, interests, or welfares than species do. As Sandler adds, "They do not have minds or intentions; they are not alive; they are not sufficiently internally organized and goal-directed to have a good of their own."[15]

What binds the moral arguments in favor of de-extinction together is an inchoate sense that something has gone terribly wrong with humans causing a high number of extinctions. It is understandable why this scenario might cause alarm and thus why some believe there ought to be measures taken to rectify it. Some analogs from interpersonal ethics come to mind in this case, such as some cases of homicide and slavery, where both perpetrators and victims of injustice no longer exist. Nonetheless, it seems restorative justice doesn't really work as a justification for de-extinction.

What might be done instead is to acknowledge the alarm that so many species are going extinct and the reasons for this. Then the message may be to be careful in how we care for species in the future. The obligation, in other

---

14 See Sandler, "The Ethics of Reviving," p. 355.

15 See both Sandler, "The Ethics of Reviving," p. 355 and *The Ethics of Species*, p. 212.

words, becomes operationalized in trying to keep further species from becoming extinct. If need be, institutions should be reformed to reduce the chances of future harms being done. In the context of conservation biology, we would have what is more appropriately called a *rehabilitative* role in reducing as much as possible the effect of species extinctions on ecological systems and on human health and welfare. This would also call for measures to stem the tide of species extinctions, especially given we are in the sixth major period of species extinction in the history of our planet.

With respect to arguments for de-extinction based on the idea that we can reestablish value we have lost due to species extinctions, there are limits to their effectiveness. First, it is not at all clear that de-extinction will bring back the sorts of relationships that constitute that value. The species emerging from de-extinction will only strongly resemble their ancient counterparts. The de-extinct wooly mammoth will not be an exact genetic match to the wooly mammoth of the Pleistocene period. Even if we were to overlook this fact of de-extinction, these species will not be independent from humans. If it is the perceived wildness of certain creatures or their independence from humans that ground their value, it is unlikely that de-extinction will really reestablish that value. After all, the ecological context into which the wooly mammoth would reemerge if de-extinction works will be quite different, and this is correspondingly true for any revived species that has been extinct for a longer period of time. Presumably, the cloning technique itself will diminish the possibility of reestablishing any value for a species we could bring back on the grounds that we are bringing back a "natural" species.

For some species that have not been long extinct, such as the passenger pigeon in the eastern half of the United States or the thylacine in Tasmania, the ecological context may not have changed enough to stymie their successful reintroduction into their original ranges. However, even in these cases, it is not just reviving such species that will sufficiently resurrect their value. It is hard to believe that reviving thylacines just to be held captive as zoo animals will reestablish their value for some admirers. But for revived species to survive in the wild would mean having a habitat conducive to their flourishing. The restoration of the thylacine must be done in the context of a restored ecological system to support them. This may prove to be logistically difficult, as revived species will have a lack of genetic biodiversity and a dearth of behavioral models.

To the arguments that de-extinction may *create* value for extinct species, a stronger case may be made. It would be hard to deny that bringing back an extinct species will likely create *some* new value, though it may not be for the species itself. Such revival would highlight the kind of novel technological advancements revealing science in its best light. The fields of synthetic biology and genetics, along with other disciplines, could revel in meeting important goals and perhaps in the process would further profit from gaining some unexpected knowledge of how species function. Perhaps the work would reveal the functions of genes in a wide assortment of animals. This line of argument might be the best hope for making the case for de-extinction.

Finally, with respect to the argument that de-extinction offers promise as a *last resort for conservation*, there may be problems. The main one is that de-extinction doesn't seem all that promising in reducing the chances endangered species will go extinct. De-extinction also doesn't really address the primary factors for why species vanish in the first place. Reviving an extinct species by itself does not alter loss of habitat, climate change, pollution, and human predation. Furthermore, de-extinction does nothing to reestablish the ecological context of extinct species. Even if you revive an extinct species, this in itself will not ensure that there will be an ecosystem for it to re-enter, especially with long extinct species. For many extinct species, their habitat simply no longer remains. The ecological costs might be too high to reintroduce a species (they may become an invasive species if introduced into an ecosystem that has not had the species before as a competitor). Also, the revived species may not be able to build a sustainable population size due to lack of species biodiversity or behavioral elasticity.[16]

Again, long extinct, revived species will not be comprised of organisms genetically identical with their extinct counterparts. They will be at best genetic hybrids, coming into a very different ecological context. We also don't know if they will behave like their extinct counterparts. In cases of species that went extinct long ago, we likely have little to no knowledge about how they acted, and thus wouldn't even have a template with which to compare resurrected species. The value of extinct species and the reasons why people generally want to see them revived goes well beyond their genetic make-up or simply having the opportunity to see them in captivity. What people value, as Holmes Rolston notes, comes from seeing salmon swimming up rivers to their

16 See Sandler, "Ethics of Reviving," p. 357, David Ehrenfeld, "Transgenics and Vertebrate Cloning as Tools for Species Conservation," *Conservation Biology* (2006) 20:723–732, Ferris Jabr, "Will Cloning Ever Save Endangered Animals?" *Scientific American*, 2013. Available at http://www.scientificamerican.com/article.cfm?id=cloning-endangered-animals (accessed December 10, 2014), and Stuart Pimm, "The Case Against Species Revival," *National Geographic*, 2013. Available at http://news.nationalgeographic.com/news/2013/03/130312--deextinction-conservation-animals-science-extinction-biodiversity-habitat-environment/ (accessed December 15, 2014).

spawning grounds, passenger pigeons blackening the sky in their mass migration, and polar bears roaming the Arctic ice.[17]

More worryingly, conservation cloning works on a species-by-species basis and the progress made will likely be too slow to retard the conservation crisis we are facing, with dozens of species of animals going extinct every day and 30 to 50% of all species vanishing by 2050.[18] It is expensive to do this research. With scarcity of resources, there is a moral impulse to steward them as best as possible. This calls into question whether, despite its allure, the focus on de-extinction is really the best use of time, treasure, and expertise for conserving species overall.

Now that we have surveyed several possible arguments in favor of de-extinction and points of contention against such arguments, we need to turn to arguments that could possibly be made against the practice. The first one has already been touched upon—that there is something unnatural about de-extinction. This is not a surprising claim, as cloning and other sorts of genetic modification have long been criticized on the basis that they and their products are artificial and consequently morally tainted. This charge could come from a religious perspective. If one believes in a monotheistic notion of God, and furthermore thinks that such a being is a creator of the natural world with sole domain over that creation, any attempt at using human scientific innovation to conduct acts resulting in the creation of animal species would transgress the laws and providence of God.

But even without this set of religious beliefs, some argue nature is its own steward, best left without human interference. They worry that science in the service of genetic modification circumvents and/or subverts the natural order of things. Surely, in many cases of de-extinction, the process will involve the mixture of genetic material from different species that could not have bred naturally or hybridized without human intervention. Since de-extinction would certainly use cloning and combined genetic material from multiple species, some critics will object to these methods even if the purpose is well-intentioned.

Is there something about species boundaries (that is, are they ethically significant in some way) that suggests transgenic biotechnology would violate them? As Sandler points out, species created with the use of transgenic biotechnology resulting in de-extinction would not be as natural as species

---

17 Holmes Rolston, "Duties to Endangered Species," *Encyclopedia of Environmental Biology* (1995) 1: 517–528.

18 See Elizabeth Colbert, *The Sixth Extinction: An Unnatural History* (New York: Henry Holt and Company, 2014).

that do not undergo this process.[19] But, the lines between species are often much more blurred than is commonly held. According to Sandler, interspecific genetic hybridization occurs in nature. He thus concludes, "[T]he mere existence of transgenic organisms is neither unusual nor ethically problematic."[20] It is not obvious that just because something has less natural value as the result of human intervention that it is ethically problematic. There are simply different species that may be labeled more or less natural. If it did matter, as Sandler quips, "then vaccines and low-fat yogurt would be ethically problematic."[21] As such, the unnaturalness argument isn't robust enough to justify stopping or even curtailing de-extinction.

One might object to de-extinction on the grounds that the procedures used to produce offspring can cause animal suffering. There is now a substantial literature on concerns about animal welfare, with the work of contemporary utilitarian philosopher Peter Singer at its vanguard. Interspecific cloning has a low success rate thus far and can lead to adverse health abnormalities for donors, surrogates, and offspring. In cases where the species to be revived has been long extinct, these abnormalities can be more frequent as there is less information about the species and greater possibility of underlying genetic imperfections. With Bucardo, after several attempts (only 7 goats became impregnated with Bucardo) there was finally a successful birth, but the animal only survived for about three days. Consequently, the concern about animal suffering from the processes used for de-extinction is legitimate.

Essentially and ecologically there are also questions about what the results of de-extinction might be. Ross McPhee, a Curator at the American Museum of Natural History, wonders whether the yield of synthetic biology is really the species that scientists are trying to revive. Is the animal that results from back breeding of modern cows with the traits of an auroch really an auroch? Is a constellation of traits scientists select for really an auroch? After all, we don't have a living auroch with which to compare a back-bred one to know if what has been created is close physiologically to an aurochs pre-extinction. As was pointed out earlier, we also won't know if the behavior of a back-bred aurochs is like that of an original aurochs because its temperament does not fossilize.

All of these issues may point to a final larger consideration with respect to nature. One might argue that animal extinction is a rather normal phenomenon, as about 99% of species ever to appear on Earth have gone extinct. Perhaps this situation is simply "the way of nature." On the other hand, if

19 Sandler, "The Ethics of Reviving," p. 357.

20 Sandler, "The Ethics of Reviving," p. 357.

21 Sandler, "The Ethics of Reviving," p. 357.

humans have primarily caused this latest mass extinction, what sort of obligations might this fact entail?

Review Questions

1) Many of the species seen as candidates for de-extinction are fauna (animals). In fact, at the TEDx de-extinction conference cited above, *no flora* (plants) were noted as candidates for de-extinction at all. Why might this be? Do you think any extinct plant species should be revived? Why or why not? Does the lack of discussion about the possible de-extinction of flora species tell us anything about how they are valued?
2) Some have argued that de-extinction is a bad idea from the perspective of ecological conservation because it serves as a distraction and diversion of resources away from efforts to keep endangered species from going extinct. Do you agree with this reasoning against de-extinction? Why or why not? Do you think it is a strong enough moral argument to try to persuade scientists from pursuing de-extinction? Why or why not?
3) What is genetic rescue? In what way is it related to de-extinction? Does genetic rescue raise any of the same concerns as de-extinction? Why or why not?

Case 2 – "Natural" Spring Water? –

Study Questions

1) What are some potential hazards of making the claim that something is "natural"?
2) What services does the National Sanitation Foundation (NSF) perform?
3) What is the US Food and Drug Administration's (FDA's) role in the regulation of bottled water? What is the role of state regulatory agencies?

A company called National Sanitation Foundation (NSF) International (based in Ann Arbor, Michigan) tests bottled water products to certify they meet United States federal guidelines for water safety. This is part of a voluntary, annual recertification process for beverage businesses that wish to be able to advertise their water as not only "safe," but also "natural." NSF tests for the possible introduction of some 160 inorganic, chemical, radiological, and microbiological contaminants in bottled water sources. NSF's testing program includes unannounced on-site visits where inspectors ensure that water companies are complying with guidelines for the treatment and handling of water products. Perhaps not surprisingly, many companies whose water meets the requirements for NSF certification do not merely state that their water is free of these separate chemical agents. They also claim that their water is "natural" with the implication that for something to be natural is somehow "better" than other products. While the federal certification NSF International bestows on its clients whose products meet its requirements does not include any mention of "natural," the company's marketing brochures imply that they test to make sure their clients' water is "natural."

NSF International was founded in 1944 and as its mission statement indicates, its aim is to protect and improve human health. As gleaned from their website, the company mission goes on to state that "Manufacturers, regulators, and consumers look to us to develop public health standards and certifications that help protect food, water, consumer products and the environment. As an independent, accredited organization, we test, audit and certify products and systems as well as provide education and risk management."[22]

Beyond the certification services that companies such as NSF International provide, there is a background layer of regulation on bottled water with respect to its safety and labeling. Along with US Food and Drug Administration's general safety standards for bottled water, state and local governments regulate these products. The FDA depends on state and local government for ensuring sanitary and safety quality control, as specified in Title 21 of the *Code of Federal Regulations* (21 CFR) in part 129.3(a). So, approval of bottled water sources is the responsibility of the state from which it is sourced. For example, in the state of Pennsylvania about 75% of bottled water originates from approved natural sources such as springs or wells. Pennsylvania state law requires that these sources be protected from contaminants. The remainder of the bottled water sourced in Pennsylvania comes from municipal wells approved by state agencies—these sources are simply municipal wells that also serve drinking water to their respective communities.[23] Some states have regulations that differ from the FDA's in both content and coverage. The state of Texas requires that water transporters who haul water in tanks or trailers must keep a minimal

22 NSF International, "Mission," (website), 2015. Available at http://www.nsf.org (accessed April 22, 2015).

23 Penn State University Extension, Colleges of Natural Sciences, "Water Facts 9: Bottled Water," 2015. Available at http://extension.psu.edu/natural-resources/water/drinking-water/water-testing/pollutants/bottled-water (accessed April 22, 2015).

chlorine residual of 0.5 mg/L in the water. However, the FDA has no such requirement.[24]

The FDA has labeling definitions to help consumers sort out different types of water, making distinctions between artesian, drinking, mineral, purified, sparkling, spring, and well waters. In the case of bottled waters, its source and other information may be available on its packaging, but this is not a requirement.

Bottled water is often treated by methods such as distillation, reverse osmosis, filtration, deionization, ozonation, or even ultraviolet light disinfection. According to the FDA's bottled water classification system, what is labeled as "bottled water" cannot contain sweeteners or chemical additives, and must be calorie and sugar-free. Yet, this doesn't mean there can't be any additives in bottled water as such. For instance, bottled water can contain extracts and "essences" derived from fruit and spices if they constitute less than one percent by weight of the final product. Some bottled waters may be carbonated either naturally or by additional treatment.[25]

In Pennsylvania, state authorities do not only regulate bottled water sourced in the state, but also oversee any foreign brands as well. Unlike other kinds of water regulated by the US Environmental Protection Agency, bottled water is placed under the purview of the FDA. The agency sets requirements on water quality, labeling and manufacturing practices. For the most part, the FDA quality requirements for bottled water follow those the EPA imposes on drinking water that flows through taps. Other characteristics of bottled water, regarding whether they can be labeled as "spring water," must fit the specific FDA definition of that term. The FDA identifies "spring" water as that which "flows naturally to the ground surface or through a bored hole tapping an underground formation that feeds a spring. Spring water may undergo some treatment, but it must maintain the same properties as the natural spring water."[26]

The FDA also sets and enforces what they call "Good Manufacturing Practices." These govern facility design and function, process controls, record keeping,

---

24 Lauren M. Posnick and Henry Kim, "Bottled Water Regulation and the FDA," US Food and Drug Administration, Center for Food Safety and Applied Nutrition, *Food Safety Magazine*, 2002. Available at http://vm.cfsan.fda.gov/~dms/botwatr.html (accessed April 28, 2007).

25 Lauren M. Posnick and Henry Kim, "Bottled Water Regulation."

26 Lauren M. Posnick and Henry Kim, "February/March 2002 Ask the Regulators—Bottled Water Regulation and the FDA," (website), 2002. Available at http://www.fda.gov/Food/FoodborneIllnessContaminants/BuyStoreServeSafeFood/ucm077079.htm (accessed April 24, 2015).

and sanitation conditions of fixtures in buildings. More specifically, bottled water must be processed, bottled, held, stored, and transported under sanitary conditions. The FDA is also charged with requiring bottled water producers to protect water sources from bacteria, chemicals, and other contaminants. Bottlers are required to utilize quality control processes to maintain the chemical and bacteriological safety of the water. Finally, the FDA requires water bottlers to sample and test not only the source water for contaminants, but also the final product.[27]

These regulations are generally enforced via site inspections at bottling facilities. During inspections, the agency examines the plant's product water and operational water supplies to ensure they are from approved sources. FDA also inspects washing, sanitizing, and bottling operations. Finally, they determine whether bottling companies have analyzed their source water and final water product for contaminants.

Bottled water is big business in the United States. According to the International Bottled Water Association, bottled water was the second most popular beverage in the US in 2005. Americans consumed more than 7.5 million gallons of it, translating into an average of 26 gallons per person. Only carbonated soft drinks outsell bottled water.[28]

Stepping back from the procedures of regulating water, we can turn to questions about what this means to consumers. Many consumers drink bottled water because they claim it is "better" than tap water. In *what* ways is it better, if it indeed is? We might need to take an "all things considered approach" to answer this question. That is, not only might we compare bottled water to tap water according to its contents, we might need to think about how each type of water is collected and processed. Tap water is just that, the stuff that flows from the tap. Bottled water involves packaging and there is the question of what is the end point for the bottles themselves. Some bottles will surely make their way into landfills and others will be recycled. What are the costs of these results in terms of resources and environmental impact?

It is hard to know what the "properties" of "natural spring water" are exactly. One might worry that the definition is circular—that is, spring water is that water which has the properties of natural spring water. Also, if spring water is treated, in what way is it still "natural"—assuming that natural means

---

27 United States Food and Drug Administration, "FDA Regulates the Safety of Bottled Water Beverages," *Food Facts*, (Fact Sheet). (Washington, DC: US Government Printing Office, 2009). Available at http://www.fda.gov/downloads/Food/FoodborneIllnessContaminants/UCM239631.pdf (accessed April 24, 2015).

28 United States Food and Drug Administration, "FDA Regulates."

"unaltered." We might also point out that it is not obvious why water tapped from an underground rock formation is "spring water." How is spring water then differentiated from water drawn up by a municipal water source? In what sense is tapped water "from a spring"? Furthermore, one who thinks natural spring water is unaltered will likely be confused when she finds out that in the US it or may not be treated. It would seem that treated water would have to be altered in some way by definition.

Answers to these questions of definition are likely important because they seek to clarify the ethical extraction and use of water. The normative questions involve whether it is ethical to advertise bottled water as natural. But there are also questions regarding other claims from some water bottling companies that they are "ethical brands" to buy. For example, Vinicius Brei and Steffen Bohm noted in a 2011 paper that some companies promote being able to serve "poor African people" with potable water by way of Western consumers purchasing bottled water.[29] Brie and Bohm conducted a study on 10 bottled water companies that claim to be "ethical" brands. Yet, their results demonstrated that bottled water companies try to influence consumers' tastes by managing the cultural meaning of bottled water. Thus, they aim to produce what the authors conclude is "a more 'ethical' and 'socially responsible' perception of their products/brands."[30] Brei and Bohm also argue that using this marketing strategy allows bottled water companies the ability to stand out in a saturated and highly competitive market. According to Brei and Bohm, these strategies allow companies to distract consumers from any attacks on the companies' environmental performances. Cultivating this image of environmental stewardship is the motivation for bottled water company campaigns to draw attention to their efforts in developing countries.

Of course, accepting this conclusion will likely depend on whether we agree that the meanings of customer goods can be culturally constructed. We have to think about whether (or to what degree) marketing campaigns can turn a seemingly mundane item such as bottled water into "a meaningful, cultural product that is not merely a utilitarian good but a symbolic message about who[m] we, as consumers, are."[31] For example, to the charge that companies merely advertise environmental commitments in developing countries as a distraction, industry advocates might respond by thinking it is a very odd and

29 "Corporate Social Responsibility As Cultural Meaning Management: A Critique of the Marketing of 'Ethical' Bottled Water," *Business Ethics: A European Review* (2011) 20(3): 233–252.

30 "Corporate Social Responsibility As Cultural Meaning Management," p. 233.

31 "Corporate Social Responsibility As Cultural Meaning Management," p. 248.

likely self-defeating tactic to distract consumers *away* from their overall environmental record by referring to an environmental justice commitment that could easily be confirmed or denied by a bit of research. If the point is to throw a *red herring*[32] out to consumers to lead them away from focusing on the environmental records of water bottling companies, it would seem a better strategy would move them as far away from environmental concerns as possible.

Ultimately, investigation of even as common a consumer good as bottled water invites us to consider what truth in advertising means and why it matters. Questions about whether the concept of the natural is merely "a term of art" that can be used to manipulate, or alternatively has some grounds in reality, will continue to arise in both business and environmental ethics.

Review Questions

1) What does it mean for water to be "natural"—especially since a good deal of water people consume comes from municipal wells containing any number of minerals? Once more, what does "natural" mean when water companies filter out materials from water extracted from their own wells?
2) Should we assume that water advertised as "natural" is "better"? Do you think this claim of "better" is merely a matter of taste (an aesthetic judgment) or instead a moral judgment? Is there anything wrong or unsavory about water companies advertising themselves as "ethical brands" or as good environmental stewards? Why or why not?
3) If given a choice, many environmentalists would likely choose "natural" products over those that are "artificial" or "processed." Given the description of bottled water above and the possible ambiguity of what it means for something to be "natural," or "spring water," are water companies who designate their water as "natural" after receiving NSF certification engaging in deceitful marketing practices? Why or why not?
4) Let's assume that no matter how it is otherwise labeled, bottled water on the market is safe. Do you think consumers should have the right to know how the water was sourced and whether anything was added to it? Why or why not? Should bottled water companies be able to call their water "pure," "natural," or "spring" no matter how it is sourced just so long as the product is safe?

---

32 A red herring is something that is intended to mislead or misdirect someone to a false conclusion.

Case 3 – And It Eats Oil! –

Study Questions

1) What did General Electric and Dr. Chakrabarty eventually receive for the latter's work on a bacterium that eats oil? In what way was *Diamond vs. Chakrabarty* a landmark United States federal court case?
2) What are the purposes of patenting? What are supposed to be the characteristics making some thing (or method) patentable?

While working for General Electric, Dr. Ananda Chakrabarty genetically engineered an organism (a pseudomonas bacterium) that has an interesting characteristic—it eats oil. While there were four existing oil-degrading bacteria already in existence when Dr. Chakrabarty accomplished his feat, they were not the most efficient cleaners of environmental messes. They would compete with each other when introduced to a spill, decreasing the total amount of crude they could degrade. Chakrabarty's bacterium outperformed these other strains by consuming more oil while doing so twice as quickly. For his work, Dr. Chakrabarty (or more accurately, General Electric, though Chakrabarty was listed as the inventor) was granted a patent on a living organism. His bacteria have since been used to help clean up a number of oil spills, including the 1989 Exxon Valdez spill in Alaska.

Patents are important because they allow the patentee exclusive rights to products, techniques, or procedures they invent for a certain period of time. While a patent is active, no one else is allowed to copy the invention without a license. This allows for a company to exclusively market their invention or to license others to do so in exchange for royalties. Thus, the patent system is supposed to provide an incentive to innovators and companies to invest the time, effort, and treasure necessary for developing new products and procedures. This is especially important with research that is financially risky—you don't always know if your best efforts are going to pay off with an invention and it may involve a great deal of resources being burned through during the research and development process.

The road to gaining a patent on Chakrabarty's oil-eating bacterium was not always easy. Initially, General Electric's patent application was rejected by an examiner on what seemed as rather straightforward grounds—it was not permissible to patent living organisms under the law at that time, according to Section 101 of Title 35 of the United States Code. The Board of Patent Appeals and Interferences agreed with the original judgment by the patent examiner.

But things turned interesting when the United States Court of Customs and Patent Appeals overturned the case in Chakrabarty's favor. They did so on the grounds that "the fact that micro-organisms are alive is without legal significance for purposes of the patent law."

The story does not end here as in 1980, Sidney Diamond, the Commissioner of Patents and Trademarks at that time appealed the case to the US Supreme Court. The case was argued on March 17 and decided that same year on June 16, with the patent being granted by the US Patent and Trademarks Office on March 31, 1981. The main provision examined to adjudicate the Supreme Court case is from Title 35 U.S.C., Section 101 that "Whoever invents or discovers any new and useful process, manufacture, or composition of matter or any new and useful improvement thereof, may obtain a patent therefor, subject to the conditions and requirements of this title." The main question for the Court to decide was whether a genetically engineered microorganism is really either "manufactured" or a "composition of matter"

The Court took the term "manufacture" in Section 101 to mean something very close to the dictionary definition of the term. That is, to manufacture is "the production of articles for use from raw materials prepared by giving to these materials new forms, qualities, properties, or combinations whether by hand labor or by machinery." With respect to "composition of matter," the term was also regarded by the Court as we do in common usage as "all compositions of two or more substances and all composite articles, whether they be the results of chemical union, or of mechanical mixture, or whether they be gases, fluids, powders, or solids."

With these terms understood in their broad senses, the Court concluded that patent law should be interpreted to mean that "anything under the sun that is made by man" can be patented. As applied to Chakrabarty's bacterium, the Court held that it had significantly different characteristics from any others found in nature. The Court also concluded that Chakrabarty's invention met the "utility" standard; as the bacterium would also have wide and important applications useful in the gas and petroleum industry for spill cleanups. Thus, since it was not simply found in nature, but of Chakrabarty's creation, the bacterium was patentable under Section 101. Chakrabarty's bacterium is distinguishable, Chief Justice Warren Burger argued, from things that are not patentable, such as laws of physics, physical phenomena, and abstract ideas—all of which cannot be patented on the ground that they are "simply there to be discovered" and could have been, hypothetically at least, unearthed by anyone. The majority opinion in the cases also noted that the bacterium is more than a new mineral discovered in the earth, or a new plant found in the wild. Chakrabarty's bacterium was not a mineral and was not found in nature, but a genetically manipulated new entity with a significant use in degrading several components of petroleum, and therefore a significant boon to oil spill cleanups.

But the patent on Chakrabarty's bacterium was not even the first ever granted on a living organism. The USPTO granted patents on single-celled organisms well before then on several occasions. One example is from 1873, when Louis Pasteur applied for and was approved a patent (US Patent No. 141,072) on a purified yeast cell. While the majority opinion in the Chakrabarty Supreme Court decision did not make reference to this fact, one might argue they could have used these earlier cases as legal precedents to rule in favor of granting the patent.

Of course, devices and techniques have also been patented over time (with several associated with living organisms). However, some have serious concerns about whether the legal criteria which the US Patent Office are supposed to use are interpreted too loosely. Nevertheless, we still need to broach the question of whether any organism *should be* patented, or if there are some types of organism and/or its parts that shouldn't be patented. One point that we likely need to keep in mind is that if part of the purpose of patenting rights is to create an environment conducive to innovation, invention, and productivity, then the Chakrabarty case is likely to be very important. Some argue that the patenting of organisms such as Chakrabarty's have had a tremendous impact on a flourishing biotech industry in the United States. Of course, it may be impossible to understand how much the biotech industry would have been stunted if Chakrabarty had lost his patent case. Yet, his win did allow a very new industry some necessary incentives for investment in a wide range of medical treatments including new pharmaceuticals, therapies, and diagnostic tools. These incentives extended into other areas such as agriculture, with the expansion of biotech companies developing a bevy of stronger and disease-resistant crops. In turn, these crops have arguably had an enormous effect on our ability to feed an ever-increasing number of humans worldwide.

But others see the patenting of genetically manipulated organisms as a Pandora's box opening to what they see as much more worrisome extensions of genetic engineering, such as the patenting of transgenic animals and human gene sequences. Public interest groups, as well as a number of religious leaders, have condemned these extensions on moral grounds. The first patent on a human gene came in 2000, when a Maryland biotechnology company, Human Genome Sciences (HGS), was granted a patent on a gene it discovered and sequenced believed to be responsible for preventing AIDS. As bioethicist Kristen Shrader-Fredette writes, "HGS discovered that people who have a mutation of their patented gene appear not to contract AIDS, even after exposure to HIV, the AIDS virus. Within five months of that patent, HGS had received more than 100 additional patents and had 7,500 pending patents on human genes. The patent on the AIDS receptor alone will give the Maryland company a monopoly over the gene, its protein, and all AIDS drugs developed through them."[33]

There is even some question of how inventive Dr. Chakrabarty was in developing his bacterium. This was an important question in *Diamond v. Chakrabarty*, since this would go a long way in determining whether Chakrabarty met the inventiveness criterion for a patent. After all, some geneticists believed that all Chakrabarty did was simply "manipulate some genes" of an already existent organism and this shouldn't pass muster with the US Patent Office as a truly "inventive step." Upon learning that the Supreme Court upheld the decision by the US Court of Customs and Patent Appeals to allow for the patent, even Chakrabarty, surprised by the Court's decision said, "I simply shuffled genes, changing bacteria that already existed. It's like teaching your cat a few new tricks."[34] On these grounds, some have argued Chakrabarty's bacterium ought not to have been patented. By extension, this would suggest that any like example of bioengineering should not be patentable either.

Other worries obtain with gene patenting beyond whether it meets the criteria of patent law. Some think gene patenting is part of the commodification of humanity that should be prohibited. Most people think markets in humans (treating them as if they are goods on the market) is objectionable. For example, slavery is in part wrongful because humans are treated as property and this degrades what it means to be human. It may even degrade the characters of people who do the buying and selling. Presumably baby selling is morally problematic for similar reasons. Perhaps more controversially, many see prostitution and organ selling not merely as unsavory, but as immoral because they commodify the human body.

Another major complaint from opponents of gene patenting is that the monopoly control companies have over patents actually stymies research by making genes less accessible and/or prohibitively expensive to work on. A patent on a gene itself, as opposed to a patent on a technique or a procedure, makes it virtually impossible to do appropriate research and development on

33 "Gene Patents and Lockean Constraints," *Public Affairs Quarterly* (2006) 20(2): 135–161 (at 135).

34 See Philip Bereano, "Patent Nonsense—Patent Pending: The Race to Own DNA," *Seattle Times*, August 27, 1995, p. B5. Also see Dale Murray, "A Lockean Argument against Gene Patenting," *Business and Professional Ethics Journal* (2001) 20 (3&4): 129–143.

the gene.[35] This can consequently render genetic tests much more expensive. Also, to be effective in conducting cutting edge research in biotechnology and developing innovative products and techniques involves many lawyers to defend patents. This funnels money that could be spent on research and development away to law firms.

Opponents of gene patenting point out that "patent trolls" further exacerbate the slow down of novel work on genes. Patent trolls are companies that accumulate patents, but then do no further work on the genes they patent. In fact, this is part of the business model. Such companies simply profit by legally enforcing patents they own instead of developing products. Biotechnology patents can be very broad and some firms purchase a vast portfolio so they can look for targets to sue for patent infringement.[36]

Others think commodification arguments are overblown. Once more, by allowing for markets in a wide variety of things, including treating human parts (goods) and capabilities (services) as commodities allows for increased routes to economic survival while providing essential goods and services to those willing to pay. This, say its proponents, extends to patents as well. Interestingly, a number of market advocates also endorse monopolies caused by patents, at least for a certain period of time. This may be surprising since it would seem that market enthusiasts would conclude that markets should always be competitive, as the drive to beat out others on the market is one of the basic principles (and purported virtues) of capitalism. But in fact, even laissez-faire capitalists believe the whole point of patents is to provide incentives to inventors by ensuring that if they are the first to develop a new product, the technique or product they devise cannot simply be stolen away. Patentees are assured that for a limited amount of time, they will have exclusive control over the fruits of their labor. Thus, some proponents of gene patenting think that the market is the most efficient way of promoting the sort of risk-taking that bringing products and techniques to consumers involves.

Review Questions

1) Some think no live organisms should ever be genetically engineered. Do you think their case is weakened when such modification is aimed at

35 "Why Are Gene Patents Controversial," *The Economist*, April 18, 2013. Available at http://www.economist.com/blogs/economist-explains/2013/04/economist-explains-why-gene-patents-controversial (accessed April 28, 2015).

36 Erika Check Hayden, "'Patent Trolls' Target Biotechnology Firms," *Nature* (2011) 477: 521. Available at http://www.nature.com/news/2011/110928/full/477521a.html (accessed April 28, 2015).

some socially valuable goal, such as cleaning up oil after a spill? Should the environmental and social usefulness of the organism be the sole justification for awarding patents on life forms? Why or why not?

2) Should Dr. Chakrabarty have been granted a patent on the bacterium he developed? Or, assuming that his technique was novel, should he have only been awarded a patent on the *technique* of deriving the bacterium? Do you think a living thing can ever be a human innovation? Explain your answer.

Case 4 – A Natural Aesthetic? –

Study Questions

1) What does the term "aesthetic" mean?
2) What does Kant mean by thinking that beauty is "disinterested pleasure"?
3) What does Kant think we mean when we ask someone, "Isn't that a beautiful sunset?"

Does it really make sense to say that Mount Everest and K2 are beautiful? If they can be, in what way; if they can't be, why not? How about professing that they are works of art? If you can't call them works of art, does that mean you can only consistently say they are not beautiful (or that we would have to suspend judgment)? These are questions that have arisen in a subfield of philosophy known as aesthetics. Philosophers working in this area take on questions concerning what is a work of art. Is art to be evaluated objectively? Or is beauty subjective, that is, merely in the eye of the beholder? Along with these questions, those working in aesthetics have also considered whether the natural world can have aesthetic properties.

Traditionally, many art critics and philosophers of art have balked at the notion that our natural surroundings can be considered works of art. This hesitance is mainly due to the idea that what defines art is that it exhibits the creative imagination of human beings and is a cultural expression—taking natural (and unnatural) objects and imbuing them with a value they previously didn't have. After all, the word "art" even derives from "artifice." This connotes elements taken out of nature and made into something unnatural by human intervention (an *artifact*). This suggests some sort of artist; someone who has a vision and is able to bring it into existence as a new artistic work. Tied up with this is the notion that the artist intends to have some aesthetic effect on the viewer. This intent could be to please the viewer or in some other way arouse her emotions. Or perhaps the artist creates a work not intending to cause a particular emotive response, but instead to convey some sort of message to the audience. In either case, it is the purposeful intention of the artist communicating with or provoking the audience member that is a key feature to artistic endeavor.

Yet, since at least around the time of Immanuel Kant (1724–1804), philosophers have seen links between the aesthetic and the natural. Kant himself spoke of how a sunset could evoke in rational beings a deep appreciation of beauty. This appreciation is a judgment he thought we could share with other similarly equipped beings. After all, when I say, "Isn't that a beautiful sunset?" I am communicating this because I think any other creature like me will be able to perceive *and* appreciate it. There are some natural phenomena we could perceive that are so awe-inspiring they challenge what Kant called the *concepts of the understanding*. The concepts of the understanding are things such as time, space, causality, and other ideas that allow us to categorize the perceptions we have of the world. In Kantian terms, experiences that challenge those concepts are those of the *sublime*. Thinking of huge violent storms or immense buildings that overwhelm our senses are instances of the sublime. When encountering something sublime, there is a sense that we can't get our head around the experience.

Kant thought aesthetic experience was to be *disinterested*. That is, instead of finding beauty in something because it pleases us, we are pleased by it *because* we find it to be beautiful. According to Kant, the judgment that we find something beautiful because it pleases or amuses us makes it agreeable, but not aesthetically satisfying.[37] I may find the taste of freshly toasted waffles to be agreeable, but I would not consequently put them up as candidates to be included in an art collection. Disinterestedness is a kind of aesthetic attitude. To be disinterested means one perceives objects in ways that are not merely in her personal interest, or to simply fulfill some purpose for her. Looking at a painting not as a commodity that can fetch a fortune, but instead examining its color, composition, meaning, and so forth, would mean that she is assessing the work aesthetically, from her disinterested perspective. Note that having a disinterested attitude towards an object doesn't mean one can't find the object interesting. The art collector only contemplating a painting in regard to its composition could still find the interactions between its subjects to be arranged in a meaningful way. So, while she is *interested* in the figures, she is still taking on a *disinterested* aesthetic attitude.

---

37 Douglas Burnham, "Immanuel Kant: Aesthetics," *Internet Encyclopedia of Philosophy*. ISSN 2161–0002, Available at http://www.iep.utm.edu (accessed April 29, 2015).

But disinterestedness allows for nature to be an object of aesthetic experience. For one thing, it means aesthetic experience does not merely need to be attached to any religious, social, personal, economic, or utilitarian meanings or interests. In fact, any of these types of interests might impede the ability to have an aesthetic experience. Kant's sense of disinterestedness in aesthetic experience allows for experience of the sublime. Instead of being so overwhelmed by the awe of seeing the Grand Canyon or the immensity of the Great Wall of China, a disinterested attitude allows us to say, "wow, that is quite a sight" in an aesthetic way.[38]

The recent interest in the connection between art and nature has risen to the point where there is burgeoning field of "natural aesthetics."[39] Ronald Hepburn, who wrote a groundbreaking essay in 1966 entitled, "Contemporary Aesthetics and the Neglect of Natural Beauty" was a pioneer in this field.[40] Hepburn criticized the artworld for failing to appreciate the aesthetic experiences emerging from the natural world, particularly experiences of natural beauty. Even so, Hepburn thinks there are two distinct differences between experiences of nature and art. As Stephanie Ross puts it, "First, the appreciator is often surrounded by, immersed in, the natural world in a special way that encourages a blending of spectator and object. Second, the natural world is itself unframed and unbounded, challenging the spectator to integrate his or her own experience. One result, according to Hepburn, is that any aesthetic quality in nature 'is always provisional, correctable by reference to a different, perhaps wider context, or to a narrower one realized in greater detail.'"[41]

With respect to the naturally aesthetic, we might also consider the work of Arnold Berleant. He thinks that even if we grant the differences between art and nature, this doesn't mean nature has less of a role to play in aesthetic

---

38 Allen Carlson, "Environmental Aesthetics," *Stanford Encyclopedia of Philosophy*, (website), January 26, 2015. Available at http://plato.stanford.edu/entries/environmental-aesthetics/ (accessed April 29, 2015).

39 This includes books and scholarly articles studying the subject such as Allen Carlson and Arnold Berleant's *The Aesthetics of Natural Environments* (Peterborough, ON: Broadview Press, 2004). Also see Sam Godlovitch's "Valuing Nature and the Autonomy of Natural Aesthetics," *British Journal of Aesthetics* (1998) 38(2): 180–197.

40 In Bernard Williams and Alan Montefiore, eds. *British Analytical Philosophy* (London: Routledge and Keegan Paul, 1966).

41 Stephanie Ross, "Review of *The Aesthetics of the Natural Environment*," *Aesthetics*, (website), 2006. Available at http://www.aesthetics-online.org/reviews/index.php?reviews_id=44 (accessed April 29, 2015). Hepburn is quoted in Carlson and Berleant, *The Aesthetics of Natural Environments*, p. 47.

appreciation. Instead of giving priority to art appreciation, he believes we should start with the appreciation of nature as the way to understand how to appreciate art. Berleant uses the idea of total engagement, which he describes as "a sensory immersion in the natural world that reaches the still uncommon experience of unity. Joined with acute perceptual consciousness and enhanced by the felt understanding of assimilating knowledge, such occasions become clear peaks."[42] It is through total engagement that we come to understand the knowledge and perspective necessary to judge whether objects are beautiful or otherwise possess aesthetic characteristics.

One implication of this work is that some environmental ethicists see a possible justification for environmental protection arising from the notion that our natural surroundings possess deep, aesthetic value. In the same way we might think it is (perhaps especially) wrongful to destroy a Van Gogh, it may also be morally wrong to destroy the Grand Canyon, Mount Everest, or some other example of nature's beautiful and awe-inspiring sights. Notice that the wrongfulness goes beyond destroying one's property. An art collector deciding on a whim to set his own personal copy of a Kandinsky painting ablaze would leave the artworld aghast. Its members would believe the collector had committed a great moral wrong, and would likely make an attempt to press legal action. This example points to some much deeper and meaningful relationship between humans and the objects of the aesthetic world, whether we create them or not.

Review Questions

1) What do you think art is? Do you think a canyon should be considered a work of art? Why or why not?
2) Can those natural formations that some call "natural wonders" have aesthetic properties and yet not be considered works of art?
3) Let's suppose sunsets and snowcapped mountains do have aesthetic value. Does such value give us compelling *moral* reasons to protect them? Should we do so with the same level of justification we might grant to what we conventionally think of as works of art such as sculptures, which are consigned to museums to preserve them from excessive handling, adverse climate conditions, or other sorts of destructive elements?

---

42 Carlson and Berleant, *The Aesthetics of Natural Environments*, p. 83. See also Ross, "Review of *The Aesthetics*."

Case 5 – To Be a Weed –

Study Questions

1) What are the four essential items plants need to promote their growth?
2) What is projected to happen to weeds in conditions where CO_2 concentration levels and temperature are greatly increased? Specifically, what are some possible worries for farmers, foresters, and land managers in the future with respect to weeds if atmospheric CO_2 concentration levels and temperatures increase?
3) What are some possible definitions of a weed? Do they all share some common factors? In what ways might definitions of a weed be subjective?

There are several definitions for a weed. Not surprisingly, farmers simply see a weed as any plant interfering with the flourishing of their desired crops and profits. Others have been kinder in their assessment. Consider Ralph Waldo Emerson's definition that a weed is "a plant whose virtues have not yet been discovered."[43] The central theme in all of these definitions, however, is that a weed is a plant held in disfavor by humans. If a plant is disliked by a significant number of us, it's deemed an intruder on the natural landscape.

However, if we take Emerson's definition more seriously and consider the historical relationship between humans and weeds, we find that judgments that some plants are "weeds" are rather subjective. Consider that many of what are now generally termed as "noxious" weeds ("noxious" is a legal definition for plants marked for destruction by federal, state, and local authorities) were originally introduced because of their perceived beauty or usefulness. For example, Chinese immigrants who recognized the ailanthus tree's medicinal purposes brought them to the eastern United States and California in the 1800's as fast growing shade trees. Now it is considered to be a very bothersome weed in urban areas of many US states.

Lewis Ziska has been conducting research to see just how bothersome weeds could be in the future, especially in the context of a changing climate. Ziska is a weed ecologist for the US Department of Agriculture, who in 2002 planted weeds in what is now a reclaimed urban, industrial landscape in downtown Baltimore. It is an area that didn't look ripe for growing much of anything, but Ziska saw an opportunity to model what much more of the Earth will be like in the years to come with intensifying climate change. Ziska's aim

43 Quoted by Tom Christopher, "Can Weeds Help Solve the Climate Crisis," *The New York Times Magazine*. June 29, 2008, p. 44.

was to create a laboratory setting mimicking the heightened concentrations of CO_2 and elevated temperatures predicted for the mid-21st century. The Intergovernmental Panel on Climate Change (IPCC) has forecast that if we engage in "business as usual" carbon emitting activities, the global average concentration of CO_2 in our atmosphere will be around 550 parts per million (ppm) by 2050. The IPCC also predicts by the same model that the global average temperature will rise by 2.6 degrees C by the middle of this century. There are several worrying implications of this, but not all life will necessarily lose out. Plants, for example, could gain from the increased carbon dioxide levels. Carbon, water, light, and nutrients are vital resources that promote flora growth. But Ziska wanted to study the complex relations between plants and the new environment with a focus on the possible implications carbon loading could have on weeds. Hence, he went to work in Baltimore.

Noticing that the labs available to him were woefully insufficient for studying the effects of climate change on plants, particularly weeds, Ziska got creative. He noticed after hearing several complaints from Baltimore residents in the summer of smog and intense heat, that in a sense, the future is now in the city. Baltimore, like a number of other urban areas, is a "heat island." The concentration of buildings and the sprawl of pavement hold in heat—Baltimore's average temperatures are about 3 to 4 degrees Fahrenheit higher than they are in nearby rural areas. The city's air, at least when Ziska began the experiment, had a carbon concentration level of roughly 440–450 ppm (in the early 2000's that was well above the global average). These conditions were relevantly similar to those predicted by a climate model projected for what the Earth's atmosphere could be like in 2050, with continued emissions increases but with some abatement.

Ziska compared three sites. One was an organic farm in western Maryland, the second a suburb of Baltimore, and the third the downtown plot. Each site was used to study one of three conditions. The organic farm would serve as the present condition, the urban site would be the scenario predicted for Earth's future by 2050, and the suburban site would stand in as an intermediate scenario. Using soil from the organic farm that already contained seeds from 35 weeds, Ziska built uniform beds at each of the sites so that the growing medium and weed population would be common to all of them.

Ziska found that the weeds grew much larger in the warmer, high CO_2 concentration plots, while also producing much more pollen. In fact, one weed species, *Chenopodium* (more commonly known as lambs-quarters) grew to 6 to 8 feet on the farm, but these weeds were dwarfed by the 10 to 12 foot members of the same species that sprouted out of the downtown Baltimore plot. Several other worrying effects on the urban site weeds could show possible negative climate changes in the general region in the future. Usually, when an area is cleared of forest and other vegetation along the Eastern Seaboard of the US, it returns to native woodland if left to its own devices. While these effects can vary due to myriad different circumstances, fast-growing annual weeds typically cover the soil first after clearing, serving as "pioneer plants." But, these generally give way to a second wave of perennials that live much longer. Trees and other shrubs then replace these second-wave perennials. Typically, all of these plants and trees are indigenous to the area, with the woodland's restoration taking decades.

This gradual transition through these stages did not occur in the urban test plot. Instead, within a short five-year period of time, there was a rapid succession of trees, with invasive weed trees dominating. Ailanthus, Norway maples, and mulberries, some of the most vexing invasive species to forest ecologists, sprang up at an alarming rate. In just five years, a 20-foot ailanthus grew in the downtown plot, completely dwarfing a five-footer that was the largest ailanthus grown at the rural site. As one might predict, weeds in the suburban site showed similar growth characteristics to those at the urban site, but were simply a few years behind.

If this is the future, there is reason to worry. First, farmers will need to be prepared to spend a great deal more money on weed control or face falling crop yields. US farmers on average lose about 12% of their crop due to weeds annually, translating into $33 billion. Herbicides already cost farmers worldwide some $10 billion each year. It is anticipated farmers will have additional costs in the future in either trying to control weeds with more chemicals, use of other weed control methods, or through crop loss. Foresters, land managers, and gardeners will need to make much more aggressive and likely costly efforts to manage weeds. That cost will likely be increased because it will be much harder to control weeds under high CO_2 concentration levels. Research has shown that some weeds (e.g., Canada thistle and quack grass) are more resistant to herbicides when grown in high CO_2 concentration conditions. This likely means farmers will have to spend even more time, effort, and money in controlling tomorrow's weeds.

It isn't just that weeds will likely become larger, more prolific, and more herbicide resistant if we continue on our current course. Their increased size and bulk changes their chemical composition in high CO_2 conditions. Ragweed grown in 600 ppm CO_2 (the concentration of CO_2 expected in mid-range climate projection models from the IPCC by 2100) produced twice as much pollen as it does in current average concentrations. This is, of course, unwelcome news to allergy sufferers. This greater volume of pollen will also be far richer in the protein that causes allergic reactions. For example, poison ivy grown in high CO_2 concentration conditions likely possess a more potent form of urushiol—the oil found in the plant's tissues that causes a rash.

The fluidity of what is called a weed is telling in a number of ways and has consequences for different environmental arguments. One issue is whether plants labeled as weeds are merely a product of an anthropocentric (human-centered) perspective. What a weed is then just seems to be subjective, opening the door to possible changes in judgment. The weeds of yesterday could potentially be the desired plants of the future. The past already gives us some idea of how our views of plants might change from desired crops to weedy pariahs. Consider, for example, wild grains such as red rice and wild oats. For hunter-gatherer societies years ago, these were valued food sources. But that time has clearly passed as both red rice and wild oats are considered to be among the worst of the agricultural weeds we face today. While the vast majority of human introduced plants do not become regarded by us as weeds (most of these invasive species do not naturalize), we know that the ones that do proliferate cause a great deal of economic and ecological damage.[44]

Some environmentalists have argued that destroying species of both flora and fauna is likely a fool's errand as the seemingly useless, annoying, or even economically damaging plants could serve useful purposes some day. This is especially true in a world where we just seem to be burning through our "natural capital" as quickly as possible. Advocates of the Amazon rainforest and the vast Boreal forests of Canada and Russia point out that these areas allow for carbon sequestration and wildlife biodiversity. Both obviously present vegetation biodiversity, that at some point will likely yield useful plants. This could either come directly (as in the creation of medicines and other innovations in biotechnology) or as sources of inspiration in biomimicry. *Biomimicry* is an approach to innovation that seeks sustainable solutions to human challenges by copying nature's patterns and strategies. Its goal is to create new products and processes well adapted to sustainable living. Its advocates believe biomimicry is a viable approach to solving a range of human problems in part because it is time-tested; we can examine a wide span of evolutionary history to provide evidence-based results to support products and processes. The second reason is that biomimicry presents us with a new, sustainable way of living over a long period of time. Humans have engaged in amazing technological feats. While these achievements warrant praise, we too often fail to account

---

44 Michelle A Marvier, Eli Meir, and Peter M. Kareiva, "How Do the Monitoring and Control Strategies Affect the Chance of Detecting and Containing Transgenic Weeds?" in Klaus Ammann et al. (eds.), *Methods of Risk Assessment of Transgenic Plants*. Vol. 3. (Basel: Springer, 1999), p. 110.

for negative externalities of the products and processes we use. We ignore certain design flaws, such as what we are to do with products at the ends of their lifecycles. Our industrial methods are often dirty and inefficient. But, biomimicry isn't just an exercise in making the processes we use more efficient. It involves a closed-loop system where waste is regenerated into another production stream. Yet, while the promise of biomimicry underscores the possible values of conservation, it obviously unsettles our notion of what a weed might be.

Review Questions

1) There is a sense in which all plants (including what we now call "native" flora) that have existed in some areas of the world since glaciation are invasive. After all, glaciation destroys or displaces all plants, leaving available habitats for replacements plants. So, what does it mean for something to be an "invasive" species? Is this just a rhetorical term used to describe flora or fauna that have perceived deleterious effects on more desired species?
2) What kinds of obligations do you think we have as individuals for weed control? Think about when you have been asked to engage (or not engage) in an activity or practice designed to protect vulnerable ecosystems, such as cleaning your boat before moving it to another body of water, or brushing off your boots before entering a nature preserve. Have you taken these precautions? Why or why not? Do you think you *should* take these precautions? Why or why not?

Case 6 – To Save a City –

Study Questions

1) What is the relationship like between the city of New Orleans and the United States Army Corps of Engineers (USACE)? Through what legislation is USACE authorized to construct and maintain levees in the New Orleans area?
2) What is the point of contention between the USACE and Ivor van Heerden?

The shockwaves from the catastrophic effects of Hurricane Katrina are still felt throughout the Gulf region of the United States. Even though the storm hit over a decade ago in 2005, much of the area is still lagging behind economically. A substantial number of New Orleans' former residents have not returned. The storm itself caused around $200 billion dollars in damages, with property damage over $108 billion (in 2013 dollars) and insurance compensation alone standing at $45 billion. Furthermore, some argue even though New Orleans has received much more attention than other Gulf Coast communities since the storm hit, the rebuilding effort there has been particularly slow, with gaps in public services and a surge in crime. In this context, some question if the city will ever "come back" to its former grandeur.

Though Katrina was a natural disaster, this has not stopped New Orleans' citizens and others from finger pointing, claiming the levees were not built sufficiently strong to hold back tidal water. Some have blamed the United States Army Corps of Engineers (USACE) for not demanding enough federal support for levee construction. Many American citizens were outraged by what they perceived as a slow response from the Federal Emergency Management Agency (FEMA) in the immediate aftermath of the storm.

Beyond questions of whether the Corps of Engineers is culpable for the failure of the levees and to what degree and at how much cost levees should be rebuilt, other environmental issues have sprung out of the storm. For years, environmentalists have worried that building along the Mississippi River delta is not only destroying habitat for several species of Gulf animals, but also deteriorates swampland serving as a natural barrier to storm surges.

In 1879, Congress created the Mississippi River Commission (MRC), which took over for the Louisiana State Board of Levees Commissioner in controlling and maintaining the Mississippi River. Building projects were then to be facilitated through the USACE and effectively made the protection of homes and commerce from floodwaters a federal as opposed to a state matter. While there were both civilian and military engineers on the original MRC, it was dominated by members of the USACE. Beginning in 1885, the MRC instituted a "levees only" policy for flood control.[45] A *levee* is an earthen wall built to keep out river or seawater. Using only levees meant that other flood protection ideas, such as use of natural outlets, cutoffs, spillways, or other means to drain high water were not utilized. The initial projects under the MRC and USACE focused on dredging the river to ease ship passage and leave it less flood prone.

45 Organization for American Historians, "Through the Eye of Katrina: Past as Prologue?" *The Journal of American History* (special issue), (2007) 94: 693–876 at 693–694. Available at http://journalofamericanhistory.org/projects/katrina/resources/levee.html (accessed June 25, 2015).

Interestingly, the goals of the MRC and the higher echelons of the federal government have often been misaligned, even from the MRC's inception. The MRC's more extensive flood control plans were rejected, as from 1881–1892 federal law prohibited the use of funds to build or repair levees for the sole purpose of protecting private property from river overflow. Instead, all expenditures for levees were geared toward navigational improvements.[46] The building projects were extensive not only in southeastern Louisiana, but throughout the Mississippi River system with levees constructed all the way from New Orleans to Cairo, Illinois, by 1926.[47] By April of that year, the USACE formally announced the levee system along the Mississippi was complete and would prevent future flooding of the river. This levee system cut off access from most natural outlets to the river.

In 1927, usually copious rainwater volume made the Mississippi River swell to historically high levels. To spare the city from catastrophic flooding, New Orleans officials ordered the blowing of Caernarvon levee at the small town of Poydras to divert water.[48] This episode is still remembered by the descendants of those living along the Caernarvon levee 13 miles downriver from New Orleans, as homes and business in low-lying St. Bernard Parish were wiped out in what many residents there saw as an unjustified sacrifice of their area for New Orleans. In many ways, this decision to dynamite the levee was based on utilitarian grounds. New Orleans' population at that time was over 400,000 residents, while in rural St. Bernard and Plaquemines parishes combined there were ~15,000 people. When politicians, bankers, business owners, and other city elites met at City Hall to devise a plan for how to protect New Orleans from the rising waters, there were no representatives from either St. Bernard or Plaquemine parishes attending. The leaders of New Orleans were able to convince then President Hoover and the USACE to dynamite the levee.

By 1955, the USACE had moved from an emphasis on flood protection to storm surge protection and began conducting hurricane impact studies. In 1962, the New Orleans District of the USACE submitted the Lake Pontchartrain and Vicinity Hurricane Protection Plan (LP&VHPP). This plan was designed to protect residents living between the Lake Pontchartrain and Mississippi River levees. Surge barriers were planned for along the lake to keep storm surges caused by hurricanes from entering Lake Pontchartrain.

---

46 United States Army Corps of Engineers, Mississippi Valley Division, "MRC History," (website), 2015. Available at http://www.mvd.usace.army.mil/About/MississippiRiverCommission(MRC)/History.aspx (accessed July 16, 2015).

47 Organization for American Historians, "Through the Eye of Katrina."

48 Organization for American Historians, "Through the Eye of Katrina."

But the LP&VHPP was not enacted until after Hurricane Betsy landed in the area in 1965, putting all engineering developments at that time to the test. In reaction to the storm, Congress passed the Flood Control Act of 1965, authorizing the LP&VHPP and granting the USACE sole responsibility for levee design and construction in the New Orleans area. However, in 1977 there was a setback in construction as the US District Court of Louisiana placed an injunction against the LP&VHPP project, citing that its environmental impact statement did not comply with standards set down in the National Environmental Policy Act.[49] By 1985, modifications of the LP&VHPP called for the installation of higher levees in lieu of surge barriers. With those alterations, construction continued under the guidance of the USACE.[50]

The USACE has been accused of trying to spin the story of the levee failure after Hurricane Katrina by insisting the levees didn't fail because they were inadequately constructed, but simply because the storm was too strong. They claimed that since the Katrina was a category 4 hurricane and the levees were only built to sustain a category 3 storm, this was the main reason why the city flooded.[51] In fact, the USACE argued that Congress only authorized them to build flood protection up to the level of a category 3 storm, that the levees were overtopped, and thus didn't really fail to hold up to the pressure of the storm surge.

However, Louisiana's top hurricane experts quickly challenged these claims, including Ivor van Heerden, then the deputy director of the Louisiana State University Hurricane Center (he still serves as Director of the LSU Center for the Study of Public Health Impacts of Hurricanes). In his response to the USACE's position, he said "We are absolutely convinced that those flood walls never overtopped."[52] Instead, he is convinced the levees were poorly designed by the USACE and simply collapsed.

Eventually the USACE agreed the levees were breached due to flawed engineering and their construction from fragile materials. Sheet pilings, which are metal planks lodged into the ground to reinforce levees and flood barriers, were not driven in deeply enough. Evidence suggests the USACE and other

---

49 Organization for American Historians, "Through the Eye of Katrina."

50 Organization for American Historians, "Through the Eye of Katrina."

51 Michael Grunwald and Susan B. Glasser, "Experts Say Faulty Levees Caused Much of Flooding," *The Washington Post*, September 21, 2005. Available at http://www.washingtonpost.com/wp-dyn/content/article/2005/09/20/AR2005092001894_pf.html (accessed June 25, 2015).

52 Grunwald and Glasser, "Experts Say Faulty Levees."

local levee boards decided to save costs where they could when building the levee system.[53]

A sign posted on a tree in the Bywater neighborhood of New Orleans serves as a makeshift memorial to those who died as a result of Hurricane Katrina. On it reads a quote from van Heerden, "If we had the will and one month's money from Iraq, we could do all of the levees and restore the coast."

There are persistent worries about the extent to which the levees could withstand another category 4 storm. After Hurricane Katrina, the federal government spent some $14.5 billion dollars in reinforcing older levees and building a more extensive flood control system, including a huge new surge barrier. The flood protection standards are now supposed to be raised to handle a 100-year storm. But some have claimed the 100-year storm standard is "a term of art that refers to storms that have a 1% chance of striking in any given year."[54] Of course, the USACE concluded that Katrina was less like a 100-year storm and much more like a once an every 400-year storm.

Katrina was a major disaster—one of the worst in American history. Over 1,800 people were killed in the storm (making Katrina the US's third deadliest hurricane). Even for those who didn't perish in New Orleans during the storm suffered greatly immediately after it, as the city was cut off from ready access to food and water. Thousands were forced to migrate to escape the brunt of the storm—many of whom have not returned to the Gulf region (about only half of New Orleans' residents have returned, with only about 15% repopulating the hardest hits areas such as the Lower Ninth Ward). Jobs were lost and some of the positions were never refilled. The Lower Ninth Ward still looks like a ghost town and nature is reclaiming the land as weeds and trees entangle abandoned properties. Stray dogs roam the streets there and in some areas of Gentilly, a neighborhood just across the river that also has not entirely bounced back from the storm. Though the Lower Ninth Ward received most of the press, other areas left with uninhabitable homes include the Seventh Ward, Lakeview, Central City, the Upper Ninth Ward, and New Orleans East. Homes in these neighborhoods are still standing, but would need to be razed and replaced. The stray dogs are the result of many pet owners releasing them before

53 Jed Horne, "Five Myths about Hurricane Katrina," *The Washington Post*, August 31, 2012. Available at http://www.washingtonpost.com/opinions/five-myths-about-hurricane-katrina/2012/08/31/003f4064-f147-11e1-a612-3cfc842a6d89_story.html (accessed August 22, 2016).

54 Horne, "Five Myths."

fleeing the storm.[55] Some thought they would be back in just a few days, left their dogs, but never returned. It is estimated that over 600,000 animals were killed or stranded due to Hurricane Katrina. Dogs continue to be abandoned in the Lower Ninth Ward as the area is now often used as an easy dumping ground for anything unwanted.

Obvious questions arise about what should be done since most of New Orleans still sits 6–8 feet below sea level and remains susceptible to flooding from large storms. New flood and storm surge protections are in place, but until this system passes the test of the next substantial storm, there seems to be little public confidence in the USACE's ability to protect the city.

Residents and businesses can purchase flood insurance from the US Federal Emergency Management Administration. But some believe there are other, more drastic levels of precaution that residents of the Gulf region (and other low-lying areas prone to flooding) could take, such as not building on flood plains. This leads us to some moral questions. Is there a personal responsibility not to live in flood plains? Is there a point where the state should use its coercive powers to close off some areas to human habitation? Libertarians think taxation is coercive and should be avoided except in specific cases such as for protection of life and property from outside influences (thus the need for a military) and domestic influences (thus the need for police protection). There is also the need to adjudicate property disputes (thus necessitating a standing judiciary, presumably publicly funded). Given this view and a scarcity of public resources, tax dollars shouldn't be wasted on anticipated future recovery efforts for future storms. On this line of reasoning, if people want to live in areas prone to floods, they should do so at their own risk. It follows from this position that tax dollars should not be used to subsidize expenditures designed to mitigate damage from floods. Since access to federal flood insurance would only encourage people to build in flood prone areas, such publicly funded policies should not be on offer. One might even go so far as to add that no public monies should be used to go to post-disaster relief, but that private citizens could, of course, supply such aid if they wished to do so voluntarily.

There are possible replies to these libertarian arguments. Those endorsing the obligation to publically fund flood protection measures for all US municipalities might point to flooding that occurs outside of the Gulf region and recent storms in other US locales. Would anyone suggest people should not

55 Wendi Jonassen, "7 Years After Katrina, New Orleans is Overrun by Wild Dogs," *The Atlantic*, August 24, 2014. Available at http://www.theatlantic.com/national/archive/2012/08/7-years-after-katrina-new-orleans-is-overrun-by-wild-dogs/261530/ (accessed July 17, 2015).

inhabit Manhattan or the Jersey Shore due to evidence of highly destructive storms such as Superstorm Sandy? What about other states suffering destructive storms, such as Texas or Florida? Furthermore, if the projections of climatologists are correct, with climate change a number of US cities besides New Orleans will be under threat of inundation from seawater. Some appear to be in even more peril of sea level rise than those in the Gulf of Mexico, such as Miami, which lies right at sea level, but has virtually no flood protection infrastructure compared to New Orleans. Also, one might argue that once a legitimate state has taken on the commitment of offering protective services, it has the obligation to follow through with that service as a matter of *legitimate expectations*. Recall that the federal government replaced Louisiana authorities to be solely responsible for managing the Mississippi River in 1879. The reason for this may have had little to do with the benefit to local residents and more to do with keeping an important conduit for commerce open for country's benefit. But once such control was taken, the government shouldn't suddenly fail to execute those duties because it finds them to be too costly or too burdensome administratively.

So then what are the obligations of the federal government to any city? Assuming that there is some level of obligation, what would satisfy the criteria for meeting said obligations? Is the 100-year test enough for storm protection? Do such duties entail marshaling as many resources as possible for the highest level of protection possible with current funds and technology? How do we weigh these obligations against those presumably owed to other municipalities threatened by wildfires caused in large part by severe droughts? If there is an obligation to protect citizens using preemptive means from one sort of disaster (hurricanes), why wouldn't other citizens be offered equal protection from other types of disasters (such as wildfires or earthquakes)? The federal government might be duty-bound to build firebreak canals, divert more water to the American West, or perhaps develop an extensive system of earthquake surveillance and infrastructural capacity to withstand exceptional seismic activity. These are important considerations and we are likely to encounter the issue of what constitutes adequate protection of citizens from natural disasters in the future.

Review Questions

1) What do you think should be the obligations (if any) of the government to private citizens with respect to protecting them against natural disasters? On what grounds might the government have such a duty?

2) The Mississippi delta region has been extremely vulnerable to storms for a long period of time. Presumably, those who have lived in this region know of the potential for devastating hurricanes given information collected over the years. What are the responsibilities of those who want to live in this region? Should they be expected to live somewhere else? Should they be offered flood insurance, and if so, should it be public, private, or both? Do you think there should be a *right* to access affordable flood insurance? What sort of right might this be and what might the corresponding obligations be? Conversely, should people living in flood plains be required to buy flood insurance? Why or why not?
3) What does this case study tell us about nature, if anything? Is it simply that south Louisiana is, in its natural state, a swampland unfit for human habitation? And if the answer is "yes" should the area be allowed to return back to its natural state without rebuilding the levees?
4) Most insurance schemes (including private indemnity plans) operate on a "shared risk" model. That is, those of us who pay into an insurance pool and don't need to make a claim subsidize others who do make a claim. But most insurance companies are for-profit businesses and they will pass along the costs of covering those in need by charging higher premiums to those who do not. So, in a flood insurance pool, those whose properties don't get flooded pay higher premiums to ensure company profits and coverage of those who are flooded. Is there anything wrong with this and what are the implications for this case study?

CHAPTER 4

Business vs. Environmental Protection

Case 7 – To Frack or Not to Frack? –

Study Questions

1) What is hydraulic fracturing and where does it occur? What are some of the benefits of fracking? What are some of its problems?
2) What were some of the recommendations the New Zealand Commissioner for the Environment revealed in the Commission's Report on fracking? Do you see any background moral considerations that might justify the recommendations in the report?
3) What are "resource consents" in New Zealand and how would they be changed if the New Zealand Commissioner for the Environment's recommendations are implemented?

There is an energy revolution occurring that is rapidly changing the economic and even the geopolitical landscape worldwide—hydraulic fracturing (more commonly known as fracking). Perhaps the biggest winner in this bonanza is the United States, which appears to be well on its way to energy independence. As recently as 2005, domestic oil production in America was in steep decline and conventionally extracted natural gas was scarce. However, today 7.7 million barrels of oil are produced daily in the US, a 50% increase from 2006 and the highest in twenty-five years.[1] This yield could increase to 11 million barrels a day by 2020, poising the US to pass Russia as the world's top energy producer. As is, the US vaulted ahead of Russia just in terms of natural gas and oil production by the fall of 2013. The US Energy Information Administration (USEIA) estimates that with respect to dry natural gas from shale deposits, America produced 7.85 trillion cubic feet (tcf) in 2011. With that said, China, which holds the largest natural gas reserves in the world, would like to get in on the action. Its government has recently invested $13 billion in oil and gas exploration and development.

But internationally hydraulic fracturing is controversial. Some countries have banned the practice, such as Tunisia, Bulgaria, and France. In the US, the

1 Gregory Zuckerman, "Breakthrough: The Accidental Discovery That Revolutionized American Energy," *The Atlantic*. November 6, 2013.

 | DOI 10.1163/9789004339996_006

states of Vermont and New York (which has significant natural gas holding shale deposits) have also banned fracking. Other places have placed moratoria on fracking, such as the province of Quebec, Canada, and in some individual US states. But the practice occurs in other areas of the US, Canada, Germany, China, and Poland. Smaller scale fracking occurs in New Zealand and Romania. While over 200 wells have been fracked in the Netherlands, a temporary moratorium on fracking was imposed in November 2013 to allow for further research on its safety. In the United Kingdom, hydraulic fracturing has been used for offshore wells for oil and gas since the 1970's and for onshore oil wells since the 1980s. But only one well in the country has been fracked for shale gas.

Hydraulic fracturing is a process of extracting natural gas and oil from previously inaccessible locations, mainly in shale rock. The technique involves drilling a vertical well roughly a mile-and-a-half deep and then with an amazing feat of engineering dexterity and the use of the latest technology, the drill bit pivots to tunnel an additional mile horizontally. This greatly increases the surface area for gas extraction. The wells are generally drilled far below the level of both individual and even municipal aquifers. There are swallower frack wells, which are more controversial since they are closer to drinking water aquifers. The vertical section of the well that extends from the surface down past the level where aquifers lie are lined with steel and cement to prevent leakage of gas, oil, and fracking fluid into the adjacent ground and water. Explosive charges are then detonated in the horizontal portion of the well in order to cause hairline fissures in the shale rock where there are layers of natural gas. Millions of gallons of water, mixed in a slurry of proppant[2] (consisting of what is called frac sand, which are granules of crystalline silica) and biocides are injected at extremely high pressure to further open the fractures. The pressure of the slurry deposits frac sand deep into the shale fractures to keep them open so gas can seep up the well and out to the surface to be collected once the water is removed. Wells may be fracked multiple times over their lifespan.

Fracking has been long established in the Barnett shale in Texas, the Haynesville-Bossier shale in Louisiana, the Fayetteville shale in Arkansas, and in the state of California. Fracking is also booming in Colorado, Wyoming, Pennsylvania, Oklahoma, Kentucky, West Virginia, and Ohio. It is expanding in the Marcellus shale of Virginia. Additionally, fracking occurs in Alabama, Mississippi, Kansas, Nebraska, Illinois, Michigan, Montana, Utah, and New Mexico.

Some are concerned that fracking can lead to earthquakes, especially in areas with little previous seismic activity such as Oklahoma and Ohio. There is

2 See Case 8—"There's Silica in Them Thar' Hills!".

some evidence supporting the link, though not for the reasons that were initially thought. The usual worries has been that the injection processes in fracking were causing seismic activity, especially the injection of used fracking fluid into deep underground wells for its disposal. But a University of Texas research team found that the slippage along faults might actually be caused by the *loss* of water, gas, and oil during the extraction process. It is still believed that any quakes possibly linked to fracking will be relatively minor, though the largest, which occurred in rural Flashing, Texas, registered 4.8 on the Richter scale. While this occurred 47 miles outside of the largest urban area in the region (San Antonio), had it happened in the city there would likely have been some building and other infrastructural damage.[3]

Another often cited issue with fracking is the possibility of well water contamination. Recent studies have led to inconclusive findings. The US Department of Energy conducted a study in Greene County, Pennsylvania—an area that has been known to have a large number of fracking wells.[4] The study examined whether fractures caused by some fracked mines in the area exhibited any upward growth of cracks through the shale that could possibly become a pathway for gas or fracking fluid to enter water aquifers. Using down-hole geophone arrays, micro-seismic signals were detected during regular fracking operations. All of the signals that were recorded were at least 2,000 feet below the Upper Devonian/Lower Mississippian gas field, and some 5,000 feet below drinking water aquifers.[5] Natural and man-made tracers were utilized to investigate whether fluid and gas from fractured Marcellus shale had migrated 3,800 feet upward to a gas producing zone of Upper Devonian/Lower Mississippian-aged shale, midway between the Marcellus shale and the surface.[6]

The combined team of governmental, industry, and academic scientists found that the crack growth stopped some 5,000 feet below drinking water aquifers in the area. There was also no discernible upward migration of gas or fluids into the drinking water from the fracked Marcellus shale.[7] Gas samples

---

3 Jordan Weissman, "Two Ways We Now Know the Fracking Boom is Causing Earthquakes," *The Atlantic*. August 27, 2013.

4 Richard W. Hammack et al. *An Evaluation of Fracture Growth and Gas/Fluid Migration as Horizontal Marcellus Shale Gas Wells are Hydraulically Fractured in Greene County, Pennsylvania*; NETL-TRS-3–2014; EPAct Technical Report Series; U.S. Department of Energy, National Energy Technology Laboratory: Pittsburgh, PA, 2014, p 76.

5 Hammack et al. *An Evaluation of Fracture Growth*, p. 76.

6 Hammack et al. *An Evaluation of Fracture Growth*, p. 76.

7 Hammack et al. *An Evaluation of Fracture Growth*, p. 76.

were taken from the Upper Devonian/Lower Mississippian gas field 2 months prior to hydraulic fracking and then up to 8 months afterward. No gas migration was detected. Further monitoring of the Upper Devonian gas field up to five months following hydraulic fracturing also showed no evidence of fluid migration. Of course, this is one study of one location, so follow up studies over a more expansive area will be necessary to set the minds of citizens at ease.

Another worry about fracking is that while there may be immense profits for shareholders and new jobs for local residents for a while, the energy economy tends to be boom or bust. The bustle of productive and profitable mining can vanish relatively quickly. As recently as 2012 in Wellsboro, the county seat of Tioga County in northern Pennsylvania, it was hard to have a conversation outside in town due to the rumble of frac sand trucks delivering proppant to mining sites. By early 2015, the town was relatively silent.[8] From a peak of drilling activity where the rolling Pennsylvania hills were filled with natural gas wells, there is only one Shell well still in operation. Eight hundred wells were drilled in the area over just a five-year period, likely making natural gas a bigger boom for the community than the lumber and coal industries had ever been. Money flowed through the town from royalties gained by local landowners for their mineral rights. Residents were able to buy new farming equipment and donations swelled to local charities. The industry also brought monies to the coffers of local government.[9]

But endemic to boom and bust economies, the high prices of natural gas that fed gas company profits and filtered money into communities with frack wells, quickly fell back to Earth. Back in October 2005, the price of natural gas hit an all-time high of $13.42 per million BTU, remaining high for about three years. However, the price began to slide rapidly, bottoming out at $1.95 in April 2012. By the end of 2014, the price of natural gas was at $3.49 per million BTU. Accordingly, the low profit margins led Shell to pull back its production capacity, despite having purchased most of the leases in Tioga County. This left just the one active rig mentioned earlier, down from the dozen that were operating not long ago.

The collapse came as a shock to the community. Jim Weaver, the Tioga County planner put it this way, "With really no warning at all, the bottom fell out of that [hydraulic fracturing activity]. In hindsight, looking at boom and bust cycles that have gone on forever, we should've known that. But when the

8 Lidia DePillis, "The Boom-Proof Economy: How to Handle a Fracking Bust," *Washington Post*. January 15, 2015. Available at http://www.washingtonpost.com/news/storyline/wp/2015/01/15/the-boom-proof-economy-how-to-handle-a-fracking-bust/ (accessed July 13, 2016).

9 DePillis, "The Boom-Proof Economy."

dollar's dangling in front of you and you're chasing the carrot, before you know it you're out on a limb, and the limb gets sawed off."[10]

For communities no longer feeling the positive economic benefits of the fracking boom, it is not always easy to simply recover their economies overnight. Once money is invested into certain types of projects, there may not be new capital to invest in diversifying the local economy and there is a pressure to profit from the investments already made. Also some decisions about land use can cause damage that doesn't allow for other uses of it over a long period of time, if ever. What some researchers call the "resource curse" can lay claim to communities that overinvest in a cash cow while ignoring other economic opportunities that may be more viable over the long term. By investing in the extraction economy, land may be despoiled that keeps the area from being a tourist attraction for years to come. Also, long-term manufacturing may be sacrificed for short-term financial opportunities in mining.

Concerns over the boom/bust cycle of the hydraulic fracturing economy and its potential environmental and health effects do not lie just in the US. New Zealand is an interesting case study, as fracking has been used as an oil and gas extraction procedure there for about a quarter century. While the number of wells fracked is relatively small when compared to the US, the history of fracking in New Zealand is also relatively much longer. Additionally, given that New Zealand is a small country in both land area and population size, a relatively small number of wells have had a huge economic effect there. The New Zealand government has been able to secure about 3 billion dollars (NZ$) in royalties from existing exploration and extraction deals with gas and oil companies. If future drilling yields are as large as expected, this revenue total looks to exceed 13 billion (NZ$). Considering it has powerful, opposing coalitions that are pro-fracking and anti-fracking, many of the tensions that have recently emerged in the US have already played out in New Zealand for some time. And with possible expansion of the practice in both countries, they share fracking as a hot topic of public debate.

At this point, fracking in New Zealand has been confined to the Taranaki region along the west coast of the country's northern island. Over sixty wells have been built in the area during the past 20 years as of 2015, with another 30 set to come online. But there is exploration on the east coast of the north island as well (in an area known as the East Coast Basin). According to energy speculators, this area may actually hold richer reserves than the New Zealand

10 "Tioga County 'A Cautionary Tale' about Fracking Busts," *Washington Post*, January 16, 2015, Available at http://triblive.com/business/businessprojects/onthegrid/gas/7583745-74/gas-pennsylvania-county (accessed August 22, 2016).

shale that is currently being tapped. TAG Oil, which is backed by the Canadian gas extraction company Apache, has invested at least $100 million into New Zealand fracking projects. They have targeted exploration and extraction sites over 1.7 million acres of land on the east coast the country. Some gas executives have seen great potential for fracking in this area, calling it the "Texas of the South" (i.e., Southern Hemisphere) and say the shale in the East Coast Basin is literally leaking oil and gas. In a survey of 500 international oil and gas executives as to what they thought were the most promising areas for petroleum exploration, New Zealand came in fifth. Natural gas is already New Zealand's fourth largest export. Its energy industry could be primed for further growth as the East Coast Basin's potential output of oil from its shale may be comparable to the abundant Bakken oil fields of North Dakota or the Eagle Ford oil fields of Texas.[11]

The same nascent worries about fracking in the US persist in New Zealand. There are the concerns about the possible adverse effects on human and non-human health from the practice. Again, the chief worry is that fracking chemicals and gas will infiltrate drinking water aquifers, either through fractures that allow seepage of fluids and pathways for gas, or as a result of fluids seeping back into ground water from well blowouts (that can be caused by leaks in the well). Advocates of fracking claim that the notorious cases where residents are able to light their water on fire (sometimes seen on YouTube) are not caused by fracking, but instead from the natural seepage of methane into water wells.

There are also concerns that the components of fracking fluids are not well known. While energy company executives and other fracking proponents declare these chemicals are harmless and can be generally found in the cleaning supplies of most homes, those using fracking to extract oil and gas have also been very tight-lipped about disclosing the exact ingredients of fracking fluids. Their rationale is that the fracking fluid components are proprietary and disclosure would destroy their competitive advantage over rivals. But what this means is that since the public (and it seems a number of governmental agencies in both the US and New Zealand) doesn't know what is in fracking fluids, there is distrust of the industry's willingness to keep people safe and skepticism that governmental agencies can properly protect ecological and human health.

---

11 "Meet The Frackers—Part 1," *60 Minutes New Zealand* (Television series episode), March 18, 2012. Available at https://www.youtube.com/watch?v=K-aN5DatUmE (accessed July 25, 2016).

Just as in the US, New Zealand's opponents of fracking have cited seismic activity induced by fracking as being a significant concern. There is the specific concern that the concentration of a large number of wells in a confined area can lead to significant seismic activity even if each individual well contributes little. The way this plays out in New Zealand is the worry that fracking wells might eventually operate close enough to areas already prone to earthquakes, such as near Christchurch on the east coast of the southern island. A recent earthquake there (not linked to hydraulic fracturing) caused extensive damage to the city and came as a surprise as there were previously undetected fault lines running along its metropolitan area. The big question is obviously whether fracking activities (both fracking itself or through the injection of disposed frack waste fluids) might set off another quake? And considering there was no previous information suggesting Christchurch is vulnerable before the recent event, is there enough information to determine whether fracking might trigger another quake despite the studies cited earlier that some fracking may cause only relatively small tremors?

Proponents of fracking claim that the worries of induced seismicity are overblown, as the magnitude of any quakes would be just 2–2.5 on the Richter scale. They add that the risk of inducing quakes in Christchurch or any other area prone to them is minor since shale fracturing would have to be done extremely close by to have such an effect. In New Zealand, there has also been little evidence found that fracking has caused earthquakes in the Taranaki district over the course of the last few decades. Of 3,300 quakes investigated in the area (note that quakes here include very minor seismic activity that wouldn't be felt on the surface), only one of them had a slim chance of having been caused by fracking. Yet, opponents of fracking cite a case that suggests some larger quakes could be caused by the procedure, such as in Basil, Austria where a rig was shut down after a 3.4 earthquake was linked to its operation. In areas where intensive fracking can be done, there may be roughly one well per square kilometer. The worry is this high concentration of wells might exacerbate localized seismic activity.

Proponents of hydraulic fracturing note that the benefits of hydraulic fracturing are mainly economic and this should not be overlooked. 3,500 people work directly in the hydraulic fracturing industry in New Zealand and perhaps two-to-three times more are indirectly supported by the industry. The claim that the fracking industry is a major boon to the economy is also the chief argument in its favor in the US. However, in America some of the industry's advocates will add a national security justification for the practice, as the US has been heavily dependent on fossil fuels from countries with which it has (to put it mildly) difficult political relations. In this context, any widespread and ubiquitous energy source that can be exploited domestically might be considered vital from a national security perspective.

Due to the several concerns noted above (besides national security, at least) the New Zealand Commissioner for the Environment, Jan Wright, headed up an investigation that concluded in 2014 concerning hydraulic fracturing in the country. The report detailed the findings of the Commissioner including six recommendations, as well as very brief comments on the association between fracking and climate change.[12] Commissioner Wright determined that while she believed a moratorium on fracking was not necessary since current law, if tweaked, could yield safe oil and gas extraction practices, the six recommendations were geared toward "plugging known gaps" in industry regulations. As Commissioner Wright put it, "During the course of this investigation I have come to a similar conclusion to the [British] Royal Society which is that fracking is safe if it is properly regulated and managed. However, I have significant concerns about how fragmented and complicated the regulatory environment for fracking is and about how these rules are being applied."[13]

It will be beneficial to look more closely at the recommendations, as they point out some of the environmental and moral issues involved with hydraulic fracturing "on the ground." The first two recommendations are interconnected. Commissioner Wright set the background for the recommendations noting that New Zealand had an opportunity to "get ahead of the curve" with respect to safe fracking. The first recommendation was to put in place a standard national plan for gas and oil exploration. Wright pointed out that in many countries lack of preparation combined with rapid expansion of fracking wells had caused numerous problems. A national plan would allow for local councils to properly prepare for new well applications. Wright added that local rules are inadequate to regulate the industry in a safe way, especially when considering that some wells in Taranaki never even required "resource consents" which in essence serve as the permitting function for fracking and all other land use in New Zealand, stipulating where wells may go and the conditions under which they may operate. Rules on permitting vary widely from district to district and with such differences there is no assurance fracking companies are

---

12 New Zealand Parliamentary Commissioner for the Environment, "Drilling for Oil and Gas in New Zealand: Environmental Oversight and Regulation," (website), June 2014. Available at http://www.pce.parliament.nz/assets/Uploads/PCE-OilGas-web.pdf (accessed April 10, 2015).

13 Issac Davison, "No Need for Fracking Ban in New Zealand," *New Zealand Herald*. November 27, 2012. Available at http://www.nzherald.co.nz/business/news/article.cfm?c_id=3&objectid=10850290 (accessed April 9, 2015).

employing best practices. Therefore, the first recommendation was for the Ministry of the Environment to develop a National Policy Statement (NPS).[14]

The second recommendation was to improve regional plans to align with the new national policy on fracking. Since at the time the report was written a number of councils were already revising their plans in any case, Wright thought it would be a good opportunity for them to overhaul their rules for the gas and oil industry. This would be particularly important for districts where fracking is already occurring or proposed to happen soon, such as in Taranaki and in the East Coast Basin. The Commissioner noted that even when resource consents are sought, this is often done without public oversight. Exploratory wells are being drilled in Manawatu, Gisborne, and Hawke's Bay, yet resource consent hearings were held despite the fact that the public had not been notified. Thus, there was no forum for the public to express any concerns it may have with new exploration. Among the factors the regional councils need to consider is how many wells they wish to grant in a given area and whether there are some environmentally sensitive areas or critical aquifers where no drilling should occur.

The third recommendation addresses well integrity, which means the structure, strength, and stability of wells. Specifically, Wright recommended that the Minister of Labor amend the Health and Safety in Employment (Petroleum Exploration and Extraction) Regulations of 2013 to include provisions for ensuring that well integrity and environmental assessment are chief considerations.[15] Ensuring that blowouts do not occur is a vital component of well safety. Part of this is obviously to protect workers, but there is also concern for protecting drinking water aquifers and this recommendation is meant to address both those issues. The regional councils in New Zealand are charged with protecting the environment and must be given the legislative tools to aid them in this respect. For local councils, this would mean that as a condition of obtaining permits to open new wells, the High Hazard Unit would be required to inspect wells to ensure the casing is sound (especially in the section that runs through the freshwater layers underground). A like inspection would be required of liquid waste pumping wells.[16]

---

14 New Zealand Parliamentary Commissioner for the Environment, "Drilling For Oil and Gas in New Zealand: Oversight and Regulation." (online video), June 2014. Available at http://www.pce.parliament.nz/publications/all-publications/drilling-for-oil-and-gas-in-new-zealand-environmental-oversight-and-regulation/ (accessed April 8, 2015).

15 New Zealand Parliamentary Commissioner for the Environment, "Drilling For Oil and Gas in New Zealand: Oversight and Regulation." (online video), June 2014.

16 New Zealand Parliamentary Commissioner for the Environment, "Drilling For Oil and Gas in New Zealand: Oversight and Regulation." (online video), June 2014.

The fourth recommendation acknowledges that even despite best efforts hydraulic fracturing operations can go wrong. Cleaning up after the fact can be extremely expensive. This means provisions need to be made *before* well installation to pay for clean up costs of any accidents that might occur (think of an escrow account). Wright admitted there had been few accidents reported from existing frack wells in New Zealand, however with a rapid expansion of wells anticipated (especially in the East Coast Basin), the need for some kind of fund from which to draw would be necessary in the future. The Commissioner concluded that while the well is active, companies owning or leasing it should have sufficient public liability insurance.

The fifth recommendation was to give regional councils more control over regulating hazardous materials associated with hydraulic fracturing. Many of the substances brought to and transported away from gas and oil wells qualify as "hazardous" under New Zealand law, including chemicals utilized in the fracking process. More specifically, under the country's Hazardous Substances and New Organisms Act (HSNOA), the Environmental Protection Authority places a host of controls on these substances. Yet, the Authority has no direct role in ensuring compliance with its own standards.[17] Instead, compliance assurance falls to inspectors from WorkSafe New Zealand and particularly the aforementioned High Hazard Unit. The problem with the status quo, according to Wright, is that the Unit's inspectors are highly trained specialists who are prepared to look only at very technical aspects of the well construction and design. Also, WorkSafe New Zealand is primarily focused on worker safety and thus its inspectors are not as attuned to the larger environmental and overall public health concerns outside of the worksite that also need attention. Given this regulatory gap, the Commissioner recommended that section 97 of the New Zealand HSNOA be amended so that regional councils can include assurances that fracking wells are complying with standards for the safe storage and handling of hazardous materials at each well site. This could be an additional charge to trained representatives of the Regional Councils who are already examining the fracking sites to see if they are meeting the requirements stipulated in the resource consents.[18] Section 23 of the Act would allow for the Regional Councils to recover the costs for enforcement of the regulations from the industry.

The sixth and final recommendation is intended to address the issue of solid waste handling at fracking sites. Where the most fracking has occurred, in the

17 New Zealand Parliamentary Commissioner for the Environment, "Drilling for Oil and Gas," (website), p. 81.

18 New Zealand Parliamentary Commissioner for the Environment, "Drilling for Oil and Gas," (website), p. 81.

Taranaki region, liquid waste has been disposed of by injecting it deep underground into wells drilled specifically for this purpose. This procedure, known as deep well injection, is standard in the industry and the Commissioner concluded it would be permissible to use in other areas of New Zealand where new fracking wells will likely come online.[19]

Yet, solid waste is another issue and is correspondingly more contentious. One disposal strategy is "mix-bury-cover." Solid waste is buried in a sump hole near the well site after being mixed with topsoil. This mixture needs to be buried below the root structures of surface plants, but above the water table. The breaking down of hydrocarbons using this method is slow because the slurry is not mixed with topsoil at the surface.[20] This method has not been used in Taranaki due to concerns that the heavy rains in that area make it difficult to isolate the waste in a high water table environment. Hence, landfarming has been used as the preferred solid waste disposal regimen. The Commissioner noted the mix-bury-cover method works better in dryer regions of the country as a way of handling solid waste from fracking. However, use of landfarming is by no means uncontroversial. In the landfarming method, the waste (of combined fracking cuttings, drilling mud and sludge, and contaminated soil) is mixed in slurry and spread on farmland. It is then mixed with topsoil and other organic matter (such as sawdust). Eventually, the land can be resown for pasture.[21] Over time microbes in the soil break down the hydrocarbons, but not the heavy metals and salts.

Despite worries about the effects of using material from fracking operations in landfarming, there is no evidence of contamination of milk from cows grazing on areas where the practice is used. Yet, New Zealand's largest dairy company will not acquire milk from any new land farms. The perception is that such milk is contaminated and this is not just a local phenomenon as New Zealand's dairy exports have suffered as those outside of the country worry about possible contamination. Wright stressed that the perception of tainted milk produced at landfarmed dairies might be as worrisome for the dairy industry as the reality of contamination.[22]

19 New Zealand Parliamentary Commissioner for the Environment, "Drilling for Oil and Gas," (website), p. 82.

20 New Zealand Parliamentary Commissioner for the Environment, "Drilling for Oil and Gas," (website), p. 59.

21 New Zealand Parliamentary Commissioner for the Environment, "Drilling for Oil and Gas," (website), p. 59.

22 New Zealand Parliamentary Commissioner for the Environment, "Drilling for Oil and Gas," (website), p. 82.

Regardless of the actual safety of milk produced from cows on landfarmed sites, the Wright suggested that a working group be assembled (convened by the Minister of Food Safety and the Minister of the Environment, and including both regional council staff and representatives from agriculture) to deliberate on two issues. The first is to "resolve the situation with livestock on landfarmed sites in Taranaki." The second suggested charge is to "consider how solid waste from oil and gas wells in the East Coast Basin should be disposed of before wells begin to proliferate."[23]

While exactly what was to be done by the proposed working group is rather vague, perhaps the recommendation was intended to allow for the widest range of policy directions. Wright followed her formal recommendations with a brief note on the association between fracking generally and climate change.[24] She noted that climate change is the biggest environmental issue for New Zealanders and the expansion of fracking must be examined from the perspective of whether the practice greatly exacerbates that problem. In 2010 in Mexico, New Zealand committed to an agreement with several other nations to only allow a maximum temperature rise of 2 degrees Celsius above pre-industrial levels. The Commissioner further explained that the country's efforts to meet the benchmark thus far have been lackluster. The Emissions Trading Scheme has been continuously weakened and New Zealand's carbon emissions are rising. Whatever targets have been met, such as those as part of the Kyoto Protocol, are owed to an intensive reforestation project in the country in the 1990's. This allowed New Zealand to be in compliance of emissions standards set for 2012. These trees will likely be harvested by 2020, making hopes of meeting the standards set for that date difficult if not impossible to meet.

Commissioner Wright pointed out that with respect to carbon dioxide emissions, there is an important distinction between oil and natural gas. New Zealand exports almost all of its oil for refining, as it is easily shipped. Thus, expansion of oil production will not lead to emissions problems in the country. But, while Wright did not make this link, it seems that New Zealand oil will still likely contribute to greenhouse gas emissions—it will just be burned *outside* of the country.

The issues with natural gas are more complicated. As compared with crude oil, natural gas does not pose as great a threat with respect to the greenhouse

23 New Zealand Parliamentary Commissioner for the Environment, "Drilling for Oil and Gas," (website), p. 83.

24 New Zealand Parliamentary Commissioner for the Environment, "Drilling for Oil and Gas," (website), p. 84.

effect. It provides more energy for each molecule of CO_2 emitted than any other fossil fuel. As opposed to coal, only about half the CO_2 is emitted when burning natural gas. If it is used to generate electricity in a combined cycle power plant instead of coal, its efficiency is even more greatly enhanced. As in the US, these facts have allowed even environmentalists worried about climate change to laud (or at least accept) natural gas as a "transitional fuel." Finding large repositories of oil and gas, combined with the enhanced technology of hydraulic fracturing, has provided an opportunity. The hope is that natural gas, which is plentiful and burns cleaner than crude oil and coal, will "buy time" for transition to lower carbon alternative fuels in the future (e.g., wind, solar, geothermal) without further high emissions. The reductions in carbon emissions experienced in the US in recent years can be significantly attributed to a shift from coal to natural gas that is accessed by hydraulic fracturing techniques.

Wright cautioned that economic complexities might limit the ability of natural gas to be a seamless transition fuel. Interestingly, since the US is not buying as much coal in a move toward greater consumption of natural gas, this has depressed coal prices to the short-term economic benefit of many other countries. Coal prices have fallen sharply and a number of countries have taken advantage by using more coal. A second barrier to the use of natural gas as a transition fuel is that fugitive emissions of methane from natural gas production still contribute to high concentrations of greenhouse gases associated with climate change. Of course, fugitive emissions of methane also come from coal mines.

Despite the hope that coal can be replaced by natural gas in New Zealand, Wright is skeptical about this possibility. New Zealanders already do not burn all that much coal. Dairy facilities using electricity generated by coal could shift to natural gas and reduce carbon emissions. However, there is an *opportunity cost*,[25] as some lower carbon, alternative energy plants that could be built now (for hydropower, geothermal, and wind) will be passed over as plentiful and cheap natural gas would mean that plants for burning it would be built instead. Natural gas from New Zealand could be exported and thus perhaps replace coal in places with high energy consumption, such as China. This might have a significant impact on lowering carbon emissions. Yet, shipping natural gas would mean first converting it in liquefied natural gas (LNG) requiring an ample supply of natural gas for it to be economically feasible.

And of course, with all of this, Wright points to one possible scenario that would greatly *increase* New Zealand's carbon emissions from fracking. Energy

---

25 An opportunity cost is the loss of potential gain from other alternatives when one alternative is chosen.

exploration in the East Coast Basin might uncover vast quantities of oil comparable to those found in the Bakken oil fields. There may thus be a replay in the East Coast Basin of what we've seen in North Dakota of *more* intensive emissions. Satellite imagery shows significant flaring in the Bakken oil fields as gas is burned off instead of captured. Oil is so much more valuable than the natural gas that could be recovered, so the gas is not collected. While flaring gas is better than simply venting it (as again methane escape would contribute to higher concentrations of greenhouse gases), there remains fear that there would be tremendous pressure to flare natural gas in the East Coast Basin.

Beyond the usual fears of excessive water use, its possible contamination, and destructive seismic activity as a result of fracking, a major moral problem in New Zealand involves the lack of public oversight and input as to both the general permissibility of the practice and the implementation of regulatory standards. In Taranaki, it falls on local councils to decide if new wells will be developed and where they will be permitted to operate. While there is some level of local input, it by and large only comes from the council. Little (or no) public commentary is permitted in the process. Where companies are allowed to drill begins with block offers from New Zealand Minerals and Petroleum. Once these offers are up for bid for drilling rights or where companies already have permits for drilling, extraction activity cannot be ceased. As Wright put it, "If you are on the land in some of these areas [where the block offers have been made], you cannot stop a well from being drilled on your land because the Crown owns the oil and the gas, not you, and that can be a bit of a shock to people."[26]

Interestingly, part of New Zealand's approach to public participation in land management might have to do with its theory of parliamentary politics. The idea is that the state is not to do "the will of the people," but instead that public input in the system comes from election of representatives, who are then supposed to use their best judgment to make decisions in the interest of the country as a whole. This political approach calls for a buffer between governmental representatives and the people. But when there are controversial issues this structure might be called into question as the public has little recourse to reverse governmental decisions that are not transparent and may harm their welfare. Local councils are limited in their control over wells and in some important ways there is little distinction in New Zealand law between fracking water wells versus oil and gas wells. While they can place some conditions on

26 New Zealand Parliamentary Commissioner for the Environment, "Drilling for Oil and Gas," (online video).

fracking gas and oil wells, local councils cannot stop wells from being drilled or operated once federal governmental officials have approved them.

With respect to resource consents, some inconsistencies are troubling—in places, the requirements are in effect non-existent and in others annoyingly constraining. In Taranaki, drilling is allowed without resource consents—which strikes many as odd and excessively lenient since that is the district with the heaviest fracking activity. On the other hand, some have criticized resource consents, saying that they are unduly required for small projects that have trivial environmental impact, such as building a deck, while not required for fracking wells where more regulation makes sense.

Beyond obvious worries for any democratic institution (even within the constraints of a parliamentary system) when important policy decisions are conducted without public input, two important specific problems occur when there are extremely limited or no opportunities for public involvement in decision-making. The first is the perception that illicit deals are being struck. This lack of transparency is counter to democracy itself, but it also breeds public distrust of both the fracking industry *and* regulatory agencies charged with protecting human health and environmental sustainability. The second is that debate is absent about the placing of wells in areas that might be of critical importance (either because they provide some essential public service or are of unique value). For example, drinking water aquifers or pristine areas might be degraded with poor placement and design of oil and gas wells.

Again, fracking in itself may not bother opponents of the practice in New Zealand (or anywhere else for that matter), but instead its rapid expansion. Correspondingly, in high-yield gas and oil fields, there may be pressure to drill too many wells in a relatively small area, leading to a worrisome concentration of fracking activity. To Commissioner Wright, it is not a single fracking well that is the problem (except of course to some who are living right next to it), but instead the *cumulative* effect of many wells lying too close together. Once a concentration of wells exists, even assuming they do not cause ecological problems, the addition of a few more might tilt the balance toward environmental harms. These will be more difficult to stop once the precedent is set. In such cases only weighty considerations will override construction of additional wells.

Review Questions

1) What do you think the role of the public should be in having input and democratic participation in discussions about environmental issues such

as fracking? Is it enough to allow elected officials to make these decisions, or should there be more extensive opportunity for public participation?

2) What level of risk should a society take to accrue economic benefits? Apply your answer to this case study.
3) Several possible policy revisions were listed in this chapter that are supposed to fill "policy gaps" with respect to hydraulic fracturing for oil and gas in New Zealand. Which (if any) of these do you think should be implemented in New Zealand and why? Are there any that shouldn't be implemented in New Zealand? Why or why not? Should measures similar to these be implemented in other countries that use hydraulic fracturing techniques for gas or oil extraction? Why of why not?

Case 8 – There's Silica in Them Thar' Hills! –

Study Questions

1) What factors make Wisconsin "ground zero" for frac sand mining?
2) What is the difference between frac sand mining and hydraulic fracturing?
3) What are supposed to be some of the benefits of frac sand mining? What are supposed to be some its problems?
4) What background philosophical principle do supporters of frac sand mining emphasize to make their case? What background philosophical principle do opponents of frac sand mining emphasize to make their case?

There is a rush on in western Wisconsin, southeastern Minnesota, and northeastern Iowa. In response to the boom in natural gas production throughout the United States and Canada, there has been an accelerated need for resources that promote oil and natural gas extraction. The main method of extraction for these valuable commodities has been by hydraulic fracturing (otherwise known as fracking).

In this process, a deep vertical well is drilled into shale deposits (~one mile and a half down) that contain numerous tiny pockets of previously unattainable natural gas. Charges are then detonated to cause cracks in the shale. To stimulate the flow of gas, a slurry of anywhere from one to seven million gallons of water (depending on the size of the well), a copious amount of proppant (3,000–10,000 lbs.), and some solvents (biocides and other proprietary chemicals) is pumped into the well at extremely high pressure. This process opens up the fractures in the shale. After the water dissolves deep into the earth or is pumped out, gas can flow out the top of the well to be collected.[27]

The proppant is where the state of Wisconsin enters the picture. As the word suggests, proppant is the material that is used to "prop" or hold open the fissures after the well has been fracked. Most fracking proppant is silica sand (or silicon dioxide (SiO_2)), which is more commonly known as quartz.

There are several virtues of the frac sand found in Wisconsin. First of all, it is from Jordanian and Cambrian sandstone deposits and the granules from these types of formations are extremely hard (with a compression strength of 6,000–14,000 pounds per square inch (psi)). This strength is necessary for withstanding the pressure to prop open the fractures in the shale rock.

Second, frac sand from Wisconsin can also be processed into the optimal size and shape to fit tightly into the fractures, but still allow for strong gas flow. Other types of sand do not suit this purpose well as they are too angular to fit properly into the fissures. But the desired frac sand is pure silica that maintains its hardness and is polished into oval-shaped granules.

It is also a boon to the Wisconsin frac sand industry that the state has a relatively robust transportation infrastructure with an extensive web of roads and railway lines that are essential for quickly handling a large outflow of traffic to busy shale plays where fracking occurs (all of which are beyond the state's borders). Wisconsin's location on the Mississippi River also allows for easy access to barges for transporting frac sand to shale deposits in the South where there is extensive fracking activity, such as in the Barnett shale in eastern Texas, the Fayetteville shale in Arkansas, and the Haynesville-Bossier shale plays in Louisiana.

All of these factors contribute to Wisconsin being "ground zero" for frac sand production. Frac sand can be found in other states, such as Minnesota, Arkansas, and Iowa (with a good deal of silica mining occurring in the last of these three). South Dakota is a new player in the industry with a massive site in the Black Hills. However, Wisconsin already is, and is primed to remain, the top frac sand producing state in the US. Also, while there has been some experimentation with use of other materials as proppants (such as silica gel in the Canadian oil fields and ceramic clay in North Dakota), they are not seen as viable alternatives to silica sand for the time being. While silica gel is thought to be a more environmentally friendlier option than frac sand, fracking companies have deemed it as prohibitively expensive. Also, the clay ceramic option

---

27 For a fuller account of the hydraulic fracturing process, see the previous case (Case 7—"To Frack or Not to Frack?").

is limited because its pellets do not have the compression strength of pure silica sand.

Finally, there is the easily overlooked, but important virtue of Wisconsin frac sand that it lies relatively close to the surface. This makes its extraction much easier as little of what is called "overburden" (e.g., clay, loam, silt, etc.) has to be removed to reach the chunks of sandstone where crystalline silica is found that can be processed into frac sand. In fact, the area along the Mississippi River has been alluring to frac sand companies for just this reason, with its exposed outcroppings near riverbanks that involve removing little to no overburden for silica extraction. Of course, steep bluffs along the river may still prove to be dangerous or completely inaccessible to excavating crews and this does put a damper on enthusiasm for extraction in these areas. Nonetheless, the overall assessment is that frac sand mining in Wisconsin is relatively cheap, making the area very attractive to companies. And of course, there are plenty of areas away from the Mississippi riverfront where silica sand can be mined.

With approximately 2,203 trillion cubic feet (tcf) of natural gas estimated to lie under US soil[28] that can be extracted, it is likely that much, much more frac sand will need to be processed and shipped to fracking sites. Given the current consumption rate of twenty-four tcf of natural gas per year in the US, there is enough natural gas to last for roughly ninety-two years. It is estimated that the amount of frac sand needed to tap these reserves would be 40–50 million tons—enough sand to fill the formerly named Sears Tower in Chicago 21 times.[29]

This means the frac sand business is extremely profitable. Private land owners who find themselves sitting on frac sand deposits are faced with opportunities for huge financial windfalls and frac sand mining companies are quickly setting up shop in western Wisconsin. To provide some idea of how quickly the industry has flourished in Wisconsin, while only five silica sand mines operated in the state in 2007, as of this writing in 2016, there are over one hundred and fifty mines and processing centers. Typically, the size of frac sand mines in Wisconsin range from a few hundred to a few thousand acres. Most of these mines operate twenty-four hours a day, for seven days a week over eight to nine months of the year.

---

28 United States Energy Information Administration, "Frequently Asked Questions," (website), 2012. Available at http://www.eia.gov/tools/faqs/faq.cfm?id=58&t=8 (accessed January 23, 2013).

29 Kate Prengamen, "Frac Sand Boom Creates Thousands of Jobs," *Wisconsin Watch*. August 19, 2014. Available at http://wisconsinwatch.org/2012/08/sand-boom-creates-jobs/ (accessed Oct. 31, 2014).

To understand the benefits and burdens of frac sand mining, we first have to understand the extraction process. After core samples are taken to determine the quality of the sand, how deep it is, and the initial removal of the overburden, large chunks of rock (called shot rock) are taken to a primary crusher, which may or may not be located on-site. The overburden is utilized to build berms around the mine to serve as a protective barrier. The berm acts as a buffer between the mine and the surrounding areas, intended to mitigate light and noise that could be bothersome to neighbors. The shot rock is crushed in a secondary crusher into "breaker rock," which is washed of impurities and refined to its final size and shape. The final product is then sometimes sprayed with a resin that further hardens the sand granules. The sand is then stockpiled for primary drying in piles. Before being shipped (sometimes by truck to a railroad spur, sometimes directly to railway cars) the sand is sent to a secondary dryer. The trucks and railway cars are supposed to be covered while in transport to reduce any fugitive dust from escaping into the ambient air. The sand is then delivered to fracking sites around the country and sometimes in Canada.

An obvious reason why some see frac sand mining as being beneficial is that it can be a very lucrative industry. With oil and natural gas companies willing to pay $200 per ton for frac sand, the mining business in Wisconsin has been extremely profitable. Proponents note that in line with the profits to be made by shareholders, there is the significant positive spillover effect of many job opportunities in an area of Wisconsin that chronically suffers from unemployment and underemployment. Many of the mining sites are located in rural areas where there is weak manufacturing capacity, seasonal jobs in the tourist industry, and a scarcity of occupations calling for technical skills. In the overall job market in these rural areas, few offer high pay and fewer employee benefits. In this vacuum, the freight hauling industry has boomed with increased frac sand production, with relatively good paying jobs available. Entry-level positions for frac sand mining companies typically pay between $15 and $20 an hour. There are not yet clear conclusions about the number of jobs the frac sand industry has created in the state, but the Wisconsin Center for Investigative Journal, using figures from the Wisconsin Economic Development Corporation, estimates that when mines and processing centers are working at full capacity there are 2,780 jobs. This is not a negligible number given the scarcity of jobs in the western portion of the state.

There has been a corresponding boom in the railroad industry. In the two-year period of 2011–2013, Union Pacific Railroad reported a 265% increase in frac sand rail transport in Wisconsin and was thus prompted to invest in the revitalization and expansion of lines that had been long in disrepair. Union Pacific committed some $35 million just to track repair in 2012. Some have argued this railroad resurgence bodes well for other small manufacturing companies that could use the rails for expansion of their shipping capacities, as well as inspire new businesses to open.

Also, to those who live on land with high quality frac sand, there is the potential for a tremendous financial windfall. Some landowners have been offered six-figure sums for mineral rights to their property. Others have been extended options to collect royalties from their property's frac sand yield, to the tune of $1.50–$3.00 dollars per ton. While only an estimate, if we consider that the average Wisconsin frac-sand mine yields 900,000 tons a year,[30] the sum from royalties would be approximately $1.35–$2.7 million per year. Permian offered $16,000 per acre (well above normal market value) for property on which it wanted to place a mine outside of Red Wing, Minnesota. These sums are quite difficult for landowners to pass up when approached by energy companies.

What role might the monetary value have in a discussion of the ethical dimensions of frac sand mining? Money may not buy happiness, but that doesn't mean financial stability has no place in consideration of the good life. Having a steady job allowing one to support herself and her family can be personally gratifying. Self-reliance can come from work and economic independence, as well as fill people with pride and self-respect. As philosopher John Rawls notes, possessing the social conditions for self-respect is part of a just society.[31] Some communities are quick to note that frac sand mining companies engage in philanthropic causes in their towns.[32] Undoubtedly, there are *some* utilitarian benefits of frac sand mining to people in communities struggling economically. Beyond that, industry advocates argue that Wisconsinites, just like other Americans, will reap the benefit of a cheap energy source—especially one that can be ramped up to serve millions of consumers relatively quickly as opposed to other non-conventional energy sources such as wind and solar.

Yet, there is a mounting worry about possible long and short-term effects of frac sand mining that are much less desirable. Mining is generally a dirty and

---

30 See Thomas W. Pearson, "Frac Sand Mining in Wisconsin: Understanding Emerging Conflicts and Community Organizing," *Culture, Agriculture, Food and Environment* (2013) 35(1): 30–40.

31 John Rawls, *A Theory of Justice* (Cambridge, MA: Harvard University Press, 1971).

32 For example, the Hi-Crush sand mining company reportedly donated $500,000 to the town of Independence, WI, to clean up Bugle Lake. See Mark MacPherson, "Frac Sand Mining Company Donates $500,000 to City of Independence," *WKBT* (website), March 6, 2014. Available at http://www.news8000.com/news/frac-sand-mining-company-donates-500000-to-city-of-indepenence/24852046 (accessed July 24, 2016).

dusty business, with neighbors being adversely affected in obvious and less obvious ways. Many are concerned there has been very little research on either the amount of fugitive dust generated from frac sand mines and crushing facilities, or the possible effects of longtime exposure to silica dust for those living nearby. We do know silica is classified as a carcinogen, as gleaned from the negative health effects endured by those who have worked directly with the substance throughout history. In the US, the Occupational Safety and Health Administration established strict worker safety guidelines for laborers in the silica processing business beginning in 1971 and these appear to be observed on frac sand mining sites as well. Among these effects is silicosis, whereby one's lungs are scarified by microscopic bits of extremely sharp silica fragments that become deeply lodged in the lungs when inhaled. The smaller the fugitive dust particles are, the more dangerous they will tend to be as they embed deeper in the lungs' cilia and cannot be extracted from the tissue. The effect is often gradual, but frequently the cause of fatalities for those with long-term exposure to silica dust. Also, the reactivity of freshly fractured silica dust seems to be higher than in aged silica, thus dust particles from frac sand processing centers are of particular concern since quartzite has been freshly crushed.

However, there is little data about the possible effects of silica on those outside of work sites. According to a 2012 report on silica sand mining from the Wisconsin Department of Resources (WDNR), this was given as a reason why there is no need to monitor silica mines and processing sites, as long as mining companies are following current state and federal guidelines for mining.[33] That is, the WDNR cites a lack of evidence for harms that come from ambient air exposure from mining sites as the rationale for not imposing a monitoring regimen. The WDNR has recently encouraged companies opening new frac sand mines and processing facilities to monitor their sites on a voluntary basis. Yet, current monitoring in Wisconsin doesn't detect particles that are under 10 micrometers. This is not a negligible concern for two reasons. The first is that obviously the smaller the silica particles are the more worrisome they are. The second is that as neighbors can clearly be in close enough proximity to mines and processing facilities that dust is visible in layers on cars, houses, and sometimes enters homes.[34] In one case a neighbor residing just downwind from a sand mind reported to a documentary film crew that dust would filter inside

33 Wisconsin Department of Natural Resources. "Silica Sand Mining in Wisconsin," (website), January 2012. Available at http://dnr.wi.gov/topic/mines/documents/silicasandmining final.pdf (accessed June 3, 2016).

34 See the case of this in Auburn, Wisconsin.

of her cabinets to the extent that she had to clean the dishes in them on a weekly basis.[35]

There are also concerns about the potential effects of frac sand mining on water. The first involves water quality for both humans and non-humans. As already mentioned, frac sand mining outfits build berms for several reasons, but there is a downside if they are not well-constructed. In 2012, there were two spills that garnered little national or even local press coverage, but they did spark worry with environmentalists.[36] The first sand spill was near the St. Croix River at a fifty-acre sand mine and processing site near Grantsburg, Wisconsin. The site is located just one hundred feet from the St. Croix National Scenic Riverway. Large amounts of water and ultra-fine sand overflowed from a pond into an environmentally sensitive area due to a faulty berm. For five days, the mixture flowed into a wetland, creek, and the federally protected St. Croix River. A hiker first detected the spill and it was concluded that it had started five days before its discovery. The main worry was that a sand spill of this magnitude could destroy the spawning ground of fish and mussel beds in the river, in addition to causing irreparable damage to the wetland. The WDNR investigated the spill, confirming it was caused by inadequate berm construction.

A second spill occurred roughly a month later at a one-hundred-and-sixty-acre mine near Blair, Wisconsin. Following a heavy rain, wet sands sluiced down more than two thousand feet onto properties near the site.[37] The first floor of a home and a garage were inundated by water and sand, as well as an adjacent creek and wetland. The mining company who owns the site did not have anything in place to prevent problems from storm water runoff. The sand may also have contained chemicals that are used in processing sand (presumably from resin added to further strengthen it). Additionally, there have been a number of accounts of sand spilling out of trucks not covered adequately in the shipment of frac sand. This is a compliance problem, as trucks are required to have such coverings in place and functioning properly.

As noted to above, there is worry that the washing and resin treating process could introduce contaminants into the water table. This is likely a low threshold effect due to the relatively small amounts of resin sprayed on the sand and that clean water is often used in the washing process. However, the presence

---

35 See *The Price of Sand*, Dir. Jim Tittle, film, Actual Films, 2013.

36 Josephine Marcotty, "Wis. Sand-Mine Spills Cause Call for Penalties against Minn. Firms," *Star Tribune*. June 12, 2012.

37 Marcotty, "Wis. Sand-Mine Spills."

of some resin contaminants is possible as well as phosphorus, both of which could flow into nearby aquifers in localized erosion from washing sand.

Even in the preparation process for mining there are future water quality concerns. As mentioned earlier, to determine the quality of the sand and the amount of overburden that may need to be uncovered to get to it, bore holes are drilled at the proposed site. These are not very large holes in circumference, but they do go deep enough to reach aquifers. State of Wisconsin mining regulations stipulate that bore holes are to be plugged after the sand quality assessment is made at the potential site. But recently, there have been cases where boreholes were left open. While there may not be a mine or other development activity soon after the bore drilling that would contaminate water, the worry is that if these holes are left unplugged (and perhaps forgotten) they can become conduits for future contamination.[38]

Contamination and wildlife habitat destruction are not the only concerns with respect to water resources in frac sand mining. The quantity of water used in frac sand processing is also an issue. Opponents of frac sand mining acknowledge that while the amount of water used in frac sand mining is small compared to that used in fracking, there are still strains on smaller local water aquifers in frac sand processing. Of course, one mitigating circumstance is that some frac sand mining companies build and utilize holding ponds which store used processing water for reuse. There is some motivation for mining companies to continue and even expand this practice.

Yet perhaps the greatest "on the ground" complaint about frac sand mining comes with respect to the noise associated with the industry. Very heavy equipment such as excavators, scrapers, and dump trucks can make a great deal of noise in the mining process, not to mention the sound emanating from constant truck traffic going in and out of mine and processing sites. Industrial scale crushing of shot and breaker rock can also be heard off-site. While the berms do lessen some of the sounds coming from machinery in mines and processing sites, they obviously offer no protection once trucks and trains transport the finished sand beyond them. Complaints from residents near frac sand mines and processing facilities have increased dramatically along with the mining boom itself. Those who live next to rail spurs have been angered by the noise of equipment used to load frac sand onto the trains.[39]

An associated concern with noise pollution is the amount of truck traffic on the roads from mines and processing sites to railway loading stations and the

38 Joe Knight, "DNR: Bore Holes Incorrectly Abandoned," *Eau Claire Leader-Telegram*. December 1, 2014. Available at http://www.leadertelegram.com/News/Front-Page/2014/08/08/DNR-Bore-holes-incorrectly-abandoned.html (accessed June 3, 2016).

39 For other issues with noise pollution, see Case 47—"Noise Pollution"?

effects this may have on road maintenance and safety. To offer some idea of the extent of such traffic, at one proposed mine in Bridgeport, Wisconsin, the estimate is that one hundred and twenty, forty-ton trucks will make as many as two hundred and fifty trips per day from the site. This kind of travel can place abnormally large burdens on transportation infrastructure, as mines are often located in areas served by small county roads. These roads can quickly strain under the pressure of lots of trips endured from trucks bearing very heavy cargo. Consequently, there is the question of who should pay the costs of these extra burdens. This worry could be reduced in negotiations during the permitting process between frac sand mining companies and counties. Typically, to open a mine, mining companies (after receiving approval from the state on storm water runoff, water use, and air pollution plans) must then submit an application for operation with the county. In other words, the day-to-day politics of regulating mining comes at the local level with towns and especially counties on the front lines. One of the issues discussed during the application decision process is what are the costs in services that will be accrued by the county and corresponding mitigation measures. It is not unusual for the county, in addition to demanding a mine reclamation plan, will also insist that the mining company set aside money (not unlike an escrow account) to fund road repair and perhaps even building of new roads.

Other worries with frac sand mining include the property value concerns of neighbors. Despite what are sometimes deep disagreements between advocates and opponents of frac sand mining, there seems to be consensus that no one wants to live next to a mine or processing facility when it is in operation. In the vast majority of cases, those who sell or lease their land to fracking outfits leave the property, but neighbors either don't have the resources or the will to relocate. This often causes deep resentment as those left behind have the burden of not only the dust and noise of a frac sand mine, they also can't sell their property (or are compensated well below normal market price) as they see its value reduced dramatically. Neighbors are left with little recourse to alter the situation and sustain heavy property value losses.

Though it is by no means uncontroversial, advocates for frac sand mining respond that its companies pay into the local tax base just like all other landholders and employers, thus neighbors receive some indirect financial support from mining in this way. Such advocates also claim neighbors' opposition toward frac sand mining reduces to envy more than of any substantive and defensible moral principle.[40]

---

40 To see the issues with using envy as a moral principle, see Justin D'Arms, "Envy," *The Stanford Encyclopedia of Philosophy*. January 22, 2009. Available at http://plato.stanford.edu/archives/win2013/entries/envy/ (accessed June 3, 2016).

The final three complaints about frac sand mining tend to be a bit more social and perhaps intangible. Some worry that since those who live next to frac sand mining and processing operations are exposed to constant noise, this causes stresses that can result in psychological problems. They go on to argue that mental health providers are scarce in the locations most associated with frac sand mining—rural areas. This is a nascent concern and it has not been adequately studied. Yet, in other instances there have been strong correlations between constant exposure to inordinately loud noise and adverse mental health outcomes in people residing in large-scale industrial settings.

In some sense, all politics is local and this seems to be the case with frac sand mining as well. Local zoning (or the lack thereof) plays a crucial role in determining whether frac sand mines and processing facilities open and under what conditions they operate. County board members and town councilors thus have heavy responsibilities with respect to regulating the frac sand industry. Most county board members in areas where frac sand activity came about quickly admit that they were blindsided by the torrent of mining permit applications. The learning curve in understanding the issue has been steep and many county boards don't have the requisite expertise to fully assess the true benefits and burdens of such mining.

There have been instances of county board members (who issue mining permits) considering applications for mining companies they later work for. In some cases this has been within an extremely short period of time, with board members ruling on a mining permit one day and resigning to join the company whose mine they had just voted on the next.[41] Needless to say, this at least raises some suspicion of possible unsavory dealings even if it is not strictly speaking an indication of conflict of interest. But other cases may involve more straightforward conflicts of interest. There have been allegations of board members who have voted on mining permit proposals from companies in which they have financial interests. Two current cases are being investigated (one in Trempealeau County and the other in Crawford County) of board members who voted on mining permit proposals, but failed to disclose their

41 See Tony Kennedy, "Local Officials Dealing Themselves a Piece of Frac Sand Boom," *Star Tribune*. December 26, 2012. Available at http://www.startribune.com/local-officials-dealing-themselves-a-piece-of-frac-sand-boom/184771921/ (accessed June 3, 2016). Note that public officials have been careful to recuse themselves from votes on frac sand permits proposed for their own land.

financial ties to the same mining companies on whose permits they voted.[42] If these allegations are true, these are classic cases of inappropriate conflicts of interest.

Finally, in interviews with residents of areas where the heaviest frac sand mining activity is occurring, there is a pervasive feeling of dismay that the industry has polarized groups of community members who had previously been much more tightly knit. These could be called "community fabric issues" and some think that even if one were to employ a utilitarian cost-benefit analysis, it is unclear if the majority of citizens in communities with large-scale frac sand mining are happier than they were before the boom. While in the short-term jobs are created in the current boom, heavy investment in frac sand mining adversely affects other important economic sectors. For example, frac sand mining is increasing in areas of Wisconsin, Minnesota, and Iowa that have depended heavily on tourist dollars for decades. In McGregor, Iowa, sand-filled trucks travel through this historic town on their way to railway distribution centers much to the chagrin of tourists and the businesses that cater to them.[43] The Great River Roadway along the Mississippi River on the Wisconsin side has attracted motorcyclists for years to the scenic route. Yet, there is a fear this road will become inundated with frac sand hauling trucks making the route undesirable for recreational bikers.

A common feature of boom economies is that they often go bust just as quickly. As one resident in a town close to the sand mines noted, "There is an incentive for the oil companies to develop alternatives to frac sand closer to the fracking sites and they won't need frac sand. So, twenty, thirty years from now we will be left with holes in the ground and no jobs."[44] There is the obvious concern that carved off hilltops and bluffs from open-pit frac sand mining will not be attractive to tourists and that in the long run both the mining and tourist industries will be left worse off by the current boom. On the other hand, those supporting frac sand mining retort that there exist mining reclamation laws designed to ensure that defunct mines are returned to useful and pleasant pasture land eventually, so the cause for concern is overstated.

Yet opponents point out that while mines are to be reclaimed after their use, this may not happen for decades in cases of particularly productive mines,

---

42 Thomas Barton, "Frac Sand Mining Divides Bridgeport," *Dubuque Telegraph Herald*, June 11, 2014. Available at http://www.thonline.com/news/tri-state/article_30eae459-9a93-5bf5-9dc3-31e281f2c3cf.html (accessed September 10, 2016).

43 *The Price of Sand*.

44 *The Price of Sand*.

especially if the market for frac sand remains robust. Even for relatively short-lived mines, reclamation of the site only restores the land to a 4:1 incline, which will obviously not have the same aesthetic characteristics of taller hills and bluffs. It seems safe to say communities with frac sand mining are dramatically split between those with green signs in their yards inscribed with "Sand=Jobs" and those who are vehemently opposed to the industry.

While it is easy to see all of this as just a practical real world debate over economic benefits and its potential cost in health, land aesthetics, and community relations, there is a bigger background disagreement over which values each side prioritizes. Those who argue in favor of frac sand mining, beyond their confidence in economic benefit, appeal to a much more basic principle of *liberty*. Many on this side argue that landowners should have the right to do with their own property whatever they wish, with no or minimal constraints. They seem to appeal to arguments akin to those made by the classical liberal philosopher John Locke and the more recent libertarian philosopher Robert Nozick. Locke argued that humans have a natural right to self-ownership bestowed upon us by God. Resources are "out there" in nature to be taken by those with the will and initiative to claim them. That is, each of us owns him/herself (each of us literally has a property right in ourselves). But each of us, through our own labor (since one's labor cannot be separated from the mind and body that labors) can acquire other previously unowned things. Just so long as one makes the property useful and doesn't leave it in waste, one becomes entitled it.[45] The only caveat (known commonly in political philosophy as the "Lockean Proviso") is that one must "leave enough and as good for others" to acquire. If this requirement is met, one is allowed to own as much property as he or she wishes and do with it what he or she wishes.

Nozick is unsure of what we could call the "metaphysical"[46] argument from Locke for robust property rights. That is, Locke is making claims about how the

---

45 It is worth mentioning that it would seem consistent with Locke's views on property that even if others own unused land, by acquiring it in fair exchange one might also be making the land more useful through developing its productive capacity in some industry. We would want to think about how this might apply to the case of frac sand mining outlined here.

46 In philosophy, metaphysics is the study of reality, which raises questions such as "What is the nature of reality?" or "Are the things that we see in front of us the most real things, or is reality something that lies beneath and not obvious to our sense perception?" In the context of Locke's metaphysical argument for property rights, he suggests its strength derives from the kind of entity God is. Locke thought God has the special feature of bestowing on humans the right to appropriate previously unowned things by claiming them and making them useful. Nozick is unsure this is truly the basis for such rights.

world is in reality and using this as grounds for legitimating property rights. (In his book, *Anarchy, State, and Utopia*, Nozick asks (likely rhetorically) that if one released the contents of a tomato soup can into the ocean, would this mean that he would thus acquire the ocean?) Beyond that, Nozick is in league with the basic Lockean attitude toward property rights and supports them in their most robust form.

Nozick has three principles of property. The first is *justice in acquisition*; as long as one finds property not owned by anyone else, he or she has the right to acquire it. The second is *justice in transfer*, that states that one may trade, sell, or gift property to another. This is legitimate just so long as any preceding link in the chain of property transfer occurred without coercion, theft, or deceit. The third is *justice as rectification*. This principle stipulates that one can seek redress if either of the first two principles is violated. Nozick even goes so far as to accept a revised version of the Lockean Proviso. A key point in Nozick's theory of property is that he doesn't think the justification for acquiring, transferring, or retaining property has anything to do with *desert*. For example, it doesn't matter if the ne'er-do-well son who gains a large inheritance from wealthy parents works hard, is a good person, or is creative or inventive. None of these traits need to be appealed to for the son to become the legitimate beneficiary. His parents (assuming they came about their wealth without deceiving, coercing, or stealing from others, and like conditions apply to its prior acquisition and transfer) are entitled to their property and have the right to do with it what they want—to bequeath it to the deserving or the undeserving, the worthy or the unworthy.

As applied to the case of frac sand mining, Locke and Nozick would likely think companies should be permitted to acquire land and the resources in it by purchasing it from other landowners. Landowners, if they voluntarily contract to make the sale, have done nothing illegitimate. In like kind, had this landowner not had frac sand deposits on her land and her neighbor did, then if the neighbor contracted with a frac sand company to sell the property to allow for mining on her land, that would have to be legitimate as well. (In fact, this is a common refrain by those supporting frac sand mining "on the ground.") To respect property rights means, to advocates of frac sand mining, an entitlement to do with something what one wishes. Liberty is intricately connected with what one can do with his/her property.

It seems as though those opposed to frac sand mining embrace a separate value that takes precedent over liberty and property rights. The idea comes from John Stuart Mill of utilitarian fame. He suggested that, yes, individuals have liberty, but that it is limited by the effects expressions of it have on other people. This leads to consideration of what philosophers call the *harm principle*. Put euphemistically, my liberty to do what I wish extends to the end of my fist and the beginning of your nose. That is, a person or company has the right to do what he/she/it wishes, as long as no one else is harmed. Those opposed to frac sand mining insist that others are being potentially harmed in the most obvious cases by adverse public health effects. Interestingly, (since this principle is also used by those supporting frac sand mining) they add there are already harms suffered by neighbors of frac sand mines due to loss of value in their property, noise pollution, and inconvenience. This means that by appeal to the harm principle, land should be regulated to minimize the actual (and perhaps potential) adverse effects on citizens.

So, liberty and its limits certainly come to bear on this issue. It is worth noting that the clash between liberty and community well-being are put to the test by many other issues besides frac sand mining. However, the links between the local debates of frac sand mining and the national debates on resource extraction in general make this a compelling case study to watch in the future.

Review Questions

1) Frac sand mining appears to be a classic case of "jobs versus animals" that often arises in debates concerning the environment. Do you think this case is so easily classified? Why or why not?
2) It was mentioned that some of the moral thinking about frac sand mining concerns calculations inspired by utilitarianism. How do you think cost-benefit analysis might be applied to frac sand mining? Do you think it is possible to measure different kinds of goods such as financial security versus mental health? Ultimately, do you think it is possible to do these sorts of calculations?
3) What do you think the relationship between liberty and property should be? Do you agree with the idea that self-ownership is the ground for ownership of other goods such as natural resources?
4) What is the harm principle? In what ways might the harm principle put limits on liberty? How does this play out in the case of frac sand mining? Should frac sand mining proceed without any regulation? If so, why? If not, why not? What sort of regulation do you think would be morally appropriate?

Case 9 – Drilling in the ANWR –

Study Questions

1) If drilling for oil and gas were to happen in the Arctic National Wildlife Refuge (ANWR), where would it occur? How much of the refuge would be developed to gain access to energy resources within it? Which groups support the drilling? Which groups oppose it?
2) What are some of the potential positive effects of drilling for oil and natural gas in the ANWR? What are some of the potential negative effects?
3) Why is the ANWR looked upon as being important to both wildlife protection and human cultural preservation?
4) How much oil is estimated to be in the ANWR? How much natural gas is suspected to be there?

There always seem to be fluctuations in the world fossil fuel market. Also, sometimes these fuels are in competition with each other, such as the case of crude oil versus natural gas. If a country (such as the United States) has access to large quantities of both lying in previously undeveloped areas, there is pressure to tap these new locations. Energy companies suspect an abundant supply of oil and natural gas can be found in the Arctic National Wildlife Refuge (ANWR), an immense area of federally protected land in the northeastern corner of Alaska, near the Arctic Circle. This is also a place that possesses critical habitat to a diverse cast of non-human animals and plants.

Since at least the 1970s, beginning with the US oil crisis, there has been persistent debate about whether drilling should be permitted in the ANWR. The crisis began in October 1973 when the Oil and Petroleum Export countries (OPEC) decided to stop sending oil to countries that supported Israel. Consequently, there were gas shortages and long lines for fuel, with gas prices skyrocketing in the US until the end of the embargo in March 1974. To avoid a repeat of the crisis, exploration for domestic sources for fuel increased, with the ANWR being seen as an attractive candidate for oil and gas extraction.

The US Geological Survey estimates there are about 7.6 billion barrels of oil lying in the ANWR, which would be enough oil to meet US consumer fuel demands for about one year. However, the 7.6 billion barrel figure is a mean estimate between the lowest of 4.25 billion barrels of crude oil and the highest of 11.8 billion barrels. Additionally, there are estimates from the US Geological

Survey to be anywhere from 0 to 10.02 trillion cubic feet (TCF) of non-associated, technically recoverable gas in undiscovered accumulations within the refuge, with a mean average of 3.48 TCF.[47]

The ANWR contains 19.2 million acres of land, with 8 million of it designated as wilderness. This makes the refuge about the size of South Carolina in land area. The ANWR contains the Brooks Range, 3 rivers designated as wild and scenic, and an ecologically diverse coastal plain. Roughly 8.9 million acres of land was set aside in 1960 to establish what was then called the Arctic National Wildlife Range, not long after Alaska was admitted into the Union as the 49th US state. It was to serve as a protected area for wildlife and fish or as President Dwight Eisenhower put it at the time, the reason for establishing Range was due to its "unique wildlife, wilderness and recreational values."[48] Though not declared as such by Eisenhower, the ANWR was also seen as a way to preserve the native cultures of the Inupiat and Gwich'in by discouraging development that might cause environmental harm interfering with traditional hunting and fishing areas used by these indigenous groups for subsistence. Some 9.61 million acres were added to the refuge in 1980 under the auspices of the National Alaskan Interest Conservation Act. The ANWR is the only conservation area in the US that encompasses an entire Arctic ecosystem, making it a particularly special place for ecologists.

If drilling were permitted in the ANWR, oil and gas would be extracted from what is known as "Area 1002," the coastal plain section of the refuge of about 1.5 million acres that lies along the Beaufort Sea.[49] Kaktovik, a tiny village with a population of about 250, is located in the coastal plain and surrounded by 94,000 acres of land owned by the Gwich'in and Inupiat. Area 1002 contains the largest onshore, unexplored, and possibly productive geological basin in the US and makes up about 8% of the land area of the ANWR.

47 E.D. Attanasi, "Undiscovered Oil Resources in the Federal Portion of the 1002 Area of the Arctic National Wildlife Refuge: An Economic Update," Open-File report. (Reston, VA: United States Geological Survey, 2005). Available at http://pubs.usgs.gov/of/2005/1217/pdf/2005-1217.pdf (accessed June 6, 2016).

48 Virginia Kramer, "Conservation Groups, Native Organizations Celebrate Wilderness Recommendation for the Arctic National Wildlife Refuge," *Sierra Club*, (website), January 25, 2015. Available at http://content.sierraclub.org/press-releases/2015/01/conservation-groups-native-organizations-celebrate-wilderness-recommendation (accessed May 27, 2015).

49 United States Energy Information Administration, "Potential Oil Production from the Coastal Plain of the Arctic National Wildlife Refuge: Updated Assessment," (website), 2002. Available at http://pubs.usgs.gov/of/2005/1217/pdf/2005-1217.pdf (June 6, 2016).

The relatively undisturbed coastal plain contains habitat for a wide variety of flora and fauna. Forty-five species of marine and land animals live in the ANWR, with some of the smallest, the pygmy shrew, to some of the largest, the bowhead whale, under the protection of the US Fish and Wildlife Service.[50] Caribou, polar bears, grizzly bears, wolves, migratory birds, and a plethora of other species make up what some call "America's Serengeti." This vast wilderness area is unique and if the ANWR were linked to two national parks in Canada (Ivvavik at ~2.51 million acres and Vuntut at ~1.1 million acres), there is the possibility for establishing a huge international park. This would seem to be the next logical step for conservation in the region, especially since there are already agreements and treaties in place to protect whales, migratory birds, polar bears, and caribou that inhabit the area.[51]

In 2015, President Obama in effect froze oil exploration on around 22 million acres of federal land and waters off Alaska, including 12 million acres in the ANWR and an additional 10 million acres in the Beaufort and Chukchi Seas.[52] In doing so, the Obama Administration recommended the declaration of Area 1002 as a wilderness area, which would preclude any drilling or other development. This recommendation to protect offshore lands is significant, as some think the reserves of the Beaufort and Chukchi Seas are the real prizes for fossil fuel extraction, likely containing much larger reservoirs than what is found in the ANWR. The move to remove more land in the Arctic from oil and gas development was also a bit of a surprise as the Obama Administration had previously opened millions of underwater acres to oil exploration off the Atlantic coast (from roughly Virginia to Georgia) and in the Gulf of Mexico to energy companies.

Proponents of drilling in the ANWR include the Alaskan state government, the Alliance for Energy and Economic Growth, ExxonMobil, and other energy companies, basing their arguments on national security concerns, energy price stability, and job growth potential. These three issues tend to intersect as

50 United States Fish and Wildlife Service, "Overview: Arctic National Wildlife Refuge," (website), 2015. Available at http://www.fws.gov/refuges/profiles/index.cfm?id=75600 (accessed May 26, 2015).

51 M. Lynne Corn, Bernard A. Gelb, and Pamela Baldwin, "Arctic National Wildlife Refuge: Controversies for the 109th Congress," *Congressional Research Service Issue Brief for Congress*, IB10136. 2015. Available at http://fpc.state.gov/documents/organization/50817.pdf (accessed May 26, 2015).

52 Joel K. Bourne, Jr., "What Obama's Drilling Bans Mean for Alaska and the Arctic," *National Geographic*, February 5, 2015. Available at http://news.nationalgeographic.com/news/2015/02/150205-obama-alaska-oil-anwr-arctic-offshore-drilling/ (accessed May 26, 2015).

worries about any political or military crisis in the Middle East can cause strong market fluctuations in the cost of fossil fuels. Proponents of drilling claim that opening domestic oil and gas reserves in the ANWR would provide a buffer zone from price fluctuations. It would also boost North Slope oil production to provide needed economic stimulus from private sector sources. ANWR drilling advocates also argue that tapping its resources would lower oil prices overall and extend the economic life of the Trans Alaska Pipeline System. This would provide many jobs not only in Alaska, but elsewhere in America as well. Advocates of drilling believe that with new extraction technologies available to the industry, these resources could be exploited with minimal environmental damage. The U.S. Congressional House Committee on Natural Resources asserts that only a footprint of 500,000 acres would be necessary to access most of the oil and gas resources in the ANWR, constituting less than 3% of the refuge's acreage.[53]

Some say the economic realities of Alaska call for opening more areas to drilling. With the completion of the Trans Alaska Pipeline in 1977, Alaska increased its production to a peak in the 1980s when about a quarter of US production came from the state. Oil from Prudhomme Bay sparked that increase. But more recently, oil production from the area (on state land west of the ANWR) has declined by about 10% per year.[54] Alaska is now only responsible for about 7% of the nation's oil production, with large portions of land in the state having been protected from energy industry development for decades. By opening the ANWR to drilling, the hope is that with the construction of infrastructure, shipping, and drilling, thousands of new jobs will be created. Estimates suggest that opening the ANWR to drilling would create anywhere from 55,000 and 130,000 new jobs.[55] Proponents also believe that billions of dollars in new revenue will become available, helping to spur both the local and national economies.

Advocates of drilling add that there are specific benefits to the state of Alaska from drilling, which is only fitting since the ANWR is located there. In fact, these advocates argue that opening Area 1002 to drilling is only a matter of the federal government keeping its promise to the state's citizens. They claim that in 1980, Congress and President Carter specifically set aside Area

53 United States House of Representatives, Committee on Natural Resources, "ANWR: Producing American Energy and Creating American Jobs," (website), February 15, 2015. Available at http://naturalresources.house.gov/anwr/ (accessed June 6, 2016).

54 Bourne, "What Obama's Drilling Bans."

55 U.S. House of Representatives, Committee on Natural Resources, "ANWR."

1002 for the purposes of oil and gas development, not designating it as a wilderness area.

Drilling advocates also point to strong support from both the Alaskan state legislature and its residents for opening ANWR to drilling. Polls indicate that Alaskans strongly support drilling, with about 70–80% in favor. Local politicians also argue that drilling in the ANWR is a non-partisan issue, with support from both Democrats and Republicans. Republican Speaker of the Alaskan House of Representatives, Mike Chenault, noted that drilling would bring much needed economic development to the North Slope. "To Alaskans it's important to not only continue to feed the work environment that we have, but it also puts together people not only in Alaska working, but all across the nation." Former Alaskan Democratic State Representative Reggie Joule used to represent a region of the state including all lands north of the Arctic Circle and from the Canadian border. His constituency was about 80% Inupiat and he reported that for years people from his district have heavily supported drilling in the ANWR. This is in large part because job prospects in his district are relatively bleak. He insists there is widespread support regardless of political affiliation. "I am a Democrat in Alaska and I support drilling in the ANWR, this is a non-partisan issue."[56]

Very local and specifically strong support for drilling comes from the village of Kaktovik, the only human community in the coastal plain of the ANWR. Again, the prospect of economic opportunities appears to be worth any risk of significant environmental degradation to the area. Alaska has never had a sales or income tax, mainly because taxes on oil revenues have filled the state government's coffers. In fact, for years, Alaskans have received an annual check from The Permanent Fund, offered to all of the state's residents as what could be seen as a royalty share of the state's resources from investing in the energy industry. However, the state has recently experienced budget deficits and declines in Permanent Fund payments. For example, in 2008, each dividend payment was $2,069.00, but in 2013, this amount decreased to $900.[57] Given all of this, there is an incentive for the state's citizens to support increased drilling to secure more tax dollars from the energy sector.

Citing estimates of around 10 billion barrels of oil available in the North Slope, advocates of drilling point out that these reserves are larger than those of countries from whom the US imports oil, such as Mexico, Angola,

---

56 U.S. House of Representatives, Committee on Natural Resources, "ANWR."

57 Alaska Department of Revenue, Permanent Fund Dividend Division, "Summary of Dividend Applications and Payments," (website), 2013. Available at https://pfd.alaska.gov/DivisionInfo/SummaryApplicationsPayments (accessed May 28, 2015).

Azerbaijan, Norway, India, Indonesia, Malaysia, Egypt, Australia, New Zealand, Turkmenistan, and Uzbekistan.[58] At peak production, the ANWR could yield 1.45 million barrels of oil a day, about as much as the US imports daily from Saudi Arabia and over one quarter of what it imports from OPEC each year. This ties into national security concerns, as more domestic production means decreasing the amount of oil that is imported from countries hostile to America. According to the US Energy Information Administration, there is a one-to-one relationship between ANWR production and foreign imports; for every barrel of oil extracted from the ANWR, crude oil imports will be reduced by one barrel.[59]

Finally, advocates of drilling in the ANWR claim that extraction techniques have improved significantly since the wildcatting days of early exploration. Advances in technology have allowed for greatly minimized environmental impact. Extended reach drilling is a major advance in oil extraction, as one drilling platform can now be used to cover a 28,000 square foot radius, an area larger than Washington, D.C.[60] Advocates believe these technological advances allow for the safe extraction of oil and gas. While accidents do happen, all industries carry some risks. However, drilling advocates conclude that the benefits greatly outweigh the harms of drilling.

On the other side, opponents of drilling in the ANWR include several non-governmental environmental groups such as Greenpeace, governmental agencies such as the US Fish and Wildlife Service, and organizations advocating tribal interests, such as the American Congress of American Indians, which represents the Gwich'in Athabascan Indians. Collectively, they point to many objections to drilling in the ANWR, but they tend to coalesce into some general areas including harms to a unique ecosystem, harms to animal species, erosion of traditional human cultures, and exacerbation of climate change. The ANWR serves as critical habitat to polar bears, muskoxen, and wolves. It also contributes to the health and sustainability of minority human cultures, such as the Gwich'in and Inupiat peoples.

Opponents of drilling cite the potential for extensive damage to a fragile ecosystem. Arctic areas are much more susceptible to disturbance than tropical areas. Tundra also takes a long time to recover from environmental

58 Alaska Department of Revenue, Permanent Fund Dividend Division, "Summary of Dividend Applications."

59 Alaska Department of Revenue, Permanent Fund Dividend Division, "Summary of Dividend Applications."

60 Alaska Department of Revenue, Permanent Fund Dividend Division, "Summary of Dividend Applications."

degradation, from tens to hundreds of years. Opponents of drilling also worry that researchers and engineers do not know about the possible effects of frigid waters on oil if there were to be a spill. They argue that before any plan for oil extraction in the ANWR could even be considered, much more study of the possible environmental impacts of possible pollution in Arctic areas should be done.

The drilling itself would do obvious damage to the ANWR land. But beyond scarification and land degradation, there is also the worry that related development will have a large and negative impact on the land. Drilling activity will leave what critics call a "spider web of disturbances" from production facilities. This includes landing strips, roads, and machines for transporting oil to other locations. Refineries can be added to that list if the oil is going to be processed at coastal sites.

There will likely be destruction of vegetation in what some have called "America's Last Great Wilderness." Sara N. Pasquinelli thinks that drilling in the ANWR should be averted so that its "unique wilderness is not jeopardized due to political and corporate greed."[61] There are hundreds of species of plants within the refuge of various types. The vegetation in Area 1002 consists of mosses, grasses, wildflowers, and shrubs as well as many other species of flora. Opponents of drilling contend that these species will be imperiled by oil and gas production. Even with a limited footprint where drilling occurs, the necessary supply network to support it would likely be spread out across Area 1002, causing more disturbance than what is estimated by proponents of drilling. Thus drilling opponents conclude that ecological harm will come as soon as facilities are constructed, including roads, pipelines, power plants, landfills, reserve pits, waste sites, and solid waste treatment systems. Even without spills, oil leaks and runoff from the drilling sites are toxic to plants. If leaks last long enough, they can leach into the soil and have long-term negative effects on plant life. Drilling opponents add that the quality of the vegetation in the area contributes to the aesthetic value that oil extraction will taint or destroy. As an unspoiled area, it would be a particularly terrible loss of wilderness if production were permitted.

Habitat for several endangered species would also be at risk. The Porcupine Caribou population is likely the most well known of imperiled species that use the ANWR for its habitat. There are only about 129,000 living today, down from

61 See Pasquinelli, "To Drill or Not to Drill: The Arctic National Wildlife Refuge v. the 'Need' for U.S. Energy Independence," *Golden Gate University Law Review* (2010) 33: 554–555.

180,000 in the 1980's.[62] While the population is found throughout Alaska, the caribou migrate to Area 1002 to have their young. The coastal plain is conducive to birthing and nursing as it is relatively free of predators. Consequently, the herd has migrated to Area 1002 for this purpose for thousands of years. Obviously, opponents to drilling worry that oil and gas extraction activities will effectively cause the Porcupine Caribou to go extinct. Proponents counter these claims by noting that the Central Arctic caribou herd not only co-exist, but thrive next to drilling operations in Prudhomme Bay. The herd is at its highest population ever, having grown seven fold since the 1970s.[63]

Thinking about the possible extinction of the Porcupine Caribou takes on another dimension when we add that the Gwich'in people depend on the herd for its subsistence. The group has used the species for food and clothing for thousands of years. Until the 1870s, the Gwich'in were a nomadic people, following the herd as its main source of protein. In fact, the word "Gwich'in" can be roughly translated to English as "the people of the caribou." According to oral tradition, the reliance of the Gwich'in on the Porcupine Caribou could go back as far as 20,000 years. With its population decline, there are limits on the number of caribou Gwich'in hunters are allowed to take each day.

Opponents to drilling also worry that polar bears, while they do not live in Area 1002, could experience greatly disrupted birthing patterns that could threaten the species if drilling were allowed. For two months of the year, about half of all female polar bears in the region seek birthing areas on land, and of them about a third do so in the ANWR. The coastal tundra region in the refuge is the "most important land denning habitat for the Beaufort Sea Polar Bear Population."[64]

More specifically, opponents of drilling take very seriously that the ANWR is unique in that it is undisturbed. It thus serves as a laboratory to study how the climate responds over time without direct interference from human development. The area is also unique due to its remoteness. This leads to another problem with having extensive drilling operations in the ANWR. The largest North Slope town is Barrow, which is very small, with about 4,000 people. There are

62 Rick Bass, *Caribou Rising: Defending the Porcupine Herd, Gwich-'in Culture, and the Arctic National Wildlife Refuge* (San Francisco: Sierra Club Books, 2004).

63 Arctic Power, "Faces of ANWR: Caribou," (website), 2015. Available at http://www.anwr.org/features/pdfs/faces-caribou.pdf (accessed June 6, 2016).

64 United States Fish and Wildlife Service, "Arctic National Wildlife Refuge: Potential Impacts of Proposed Oil and Gas Development on the Arctic Refuge's Coastal Plain, Historical Overview and Issues of Concern," (website), 2001. Available at http://library.fws.gov/Pubs7/arctic-oilandgas-impact.pdf (accessed May 28, 2015).

concerns about how much of an influx of oil and gas industry workers that the area could accommodate. Another worry with the remoteness of the site is what happens if a spill occurs and what kind of response would be available. When the Deepwater Horizon spill occurred in the Gulf of Mexico in 2010, it was about 40 miles off from the Louisiana coast. But there was still extensive damage even with the good fortune of having several port cities surrounding the area of the explosion and the spill. New Orleans, Biloxi, and Mobile were relatively close by (and even Houston further away, but still in the Gulf region), serving as staging areas for rescuing workers from the rig, putting out fires, and eventual cleaning up the spill. The cleanup effort would have been much less successful without the number of ports and marine vessels available for use in the area. Comparatively, such a concentrated constellation of transportation centers simply does not exist in northern Alaska and northern Canada. The likely response time to a spill similar in proportion to Deepwater Horizon would be, in the determination of ecologists, catastrophically slow. Another recent oil spill off the coast of Santa Barbara, California, was a reminder that accidents do occur despite industry best practices and makes environmentalists distrust the industry's ability to extract oil safely. And of course, there have already been oil spills in Alaskan waters, such as from the Exxon Valdez in Prince William Sound off the state's southern coast; an event that has left a lasting scar on environmentalists.

There is also debate about how much oil is actually in the ANWR. Opponents to drilling are skeptical of claims that drilling in the refuge will really allow the America to be energy independent. They charge that the amount of oil found there will not be sufficient to give economic security worth destroying the value of the refuge. While Area 1002 contains a great deal of oil, from the perspective of a national energy solution there are other areas more suitable for drilling, especially given the unique ecological characteristics of the ANWR. The oil in the ANWR only "provides a short-term patch to ... significant energy concerns."[65]

Drilling opponents also claim the job creation projections from exploration of oil and gas in the ANWR are overestimated and misleading. Oil production in the region will create some jobs and will likely enhance the profits of some oil executives, but the question is whether it will be enough to really provide the level of economic and national security that proponents claim it will.

---

65 Nicholas J. Monaghan, "'Drill, Baby, Drill!' The Arctic National Wildlife Refuge and America's Energy Reckoning," *Notre Dame Journal of Law, Ethics & Public Policy* (2009) 23: 649–672 at p. 649. The article is available at http://scholarship.law.nd.edu/cgi/viewcontent.cgi?article=1110&context=ndjlepp (accessed May 28, 2015).

Finally, opponents of drilling in the ANWR assert that expanding domestic oil extraction diverts attention away from the real problem—that Americans are addicted to oil. Trying to fix the issue by drilling for more oil is a short-sighted approach that ultimately will not work. Both sides agree drilling in the ANWR will not solve *all* of the nation's energy needs and doesn't constitute a comprehensive energy policy solution. As Lincoln Chafee adds, "[a]llowing oil and gas development in the coastal plain promises only short-term benefits that may irreparably damage the wildlife values and unique vitality of the Arctic Refuge."[66] With that said, advocates of drilling believe that until the global marketplace can either promote some form of alternative energy or otherwise get consumers to use less, oil shortages will continue.

In some ways, both opponents and proponents engage in some sort of *cost-benefit analysis* to try to make their respective cases regarding drilling in the ANWR. As noted in the opening chapter, cost-benefit analysis can be seen as an application of utilitarianism. Utilitarianism is the view in moral and political philosophy that the appropriate action or rule is that which secures the greatest good for the greatest number, usually measured in happiness or pleasure.[67]

Advocates for drilling tend to use cost-benefit analysis more straightforwardly and as the primary approach to support their position. Those opposed to drilling, while sometimes employing cost-benefit analysis (or at least some sort of utilitarian calculation) to supplement their position, tend to lean on deontological arguments such as that it is wrongful to harm animals, the environmental, and other humans. They believe that no matter the possible economic benefits of drilling, allowing it violates certain obligations to the environment, or to animals, and perhaps even to humans who greatly value intact, wilderness areas. Yet, when they are trying to buttress their position against drilling, even its opponents usually enter into a utilitarian analysis, suggesting that all things considered, there is not enough oil and gas in the ANWR to fulfill energy needs. They will add that overall, with possible destruction from drilling (perhaps catastrophic if there is a spill), the costs will be much greater and last much longer than the benefits.

Arguments that tend to rely on cost-benefit analysis do enjoy a common sense feature making them attractive. There appears to be a definite answer to a policy question depending whether the "pros" outweigh the "cons." Also,

66 Quoted in Monaghan, "'Drill, Baby," p. 659.

67 For a more detailed discussion of utilitarianism and cost-benefit analysis, see the section on "Utilitarianism" in Chapter 1 on general moral theories.

using cost-benefit analysis seems to give credence to goals agreed upon by the majority, and hence is attractive to proponents of democratic forms of decision-making.

But several critics pounce on this style of reasoning. It seems like some important values get left to the sidelines or are incommensurable when using cost-benefit analysis. How much is the sight of majestic polar bears or caribou worth? How do we put a price on a mountaintop? Even if actuaries believe that a price can be put on anything (even human life) does that merely capture what people are willing to pay on the market for clean water or the sight of muskoxen, or does it reflect the *true value* of those entities? On the other hand, will more people be made happier than not by drilling in the ANWR and is money a good indicator of that? After all, it is interesting that even though many strive for wealth, the level of happiness that people have is not directly correlated to it as recent studies have indicated. And how do we assess what the relevant moral community is in the first place? Is it all Alaskans? Is it all citizens of the United States? Considering the effects of climate change, is it the world?

Additionally, as with utilitarianism, there are criticisms that in using cost-benefit analysis, while perhaps everyone's interests are taken into account, there is nothing to prevent some from being forcibly sacrificed while others get to enjoy the benefits. Think about eminent domain. Some who believe in strong property rights would likely balk at the notion that family farmer (perhaps someone whose family had owned the property for generations) is forced to give his land to be used for a highway. Likely the argument against the government using eminent domain is that even if the highway makes the most people happy, the government is still violating the property rights of the farmer. Taking one's land is forcing an individual to sacrifice far too much to satisfy the needs of society. One might further argue that even if cost-benefit analysis revealed drilling in the ANWR would generate more financial gains and greater national security, certain species of plants and animals (and perhaps even members of indigenous groups in the ANWR) would incur too great a loss.

All of these questions and concerns arise with cost-benefit analysis, which calls into question whether it can really serve as a morally appropriate way to determine whether any policies (environmental or not) ought to be adopted. However, given that many also question the inherent value of ecosystems, it is hard to see exactly how some weighing of competing interests can be avoided. And if the debate over drilling in the ANWR will be decided by the values of the environment to humans, this will likely mean its participants will have to re-enter the murky waters of weighing the costs and benefits of the practice.

Review Questions

1) Do you think drilling should be permitted in the ANWR? If not, why not? If so, why and under what conditions? Should there be regulations in place governing drilling?
2) At times, environmentalists are accused of being paternalistic; though they are supporting certain policies because they are looking out for vulnerable groups, they may not understand that those groups do not agree with them. What do you make of the disconnect between environmentalists and indigenous people in this case study? On the one hand, some environmentalists argue against drilling in the ANWR because it will harm an ecosystem sustaining indigenous people. Yet, many indigenous people argue that drilling in the ANWR should be permitted to create jobs (especially crucial to young Gwich'in and Inupiat people). Do you think paternalistic arguments are good grounds for policy adoption? Why or why not? Even if you do not agree with paternalistic arguments, should they have some role to play in whether drilling is permitted in the ANWR?
3) Some problems with using cost-benefit analysis to decide public policy issues, such as the permissibility of drilling in the ANWR, were outlined above. What are some of these problems and do you think there are ways around them? Overall, what do you think the likelihood is of coming to a morally acceptable conclusion about oil and gas extraction in the ANWR? Explain your answer.
4) One of the reasons why some groups oppose drilling in the ANWR is due to the possible contamination of soil and water along the coastal plain and loss of habitat for the Porcupine Caribou, an endangered species. However, also remember that the Gwich'in depend on the species for its survival. Should the Gwich'in be allowed to continue to hunt this species? If the Gwich'in should be able to hunt caribou, should it also be permissible to block oil and gas companies from drilling to protect Porcupine Caribou?

Case 10 – The Lakes and Phosphorus Twins: Monona and Mendota –

Study Questions

1) What is the phosphorus cycle? How does phosphorus enter lakes and rivers and what is its origin? What are some consequences of large amounts of phosphorus infiltrating waterways?
2) What is the difference between dissolved and particulate phosphorus? Do these two forms of phosphorus have the same effect on lakes? What is the ecological significance of these forms of phosphorus?
3) What are some moral questions that arise from increased amounts of phosphorus in lakes and rivers?

The city of Madison, Wisconsin, is often lauded for its picturesque setting. Its downtown portion lies on an isthmus, which is a strip of land between two bodies of water: in this case Lake Mendota to the north and west, and Lake Monona to the south and east. This geographical layout provides several residents of, and visitors to, Madison alike with stunning lake views and easily accessible recreational activities on the water—boating, swimming, and flyboarding in the summer; ice skating, ice fishing, and ice sailing in the winter. With this readily available natural setting, when combined with its status as state capital, its serving as home of the world-class flagship University of Wisconsin campus, and its possession of a relatively good economy, Madison routinely sits at or near the top of many lists for most livable cities in the United States. The regionally famous farmers' market occurring downtown every weekend from early spring though late fall evokes nostalgic Americana, with a bevy of locally produced goods making it one of the most successful in the nation. The proximity of the city to agricultural bounties also fuels the farm-to-table movement whereby locally sourced fruits, vegetables, meats, and baked products are supplied to regional restaurants and residential kitchens. This prevalent supply of goods has created a restaurant and specialty shop scene that is rapidly turning Madison into a "foodie" destination.

But note that even desirable places to live can have significant problems, and sometimes the very features that help make a location attractive can also work against it. From one perspective, living on an isthmus is a turn of good fortune, with natural beauty close at hand and lakeshore lifestyle available to many. From another viewpoint, this geography is also a recipe for urban sprawl with its attenuating problems of traffic congestion and harmful spill off that easily taints local waters. Being an appealing area serves as a magnet for more people, placing greater demands on local infrastructure and contributing to pollution. Likewise, while having a heavy agricultural presence has left the city with a cornucopia of products, it also has a cost as farm runoff and soil erosion loads local waterways with contaminants even though the sources are many miles away.

Of course, part of this pollution is quite local, originating in runoff from storm drains that eventually finds its way into the lakes. Motor vehicles leak oil, gas, and a host of other lubricants that wash into the storm water system. A high concentration of businesses and people produce waste that needs to be dealt with and is a massive burden that sometimes cannot be properly controlled before polluting water. Trash from people and fluids from motorboats find their way directly into the lakes. All of these factors contribute to supplying contaminants into soil and water that build up over time.

But for all of these issues, the greatest contributor of pollution in the Madison area lakes is phosphorus, which comes from large-scale farming operations surrounding the city, and its negative effects have become more noticeable over recent years. Lakes Monona and Mendota are part of a much larger Yahara watershed that spreads for miles outside of the Madison metropolitan region and includes a wide swath of agricultural land. A large, modern agricultural industry means a great deal of fertilizer and with several farm animals come piles of manure. For lakes fed by portions of the Yahara watershed near these farms, this results in exposure to copious amounts of phosphorus. Over time, through the erosion of phosphorus-rich soils, not to mention the direct spill off of fertilizers and manure that flows into the watershed, excessive phosphorus builds up in Madison area lakes.

Phosphorus in and of itself is not a bad thing. It is naturally occurring in soil, water, and all living organisms. Phosphorus is critical for the health of plants; it is one of the sixteen critical elements that promote plant growth, so its appearance in soil is important. But what is good for plants is not always good for bodies of water, as phosphorus, especially in high concentrations, impairs water quality by stimulating excessive growth of algae and aquatic vegetation.[68]

Phosphorus attracts cyanobacteria (also called blue-green algae), which feed on nutrients and quickly experience explosive population growth. This process has a dramatically negative effect on aquatic life in lakes, ponds, and streams. Cyanobacteria blooms are toxic and can cause fish kills that reduce the recreational value of a body of water. They can also be harmful to humans

68 University of Wisconsin-Madison, Nutrient and Pest Management Program, "Phosphorus Movement from Land to Water," *Wisconsin Crop Management*, November, 4, 2010. Available at http://ipcm.wisc.edu/download/pubsNM/P_Movement_landwater.pdf (accessed January 23, 2015).

as some blue-green algae produce neurotoxins that, if consumed, can cause vomiting, headaches, stomach cramps, muscle weakness, and fever.[69] Contact with blue-green algae can lead to asthmatic symptoms as well as skin irritation. In extreme cases, such contact can even be fatal to humans and pets. As blue-green algae are bacteria, they can also cause gastrointestinal issues. Children, due to their lower body weight, are at greater risk of illness if they ingest toxins released by blue-green algae.

Large amounts of cyanobacteria can further diminish the recreational value of lakes by forming algae blooms making swimming unpleasant. If the blooms are excessively large, beaches need to be closed. Algae blooms caused by excessive phosphorus also deteriorate the quality of drinking water.

To better comprehend how lakes become contaminated we need a clearer understanding of what is called the phosphorus cycle. That is, we must look closely at sources of phosphorus and how it is transported from land to water. Soil, fertilizer, manure, vegetation, other soil amendments, as well as municipal and industrial discharges are all sources of phosphorus. The primary transport vehicles of phosphorus are soil erosion and water runoff.

Like other nutrients, phosphorus runs through a cycle of interactions from soil to plant to animal. The phosphorus cycle is composed of a complex relationship of chemical and biological reactions that determine the availability of phosphorus. For example, the weathering of minerals and all of the other sources listed above are points of origin for soil phosphorus. To remove it from soil involves crop uptake and harvest, runoff, erosion, and sometimes, by leaching. Most phosphorus can be found in topsoil as a combination of organic and inorganic matter. Again, phosphorus in both forms is important to plant growth, but how much is available depends on environmental conditions as well as soil characteristics. Most phosphorus is tightly bound to soil (called "fixation") that makes it unavailable to plants.

In plants we also have a significant source of phosphorus from both native and agricultural vegetation. In spring runoff, it is transported from alfalfa, forest litter, grasses, and crop residues. Ruptured plant tissues, as a result of freezing or thawing, can also release phosphorus. Plant derived phosphorus can also be transported as a result of heavy rains or during the spring melt where it is dissolved in runoff water.

Manure is another significant source of phosphorus. When it is applied to land for agricultural purposes, manure recycles vital nutrients to promote plant growth. Manure can also be an important soil additive, enhancing the

---

69 Wisconsin Department of Natural Resources, "Blue Green Algae," (website), 2014. Available at http://dnr.wi.gov/lakes/bluegreenalgae/ (accessed January 29, 2015).

physical properties of the soil and leaving it as an improved seedbed for plant growth. Unfortunately, despite all of these advantages of manure use as a soil amendment, if too much is used on the land for too long, it can also become a source of excessive phosphorus.[70]

Livestock feed is another source of phosphorus as it relates to manure. Cows need a high nutrient diet to achieve their growth goals, but even then their dietary rations often involve excess nutrients, including high levels of phosphorus. The amount of phosphorus in farm animal manure comes directly from dietary inputs. Excess dietary phosphorus thus accelerates the amount of it in soils and, in turn, obviously increases the amount of it that can be displaced into bodies of water via runoff. Consider that according to the National Research Council, the recommended range for phosphorus levels in feed for lactating dairy cows is from .35 and .38 percent. In cows consuming phosphorus at the high end of that range, the fecal phosphorus emitted is 47lbs. / per cow / per year. But if the phosphorus intake of lactating cows is increased to .48%, the amount of fecal phosphorus emitted increases to 65 lbs. / per cow / per year. The amount of phosphorus increases even more if the consumption level is .55%, with 78 lbs. / per cow / per year of fecal phosphorus emitted.[71]

However, the largest source of phosphorus comes in the form of agricultural fertilizer. Soils provide some phosphorus naturally, but they usually need to be augmented by many other inputs in agricultural settings. Since in nature soils have little phosphorus and much of it is already "fixed," there is usually too little available for plant use to yield desired levels of crop production. Thus, to promote plant growth, farmers have incentives to add phosphorus to their cropland via fertilizer. This supplementation is necessary because of the rapid rate of bonding of phosphorus to soil particles.

Over time large applications of supplemental fertilizer saturates agricultural soils with phosphorus to the point where it is being lost either through runoff or erosion. This is because if too much phosphorus is added to the soil and is not removed quickly enough, this will in turn raise levels of phosphorus above what is needed to optimally grow plants. In most cases, soils in this condition are susceptible to heavier losses of phosphorus via surface water runoff.

Finally, industrial and municipal water discharged into lakes introduces phosphorus into these bodies of water. This, of course, is not surprising as these discharges typically serve as sources of a number of pollutants infiltrating

---

70 University of Wisconsin-Madison, Nutrient and Pest Management Program, "Phosphorus Movement from Land to Water."

71 University of Wisconsin-Madison, Nutrient and Pest Management Program, "Phosphorus Movement."

lakes. Storm water runoff coming from streets, driveways, parking lots, roofs, and lawns all contribute to feeding phosphorus into surface waters. Sometimes phosphorous filtered out by wastewater treatment still ends up in lakes later through other means. Biosolids, including sewage sludge, come from the separation of liquid from solid waste in municipal wastewater treatment plants. These solids are rich in phosphorus, but are often applied to nearby agricultural land as a nutrient source. As is the case with animal manure, phosphorus build up is likely after many years of biosolid application, eventually resulting in excessive levels reentering the watershed.[72]

With respect to the transport of phosphorus, two interrelated processes mentioned earlier deliver it from land to bodies of water—runoff and erosion. Runoff is the movement of water over the land's surface. Erosion occurs when runoff strips soil particles from the land surface and relocates them to a different area. As runoff water interacts with soil, the water itself can pick up phosphorus from the soil, plant material, and manure with which it comes into contact. Runoff water can contain both *dissolved* and *particulate suspended* forms of *phosphorus. Particulate phosphorus* is bonded with soil and other organic materials in the runoff. 60–90% of phosphorus stripped from agricultural land in the runoff process is of the particulate type. Researchers have found that as the erosion rate increases, the particulate phosphorus rate also increases, while the *soluble form of phosphorus* decreases dramatically.

Particulate phosphorus is not readily accessible to plants to use for nutritional purposes. Yet, *dissolvable phosphorus* entering the lake through runoff is instantaneously useable to plants and promotes the growth of algae and aquatic vegetation. The eroded soil picked up from runoff eventually settles to the bottom of the lake, taking with it bonded phosphorus. We can think of this as a reservoir for phosphorus that can be released in the future. So, while there may not initially appear to be a problem with phosphorus at the bottom of a lake, it can cause difficulties from its future release. One way soluble phosphorus in sediment at the bottom of the lake or lying on organic matter can be released is by biological interactions. Some of this soluble phosphorus becomes bound to surfaces in the lake bottom before it reaches the overlying water. However, some soluble phosphorus dissolves into lake water.[73]

Unfortunately, it doesn't take much dissolved phosphorus to decrease water quality. In fact, when compared to the amount of phosphorus needed

---

72 University of Wisconsin-Madison, Nutrient and Pest Management Program, "Phosphorus Movement."

73 University of Wisconsin-Madison, Nutrient and Pest Management Program, "Phosphorus Movement."

for optimal plant use, or the amount found in manure and fertilizer, the amount necessary to significantly diminish water quality is very small. For example, surface water application of phosphorus that can damage water quality can be as small as 0.01 parts-per-million (ppm).[74] Note that this is ten times lower than the amount of phosphorus usually found in soil necessary for plant growth, which is typically 0.2 to 0.3 ppm.

Another possible conduit for phosphorus movement is by leaching, which is when it infiltrates groundwater from soil. Generally speaking, leaching can occur at sites with a high water table, soils that are heavily saturated with phosphorus, or in sandy, highly permeable soils. Sometimes, though it is rare, phosphorus can be carried by groundwater to surface water as well. Groundwater contributions to phosphorus in lakes are possible, but in Wisconsin the amount is negligible.

A further distinction that should be made in relation to phosphorus is between *point* and *non-point sources of pollution* (also called *runoff pollution*). *Point sources for pollution* are easily (or at least more easily) identifiable, such as the end of a pipe or a smoke stack. *Non-point sources of pollution* are more diffuse and harder to identify as they come from runoff and erosion over a widespread area. While one might tend to associate point source pollution with urban areas and non-point source pollution with agricultural activities in rural areas, these categorizations are too simplistic. While these descriptions often fit, there are notable exceptions. For example, in urban areas, runoff entering storm drains during heavy rainfalls can pick up sediments from construction sites, parking lots, streets, and lawns. Pollutants can often be found in these transported sediments and thus, indicate non-point sources of pollution. Conversely, the agricultural industry can supply point sources of pollution. Consider the pesticides, manure, and fertilizer from a storage facility that can spill directly into a nearby waterway.

Due to stricter regulations and heavy capital investments into upgrading infrastructural capacities, pollution from industrial and municipal sources has decreased dramatically. Given the success of point source pollution reduction and the substantial cost of continued reduction of this source of excessive phosphorus infiltration of water sources, there has been more public interest in turning to the problem of reducing pollution from non-point sources to improve water quality. In Wisconsin, there are many more agricultural watersheds than urban watersheds. However, many of the state's watersheds span over *both* urban and agricultural areas. Non-point sources of pollution in

74 University of Wisconsin-Madison, Nutrient and Pest Management Program, "Phosphorus Movement."

urban areas can yield significant negative impacts in small, localized areas. Yet, given that agricultural non-point sources of pollution are spread over a much wider area, their overall impact on lakes, streams, and rivers is much greater, and consequently, recent efforts to improve water quality have targeted these sources of pollution in Wisconsin.

While non-point sources are responsible for most of the pollution of Wisconsin's lakes and rivers, this doesn't mean this type of pollution cannot be concentrated. For instance, 90% of the phosphorus coming from non-point sources of pollution that reaches the state's lakes comes from 10% of its agricultural area. There are many factors that go into whether a location contributes phosphorus into bodies of water, including exposure to heavy rainfall, soil moisture, soil erodibility, topography, and soil water storage capacity. Some locations with significant amounts of phosphorus do not have the features necessary for transporting it to lakes. On the other hand, areas one wouldn't expect to be major sources of phosphorus contribute to its concentration in lakes because they are subject to significant runoff, soil erosion, or have other traits making them conducive to phosphorus loading. All of these factors indicate that the dynamics of land-to-water transport of phosphorus compose a rather complex process.

Despite intensive efforts to try to do so, including implementation of improved land management practices, Madison lakes are not measurably cleaner than they were thirty years ago.[75] Limnologists say these efforts have been counteracted by the increased frequency of heavy rainfalls caused by climate change that wash several pollutants into the lakes via storm drains.

While it is important to understand what factors drive excessive phosphorus being spread on the land and how it enters lakes, it is also imperative to know how it affects aquatic life in more detail. One worry is *eutrophication*, which is when too much phosphorus in water triggers excessive growth of algae. Blue-green algae can collect in think mats (what some call "pond scum") that hugs close to the shorelines of lakes. After the algae bloom ends, it will die off and sink to the bottom of the lake to decompose. In turn, during the decomposition process of blue-green algae, other bacteria will flourish and their metabolism will require an abundance of oxygen. The unnaturally high rate of oxygen consumption will deplete the amount of it needed by aquatic fauna. The water aeration rate will not be able to restore oxygen quickly enough to offset the amount consumed by bacteria, and the lake will suffer low oxygen concentration,

75 Kate Golden, "Yahara Beach Closures Highlight Algae, Bacteria Threats Statewide," *Wisconsin Watch*, April 23, 2014. Available at http://wisconsinwatch.org/2014/04/yahara-beach-closures-highlight-algae-bacteria-threats/ (accessed January 27, 2015).

leaving it *hypoxic*. This causes stress for fish that cannot get enough oxygen and if conditions leave a body of water with no oxygen, it becomes *anoxic* and is lethal to aquatic life. Areas where there is no oxygen left for animal life are called "*dead zones*" with perhaps the most famous example being a substantial portion of the Gulf of Mexico.

Blue-green algae blooms also imperil aquatic plants. Plants living on lake bottoms can have their growth impaired by algae flourishing on the water's surface. Algae effectively blocks light vital to the health of lake bottom plants, as they cannot photosynthesize and thus experience stunted growth. This causes yet another stress factor for fish and other aquatic fauna as they often use lake bottom vegetation to hide from prey. These negative features of blue-green algae blooms, especially given the frequency with which they occur on Madison lakes, are changing the local ecosystem and decreasing its biodiversity.

Beach closures are also common on the Madison lakes due to blue-green algae blooms, which are symptomatic of low water quality. When combining the totals from Lakes Monona and Mendota with two other prominent area lakes (Waubesa and Kegonsa), the number of beach closure days were highest in 2008 and 2009, when they were closed for 156 and 196 days respectively. 90% of those closures were due to harmful blue-green algae blooms, though most recent closures have been attributed to other bacteria fed by algae blooms. Either way, phosphorus additions to the lakes contribute heavily to this issue. The average number of days Madison area beaches have been closed by Madison and Dane County Public Health Departments each summer since 1996 is 75. While the number of days of beach closures were appreciably lower more recently, with 101 in 2012 and only 48 in 2013, the lower figure in 2012 was caused by drought whereas in 2013, cooler temperatures prevented large areas of algal scum. But these lower numbers appear to be caused more by anomalies in weather than indicative of any real improvement in water quality.

Finally, shorelines are made aesthetically unpleasant by algal blooms that can eventually have a real economic impact. Severe blooms cause some types of phytoplankton to release a secretion that, when whipped up by wind and wave action yields unappealing looking yellowish-green foam. Blue-green algae blooms not only discolor water, but also can give off a bad odor.

Steve Carpenter, Chair of the Limnology[76] Department at the University of Wisconsin-Madison, thinks phosphorus management is what he calls "a keystone issue" for the health of lakes. Phosphorus control is so important because it is the first step toward solving any other problems lakes might have. Take, for

76 Limnology is the study of lakes.

example, the problem of invasive species. They tend to penetrate phosphorus-rich areas. In this unbalanced nutrient environment, invasive species are able to outcompete native ones and crowd them out.[77]

Some farming practices can help reduce the erosion that leads to high concentrations of phosphorus in area waterways, such as contour plowing and use of winter cover crops. But the high level of phosphorus in Madison lakes now likely makes these solutions impractical. Restoration measures thus have to be more dramatic.

One strategy involves directly combatting the overabundance of phosphorus that creates algae blooms by trying to recycle it.[78] To this end, Madison began recycling phosphorus in the summer of 2014 in Wisconsin's first phosphorus recycling facility. The goal is to filter as much phosphorus out of the process as possible before it reaches the lakes, removing it from wastewater at the facility and then placing it in "lake-friendly" fertilizer that can be used in urban lawns and gardens. To do this, the fertilizer is run through a system whereby there are two outputs. One is a sludge mainly made out of bacteria and the other is wastewater containing high levels of phosphorus.

The Madison Metropolitan Sewage District (MMSD) then uses a biological removal process to take phosphorus out of wastewater to feed micro-organisms. Once the phosphorus is concentrated in the micro-organisms, they are removed from the system as waste sludge. The micro-organisms take most of the phosphorus with them. The wastewater then filters into the rest of the treatment cycle before being jettisoned into the lake. The waste sludge is sent to a tank where it is gently mixed with a small amount of acid-phase sludge, a procedure promoting release of the phosphorous. The combined waste sludge and dissolved phosphorus are then directed to a gravity belt thickener where the solids are thickened prior to introduction to an anaerobic digester. In the meantime, the liquid phosphorus-rich filtrate is captured. Reactor feed pumps move the phosphorus-rich filtrate to the phosphorus harvesting reactors. The filtrate is then turned into a solid particulate called struvite. Struvite contains an equal ratio of magnesium, ammonia, and phosphate.

A bed of struvite particulates (called prills) are lifted through the reactor and chemicals optimizing the formation of struvite are added. Struvite collects on the outside of the prills, sort of like the formation of a pearl, until they reach the desired size for harvest. The specific size selected for harvest is

---

77 For more on invasive species, see Case 42—"Beware the Invaders!".

78 Eleanor Nelsen, "Will Recycling Phosphorus Help Stop Algae Blooms?" *KQED Science*, (website), October 7, 2014. Available at http://science.kqed.org/quest/2014/10/07/will-recycling-phosphorus-help-stop-algae-blooms/ (accessed January 23, 2015).

determined by demand from the fertilizer industry.[79] When the prills are ready, they are removed from the rest of the water by a screen and dried for storage (the Madison facility actually supplies the dryer heat via a hot water system fueled by biogas to save energy).

While much of this case study has been very technical, there are a couple of philosophical issues of interest here, but we need some background context to draw them out. Notice the regulatory requirements that some think are necessary combat the problem. The worry is not so much that agribusiness uses phosphorus, but that it uses too much. We have a case where there needs to be a critical balance set, as solving one problem has led to overcompensation that causes yet another problem. Recall that the urge to add fertilizer containing phosphorus comes from the desire to optimize crop growth. But since in nature the majority of phosphorus is of the particulate form bonded with soil and unavailable for absorption into plants, farmers add more phosphorus. As a consequence, they often add too much, to the point where it can no longer be taken up by plants and will be carried off by runoff and erosion. This means surface waters near agricultural areas will eventually transport phosphorus to Madison area lakes.

So, there are two important points to consider here. First of all, for any individual farmer, there is a strong incentive to maximize crop growth, which will more likely happen by adding phosphorus than not. Even if too much phosphorus is applied, the farmer doesn't suffer the burdens of that choice. The farm achieves the highest possible yields even if excessive fertilizer is applied. Also, since the phosphorus is likely carried off site miles away to another part of the watershed, even the costs of dirty water affect others disproportionately. If some farmers want to go through the trouble of mitigating the application of phosphorous—then so be it, but there is no strong incentive for any individual farmer to do so on his/her farm. Admittedly, this is not what economists would call a commons (which is a resource that is open to all, but is owned by none). After all, farms in the United States are mainly privately owned. However, there are features of this environmental issue that strongly resemble what scholars call the *tragedy of the commons*.[80] That is, when actors have an opportunity to receive benefits without any immediate or long-term risks, there is little incentive for them to cease or curtail that action even when it burdens others.

---

79 Madison Metropolitan Sewage District, "Phosphorus Harvesting," (website), 2015. Available at http://www.madsewer.org/Programs-Initiatives/Phosphorus-Harvesting (accessed June 7, 2016).

80 For background on this, see Garrett Hardin's seminal work "Tragedy of the Commons," *Science* (1968) 162(3895): 1243–1248.

The second thing to notice is most of the phosphorus comes from non-point sources. Of course, one way that this is problematic is due to the brute fact that it is hard to combat this sort of pollution because there is no "smoking gun" that can be pointed out as the culprit. But the problem takes on another dimension as well. It allows for those who likely contribute to the phosphorus problem in lakes to deny they are really causing the issue. Each individual farmer might well claim that even if phosphorus leaves their property and goes elsewhere, there is no way to tell in such a diffuse system if it is erosion runoff *from their land* that is resulting in harmful phosphorus buildup. The chain of events begins at a point far too remote from where the harm is occurring to pin blame on any particular farmer.

The classic problem of *vagueness* arises here, in which there are concepts that by their nature are indeterminate. The paradigm example of a vague concept comes in trying to define a heap. How many grains of sand does it take to make a "heap" a "heap"—one more than one? This leads to another related philosophical problem known as the *sorites paradox*. A paradox is a statement or argument that in itself appears to be sound, but leads to a conclusion that seems to be senseless. In the case of sorites paradox, we have a group of arguments that are paradoxical due to vaguely defined predicates. Let's return to the heap example. If a have heap of twenty-seven stones, does it cease to be a heap by continuously removing stones? Is a twelve stone heap not a heap any more if we take one away? What happens when one keeps taking stones away? At which point do we no longer have a heap? No one stone can be identified as *the one* that makes the difference between having a heap and not having a heap. Another example of a vaguely specified concept is baldness. We seem to know what baldness is, but at what point—at the loss of what number of hairs—does it take for someone "to become bald." It appears there is no suitable number of hairs at which one becomes "bald" as opposed to "not bald."

How does this apply to the case of phosphorus in the Yahara watershed lakes? While it is not clear we have the sorites problem or vagueness in this case per se, we do seem to have similar elements with the issue of water pollution. No one in the scientific community is claiming there is one large farm degrading water quality in the Yahara watershed. But it does seem that incremental contributions from a large number of farms over time, even if each of the contributions of phosphorus are negligible, cause a major watershed problem.

This has implications for human interactions with the environment. Again, if I am a farmer and you want me to cease putting so much phosphorus on my crops, you are going to have a hard time convincing me to alter my actions. There is no way my individual contribution can be pointed to as *the one* that tips the balance over to "polluting Madison lakes." After all, my intention is not to "pollute Madison lakes," but instead to "increase my crop yield." If this is combined with what appears to be the flimsy case that I really have much of an effect on the watershed, I do not have much motivation to cease putting, or to even put less, phosphorus on my cropland.

In fact, there is even less motivation for any individual farmer to take reform measures if he/she considers what *other people* are likely doing. Imagine I am a farmer who cares somewhat about the environment. It will still be a tough sell to get me to reform my farming practices. Even if all farmers desired to have a cleaner and more robust watershed, if I think that my act is negligible, and if there were lots of other farmers willing to curb their use of phosphorus (perhaps they are small organic farmers who have low impact on the environment), I would then have little motivation to curb mine. I can simply free ride on the reform actions of the other farmers and perhaps experience a win-win-win situation—I don't have to endure the burden of foregoing my usual farming practices, I continue to get a high crop yield, *and* the watershed becomes cleaner! This paragraph is an instance of when the sorites problem is applied to a social issue, which is known in political philosophy as the *free-rider problem*.

The free-rider problem is particularly sticky with respect to social outcomes because it isn't at all clear I am being irrational in pursuing direct, short-term benefits. In fact, it would seem that if I put all of the factors together it is rational for me to free-ride. After all, changing the practices on my farm alone won't help the Yahara watershed, I risk losing crop yield if I reform my practices, enough others might reform their practices to clean the watershed, and the goal of cleaning the watershed may likely be attained without my contribution. I get to enjoy *both individual and collective benefits* of the actions of others without having to do anything. What could be more rational than that?

Of course, with many collective actions such as the one we've been discussing, if enough people reason as the farmer above does, the desired collective goal will not likely be met. Is there a way of showing that it is, indeed, rational to act collectively without being coerced? Political theorist Richard Tuck thinks there might be.[81] Recall that what first drives the failure to contribute to a collective goal is the thought that there is no single action I can take to either pollute a lake or to bring it to the status of "not polluted." Hence my contribution in either case is negligible. Tuck suggests a different way of thinking about the issue, using the example of climate change. Even though there is no precise tipping point at which global climate change occurs, at some level (though we can't give more than a general account of where) there is

---

81 See Richard Tuck, *Free Riding* (Cambridge, MA: Harvard University Press, 2008).

the equivalent of a threshold. From the standpoint of my collaborative action, it is similar to a threshold case like voting. Tuck thinks that in cases of voting, my vote does count even when it is *not necessary*; it is part of the set of votes *sufficient* to get my candidate elected. In that sense, *I* get my candidate elected. Correspondingly, with respect to taking some action to curb pollution on my farm, I have reason to do so *if* other farmers are acting to do so as well. What Tuck concludes is that we do have good reason to support *collective action*; it is thus irrational (if you care about the environment) to free ride on the collective actions of others who are trying to curb greenhouse gas emissions. He also thinks coercive measures are justifiable and that we need all the help we can get to overcome our irrational desires so that we might meet certain environmental goals we want to achieve. But we shouldn't insult people by telling them that they are being rational in not wishing to contribute, but then try to force them to comply with some collective goal.

Tuck still cautions us not to be naïve. It does seem preposterous to think that my shutting off the lights will have much of an effect on curbing global warming without like agreement from others. But, this is precisely why it is crucial to work toward large-scale international accords to commit to lowering carbon emissions.

Perhaps we could apply Tuck's work to the issues of excessive phosphorus in the Yahara watershed. There is reason, if one desires a watershed to be cleaner, to act with others *if* they are sincerely reforming their own agricultural practices. But what is more important is to try to actively promote policy changes to protect water quality. The question that remains is, if farmers do not have this impulse, and they are not convinced there is any duty to take action to curb phosphorus release into the watershed, will merely the fact that "other farmers are doing it" be enough to convince them to take like action?

Review Questions

1) What do you think about the prospects of applying Richard Tuck's observations on possible justifications for engaging in small acts when combatting climate change to small acts individual farmers could make in reducing phosphorus loads in local watersheds? What about phosphorus loads that come from the activities of residents, such as overapplication of fertilizer to residential lawns, or the application of rock salt and other deicers to driveways and sidewalks (see Case 43—"A Different Approach to Snow Removal" and Case 44—"The Deicing Dilemma")? Would these individuals simply take advantage of the morally good actions of farmers and thus lead us back to the original free-rider problem?

2) One strategy used to entice farmers to control runoff from their farms includes paying them to grow crops organically, use contour farming practices, or cultivate a barrier of plants around their fields to collect phosphorus before it runs off. Do you think there are any moral problems with paying farmers for employing these strategies to protect watersheds from excessive phosphorus loading? Why or why not?

Case 11 – A Great Lakes Diversion.... –

Study Questions

1) Why are city officials in Waukesha trying to divert water from Lake Michigan to their city? What is the diversion plan?
2) Why do Waukesha city officials think that diverting water from Lake Michigan to the city and then returning its flow back to that lake via the Root River is ecologically beneficial?
3) What are some concerns about the water diversion plan?

Officials in Waukesha, Wisconsin, continue to press for a diversion of water from Lake Michigan to their city, which is presently prevented by federal law. The motivation behind the diversion is to secure an adequate supply of drinking water for Waukesha's residents, to establish an outlet for treated wastewater, and also to provide a conduit for river freight traffic into the region. Waukesha city officials argue that a diversion would not create a negative impact on lake levels or water quality, but there is no consensus on those claims. Waukesha is considering using a Lake Michigan tributary (the Root River) as its discharge point for millions of gallons of diverted water daily for return flow into the Great Lakes. The cost estimate for building new water utility pumping facilities in Milwaukee to supply Waukesha with water and to the Milwaukee Metropolitan Sewerage Commission are in the multiple millions. Opponents of the plan say that realigning water supplies from one basin to another across a region—using one of the inter-connected Great Lakes as the source—is fraught with ecological and financial costs.

It is indeed strange to think that the diversion of water would ever need to emerge as a topic of debate in southeastern Wisconsin. In the late 19th and early 20th centuries, there was an ample supply of water in the area. In fact, what was then the town of Waukesha was well known in the Great Lakes Basin for its bounteous supply of pure water, with rich aquifers laying beneath its woods and meadows. The purity and taste of the area's water were so well regarded that spas grew up around the town, selling it (while touting its restorative health qualities) to city dwellers in Milwaukee and Chicago.[82]

But Waukesha now is much more city than town as its population has increased. Its water resources have become strained from serving a population that rapidly expanded and is consequently dealing with water quality degradation. Waukesha has 71,000 residents and suburban growth and industrial expansion have drained the deep aquifer it uses for its public drinking water. Water levels in the deep aquifer are down hundreds of feet from what they were when compared to predevelopment levels.[83] Both industrial and residential expansions have caused pollution in local water sources, with radium (a known carcinogen) levels in aquifers exceeding US federal standards. Though this supply has been supplemented by water from another swallow aquifer, the city is concerned that this supply won't last and thus seeks a long-term solution to water shortage issues. Additionally, Waukesha is mandated to find a new source of water from sources outside the radium bearing aquifer by 2018.

While calls to divert surface water from Lake Michigan have been made since at least early this century, a formal proposal was submitted in October 2013 by the city. The proposal requests that water enter through the Oak Creek treatment plant, which is 31 miles east of Waukesha. According to city officials, they want to buy 10.1 million gallons of water per day and have it transported by pipeline, which according to their estimate is about 1 millionth of one percent of the total supply of water from the Great Lakes.

But even though this seems like a negligible amount of water, there has been a great deal of controversy about the proposal. Part of the dispute can be explained with a bit of background. Eight US states (Illinois, Indiana, Michigan, Minnesota, New York, Ohio, Pennsylvania, and Wisconsin) agreed to the Great Lakes-St. Lawrence River Basin Water Resources Compact, which is an interstate, legally-binding agreement that no water in this immense water basin can be diverted into other water basins. This compact is the means by which the commitments made internationally with the Premiers of Ontario and Quebec will be enacted, which together is known as the Great Lakes-St. Lawrence River Basin Water Resources Agreement. President Bush signed

82 Kaye LaFond, "Waukesha Presses First Test of Great Lakes Water Compact," *Circle of Blue* (website), July 9, 2014. Available at http://www.circleofblue.org/waternews/2014/world/waukesha-presses-first-test-great-lakes-water-compact/ (accessed May 15, 2015).

83 Wisconsin Department of Natural Resources, "City of Waukesha Water Diversion Application," (website), May 2015. Available at http://dnr.wi.gov/topic/wateruse/waukesha diversionapp.html (accessed May 15, 2015).

the measure in October 2008 after all of the involved US states ratified it and it was passed by both the US House of Representatives and the Senate.

While GL-SLRBWRC is designed to prevent cross-water basin diversions of water from the Great Lakes watershed, exceptions *may* be made for municipalities that lie on its outskirts. Though Waukesha is within just 15 miles of Lake Michigan (and merely 1.5 miles from the bordering Great Lakes basin) it is actually in the adjacent Mississippi River water basin.[84] The Fox River (Illinois River Tributary) that runs through the city is also part of the Mississippi River Basin. However, as a community that lies in a county that "straddles" both the Mississippi River and Great Lakes basins, the city's water district is eligible to apply for an exemption that, if approved, would allow for a diversion of water. As noted by the City of Waukesha in its application, "*A proposal by a Community within a Straddling County to divert Great Lakes water is excepted from the prohibition against Diversions, provided that the Proposal satisfies all of the requirements of the Compact and Wisconsin Statutes section 281.346(4). Compact section 4.9.3; Wis. Stat. § 281.346(4).*)[85] Furthermore, according to Wisconsin State Statutes, "[A c]ommunity *within a Straddling County means any incorporated city, town or the equivalent thereof, that is located outside the Basin but wholly within a County that lies partly within the Basin and that is not a Straddling Community." Compact section 1.2. See also Wis. Stat. § 281.346(1)(d). A "'[s]traddling county' means a county that lies partly within the Great Lakes basin."*[86]

Waukesha's formal application to the State of Wisconsin is the first test of one of the "straddling" water district diversion exemptions and thus could set a precedent for whether and to what extent Great Lakes water is allowed to flow into nearby basins. There are at least seven other cities in Ohio, Indiana, and Wisconsin that are in relevantly similar situations to Waukesha's and might soon petition for water.

According to City of Waukesha data, this is not merely a ploy for the city to obtain cheap water for urban sprawl. The city anticipates that by 2050 the number of citizens will be ~76,000 or only a 0.5 increase in population. The service area will expand, with around ~97,000 residents consuming water from the Waukesha water district. The city projects that through a strategy of water

84 City of Waukesha. "Application Summary: City of Waukesha Application for a Lake Michigan Diversion with Return Flow." (website), October 2013. Available at http://www.ci.waukesha.wi.us/c/document_library/get_file?uuid=a972a2e4-d45b-4748-9948-17c0ce17b692&groupId=10113 (accessed May 15, 2015).

85 City of Waukesha, "Application Summary." (Italics in original).

86 *Wis. Stat. § 281.346(1)(tm). See also Wis. Stat. § 281.348(3).* (Italics in original.).

conservation that began formally in 2006 and light population projections into mid-century that 10.9 million gallons of water per day should meet the city's needs. In its application for water diversion, the city requests a maximum of 16.7 million gallons for days of high use by the final buildout of the service area by 2050. By that time, the diverted water will not only serve all of Waukesha, but perhaps also portions of the City of Pewaukee, and the towns of Waukesha, Genesee, and Delafield in the future.[87]

The Waukesha water diversion proposal has two main components. One is delivering water to the Waukesha water district. If the proposal is accepted as is, this would be done, as alluded to earlier, but drawing in water purchased from the Town of Oak Creek waste treatment facility. This water would then be distributed via Waukesha's existing water system. The second component is to take water after consumptive use, which includes remaining water, as well as infiltration and storm water inflow, and treat it using Waukesha's wastewater facility. This water would then be piped out to the Root River to be discharged back into Lake Michigan. According to the city's plans, most of the water would be returned to the lake after treatment.

Waukesha cites in its application that it decided to proceed with the Root River as a discharge site instead of Underwood Creek because analyses indicted that there would actually be environmental benefits from using the Root River to enhance Great Lakes fisheries.[88]

A final decision on the plan may not be coming anytime soon. Even if the state of Wisconsin agrees, the proposal would then have to pass muster with the other states that are partners to the Compact.[89] And even if all of these states approved the plan as of late 2015, construction of a new water supply service could not be completed before 2020, according to Waukesha Water Utility Manager Dan Duchniak.[90] With these delays in mind, it was expected that the City would ask for a two-year extension on finding a radium-free drinking water source.

---

87 Wisconsin Department of Natural Resources, "City of Waukesha Water Diversion Application: Background," (website), January 6, 2016. Available at http://dnr.wi.gov/topic/WaterUse/waukesha/background.html (accessed June 7, 2016).

88 City of Waukesha, "Application Summary."

89 Don Behm, "Waukesha Concedes It Can't Make Deadline for Radium-Free Water," *Milwaukee Journal Sentinel*, November 19, 2014. Available at http://www.jsonline.com/news/waukesha/waukesha-concedes-it-can-meet-deadline-for-radium-free-water-b99393845z1-283255831.html (accessed May 15, 2015).

90 Behm, "Waukesha Concedes."

This case study brings to light a number of background questions concerning water. Who should own water and how much of it? Is water something that should be owned (or at least controlled) by states, or should it be governed at the county or municipal level? This question becomes especially important given that "on the ground" in many towns, suburbs, and urban areas, US residents receive water through local water districts. When a municipality runs out of potable water, how much of its past actions with respect to use of water should be taken into account in justifying resupply? Should its history of developing industries known to pollute water be taken into account? What is happening in Waukesha now may very well be precedent setting, but the issue isn't just Waukesha's. It could be coming to a community like yours soon.

Review Questions

1) Is the water diversion plan a good solution for the Root River and for the people and communities downstream from Waukesha? Why of why not?
2) It seems that this case outlines an intriguing distinction about how ocean water (at least international waters) and fresh water are treated both legally and morally. Ocean waters are treated more as a commons, while rivers and lakes are treated more like a private property system. What do you think might explain this difference? Do you think they should be treated differently? Why or why not?
3) Because Waukesha straddles two river basins (the city lies in the Mississippi River Basin but is less than a few miles from the Great Lakes Basin), should the area be allowed an exemption from the Great Lakes Compact allowing the city to divert water? Should there be any ban on diverting water from one basin to another? Why or why not?

CHAPTER 5

The Environment, Global Challenges, and Global Economies

Case 12 – Climate Change Accords and Vanishing Islands –

Study Questions

1) What was the result of the Kyoto Protocol? Have climate change talks since then yielded any agreements on climate change mitigation or adaptation? Give specific examples to support your answer
2) What was the result of the COP21 talks in 2015? Explain its key features.
3) What are some of the issues faced by the people of the Maldives as the sea level rises? What are some of the adaptive strategies the population there will likely need to use within the next 50–100 years?
4) Why has the United States been criticized in the process of trying to reach an international greenhouse gas emissions reduction agreement? Is the US criticized more than other nations in climate talks? Why or why not?
5) What is the distinction between mitigation and adaptation strategies in response to climate change?

On October 17, 2009, the then President of the Republic of the Maldives (henceforth Maldives) called for a cabinet meeting like no other in the history of any country's government—he held it underwater. President Mohamed Nasheed and his governmental ministers, to highlight the possibly devastating effects of climate change on his low-lying island nation from sea level rise, signed a document calling for a global reduction of carbon emissions while 3.5 meters (about 20 feet) under the surface of a lagoon.[1] The participants (donning scuba gear) communicated with hand signals and by writing on whiteboards during the half-hour session. Leading up to the Copenhagen global climate change reduction talks later that year, the underwater cabinet meeting was an intentional attempt to focus attention on climate change's effects on not only the Maldives, but on an array of other small island states. Some of these nations could be completely submerged, or at least have populations facing death from

1 "Maldives Cabinet Makes a Splash," *BBC News*, (website), October 17, 2009. Available at http://news.bbc.co.uk/2/hi/8311838.stm (accessed June 8, 2016).

 | DOI 10.1163/9789004339996_007

flooding, forced emigration, and/or severe infrastructural damage within this century due to rising waters.

Nasheed's hope was to come away from the Copenhagen talks with a legally binding agreement to reduce carbon emissions significantly. His goal was to have all countries agree to decreases in emissions, with targets of allowing only a 1.5 degree Celsius rise in the mean global temperature over the pre-industrial level and of lowering the atmospheric carbon dioxide concentration to 350 parts per million (ppm). His worry, of course, was that yet another climate summit would pass with much talking, but no action. Nasheed expressed frustration that while other countries debated minor textual changes in any accord, his country was already sinking—a scenario akin to Nero fiddling while Rome burned.

Climate talks resulting in no concrete plans for reducing emissions have clearly become a habit. In January of 2008, a two-week session was held in Bali to address several issues related to global warming. No binding agreements came from the meeting. At that time, the Bali Action Plan included a provision for deep emissions cuts. It also contained a timetable for two years of further talks that at the time was thought likely to initiate the first formal emendation to the Kyoto Protocol, which had been adopted in 1997 with the Maldives being the first nation to sign. However, governments mainly pulled back their commitments to the cuts and the additional two years of talks did not yield wholesale changes to the Kyoto agreement.

Even before the debates over the Bali Action Plan, the United States had been roundly criticized for not adhering to the mandates stemming from the talks in Kyoto, especially since American industries and consumers are widely considered to be major causes of greenhouse gas emissions contributing to global warming. The US response under the Bush Administration was that curbing carbon emissions is not merely the responsibility of Americans, but that "the negotiations must proceed on the view that the problem of climate change cannot be adequately addressed through commitments for emissions cuts by developed countries alone. Major developing economies must likewise act."[2] However, the leaders of countries with developing economies, such as China and India, believe they have no obligation to adhere to the Kyoto Protocol since developed nations like the US have not done so.

Later, while each country calculated specific rates of reduction during the Copenhagen Accord talks, again the agreement was not legally binding

2 Thomas Fuller and Arthur Revkin, "Climate Plan Looks Beyond Bush's Tenure," *The New York Times*, December 16, 2007. Available at http://www.rainforestcoalition.org/documents/nytimes_12-16-071.pdf (accessed June 8, 2016).

on them. In effect, this meant there was still no replacement for the Kyoto Protocol. Along with accepting the continuation of the Kyoto Protocol, a further timetable was established at the end of the Copenhagen Accord to come to a binding legal agreement by 2012. That didn't happen, but talks continued and at the 2013 Warsaw Climate Change Conference, UN members set the groundwork with the hope of coming to an agreement to replace the Kyoto Protocol (now almost twenty years old) with a universally binding agreement in Paris for reduction in the emission of greenhouse gases (to be ratified in 2015). There was also support for all member nations to voluntarily begin cutting carbon emissions by the beginning of 2015 so that peak emissions would occur by the end of this decade. Ultimately, the goal according to United Nations Framework Convention on Climate Change (UNFCCC) executive director Christiana Figueres is to get to zero net emissions by the second half of this century.[3]

During deliberations at the Copenhagen Accord, most countries agreed that the target for acceptable temperature rise should be no more than 1.5 degrees Celsius above pre-industrial levels by the end of the decade. Yet, the 1.5-degree C criterion was formally dropped out of the text of the Accord, replaced with a more lenient threshold of 2.0 degrees C above pre-industrial levels. At the time, the possibility was left open that the 1.5-degree threshold could be revisited as part of the 2015 Paris talks. Yet meeting this threshold is still problematic as David Collings observes, "[S]ince we've already raised the average temperature by 0.8 degrees, and the temperature will rise another 0.6 due to the inertia of the world's climate, 1.4 of these 1.5 degrees are already inevitable leaving us virtually without hope of reaching the target. Even if we acted *today* [2014] to eliminate our green house gas emissions *entirely*, we'd still barely meet the goal."[4]

At least with respect to international climate agreements, COP21, held in Paris in late 2015, offers some hope in taking action on climate change. The talks resulted in the Paris Climate Pact, which is being called a "landmark

---

3 "UNFCCC: Warsaw COP 'Pivotal Moment to Step Up Climate Action'," *Clean Tech Poland*, (website), November 9, 2013. Available at http://cleantech.cleantechpoland.com/?page=news&id=113&link=unfccc-warsaw-cop-pivotal-moment-to-step-up-climate-action- (accessed June 8, 2016).

4 David Collings, *Stolen Future, Broken Present: The Human Significance of Climate Change* (Ann Arbor, MI: Open Humanities Press, 2014), p. 62. See also Mark Hertsgaard, *Hot: Living Through the Next Fifty Years on Earth* (Boston: Houghton Mifflin Harcourt, 2011), p. 70.

agreement."[5] First and most importantly, this is the first universal, legally binding, and comprehensive climate agreement within the 196-nation, UN framework. Here are some of the key features of the new agreement.

The first calls for holding the increase in the global average temperature to well below 2 degrees C above pre-industrial levels. In addition, there would be pursuit of efforts limiting the temperature increase to 1.5 degrees C above pre-industrial levels. This seems to endorse scientific predictions that an increase in the global temperature average over 2 degrees C (3.6 degrees F) would virtually "lock" us into catastrophic climate change, with substantial impacts such as greater frequency of devastating floods and droughts, more powerful storms, and widespread food shortages.[6] But even short of that, the parties to the agreement realized that warming to just 1.5 degrees C (2.7 degrees F) would likely be devastating to low-lying island nations, as the rising sea level would leave them submerged. Of course, this might not be an achievable outcome and the national plans submitted to the conference will only likely result in keeping average global temperature to 3 degrees C above pre-industrial levels.

A second key related feature is that member nations should reach a global peak of greenhouse emissions as soon as possible. A proviso to this is that developing countries will likely take longer to meet these goals. After reaching this peak, the expectation is for rapid decreases in emissions. This is supposed to send a message to the fossil-fuel industry that "business as usual" practices are unacceptable and the remaining oil, gas, and coal must stay in the ground. Notice though that the language does not include what was proposed in a previous draft for "reaching greenhouse gas emissions neutrality in the second half of the century."[7]

A third key feature is deemed crucial to poor and low-lying nations. The agreement recognizes "the importance of averting, minimizing and addressing loss and damage associated with the adverse effects of climate change." Yet, this acknowledgement does not go much further as it "does not involve or provide a basis for any liability or compensation." This means that wealthier nations cannot be held legally liable for any adverse effects of climate change.[8]

A fourth feature is that the agreement requires all countries to update their detailed plans for emissions reductions. Member nations will be required to

5 "The Road to a Paris Climate Deal," *The New York Times*, December 12, 2015, Available at http://www.nytimes.com/interactive/projects/cp/climate/2015-paris-climate-talks/key-points-of-the-final-paris-climate-draft (accessed September 4, 2016).

6 "The Road to a Paris Climate Deal."

7 "The Road to a Paris Climate Deal."

8 "The Road to a Paris Climate Deal."

make their emissions policies more stringent, showing their emissions levels by 2020 and reporting every five years thereafter.

The fifth feature of the new climate accord is that it requires what is called a global "stocktake"—that is, an overall evaluation of how member nations are doing in decreasing emissions compared to their national plans. This will begin in 2023, with re-assessments every five years.[9]

The sixth feature of the climate deal is that it requires monitoring and reporting of greenhouse gas emissions using a standardized global measurement system. Counting and verifying each country's emissions is thought crucial to any plan's success. This includes an independent body doing the assessments as opposed to merely the self-reporting of member nations. Even with this common accounting system, the details will have to be worked out. Also, establishment of an outside body to do the assessments still needs to be done.[10]

The seventh feature is the establishment of a "Capacity-Building Initiative for Transparency." This is designed to aid developing countries meet a new requirement that they submit a national "inventory report" of anthropogenic causes of emissions.[11]

The eighth feature of the Paris Climate Pact is that nations need to establish "a new collective qualified goal" of $100 million a year for financing climate change mitigation and adaptation, beginning in 2020. While it is not clear that this is an "action" item, since it only is listed under the "decisions" section in the Pact, it does stipulate that $100 billion is the bare minimum to be set aside for funding climate change initiatives.[12]

Finally, the ninth feature of the Pact is that countries will commit to the "highest possible ambition" when updating their emissions goals. It does not set a numeric target and furthermore merely notes that all nations have "common but differentiated responsibilities, in the light of different national circumstances." In this context, the agreement stipulates that rich countries need to have "absolute" reductions in emissions, while developing countries should "continue enhancing their mitigation efforts."[13]

But even with the Pact in place, there is the additional worry that even allowing the global average temperature to rise to the 2 degrees Celsius threshold will be too high. The predictions of the effects of global temperature increases have changed since when the 2-degree C goal became the projected threshold

---

9 "The Road to a Paris Climate Deal."

10 "The Road to a Paris Climate Deal."

11 "The Road to a Paris Climate Deal."

12 "The Road to a Paris Climate Deal."

13 "The Road to a Paris Climate Deal."

for avoiding dangerous climate change back in the early 1990s. The science of predictive climate models has become much better since then and there are grave concerns that what were once considered tolerable effects of a projected 2 degree C rise will now be much worse. A building consensus believes the tipping point for irreversible and dangerous climate change could come at something closer to a 1 degree C rise in global temperature from pre-industrial levels. The Paris agreement will not likely hold UN nations to that threshold.

The Intergovernmental Panel on Climate Change (IPCC) has made it clear that low-lying islands (especially the "atoll nations" in the Pacific and Indian Oceans) are the most vulnerable to "climate change, seasonal-to-interannual climate variability, and sea level rise."[14] One case in point is obviously the Maldives. The Maldives is an archipelago composed of islands in the southern Indian Ocean, most of which are at a very modest elevation of four feet above sea level, with none rising more than six feet. As such, it is the lowest-lying country in the world.

First gaining independence from Britain as a sultanate in 1965, the Maldives became a republic three years later. Only about 200 of the 1,912 coral islands that make up the Maldives are inhabited. Economically, the Maldives is heavily dependent on the tourist industry (which represents about 33% of the country's GDP), as it is packaged as a warm and sunny getaway location for the wealthy. The fishing industry (including fish processing) provides a livelihood for many of the island country's 400,000 residents. Coconuts, corn, and sweet potatoes are the chief agricultural products of the Maldives. The country's main exports are fish and clothing.[15] The island nation has mainly recovered from the severe damage caused by the December 2006 Indian Ocean tsunami. That tropical weather event killed 88 people, with 42 others missing but presumed dead. It also wiped out about half of the country's GDP.

Tourists are drawn to the extreme beauty of the Maldives' azure colored waters, pristine white bleached and golden beaches, and lush tropical forests. Yet, the Maldives is also known as an "extreme test case" for climate change. Damian Carrington of *The Guardian* went so far in his description of what the country faces in the future that it sounds hyperbolic, "Like rays of burning

---

14 Intergovernmental Panel on Climate Change, "Assessment Report," 2013, 19.3.4.1. Available at http://www.ipcc.ch/ipccreports/tar/wg2/index.php?idp=671 (accessed February 16, 2015).

15 "Maldives Facts," *National Geographic Atlas*, 8th Edition, 2008. Available at http://travel.nationalgeographic.com/travel/countries/maldives-facts/ (accessed June 8, 2016).

sunlight concentrated through a magnifying glass, almost all of the world's environmental problems come into sharp focus in the Maldives."[16]

The first issue is a rising sea level. The Maldives archipelago is not built on sand, but instead on coral reefs. Coral reefs are the most endangered ecosystems in the world. The bright white sand making up the Maldives world-famous beaches originates from fractured and ground bits of dead coral. Along with higher sea level, ocean acidification and rising ocean temperatures kill coral. This is obviously problematic for the local fishing industry, as many of the marine species harvested by Maldivians are highly dependent on healthy coral reefs. The tourist industry will take a substantial hit due to beach erosion and the degradation of dive sites from the bleaching of coral reefs.[17]

The 2014 Intergovernmental Panel on Climate Change Report on the world's latest climatic conditions and future prognosis offered even starker and more imminent dangers of taking no action on climate change. The Maldives will be greatly affected even according to the most conservative climate model forecasts. Much of the critical infrastructure in the Maldives and other small island nations lies along the coastline, as does the lion's share of their populations. These coastlines are already at sea level, which means any increase in water level will have destructive results. It is estimated the Maldives will be inundated by seawater perhaps as soon as the end of this century. This obviously means a host of problems including increased shoreline erosion, changes in aquifer volume and water quality with saline intrusion, coral-reef deterioration from sea level rise, outmigration caused by permanent inundation, social instability due to interisland migration, loss of tourist investment, increased vulnerability of human settlement due to decrease in land area, and loss of agriculture and vegetation. It is expected that with just a 90-centimeter sea level rise, the capital of the Maldives, Male, will become 85% inundated by water. Sea-flooding, coastal erosion, and saline intrusion into aquifers is already occurring in a number of the atoll nations, including the Maldives.

With respect to ocean temperature, according to the IPCC report, current models predict the Indian Ocean is likely to warm 1–2 degrees Celsius, which

---

16 Damian Carrington, "The Maldives is the Extreme Test Case for Climate Change Action," *The Guardian*, September 26, 2013. Available at http://www.theguardian.com/environment/damian-carrington-blog/2013/sep/26/maldives-test-case-climate-change-action (accessed February 16, 2015).

17 Ann Powers, "Sea-Level Rise and its Impact on Vulnerable States: Four Examples," *Louisiana Law Review* (2012) 73(1): 151–301 at 160. Available at http://digitalcommons.pace.edu/cgi/viewcontent.cgi?article=1863&context=lawfaculty (accessed February 24, 2015).

sounds like a small amount until considering that even small temperature increases cause changes in the ocean dynamic that can have a strong effect on small islands. Models indicate that seawaters surrounding the Maldives will double in CO_2 concentration levels. This, combined with already rising ocean temperatures, is a recipe for moderate temperature increases into the future. Also, tropical oceans are likely to experience between 20–30% more rainfall. Certainly, there is no consensus with the models to demonstrate there will be any increase or decrease in the *frequency* of tropical cyclones due to warming sea temperatures in the Indian Ocean.[18] However, some other studies indicate a possibility of a 10–20% increase in the intensity of tropical cyclones formed in high CO_2 conditions.[19]

The mean surface air temperature for the Indian Ocean is expected to increase by 1.64 degrees C by the 2050s and 2.61 degrees C by the 2080's in a model that accounts for both rising temperatures caused by greenhouse gas emissions and higher CO_2 levels, even with the possible cooling effects of aerosols taken into account. Using the same model, annual mean precipitation change is anticipated to increase by 1.6% in the 2050s and by 4.3% in the 2080s.[20]

The main association between the four greenhouse gases (carbon dioxide, methane, water vapor, and nitrous oxide) and rising temperature is embodied in the greenhouse effect. Of these gases, CO_2 and methane are the two most important with respect to current emissions. The basic premise is that energy from the sun reaching the Earth has a hard time returning to space. With the greenhouse effect, some of the energy that would normally be reflected back into space is instead trapped in the atmosphere, where it is absorbed and released by greenhouse gases. Interestingly, under normal conditions the greenhouse effect is natural and beneficial as it keeps the Earth's temperature above freezing.[21] Yet, as we add copious amounts of greenhouse gases to our atmosphere their cumulative strength is increasing, supplying too much of a good thing and overwarming our planet. According to NASA, about 30% of

---

18 Intergovernmental Panel on Climate Change, Working Group II, "Impacts, Adaptation and Vulnerability," *Fifth Assessment Report of the Intergovernmental Panel on Climate Change* (Cambridge, UK and New York, NY, USA: Cambridge University Press, 2014), 17.1.4.3.4.

19 Intergovernmental Panel on Climate Change, Working Group II, "Impacts, Adaptation and Vulnerability," 17.1.4.3.5.

20 Intergovernmental Panel on Climate Change, Working Group II, "Impacts, Adaptation and Vulnerability," 17.1.4.3.5.

21 University Corporation for Atmospheric Research, "Greenhouse Effect," (website), 2011. Available at http://scied.ucar.edu/longcontent/greenhouse-effect (accessed July 18, 2016).

the radiation striking the Earth is reflected back into space by clouds, ice, and other reflective surfaces. The other 70% is absorbed into the atmosphere, the oceans, and the land.[22]

Normally, solar energy that hits our Earth is reflected back out into space as heat. A greenhouse gas has a more complex structure than other gases in the atmosphere. This allows it to absorb heat, and thus the higher its concentration, the more heat it traps, meaning less heat that can escape back out into space. One might wonder why the solar energy isn't merely captured on its way to Earth. The reason is that ultraviolet rays come through the atmosphere, but simply pass through greenhouse gases. Yet, the radiated energy is reflected back as infrared energy, which is instead absorbed into greenhouse gas molecules. From there, the heat is either radiated to other gas molecules, into space, or most commonly, back to Earth, thus warming it.[23]

All greenhouse gas molecules have three or more atoms. Their bonds are relatively loose, thus vibrating when they absorb heat. This vibration allows greenhouse gas molecules to radiate heat to other greenhouse gas molecules and this action keeps the heat near the Earth's surface. Most of the gas in the atmosphere is oxygen and nitrogen, which in contrast to greenhouse gases are composed of two atoms, are tightly bound, and cannot vibrate. Hence, they do not absorb heat and don't add to the greenhouse effect.[24]

Two important greenhouse gases warrant closer investigation—carbon dioxide and methane. Carbon dioxide is composed of two oxygen atoms and one carbon atom. Even though CO_2 makes up only a fraction of the atmosphere, it can have a very large effect on the environment as a greenhouse gas. Back when the Industrial Revolution began, during the mid-1800s, the concentration of CO_2 in the atmosphere was 270 ppm.[25] However, mainly from the burning of fossil fuels, which emit CO_2, the concentration rose above 400 ppm for the first time ever in 2014. Methane emissions have also increased steadily. It is a powerful greenhouse gas that can absorb far more heat than CO_2. Methane is made up of one carbon atom and four hydrogen atoms. Like CO_2, there is only a small amount of methane in the atmosphere, yet it also has

22 Marc Lallanilla, "Green House Gas Emissions: Causes and Sources," *LiveScience.com*, February 10, 2015. Available at http://www.livescience.com/37821-greenhouse-gases.html (accessed February 25, 2015).

23 University Corporation for Atmospheric Research, "Greenhouse Effect."

24 University Corporation for Atmospheric Research, "Greenhouse Effect."

25 University Corporation for Atmospheric Research, "Greenhouse Effect."

an oversized effect on warming the Earth as a greenhouse gas.[26] Methane is a fuel and when it is burned emits CO2 into the atmosphere.

If the current emission trends continue, carbon dioxide levels will likely double within this century, which will only exacerbate global warming. Coupled with increased amounts of methane and nitrous oxide also entering the atmosphere, the greenhouse effect will only increase in its intensity. Of course, the burning of fossil fuels is not the only way greenhouse gases are added into the atmosphere. Farm animals are also major contributors to the concentration of greenhouse gases as they digest food and emit methane. This is one of the reasons why some have advocated eating a vegetarian diet. Decreasing the demand for animal meat would lessen the incentives for maintaining large-scale animal operations and correspondingly lessen methane emissions.[27]

But while there is a good deal of talk about the profound and grim *future* effects of climate change, there is evidence that some particularly vulnerable small island nations are already facing them. The Maldives is prone to natural calamities and recent storms have caused extensive damage to the country. The tsunami of 2005 caused by a massive earthquake was already briefly mentioned, with this event putting a major dent in the tourist industry for quite some time. This is important since, as we've seen, tourism is the most significant economic cog in the Maldives. Major flooding in 1987 also did a great deal of damage to the Male International Airport, which is an essential transportation hub for the nation.

Even without such storms, the sea constantly gnaws at the shores of the Maldives. The film *The Island President* documents this in scenes where Nasheed is shown travelling to yet another beach where 100, 200, or 300 feet of coastline has eroded away and fallen into the ocean.[28] The Maldives has already diverted large sums of money away from a host of other important sectors of governmental spending such as health care and education to fund the building of seawalls, which are at best only a temporary solution to the problem of rising seas.[29]

This shouldn't be surprising since the latest information on sea level rise from the IPCC finds (with high confidence) that "The rate of sea level rise since the mid-19th Century has been larger than the mean rate during the last two millennia.... Over the period 1901 to 2011, the global mean sea level rose by

26 University Corporation for Atmospheric Research, "Greenhouse Effect."

27 University Corporation for Atmospheric Research, "Greenhouse Effect."

28 *The Island President*, Dir. Jon Shenk, film, Actual Films, 2012.

29 *The Island President*.

0.19 (0.17 to 0.21) meters."[30] It is also very likely that from just 1993 to 2010 the mean average sea level has risen 3.2 [2.8 to 3.6] millimeters (mm) per year, as opposed to 2.0 [1.7 to 2.3] mm per year from 1971 to 2012, and 1.7 [1.5 to 1.9] mm per year from 1901 to 2010. Tide-gauge and altimeter data are consistent with the high levels recorded during the 1993 to 2010 time period—it is likely that similar high rates occurred between 1920 and 1950.[31]

And the association between increases in the mean rate of global sea level rise and increased glacial ice pack melt seems to be well established:

> Since the early 1970s, [there is high confidence that] glacial mass loss and ocean thermal expansion through warming together explain about 75% of the observed global mean sea level rise. Over the period 1993 to 2010, global mean sea level rise is, with *high confidence*, consistent with the observed contributions from ocean thermal expansion due to warming (1.1 [0.8 to 1.4] mm per year), from changes in glaciers (0.76 [0.39 to 1.13] mm per year), Greenland Ice Sheet (0.33 [.25 to .41] mm per year), Antarctic Ice Sheet (0.27 [0.16 to 0.38] mm per year, and land water storage (0.38 [0.26 to 0.49] mm per year). The sum of these contributions is 2.8 [2.3 to 3.4] mm per year.[32]

Current estimates suggest that sea level will not only continue to rise, but at rates higher than the eight to ten inches it has since 1880.[33] If the current rate of carbon emissions continues unabated, a recent survey of experts reported that sea levels might rise three feet by 2100.[34]

But any significant cuts in carbon emissions are difficult to imagine when considering that the Kyoto Protocol didn't yield many beneficial results. For instance, most nations participating in the Kyoto talks pledged to decrease their greenhouse emissions by about 5% of 1990 carbon concentration levels

30 Intergovernmental Panel on Climate Change, "Summary for Policymakers," in T.F. Stocker, D. Qin, G.-K. Plattner, M. Tignor, S.K. Allen, J. Boschung, A. Nauels, Y. Xia, V. Bex and P.M. Midgley, eds. *Climate Change 2013: The Physical Science Basis. Contribution of Working Group I to the Fifth Assessment Report of the Intergovernmental Panel on Climate Change* (Cambridge, UK and New York, NY, USA, Cambridge University Press, 2013), p. B4.

31 Intergovernmental Panel on Climate Change, "Summary for Policymakers."

32 Intergovernmental Panel on Climate Change, "Summary for Policymakers."

33 Carol Davenport, "Rising Seas," *The New York Times*, March 27, 2014. Available at http://www.nytimes.com/interactive/2014/03/27/world/climate-rising-seas.html?_r=0 (accessed February 18, 2015).

34 Benjamin P. Horton et al., "Expert Assessment of Sea-level Rise by AD 2100 and AD 2300," *Quaternary Science Reviews* (2014) 84: 1–6.

by 2012.[35] However, as that deadline has past we can now report there was actually a 38% increase in global carbon emissions from 1990–2009.[36] Perhaps the Paris Climate Pact will show more success.

Other factors have an impact on the moral claims made by Maldivians and others interested in climate justice. One is that the anthropogenic contribution to climate change from the people of the Maldives is practically negligible. In fact, even when combining all small island nations, they still contribute less than 1% of all greenhouse gases.[37] However, they stand to be among the first to lose (and lose heavily) from the effects of climate change. In sum, Maldivians who are able to leave will likely be the first "climate change refugees."

Even though the Maldives has contributed very little to the high concentration of greenhouse gases melting the polar ice caps and the glaciers of the Greenland Ice Sheet and raising sea level, former President Nasheed pledged (in March of 2009) that the country would be carbon neutral by 2019. During his presidency, Nasheed invested funds in solar and wind power to transition the Maldives to an energy infrastructure infused with more renewable energy sources.

But apart from the normal challenges of trying to become carbon neutral, the Maldives also has to deal with the misfortune of political realities. A power struggle occurred between a long-standing dictatorship and Nasheed's democratic movement. Nasheed claimed that he was forced to resign "at gun point" by police and army officers loyal to the country's former 30-year dictator Maumoom Gayoom. In a February 2012 coup d'état, Nasheed was forced out of power by Gayoom's half-brother, Abdulla Yameen. There is little evidence that the current regime is committed to following Nasheed's plan for carbon neutrality. Nasheed now lives in exile in the United Kingdom.

What makes climate change in part such an intriguing issue is that it came to the public's attention only a relatively short time ago. Thus, it should not be surprising that philosophers have said little about the topic. However, a burgeoning literature is developing on climate change, especially concerning moral issues surrounding it. Accordingly, several issues of global environmental justice emerge from this case study. Perhaps one way to categorize these issues is by placing some under the aegis of *mitigation* and others under *adaptation.*

---

35 Collings, *Stolen Future*, p. 60.

36 Robert B. Semple, Jr., "Remember Kyoto? Most Nations Don't," *The New York Times*, December 3, 2011. Available at http://www.nytimes.com/2011/12/04/opinion/sunday/remember-kyoto-most-nations-dont.html?_r=0 (accessed February 17, 2015).

37 IPCC, 17.2.1.

In the context of climate change, *mitigation* involves lessening its effects and strategizing institutional responses to try to stop or curb it. Yet, many who study its effects believe the way the discussion about it is framed needs to change from mitigation to adaptation. With relatively little activity on the part of states to respond to dire warnings from climatologists, there is growing consensus that we are beyond the point of trying to stop or even meaningfully curb climate change. Instead, we need to focus on *adaptation*, which calls for accurately assessing the consequences of different emission level scenarios, and to prepare as best as we can for the worst. This may involve moving populations, relocating essential infrastructure, preparing for mass migrations, and developing the services that will likely be required for accommodating them.

Several moral questions arise from considerations of mitigation. Who is responsible for climate change? How should we lessen the further effects of climate change? What obligations, if any, do we have to future generations? Should international bodies take more aggressive regulatory means to slow down emissions? Would a carbon tax more effectively lessen greenhouse gas emissions?

With respect to adaptation, there is obviously the worry of extreme harms befalling people who will not be able to escape rising waters. This was one of the reasons that among the first tasks Nasheed put to himself and his ministers when he was elected President in 2008 was to figure out a funding plan to support the relocation of his people. Part of this was trying to find a location that could hold the entire population of the Maldives and how to acquire the land necessary for that transition.

There are also the issues of whether funding should be allocated to temporary strategies such as building seawalls to stem erosion occurring on many of the inhabited islands of the Maldives. But is this merely akin to placing Band-Aids on steadily widening wounds as seawater continuously gouges chunks out of the Maldivian shoreline? Does it make sense to pour money into these protective strategies if they are going to be ineffectual in a few years' time?

While the Maldivian government's urgent pleas to world leaders to reduce the effects of climate change by reducing their profligate emission levels might be seen as morally appropriate, it still may seem futile for the island nation to itself expend resources on becoming carbon neutral. Should that funding be spent on other projects? Nasheed has actually already given a response to this criticism, saying, "At least as we are dying here in the Maldives [from rising seawaters], we will be doing the right thing." Along with the possibility that not contributing to a moral problem is the correct course of action, one might add there is a symbolic appeal to their stance. The Maldives wants to not only provide an example of carbon neutrality and make a statement, but also "puts its money where its mouth is" and accepts the burdens of carbon offsets. Of course, we can still ask whether this is the proper moral response.

Beyond the debate about whether mitigation or adaptation is the best approach to climate change, there are issues of justice. Citizens of the Maldives and other small island states might argue they are unjustly and unnecessarily suffering the burdens of living in a location most greatly effected by anthropogenic climate change with no end in sight. As philosopher James Garvey points out, considerations of unnecessary suffering and lessening it are among the most moral of reflections.[38] Since their country's contribution to the problem has been negligible, it is unfair that Maldivians and their other small island nation counterparts should endure the full cost of these burdens (not just in life and hardship, but also economically). Therefore, those in small island states appeal to the *harm principle* and *equality* (that we all need to share some of the suffering) in debates over climate justice. Corresponding moral obligations of states that are large-scale CO_2 emitters would likely call for mitigation—there is a moral responsibility for lessening CO_2 emissions in a meaningful way. There is the additional obligation of large-scale greenhouse gas emitting countries to help other countries that have not caused climate change by funding the latter's mitigation *and* adaptation strategies.

But larger producers of GHGs retort that there are factors lessening their moral responsibility for the effects of climate change. They argue that they are not the only emitters. Thus, for meaningful emissions mitigation, other CO_2 emitters will also have to commit to reducing their emissions. So, if others do not sign on to substantial carbon reductions, it isn't fair for them to be the only (or only among a small group of) major GHG emitters paying a greater cost for climate change mitigation. This was more or less former President George W. Bush's line of argument referred to earlier in this case study. There is even less support from the larger GHG emitting countries that they should shoulder the cost of the adaptation of other nations to a climate-challenged world.

Also, not all major carbon emitters came into the game at the same time. India and China are now two growing monster economies, but they have only been significant economic powers (with major emissions) for a relatively short period of time. They are just now realizing the benefits of a robust economy, and they feel they have an obligation to their own citizens to spread that prosperity throughout a wider range of their populations—to give once poor Chinese and Indians better lives. They argue that goal will be difficult if not impossible to accomplish with major emissions reductions.

---

38 James Garvey, *The Ethics of Climate Change: Right and Wrong in a Warming World* (London: Continuum Press, 2008).

Some argue that developed countries, such as the US, should carry a greater proportion of the burdens for reducing pollution that leads to climate change due to a couple of important factors. One is that they possess the greatest ability to reduce; that is, they have more luxury emissions that could be decreased. The second is that cumulatively, developed countries have contributed the greatest amount of emissions over decades and have already derived immense economic benefits from manufacturing, such as profits, jobs, and easy access to consumer goods. The moral idea is fairly elementary; those who contribute the most to a problem over a long period of time have a special ethical obligation to not only mitigate future negative environmental impacts, but also to take remediation steps and pay the associated costs. Part of these costs may also involve aiding developing countries in mitigating their carbon contributions and perhaps even funding their adaptation strategies.

Nasheed was careful to point out that rising sea levels are not just a challenge to small island nations. New York City's borough of Manhattan, Miami, New Orleans, and other US cities also face submersion to varying degrees. And certainly India and China will not be immune to the effects of more frequent storm surges and flooding. In fact, 50% of the world's population lives in coastal areas.

One challenge to the idea that any of us should care about climate change is that its worst effects will not be suffered by us, but by future generations.[39] It is indeed an interesting feature of climate change that it would involve people now making sacrifices for those not yet born. Garvey responds by saying that part of what ethics involves is putting the needs of others ahead of our own. Morally upstanding people make ethical decisions for people right in front of them. But, there isn't really that much of a moral difference between making choices for those close by than there is for those far away. Garvey goes on to argue that not only is spatial distance irrelevant to what our moral obligations are to others, but that temporal differences are not morally relevant either. He uses the famous example from the utilitarian philosopher Peter Singer that assuming we are able to do so without risk, we have a moral obligation to wade in after a child in a shallow pond to save him. You ought to do so even if you don't know the child and even if this causes some inconvenience to you (e.g., you might have to get wet and your clothes a bit muddy). In like kind, even if you face some inconvenience, Singer thinks that you have some obligation to help the child in Africa who faces starvation—the distance she is away from you is not morally relevant. Garvey simply adds that these obligations also

39 See the brief discussion of the problem of future generations in this book's introduction. For a more detailed discussion of it, see Case 14—"The Cost of the Future."

extend over time—whether you are aiding someone right now or in the future doesn't mean you don't have some sort of obligation to help those in need. Since there isn't much of a difference morally speaking between the person alive today and the one who will be live one hundred years from now, we have grounds for thinking we have obligations to future generations.[40]

But of course, one possible morally relevant difference is that the person in the future may or may not exist. I don't know how many people will exist 100 years from now (if any at all) and I certainly don't know *who* will existence. Moreover, I can only guess at their preferences even if they were to come into existence. Will they be all that concerned with the environment? Would they be more satisfied by inheriting a world in which there is pristine wilderness but perhaps little economic opportunity? Or would they be more thankful for a modern economy that while contributing some dirty emissions provides some higher degree of prosperity and more creature comforts?

Future generations concerns go even further than this. As Derek Parfit observed, there is a *non-identity problem* deeply imbedded in the problem of future generations. Personal identity is a radically contingent thing. Consider this scenario. If my father had never been in the parking lot of the drive-in theatre in 1953 and my mother had never been convinced to "stand up" her boyfriend at the time and instead go to the same drive-in with her friends, my parents would never have met, fallen in love, gotten married, and had me. I simply wouldn't have existed.

Now put this in the context of climate change in the future. Let's suppose that humans choose to adopt green lifestyles and somehow avert the worst projected outcomes of global warming. But notice that this would be a completely different set of human beings generations from now than those who might have existed if we simply continued with heavy emissions of greenhouse gases into the atmosphere. The peculiar conclusion is that people of the future in the latter condition would have no room to complain since they wouldn't exist without their ancestors making the choices they did.

Here are some possible replies to this challenge. A utilitarian could say that all things considered, since the moral goal is to maximize pleasure for the greatest number of humans (and perhaps non-human animals) in the future, it would make sense to follow a green route. This future set of people who will exist will be better off than the set who would have inherited a much more degraded environment from ancestors making fewer green choices. This alternative set resulting from ancestors making worse environmental decisions would likely suffer the burdens of the greater likelihood of droughts in some

40 Garvey, *The Ethics of Climate Change.*

places and floods in others. Given these considerations, from a utilitarian calculation of future prospects, it is pretty clear which set of humans should be brought into being.[41]

But what of the argument that the US has no obligation to reduce its emissions since it cannot stop climate change alone and other large emitters who could make meaningful reductions (such as China and India) have not done so? Garvey notes that America is the largest contributor to climate change per capita, when one considers that 4% of the world's population contributes 25% of the emissions and yet does very little to reduce. Along with feeling moral outrage at that, one could point out that the US also has the largest amount of room to reduce emissions, as it has more luxury emissions than most other countries. Even at the individual level, the average American is the beneficiary of much higher emissions per capita than the average citizen of any African country. Hence, the average American should be doing a great deal more to reduce. Citizens contributing to luxury emissions are clearly not causing the same amount of harm as a country, but there is some harm being committed.[42]

Some argue that we are not (morally speaking) "off the hook" even if we decide to reduce our impact on the environment and others fail to do so. Recall Nasheed's response to the suggestion that reductions of greenhouse gas emissions in the Maldives is not an effective expenditure of resources, given that the country contributes very little to the problem. Perhaps the same can be said of individuals who could reduce but wonder if they should because their neighbors might not. But as Nasheed seemed to suggest, the immoral actions of others don't give us a pass on doing the right thing.

Still, one who thinks climate change is occurring and is harmful, might remind us of the human ability to adapt to new conditions. Moreover, technological innovations have aided humans in overcoming a host of problems throughout history. Couldn't the same be expected in the case of climate change? Maybe economics and technology can save us from its worst consequences without having to resort to radical emission reduction measures. We might also find hope in making alternatives to fossil fuels more cost competitive to offer incentives for energy producers to invest in alternative energy and for consumers to be able to afford such a transition.[43]

---

41 Garvey, *The Ethics of Climate Change*.

42 Garvey, *The Ethics of Climate Change*.

43 See Steven D. Levitt and Stephen J. Dubner, *Super Freakonomics: Global Cooling, Patriotic Prostitutes, and Why Suicide Bombers Should Buy Life Insurance* (New York: Harper Collins, 2009), ch. 5.

But there are limitations to the "technology can rescue us" argument. Given the range of ways that climate change is expected to challenge us in the future, and the depth of the problem we are facing, there is little reason to believe technology will be the silver bullet that solves all of these issues. There is, for example, a great deal of hope for the prospects of developing a hydrogen economy. However, it is hard to consider hydrogen as an alternative energy "savior" in anywhere near the timescale it is needed by (as in "right now"), especially since it is not even a viable fuel source yet.

The moral questions become even trickier once we recognize what else may be required of humans to really face climate change head on. We haven't even mentioned the issue of world population, which continues to increase. There are currently 7.2 billion people on Earth and if the upward trend continues unabated, it is now expected (as gleaned from a 2014 study published in *Science*) that this number will likely rise to 11 billion by 2100.[44] Are there too many people on Earth? Most working on climate change issues admit there are likely already too many humans given scarcity of resources and excessive greenhouse gas emissions worldwide. That leads to a rather obvious question—do we have obligations to reduce the future number of humans? Would this mean imposing policies limiting what some have called *the right to reproductive liberty* by placing quotas on the number of children that parents can have? There have already been such policies in some countries. Perhaps the most well-known of them is the very recently rescinded one-child policy in China—though it wasn't meant as a response to climate change, but more due to general scarcity of resources.

However, having and raising children brings joy and happiness to many humans, not to mention the sense of meaningful purpose that this aspect of life might include. This, in fact, may be one of the reasons why China now allows each family to have two children as of late 2016. Additionally, if human happiness is the most important value with respect to climate change, we have to carefully consider that for different humans, there are varying conceptions of the good life. For some, the pleasures of hot baths, cheap food, travel, etc. trump the environmental costs. For others, staring at sunsets will turn them on; for still others, living the good life includes costly (economically and ecologically) vacations involving long haul jet travel.

One final ethical concern climate change adaptation involves is the question of migration of people from decimated landscapes to sanctuary locations.

---

44 Patrick Gerland et al., "World Population Stabilization Unlikely This Century," *Science* (2014) 346(6206): 234–237.

In the Maldives, some residents will find their home islands to be uninhabitable. They will likely look to move to other islands at a higher elevation. Of course, eventually these islands will suffer the same fate. Part of the adaptation process will likely involve inter-island migration. Will Maldivians seeking sanctuary on other islands in their country be welcomed by their fellow citizens already living on them? What strains will this put on the local resources of places that serve as temporary sanctuary islands? What moral obligations do Maldivians on "sanctuary islands" have to those who come to their islands? Should the Maldivian government prepare some currently uninhabited islands that may be submerged later to accommodate the first wave of inter-island migrants?

Of course, at some point, all residents of the Maldives will likely have to leave the country. Where will they go? What ethical obligations, if any, do other countries have in accepting migrant Maldivians or those from other low-lying countries escaping the worst effects of climate change? Even if some countries step forward to accept climate refugees, should there be some coordination between them to ensure that the burdens of doing so will be equitably distributed? Should the UN put pressure on states more hesitant to accept climate change migrants, especially if these are relatively affluent nations? This looks to be a humanitarian crisis as more people are displaced not only from small island nations, but also in other areas around the world facing rising seas. We have already seen the flashpoint caused by migration due to war, most recently and dramatically of Syrians fleeing their war-torn land to enter Europe.

Ethical issues with respect to climate change are thorny and in some ways different from other sorts of moral problems. For example, one might not be directly implicated in the cloning debate because even if she finds it to be interesting and of great importance to other people, she may not really worry that *she* will be cloned. Yet, if projections about climate change are correct, ultimately no one will be able to escape its effects. We are all, so to speak, "in it." And even if we don't suffer any of the costs of climate change, all of us are still implicated by our contributions to the problem. The choices you and I make have an impact of some sort in the future.

Despite the doom and gloom surrounding climate change, some thinkers believe there is still hope, at least in that mitigation will to some extent lead to adaptation. While we have already experienced some negative effects of climate change and are locked into others in the future, there is still time to mitigate some effects and make adaptations where necessary. The big question is how we should come to decisions to face a climate-challenged future?

Review Questions

1) Assuming that some larger developing countries (such as those of China and India) have never (at least not recently) experienced economic well-being, what should their obligations be to reducing carbon emissions (if any at all)? Is their level of responsibility lower than that of established countries?
2) Should the unwillingness of developing countries to lessen emissions (out of fears of stunting their economic growth) allow leaders of developed countries to act likewise? Why or why not?
3) Recall the principle that those who cause the greatest amount of harm to someone or something have the greatest share of responsibility to pay for it and to take measures to mitigate harms in the future. Do you agree with this principle? If so, why? If not, why not, and what should the responsibility be for developed countries with respect to climate change? Is there another principle or measure that should replace it?
4) Consider the moral questions raised in this case study about mass human migration that will likely be caused by climate change. What obligations (if any) do countries like the Maldives have for preparing (perhaps by spending a great deal of money) for outmigration of its population? What obligations do you think other countries have in accepting climate refugees? How should governments balance goals such as providing sanctuary to humans under threat from climate change versus national security?
5) *Beasts of the Southern Wild* is a movie set in the bayou (the Mississippi River delta area of southeastern Louisiana also referred to as "the bathtub" by the characters in the film). It depicts the dangers of living in a rural, low-lying area during and after Hurricane Katrina. Katrina was merely a warning for what could likely someday be the first coastal area to disappear in the United States due to rising sea level. This area is culturally unique to the US with a Cajun population with its own distinct traditional music and festival culture. Many Cajuns see themselves as culturally distinct from the rest of the American South and worry that social, climatic, and economic pressures will force them out of the bayou, which they think helps define them and serves as a touchstone for their cultural identity. They believe they will lose their separate lifestyle and traditional ways if they have to move. Do you think there should be any special group rights for a distinct population that would require measures to protect them and their culture from the negative effects of climate change?

Case 13 – They Kill Cell Phones Don't They? –

Study Questions

1) What are the typical steps in the lifecycle of a cell phone sold in the United States? Where do cell phones end up?
2) What is "urban mining" and what are the economic incentives that drive it? Are some kinds of urban mining safer than others?
3) What are some of the potential environmental hazards of discarding cell phones in landfills? What are some of the potential health hazards for those who do "urban mining"?

For many years, the Belgian corporation Umicore was a conventional mining company, specializing in the acquisition of copper. In an early form of global commerce (beginning in the 1800's) Umicore mined copper from the African Congo and shipped raw ore to its smelter in Antwerp. The company, however, now does "mining" of a different sort—by processing various metals from cell phones that have been discarded.[45] Umicore has in effect entered what is known as "urban mining" (or what the company has dubbed "aboveground mining"), which is part of the e-waste industry, a rapidly growing sector of green business. *E-waste* is simply any discarded electronic or electrical devices or their components (e.g., cell phones, computers, televisions, etc.).

There is certainly money to be made from these new ventures, as there is a great deal of e-waste to be mined. In fact, e-waste is the fastest growing municipal waste disposal stream in the United States and one of the fastest growing in the European Union (EU) as well. In the 15 countries that comprise the EU, over 9 million tons of e-waste was generated in 2005, and this is projected to grow to 20 tons by 2020.[46]

Overall, according to the US Environmental Protection Agency (USEPA) over 2.5 million tons of e-waste is generated in America each year.[47] More specifically, it was estimated by the US Geographical Survey (USGS) that by 2005,

45 Jon Mooalem, "The Afterlife of Cellphones," *New York Times Magazine*, January 13, 2008, p. 40.

46 European Commission, "Waste Electrical and Electronic Equipment," (website), March 5, 2015. Available at http://ec.europa.eu/environment/waste/weee/index_en.htm (accessed March 6, 2015).

47 Derek Markham, "How E-waste is Becoming a Big, Global Problem," *National Public Radio* (Radio transcript), January 11, 2013. Available at http://www.npr.org/2013/01/11/169144849/how-e-waste-is-becoming-a-big-global-problem (accessed March 3, 2015).

there were already over 500 million old, unused cell phones collecting dust in American homes. Cell phones may be the most valuable form of e-waste, as each one contains about a dollar's worth of precious metals, most of which is gold. This means that even over a decade ago, the metals combined in the phones (gold, platinum, silver, palladium, and copper) simply contained in American homes were already worth more than $500 million.[48] The rapid proliferation of cell phones shouldn't be surprising since Americans on average replace them every 22 months, and 150 million of them were junked in 2010 alone.[49]

Surely part of what is driving this trend is that the electronics industry, generally speaking, "designs for the dump."[50] That is, devices are intentionally produced so that they will become obsolete in a short timeframe and upgrading currently requires buying a whole new device. However, Umicore and its competitors only received 1 percent of all phones discarded globally in 2006. And while there has been some increase in these numbers since then, the vast majority of cell phones eventually go to landfills.

Most cell phones that are discarded are reused, often purchased by companies that then resell them to overseas buyers. While the latest cell phones are relatively inexpensive in the United States, they are quite pricey throughout the world and there are eager consumers for second-hand American phones. After use in the US and other developed countries, the lion's share of cell phones (at least of those that don't end up in the desk drawers of consumers in these countries) are sold in a secondary market in South America, Eastern Europe, Asia, Africa, and the Caribbean. In the countries of these regions, phone and data plans are quite cheap; however, since new phones are very expensive there is great demand for relatively inexpensive used ones. Accordingly, this demand drives a lucrative secondary market. For instance, a California buyer snatched up lots of old Qualcomm QT 1000 mobile phones for $11 a piece. The buyer then was able to sell them off in the Ukraine for $111 per phone for a very hefty profit.

---

48 Mooalem, "The Afterlife," p. 41.

49 Layla Acaroglu, "Where Do Old Cellphones Go to Die?" *The New York Times*, May 4, 2013. Available at http://www.nytimes.com/2013/05/05/opinion/sunday/where-do-old-cell-phones-go-to-die.html?_r=0 (accessed February 26, 2015).

50 Victoria Loewengart, "Urban Mining of Rare Earth Elements in the United States: A Win-Win Proposition," Unpublished paper, 2011. Available at http://www.academia.edu/2578513/Urban_Mining_of_Rare_Earth_Elements_in_the_United_States_A_Win-Win_Proposition (accessed June 9, 2016), p. 2.

Reuse has always been a chief virtue of green business, but the efforts of urban mining and other reuse of e-waste has its limits. The true end of the line for most cell phones comes in some of the poorest places on the planet, where "urban mining" takes on a sinister and dangerous form. Eventually, many of the phones end up in large landfills like those in China, where people burn off plastic coverings and wires to release and salvage several kinds of precious metals. On their way to the burning heaps, many cell phones and other electronic devices snake their way through a shadowy underground economy from third party dealers who acquire phones that are meant to be recycled. Or as Jim Puckett, Executive Director of the Basel Action Network (BAN) explains, "A recycler can be a recycler in name only."[51] These dealers have found that they can make a great deal more money by simply exporting the material. This is particularly easy from the US where there are no laws preventing the shipping of e-waste out of the country.

Consequently, hundreds of shipping containers filled with e-waste leave ports in North America everyday. Sometimes these so-called "recyclers" simply dump the phones in a host of countries with large, poor populations and lax environmental regulations such as Ghana, India, Vietnam, Pakistan, Nigeria, Thailand, the Democratic Republic of the Congo, and the Philippines. E-waste is often illegally shipped into ports disguised as usable second-hand electronic devices or hidden behind functional electronics in shipping containers originating from developed countries such as the US and the United Kingdom. From the ports they are trucked inland to the dumps. Some old phones and other electronic devices can also end up for a time in market districts, but those they do not sell are taken to the dumps for metal extraction.

Urban mining in poor countries often involves patently vulnerable populations, such as children who pile the e-waste to be burned and mined. The salvaging operations in China and Ghana, for instance, can be quite large with hundreds of children and young men working in the dumps. After extracting the cooper wires and bits of silver and gold thread, they sell them to recycling merchants for a few dollars. It has been reported that in India young boys use mallets to open computer batteries. There is obviously a danger in this, as flecks of cadmium they are trying to collect splash onto the boys, leaving their feet and hands covered in toxic residue.[52]

---

51 Terry Gross, "After Dump, What Happens to Electronic Waste?" *National Public Radio* (Radio transcript), December 21, 2010. Available at http://www.npr.org/templates/transcript/transcript.php?storyId=132204954 (accessed February 27, 2015).

52 Acaroglu, "Where Do Old Cellphones."

A number of organizations, including Greenpeace and the BAN, post YouTube videos of children inhaling fumes that come from the burning of casings as they take out small pieces of metal and sort out different kinds of plastics for recyclers. In Ghana, there are reports of boys (most in their early teens, but others as young as ten) burning piles of copper wire to release it from its plastic encasing. At times they also melt down old air conditioners. To start and maintain the fires, the boys light tires and foam insulation from old refrigerators. Sometimes the fire changes color, including green when copper is burned, but the overpowering sensory impression comes from the acrid smell of tires burning with billows of thick, deep black smoke. Pools of aluminum and melted plastic are left for the children to sort through to collect valuable metals.

Many boys in Ghana who engage in burning do so after school or during the weekends to earn extra money for school fees. While education is technically supposed to be free in Ghana, in reality only primary school is. Thus children in poor families have to come up with money for lunches, exams, books, and a number of other fees that add up quickly. Of course, in other reports there are also many orphans who work in the dumps who may or may not be going to school.[53]

Women also work in this mining industry. They are often found laboring over baths of hot lead, in which circuit boards are melted down for the extraction of silver and gold.[54] Puckett describes the scene of women working an e-waste dump (which he calls a "cyber-age nightmare") in Guiyu, China: "It's the only part of the world where you'll go and see thousands of women any given day who are sitting... basically cooking printed circuit boards. As a result, they are breathing in all of the flame-retardants and the lead and tin that are being heated up. You smell it in the air. You get headaches as soon as you enter this area. It is really quite sad."[55]

It is clear that this industry in developing nations is unregulated, inefficient, and dangerous. The environmental damage done in these burning operations is also significant. Descriptions of the landscape from junkyard metal extraction in the Agbogbloshie region of Ghana near the capital city of Accra are nightmarish. *Atlantic* journalist Yepoka Yeebo describes how the area used to look before the e-waste dump was created:

---

53 Gross, "After Dump."

54 Acaroglu, "Where Do Old Cellphones."

55 Gross, "After Dump."

> I've heard from my grandfather and my grand uncles that it used to be a massive swamp, right in the middle of central Accra. And it's on one side of a lagoon, which is like the drainage basin for the whole of the capital city of Ghana. And it was a colonial-era football field—it was where the national stadium was until it flooded horribly in the sixties and lots of people died. And so the government decided to turn it into an industrial area and then there was a series of military coups and that was abandoned and it slowly got taken over by this market district.[56]

But today, since the arrival of e-waste and urban mining, the picture of the place is entirely different according to Yeebo:

> It looks like hell. It's this massive blackened field, and the first thing you notice is this pool of thick, black smoke. It burns; it's this choking smoke. And I sort of coughed and splattered my way through my first few minutes right in the middle of the dump. And it's dotted with these little hills of old electronics and scrap bits, or piles of CPUs, heaps of car doors, stacks of old televisions. And hundreds of people work there.[57]

The obvious health risks to those burning electronic devices to extract the metals begin with the choking smoke from the melting plastic. Chlorine is an important component of plastics in cell phones (particularly PVC, which makes up about 30% of the material in them). Since plastics are burned off during the extraction process, chlorine exposure is a genuine concern as it can lead to tissue and cell structure damage.[58] Likewise, bromine is used in plastics (including the circuit boards) of various electronic devices as a fire retardant and is released during the burning process.[59] Bromine exposure can lead to cognitive and developmental problems in children. Moreover, it can contribute to disruption in thyroid balance, as well as lead to brain damage and cancer.

---

56 National Public Radio, "A Shadow Economy Lurks in an Electronics Graveyard," (Radio transcript), January 4, 2015. Available at http://www.npr.org/2015/01/04/374780916/a-shadow-economy-lurks-in-an-electronics-graveyard (accessed February 28, 2015).

57 National Public Radio, "A Shadow Economy."

58 Paulie Anthony, "Cell Phone Toxins and the Harmful Effects on the Human Body When Recycled Improperly," *E-Cycle*, (website), October 15, 2013. Available at http://www.e-cycle.com/cell-phone-toxins-and-the-harmful-effects-on-the-human-body-when-recycled-improperly/#sthash.NPrNRJ50.dpuf (accessed March 2, 2015).

59 Gross, "After Dump."

In a study conducted by the Ecology Center of Ann Arbor, Michigan, of the thirty-six brands of cellular phones they examined, all of them contained at least one of the following toxic elements; chlorine (as just mentioned), cadmium, lead, bromine, or mercury.[60] These chemicals are linked to an array of adverse health effects such as birth defects and impairments to learning development in young children. Harmful chemicals have been found in the soils taken from e-waste dumps in China at levels 10 to 100 times higher than background levels.[61] Cadmium is specifically known to have high acute (short-term) toxicity in studies with rats. Additionally, acute inhalation exposure to high levels of cadmium can cause respiratory problems such as pulmonary and bronchial irritation.[62] Chronic (long-term) exposure to cadmium either orally or through inhalation can lead to the toxin's build up in the kidneys that can cause disease. Lead exposure can cause reproductive, blood, and nervous system damage. Mercury, up to 2 grams of which can be found in a single phone, can cause kidney and brain damage. Arsenic can be found in the microchips of many electronic devices, including cellphones, and exposure to the substance in high doses can be fatal. Even at lower exposure levels, it can still cause skin irritation and negative effects on the liver, nervous, and respiratory systems.[63]

There are other dangers as well, such as small explosions—one of which was reported by Yeebo. She observed a blast hurl a burning aerosol can through the air, almost hitting her and the boys she interviewed in the dump. Other physical dangers are present as well, with barefooted children working with metals suffering from lacerations on their exposed hands and feet. Boys who have worked in the dumps for some time are beginning to die at young ages of respiratory ailments and cancer.

Economic desperation likely drives this activity given its dangers. In addition to the toxic fumes released in the burning process, since these are open pit burning areas, rains can wash some of the effluents into creeks and streams near a landfill site, worsening the pollution of aquifers—many of which are used as sources of drinking water by those doing the burning who may live nearby the dump. It is readily apparent that despite clear evidence of both

60 Tim Wall, "Cell Phones Toxic to Humans and Earth," *Discovery News*, (website), October 5, 2012. Available at http://news.discovery.com/human/cell-phones-toxic-121005.htm (accessed March 2, 2015).

61 Wall, "Cell Phones Toxic."

62 United States Environmental Protection Agency, "Cadmium Compounds (A)," (website), 2000. Available at http://www.epa.gov/airtoxics/hlthef/cadmium.html (accessed February 28, 2015).

63 Anthony, "Cell Phone Toxins."

human health and land health issues from the metal extraction process, little is done to regulate the burning pits.

Companies such as Umicore are trying to properly recycle materials in a way that doesn't involve harming workers or the environment. Umicore extracts 17 separate metals from old televisions, computers, and cell phones. Its smelter burns at 2,116 degrees F, with its mouth being a pit 13 feet wide and 46 feet deep. A conveyer belt feeds shredded circuit boards and other metal scrap to the pit. The metals exit the smelter's base in a melted, glowing sludge and then falls into a cauldron about the size of a house. It then proceeds to tanks filled with acid that is electrocuted. This process separates out the copper that collects at the end of the tank, ultimately creating plates of 99.9% pure, shiny copper. Bars of gold also emerge from the facility.

To keep the recycling process as green as possible, the facility has a giant gas-cleaning-and-filtration system. This collects and then neutralizes the carcinogenic and endogenic-altering chemicals to the point where the final effluent not only meets, but exceeds EU air quality standards. While even this process is admittedly not entirely clean, it is much cleaner than the "low-tech" metal extraction occurring in the dumps described in this case study. Of all of the materials that Umicore processes in its recycling facility, only about 0.5 percent can't be safely put back into usable form.

For those who are worried morally about the disposal of e-waste there may be other possible solutions to the problem. One comes from the production end, simply putting pressure on those who set the specifications for the development and manufacturing of electronic devices to reduce the number of toxins in making these products. Given the problems of dealing with toxins that we've seen once they are *in* products, it may make the most sense to try to mitigate their use in manufacturing up front. Also, consumers could reward device manufacturers who are committed to using fewer toxins in their products and/or properly recycling their products by buying from "greener" producers.

With respect to regulations on urban mining, there generally are none within the countries where most of the extraction takes place. Internationally, there is the Basel Convention, which is an agreement with the goal of reducing shipments of toxic waste from developed to undeveloped countries. Originally, this agreement (otherwise known as the "Basel Ban") didn't primarily target e-waste, mainly because when discussions leading to the agreement occurred in 1989 and when it was signed in 1992 (including by all 15 countries what is now the EU), there wasn't yet such a need. At the time, the agreement was meant more as a check on the disposal of liquid chemicals. But in effect, by January 1, 1998, the agreement banned the export of all hazardous waste, including toxins "locked into" products from 29 of the wealthiest, most industrialized countries of the Organization of Economic Cooperation and Development (OECD) to non-OECD countries. The US, however, has never signed on to this agreement.[64]

Internally, the EU put in place the first Waste Electrical and Electronic Equipment (WEEE) Initiative in February 2003, which allowed for the creation of collection schemes for consumers to return their WEEEs free of charge.[65] This approach is designed to help maximize the number of recycled or reused electronic items. Also, in December of 2008, the EU first proposed to revise the Directive given how quickly the waste stream was growing, but changes only took effect fairly recently—on February 14, 2014.[66]

With respect to greener design of electronic products and thus trying to lessen the toxic effects of e-waste by reducing the toxicity of their production, the RoHS Directive came into effect in 2003. This EU legislation restricted the use of hazardous substances in the manufacture of electronic equipment. More specifically, the measure requires that heavy metals such as lead, cadmium, mercury hexavalent chromium, and fire retardants such as (polybrominated biphenyls (PBBs) and polybrominated diphenyl ethers (PDEs)) be replaced by safer alternatives.[67] Again in December 2008, the Commission proposed to revise the agreement and it became effective in January of 2013. The US not only failed to sign on to these agreements, but has lobbied against them. When the EU developed the WEEE Directive in the early 2000s, the US fought against the measure, taking their case to the World Trade Organization (WTO) on the grounds that they saw it as a major barrier to trade.[68]

In the US, only a few individual states have taken measures to at least reduce e-waste entering domestic landfills. California and Massachusetts require that computer monitors and old TVs be recycled, but this provision doesn't include cell phones at this time.[69] But of course, laws are limited in their effectiveness as there are always issues of noncompliance. It is difficult to monitor and enforce violations of an electronics dumping ban. There is often a fee associated with recycling in states with such a ban, sometimes around $20, but have been known to be as high as $60. So, facing a fee people don't want to pay, the

64 Bill Moyers, "Toxic E-Trash: Recycling Facts and Laws," *NOW* (Television news transcript), July 7, 2001. Available at http://www.pbs.org/now/science/ecycling.html (accessed March 6, 2015).

65 European Commission, "Waste Electrical."

66 European Commission, "Waste Electrical."

67 European Commission, "Waste Electrical."

68 Moyers, "Toxic E-Trash."

69 Emily Harris, "Toxic E-Waste," *NOW* (Television news transcript), July 19, 2002. Available at http://www.pbs.org/now/printable/transcript_ecycling_print.html (accessed March 6, 2015).

inconvenience of taking devices to be recycled, and the relative ease of simply littering, there is a strong incentive to simply not comply. Thus, what waste management haulers in Oakland, California, found was that instead of recycling them, residents would simply abandon their monitors and old TVs on curbs, leaving it to the city to eventually dispose of them.[70] As Susan Kattchee, Oakland's Recycling Supervisor noted, "We now have to send an additional crew out, with a separate truck, because the TVs and monitors have to be specially handled on the streets. They can't just be picked up and thrown in a regular garbage truck. And they have to be off loaded and handled appropriately at that point. And then again, the cost of disposal."[71]

A number of charities, including Goodwill, which used to accept large donations of TVs, no longer do so because disposal costs outstrip the profits they used to make. According to Senior Vice President of Operations, Terry Fitzpatrick, of the Goodwill of the Greater East Bay, the economic incentives for the company of accepting discarded electronic devices has completely changed as a formerly significant moneymaking avenue for the non-profit is no longer worth it: "We used to get—this is the local Goodwill about $50,000 a year in revenue from the sale of TVs and computers. Well, since we no longer accept them due to the high cost of disposal not only have we lost that revenue source, but our costs for handling and disposing of those items has gone up as well. And that can be as high as $40,000 a year."[72]

Of course, just as people are willing to dump their old computers and TVs in the street, a significant number also drop their electronic devices at Goodwill's steps, thus leaving the company to dispose of them. Now having to do something with these old devices and their toxic waste, the State of California has begun to address this issue and lawmakers have bandied about a proposal to require electronics retailers to add a $30 fee so that consumers pay the disposal costs. In this way, the cost of disposal would be built into the upfront price of such devices. Some argue that this only right—that those who want and use electronic devices have to contribute their fair share of disposal costs.[73]

The electronics' industry is fighting any proposals for upfront fees, worrying that sales will be reduced due to increasing costs for consumers. They also argue that securing compliance will be difficult as some consumers will purchase devices via the internet to avoid the fee. On the other side, some environmentalists object to consumers paying for disposal fees, as they see it as

70 Harris, "Toxic E-Waste."

71 Harris, "Toxic E-Waste."

72 Harris, "Toxic E-Waste."

73 Harris, "Toxic E-Waste."

the manufacturers' responsibility to take back and handle the disposal of their products directly.[74]

From these possible policy revisions, we get some sense of the moral problems that result from in this case study. One of the largest has to do with international economic justice. As with respect to a number of environmental problems it seems that developed countries that design, produce, and sell electronic devices and their large, relatively affluent consumer bases enjoy most of the benefits of the electronics industry, while poorer countries tend to end up with an excessive share of the burdens. Many of the countries of the West get to have cool phones and other stylish items, but their disposal happens thousands of miles away, out of sight and out of mind, under shady and clandestine conditions that are unregulated by both developed and undeveloped countries. The result is that poorer countries become the "trash heap" of wealthier ones and some argue that this scenario is exploitative. So, on at least two important levels there are burdens placed—on the ecosystems of countries that have the least capacity to handle them, and on human workers and those living next to the dumps whose drinking water and soil are contaminated by toxins spilling off from the dump sites.

Surely, those who are harmed may be in different categories. For example, while both workers and those living next to dumps may incur damages due to dirty urban mining, one might argue that those not participating in the scrap collection are wronged at a higher level. They in no way asked for the contamination, they are not unlocking toxins from electronic devices for pay, and thus they likely have a direct claim to being harmed. The interesting question is who are they harmed by and with what degree of culpability? Are those burning e-waste in the dumps to be held to the highest degree of culpability by those who don't salvage metals, but suffer the harmful health effects of the process? Or should blame be placed on those who unscrupulously ship e-waste to undeveloped countries in the first place? Moreover, while there is a growing call to hold manufacturers of electronic devices accountable and to have them incur more of the cost for its disposal, is it because they harm citizens who don't salvage metals but suffer the effects of them in Ghana, China, and the Philippines?

With respect to the possible wrongs done to those who work in dirty urban mining, what might ground the claim that they are exploited? While theories of exploitation are very complex and can't be given full treatment here, we can say enough about them to give some traction to answering significant questions in this case study. In its most essential form, exploitation means taking unfair

74 Harris, "Toxic E-Waste."

advantage of someone. This would seem to imply that someone gets an advantage "off the backs" of others. While this might be done intentionally, there doesn't seem to be a requirement that exploitation only occurs when it is intentional. For example, I could gain exorbitant profits as a shareholder of a company that pays its workers extremely low wages without ever meeting them or wishing them harm. Yet, I would still be said to exploit them on the criterion for exploitation cited above. So the geographical, economic, and social separation between the CEO of Apple and the person, say an adult female extracting metals from circuit boards with acid in India, is not relevant to claims of exploitation.

Also, for a relationship to be exploitative, it has to be in some sense unfair. That is, person A derives a benefit that is undeserved *due to* a transaction with person B, who suffers a burden while creating A's benefit. In philosopher Allen Buchanan's definition of exploitation, the salient factors involve the harmful, merely instrumental use of one or her capacities by another, and that the user does so *only* to fulfill his/her own advantage or to advance her own ends.[75] This is clearly an application of the Kantian idea that one should never act to use another *merely as a means* to one's own ends.

However, those who claim that exploitative action can only occur when one is *forcing* another to do something and where the former is garnering a benefit and the other is saddled with a burden, might counter charges of exploitation in urban mining. Note that they would likely even accept part of Buchanan's definition of exploitation. A coercive relationship is clearly the instrumental use of someone because forcing him to do something against his will is using him merely as a tool. But arrangements whereby one gives *consent*, even if it is given begrudgingly, is not considered an exploitative relationship by some social contract theorists because all parties have accepted the terms of a contract or entered into an activity that they could have refused to do.[76] So, slavery would be exploitative, but deciding to melt down cellphones to find scrap metal to sell (even for an absurdly small amount of money) would not be a case of exploitation.

The likely reply to those who think urban mining is not exploitive is that for many, perhaps even most urban miners, consent is not possible. Recall that many who do the most dangerous metal extraction work in the dumps are children. Part of the reason they are regarded as a vulnerable population

---

75 Allen Buchanan, *Ethics, Efficiency, and the Market* (Totowa, NJ: Rowman and Allenfeld, 1985).

76 This is the view of Robert Nozick and would likely be accepted by most libertarian political philosophers, political theorists, and strong free market enthusiasts. See his *Anarchy, State, and Utopia* (New York, Basic Books, 1974).

is because they are not yet choice-bearing agents. They lack the capacity to gather appropriate information and make fully rational decisions. They also lack experience in fully understanding the full ramifications of their actions. This is especially true of actions that have long-term and/or systemic negative effects in the future.

Another interesting moral aspect of this case has to do with the interplay of moral motivation and knowledge of a complex environmental issue. Sometimes learning the facts about a case can *reduce* one's moral motivation to do what she thinks is right even by her own lights. For example, imagine that you believe the dirty form of urban mining is a morally bad practice. You might think the process causes harm to people and that long-term suffering and death are evil. Workers in the dump put their health at risk only because they're impoverished. Thus, they do not truly consent as they have no choice in a meaningful sense—their "choice" is to forgo survival and enhancement entirely or to labor in unsafe conditions. Furthermore, you think that spilling toxic residues into the soils and waters of developing countries is an ethical issue as it harms the land. You want to do something about this on a personal level—you don't want *your* electronic equipment to reach these landfills. But, with little investigation, you recycle your phone. The problem is that about 80% of phones recycled will still end up in landfills overseas. The moral problem becomes an epistemic one. For you are uncertain if *your* phone will really be kept out of a landfill. Once it leaves your hands, it can go through the shady streams of the e-waste economy where it will end up where you didn't want it to despite your intentions and efforts. Thus, with some evidence that your phone will not reach the desired destination anyway, you are disinclined to recycle it.

This leads to what is likely an undesirable scenario. You may be more inclined to (begrudgingly and guiltily) throw your old cell phone in the trash. You might actually believe that it is better to pollute in the US or the UK by simply throwing your phone into the trash. Why? Because it will likely not find its way into a shipping container to become someone else's problem overseas. Even though your phone will likely contribute to pollution and America's landfills are in areas close to impoverished or politically impotent people, it is less likely that children and women will be harmed since there is little economic incentive to do dirty urban mining in the US.

There will likely only be more e-waste over time as it is still the fastest growing sector of waste in landfills not only in the US, but in the rest of the developed world as well. More is destined to go overseas. And this might be only the tip of the iceberg. It is projected that in undeveloped, but rapidly developing countries, such as India and China, as more of their gargantuan populations become wealthier, their taste for electronic goods will mean they will likely soon join the ranks of nations contributing substantially to the e-waste problem.

Review Questions

1) Should there be more governmental regulation over the disposal of cell phones and other electronic devices in their countries of origin? What should these regulations look like?
2) Should US consumers be forced to sell their cell phones to companies that resell them to other consumers overseas? Short of forced sale, should governments develop incentives programs to encourage safe disposal of toxic components of cell phones and other electronic devices?
3) Should consumers refrain from buying new cell phones, or at least buy them less frequently, having them repaired instead?
4) Do companies have any obligation to build longer lasting cell phones in an attempt to reduce the number that will eventually need to be disposed of? Should they be forced to collect phones for recycling purposes?
5) Can you think of other strategies to help solve or reduce the amount of e-waste? Explain.
6) Who should pay for the disposal costs of electronic devices such as cell phones? Is the current way of handling the situation in California fair? Would it be fairer to require customers to pay the disposal fees upfront, or is this something that should be a responsibility of companies designing and/or manufacturing the devices?
7) List all of those (if any) who you think are exploited as a result of the e-waste industry. Why do you think they are exploited? If you do not think anyone is being exploited, why not? Do you think it is possible to exploit someone and yet not do them a wrong? Explain your answer.

Case 14 – The Cost of the Future[77] –

Study Questions

1) How might one generation have to suffer greater burdens than prior generations with respect to the environment?

77 I would like to thank Phil Certain for bringing this case to my attention.

2) What problem is one of the most difficult to solve with regard to environmental ethics and how is it appealed to in this case?
3) What is the *non-identity problem* and what appear to be its implications for possible obligations to future generations?

We often hear people who try to argue in favor of a green lifestyle make an appeal to the future. Future generations of unborn children will suffer the direst of consequences from our short-term beneficial, but long-term environmentally destructive behaviors. We are asked to leave an environment at least as good, if not better, for our children and our children's children.[78]

However, even if we owe obligations to future generations, we need to ask *how much* they are owned. We can't simply assume that mitigation of environmental degradation comes without cost. In this context, consider the following. In 2008, United Nations projected it will cost $20 trillion over twenty years to switch over to cleaner energies.[79] The International Energy Agency revised this estimate, noting that it will likely cost $4 trillion within the next decade to make this conversion. While part of this burden may be included in the costs to switch to cleaner energy, by 2014 the UN estimated the annual cost of aiding poor countries adapt to changes in climate and convert to cleaner energy would be about $500 million per year. Japan and Europe have already spent millions of dollars in efforts to meet even the modest carbon curbing requirements agreed to in the Kyoto Protocol established over two decades ago. These costs are likely to be increased by the rising price of food (as corn and other products grow scarcer due to their use as renewable fuels).[80] Given the latest climate projections, it doesn't seem as if the food price problem will be going away anytime soon. These considerations raise important moral and economic questions such as should we sacrifice lots of resources for this generation to avert the adverse effects of global warming for future generations?

Societies are confronted with a wide spectrum of social, health, economic, and political problems, all of which are likely worthy of some sort of investment to help solve. World hunger, lack of access to potable water, malaria, AIDS, Ebola, and Zika are very pressing, deeply rooted, and serious problems

78 For more on the problem of future generation, see this book's introduction and Case 12—"Climate Change Accords and Vanishing Islands."

79 Eduardo Porter, "Are the Grandkids Worth the Money?" *New York Times*, March 14, 2008. Available at http://www.nytimes.com/2008/03/14/opinion/14fri4.html?_r=0 (accessed June 9, 2016).

80 See Case 18—"Food Ethics, Obesity, and Voting with Your Fork," Case 30—"Food vs. Fuel: Unintended Consequences of Ethanol," and Case 34—Filling Up on Kudzu."

that face us right now. Several people are dying of these conditions and ailments.[81] The question is how much ought we to sacrifice in terms of helping people today to solve the anticipated problems of tomorrow?

Climate change obviously presents a significant challenge. The latest IPCC report warns that if clean energy is not dominant in the world's energy supply by 2050, we will likely cross a threshold into catastrophic climate change. This is not just the opinion of a few scientists, but also the consensus of hundreds of experts. Fossil fuels now produce about 80% of all the world's energy, so meeting emission goals is daunting.

The problem is even deeper if we think people's rights are irrelevant to when they are born. Yet, can we assume our grandchildren's well-being is just as important as our own? In an editorial written in the *New York Times*, Eduardo Porter posits that it might: "Assuming our grandchildren's welfare is just as valuable as our own provides a metric to measure the value of investments for the future: devoting X percent of the current generation's income to forestall global warming would be a good deal if it produced a benefit amounting to more than X percent of that future generation's income."[82]

This appears to be an effort to justify intergenerational equity. The Stern Review on the Economics of Climate Change was a British study that first stressed the enormous cost of stalling on action to reduce climate change. Back in 2006, the organization cited the deep and troubling implications of climate change as a justification for spending one percent of the world's income, which would have been about $700 million at that time, to cut carbon emissions.

But if we examine human behavior, we are all too likely to sacrifice the future for the present, not worrying about what may come tomorrow. All we need to do is look at the lack of savings rates of citizens in not only the United States, but also in several other countries around the world.[83] As Porter points out, since children and unborn members of future generations can't vote, measures to protect the future have no strong advocates. Even in more socially democratic countries in Western Europe, public opinion polls show that the older the persons polled, the less likely they are willing to pay to reduce carbon emissions.[84] Porter noted that in Florida, there is a popular bumper sticker that reads, "I'm spending my children's inheritance."[85]

81 Porter, "Are the Grandkids."

82 Porter, "Are the Grandkids."

83 See Richard T. Thaler and Cass Sunstein, *Nudge: Improving Decisions about Health, Wealth, and Happiness* (New Haven, CT: Yale University Press, 2008).

84 Porter, "Are the Grandkids."

85 Porter, "Are the Grandkids."

It is important that we don't assume future generations, even if they will face challenges, will necessarily be worse off than ours. In fact, the American dream has rested on the notion that in the US there is a viable opportunity for one's children to have a brighter future than their parents did. People in the future will likely be richer and more productive than we are. With this assumption, one might question why we ought to spend heavily on the welfare of "tomorrow's child." Besides, one might argue it would actually be cheaper to cut carbon emissions in the future using advanced technology, than by trying to do it with today's blunt instruments.

With these thoughts in mind, the Copenhagen Consensus, a group of respected economists, has argued against what they see as aggressive (and unduly expensive) efforts to curb carbon emissions on these grounds. They think the costs of doing so will likely be greater than the benefits, and that alternatively money would be better allocated in addressing currently pressing social and economic problems. But of course, all of this comes in the face of deep uncertainty. The gnawing notion is that climate change heads us toward a catastrophic future. The assumption that future generations will necessarily have more money and enhanced technological capability to quickly scale back carbon emissions cannot by any means be assured.

Behind all of this are some philosophical problems. In the environmental ethics literature, one of the most perplexing issues facing those who wish to argue in favor of ecological protection is the *problem of future generations*. Many factors lead to it. While the negative impact of some environmental destruction is endured at the time it occurs, much of it is not fully witnessed by the generation that creates it, but instead will be suffered by generations of people not yet born. This means the burdens of one generation's policies and actions are often taken on by "almost certain to exist" humans, who not only fail to derive any possible short-term benefit from the policies or actions of their predecessors, but also didn't create such policies, or even have an opportunity to consent to paying possible costs. Some argue this is patently unfair.

However, this conclusion appears to be philosophically problematic. To see why, we need to think about the usual connections made between persons, moral status, and rights. Most arguments for why any sort of being possesses moral status comes from characteristics those beings have (e.g., the ability to think or feel). If a morally relevant status is present, then rights are looked upon as ways of protecting such status. Notice that what lies at the heart of the abortion debate is whether fertilized ova, blastocysts, zygotes, and fetuses have the appropriate moral status seen as requisite to be conferred rights. Yet, if it is already controversial as to whether these entities have moral status and possible attenuating rights (to life, preservation, etc.), it seems *even more debatable* that those not yet conceived could possibly have moral status, and consequently, moral rights. After all, doesn't something actually have to exist to be accorded rights? Given this context, what obligations can current generations have to those who do not exist now and may or may not exist in the future? Why should we forgo opportunities now to save for the future?

Another philosophical issue arising out of thinking about the problem of future generations is called the *non-identity problem*, which was brought to light by the philosopher Derek Parfit. It goes something like this. At the heart of the problem of future generations is the charge of unfairness. One generation gains most of the benefits and few of the costs of some activity or policy at the expense of another generation that endures many burdens and enjoys few to no benefits. Even if that is not unfair, then the fact that the burdened generation has no say in whether said activities should have occurred or policies should have be implemented because they didn't exist, certainly seems unfair—sort of like taxation without representation.

But Parfit notes that without said policy or activity, assuming that people are caused to exist by such things, those who exist in future generations would not have existed if those in prior generations hadn't acted as they did. To see this, consider the following thought experiment involving two societies; "green society" and "brown society." "Green society" adopts policies to try to ensure that the future generations have a reasonably clean environment. "Brown society" conducts "business as usual" practices; adopting development friendly policies that pollute. In the future, residents of green society enjoy cleaner land, air, lakes, and oceans than those of brown society. But notice other significant differences between these two societies over the years. They have adopted different policies, so their economies are different. Different jobs are available, which means different people meet, bringing different people into existence than if some other policy were in effect—and this is true of both societies.

We can also imagine that over time, two completely different sets of people are in existence. That is, those in future generations are non-identical to the people who would have existed instead if different policies and actions had been taken by their previous generations. However, note an implication of this. In either case, it would appear that those in existence would have no complaint about their lot, as without the policies or actions implemented by their predecessors, they *wouldn't exist*. Even if you are worse off than those in previous generations, you are better off than not having existed at all. If you are convinced by this line of reasoning, it appears that you would have to solve the non-identity problem to show why we have obligations to future generations.

Review Questions

1) Some think people's rights should not depend on when they are born? Do you agree? Why or why not?
2) Should we assume the welfare of future generations is just as important as our own? Explain your answer.
3) The problem of future generations is important because it reminds us that moral status is necessary to ground rights to life and perhaps a (at least minimally?) clean environment. But is there another possible reason why we have obligations to future generations even if they don't have rights?
4) If future generations should have rights, what kinds of rights would these be? Would they be rights as *potential* individuals? Or might they be more like potential *group rights* (such as a proposed right of peoples or nations to secede from larger political units)?
5) Perhaps one of the more intriguing, but perhaps unsurprising facts emerging from the recent climate change debate is that the older people get, the less they wish to spend on curbing carbon emissions. Do you think the elderly hold a just position (i.e., that they should simply enjoy their final years without worrying about future generations)? Or do you think they should care about their impact on the environment for the sake of future generations? Why or why not?

Case 15 – Climate Change and War –

Study Questions

1) What are some of the possible ways social scientists and policy analysts believe climate change could precipitate war? Why do other scholars think this worry is overblown?
2) How might the work of Thomas Hobbes apply to the possible connection between climate change and war?

Can climate change have what might be considered to be a substantial unanticipated adverse effect—global destabilization leading eventually to war? We can trace the idea that given conditions of scarcity of desired resources humans will fight to acquire them all the way back to the English philosopher Thomas Hobbes (1588–1679). Hobbes thought conflict arose from human nature—that we always naturally act in ways to benefit our self-interests (i.e., *psychological egoism*). He believed that psychological egoism, which also involves a healthy dose of vanity and thirst for glory, mixed with scarcity of resources, concludes in a ceaseless war of all against all. Without a centralized political authority to keep people in line, lives are left (in his famous phrase), "solitary, poor, nasty, brutish, and short."[86]

But whether we agree with Hobbes that we are psychological egoists, modern day policy makers don't need to go far to find warning signs of possible conflict—especially after witnessing what seem to be destabilizing effects of climate change on weak and vulnerable nation-states. In a 2007 report, a group of retired United States generals and admirals warned leaders in Washington that climate change should be included in America's national defense strategy (and it since has).[87]

Other international leaders are taking this call seriously. Also in 2007, United Nations Security General Ban Ki-moon noted in the *Washington Post* that the fighting and genocide in Darfur might have been attributed chiefly to an extended drought in the area. In fact, he described the fighting in the Sudan as the first climate change conflict.[88] This long time dry period caused rifts between herders and farmers as they jockeyed for control over decreasing acreage of arable land. Increased warming over the Indian Ocean has dramatically altered weather patterns throughout a wide area of East Africa—all of which has been caused by human activity. Furthermore, the Nobel Panel echoes the Hobbesian concerns of a different era by saying that global warming "may induce large scale migration and lead to greater competition for the Earth's resources … [and] increased danger of violent conflicts and wars, within and between states."[89]

The background assumption is that over time drought has been caused by rainwater patterns altered by climate change, which has reduced the amount of water available in the region. In turn, drought contributed to tensions and Moon is likely reflecting findings in the past few decades that in years of lower precipitation, there have been more instances of conflict.

In a 2009 study (called by them "[t]he first comprehensive examination of the potential impact of global climate change on armed conflict in sub-Saharan Africa"), Marshall B. Burke and a group of colleagues from several American

---

86 *Leviathan* (New York: Penguin Classic, 1981), ch. 13.

87 Stephanie Giry, "Climate Conflict," *The New York Times Magazine*, December 9, 2007, p. 62.

88 Mark Notaras, "Does Climate Change Cause Conflict?" *Our World*, (website), November 27, 2009. Available at http://ourworld.unu.edu/en/does-climate-change-cause-conflict (accessed June 12, 2016).

89 Giry, "Climate Conflict," p. 62.

universities did a regression analysis of historical data of temperature changes and internal conflict in the region.[90] They found that "there are substantial increases in conflict during warmer years." For example, the evidence gathered by Burke's team suggests that a 1% increase in temperature leads to a 4.5% increase in civil war within the same year and a 0.9% increase in the following year. If this trend were to continue (at least as projected from the time Burke's study was finished), given the average temperature from 18 different climate models, this would mean a 54% increase in armed conflict in the region by 2030. Burke et al. concluded that the main driver of conflict comes from economic uncertainties expected from temperature-related decreases in crop harvests (especially in areas heavily dependent on agriculture). This conclusion correlates with earlier work that "economic welfare is the single factor most consistently associated with conflict incidence."[91] In a 2013 study, Sol Hsiang joined Burke to conduct a meta-analysis of 50 studies on civil violence and climate change. They reported that not only do higher temperatures correlate with greater incidences of civil conflict (which matches Burke's earlier findings), but that higher levels of precipitation do as well.[92]

At first glance, it may make sense that climate change could foster conditions ripe for the destabilization of nation-states and thus exacerbate long-existing tensions between countries (or groups within states). If droughts, floods, and failed crops are caused by global warming, it is not a stretch to imagine these conditions could lead to conflicts further induced by economic inequalities, transfers of large populations, and poverty.

The Environmental Change and Security Program (ECSP) within the Wilson Center was created to examine environmental issues connected to health and population dynamics. In the work done by this group, these facets are examined with respect to their connections with human insecurity, foreign policy, and conflict. The three primary focus areas of ECSP are: 1) integrated development, including health, environment, livelihoods, population, and security, 2) Environment, conflict and security, and 3) water. These foci present us with numerous examples of how climate change will likely increase social and political divisions.

The first focus area of integrated development, livelihoods, population, and security involves the idea that poverty forces the impoverished to face many

90 Notaras, "Does Climate Change Cause…?".

91 Notaras, "Does Climate Change Cause…?".

92 Amy Westervelt, "Does Climate Change Really Cause Conflict?" *The Guardian*, March 9, 2015. Available at http://www.theguardian.com/vital-signs/2015/mar/09/climate-change-conflict-syria-global-warming (accessed May 7, 2015).

tightly bound, difficult, and systemic development problems. For example, consider urbanization, youth bulges, and migration—three types of global population dynamics. All of these factors can lead to social and political destabilization and cause conflict. ECSP can be seen as a forum to research ideas on *climate change adaptation*. As the challenges from climate change occur, nations will need to understand them and devise strategies to diffuse conflict. Practical policy options will require discussions about population-health-environment connections and demographic security—especially in developing countries.[93]

The second area of focus for ECSP's work, involving environment, conflict, and security, comes from recognition of how natural resources (or the lack thereof) are increasingly causes of conflict and the insecurity of states. Climate change is thought to be a "threat multiplier" over a range of security areas. ESCP believes environmental interdependence is a way of enticing cooperation and peacebuilding, as countries see their reliance on the stability of other nations. Thus, ECSP works with policymakers and scholars to deliberate about research and develop policies aimed at ameliorating adverse environmental and developmental problems.

The third area concerns water availability, as obviously its scarcity causes fundamental challenges to human health and the stability of communities. Publications produced by ECSP examine how water supplies can cause both conflict and cooperation. The social and economic value of water is also emphasized, as well as its relationship to health and disease.

The instability of states from climate change highlights some additional problems. The poorest of countries are hit particularly hard by climate change because they are in the worst possible position to build the infrastructure to prepare for rising sea levels and encroaching deserts. Also, their populations often don't have the means to "just move" if negative effects of climate are approaching. Finally, they don't have the means of cleaning up. This adds up to poorer, developing countries of the world receiving the most structural damage during violent storms, the most deaths (because impoverished residents can't easily escape), and taking the longest to rebuild if they are able to rebuild at all.

But of course, it is still controversial whether there is a strong connection, if much of one at all, between climate change and warfare. Idean Salehyan, from the Department of Political Science at the University of North Texas, warns there is little empirical data to support the Hobbesian idea that

93 See Wilson Center, "About Environmental Change and Security Program," (website), 2015, Available at http://wilsoncenter.org/about-ecsp (accessed May 6, 2015).

scarcity necessarily leads to conflict. Interestingly enough, Salehyan notes that if anything we tend to fight over an *abundance* of resources, not a dearth of them. Furthermore, doubts surface from those who believe Burke's work misses the inherent complexity involved between climate conditions changing for the worse and conflict. Dr. Vesselin Popovski, the Senior Academic Program Officer of the United Nations University Institute of Sustainability and Peace and Security Section, thinks that if there is any link between climate change and conflict at all, it is an indirect one. Popovski notes a study from the International Peace Research Institute in Oslo where researchers concluded that the "the causal chains [between global warming and conflict] suggested in the literature have so far been rarely substantiated with reliable evidence."[94] Popovski adds that "[t]he causes of conflict are primarily political and economic, not climatic. Warlords—who foster conflict—may exploit drought, flooding, starvation, agricultural or natural disasters in their strategies, like they did in Somalia and Darfur. But what will drive their fight is not the rain, the temperature, or the sea levels—they will always fight for the same goals of power, territory, money, revenge, etc."[95]

Popovski concludes, "There is no doubt that impoverishment and human insecurity may arise as a result of climate change, if preventative measures are not undertaken. However, there is missing evidence that global warming directly increases conflict."[96] Even in considering recent conflicts such as in Syria, a country that has suffered from extreme drought, some researchers believe we shouldn't jump to the conclusion that climate conditions *cause* civil war. One major issue facing those who propose a strong causal link between climate change and conflict is how to test the hypothesis. The problem is social and scientific. As Andrew Solow, senior scientist at Woods Hole Oceanographic Institution describes it, "[I]f you want to show that climate change has contributed to an increase in civil violence, then you need to control for other factors. This is a fundamental scientific principle. But it is difficult to do."[97]

To their credit, Hsiang and Burke are cautious about their claims concerning possible causes of conflict from climate change. In a statement accompanying their study, Hsiang mentions that, "There's no conflict that we think should be wholly attributed to some specific climatic event. Every conflict has roots in interpersonal and intergroup relations. What we're trying to point out is that

94 Cited in Notaras, "Does Climate Change Cause…?".

95 Notaras, "Does Climate Change Cause…?".

96 Notaras, "Does Climate Change Cause…?".

97 Cited in Westervelt, "Does Climate Change Really…?".

climate is one of the critical factors [that] affect how things escalate, and if they escalate to the point of violence."[98]

No matter what, perhaps the common ground between those who think climate change plays a causal role in conflict versus those who don't might be that climate change is a "threat multiplier." This is the way the US Pentagon has characterized the link between climate change and conflict. As former Defense Secretary Chuck Hagel put it, "Rising global temperatures, changing precipitation patterns, climbing sea levels and more extreme weather events will intensify the challenges of global instability, hunger, poverty and on conflict."[99]

One may wonder why this debate ultimately matters. After all, both sides wouldn't deny that there is *some* connection between climate change and conflict. Does it really matter whether climate change is a root cause, a contributing factor, or something that lengthens conflict and intensifies civil threats? Or is this debate merely a pedantic philosopher's point? Solow doesn't think so. If we wrongly believe climate change is a cause of conflict when it isn't, this is a problem. "I'll put this in a crude way: no amount of climate change is going to cause violence in the state where I live (Massachusetts), or in Sweden or many other places around the world. If we want to reduce the level of violence in other places, then it would be more efficient to focus on these factors: to bring people out of abject poverty, to provide them with the technology that loosens the connection between climate change and survival, to reduce corruption, and so forth, rather than on preventing climate change. I sometimes have the feeling that some people only care about human suffering if it can be traced back to climate change."[100]

While there is the old saying that "all's fair in love and war" and over history humans have certainly behaved as if this edict is true, there has also been a long *just war* tradition. The Bible hints at possible ethical standards within warfare as well as just causes for going to war. As early as the 4th and 5th centuries, St. Augustine (354–430 AD) commented on the morality of war from a Christian perspective, mainly protesting against how warriors can acquire a love of violence in combat. But it was St. Thomas Aquinas (1225–1274) who systematically examined the ethics of war. In his *Summa Theologica*, Aquinas outlined much of what is still discussed today as the set of principles for just war.

The theory of just war includes both the moral conditions under which it is permissible to go to war and who should go to war (or more precisely, which institutions can go to war). It also involves what ethical principles should guide

98 Westervelt, "Does Climate Change Really…?".

99 Westervelt, "Does Climate Change Really…?".

100 Westervelt, "Does Climate Change Really…?".

how wars ought to be waged. Just war theory breaks into roughly two large categories. *Jus ad bellum* involves those principles governing whether a country should go to war. *Jus in bello* involves principles for how to fight in war. Since the possible link between climate change and conflict only really involves under what conditions countries should go to war, we'll only concentrate on jus ad bellum.

There are six principles or conditions that a country must meet for justifying going to war in just war theory. These principles are not only from a consequentialist position, but also come from a deontological tradition. First, a state should have just cause; that is, a nation must have good reason go to war. While most just war theorists don't think it is justifiable to simply initiate aggressive action, it is justifiable for a country to defend itself from attack. While this would seem to suggest that only a nation's right to self-defense would be a justified reason to enter war, some instances of pre-emptive strike might also be justified if there is good evidence of imminent attack.

A second condition for going to war involves *right authority*, which means that war can only be justifiably authorized by the sovereign power of the state (assuming it is a legitimate state). The third principle is *right intention*, which roughly means that a nation is waging a just war when it is doing so for a just cause and not merely for reasons of expansionist nation-building or for glory.

Next, wars should only be fought by a nation if it has a reasonable chance of success. Recognize that while this is a *necessary* condition for war, it is not a *sufficient* condition. The concept of a reasonable chance of success is consequentialist in the sense that the costs and benefits of a war must be taken into account to see if it ought to be waged. While there are difficulties with this principle (as there are with other *jus ad bellum* principles), the thrust of the claim is that nations should not waste human life and economic resources when it is clearly overmatched.

A country going to war also ought to use the *right proportion* of force to win the war. The desired end should be proportional to the means used. We can think of this as the "anti-genocidal" policy. This principle overlaps with *jus in bello*, which also has a proportionality principle. But in this case, the emphasis is not on how much force should be used to neutralize an individual opposing combatant or an opposing army. The idea with respect to how much force is necessary in the case of *jus ad bellum* is that a just war requires a goal, an end game, and that end must be proportional to other principles of just cause. This is a provision to try to minimize destruction in warfare. So, nuclear weapons should not be used in a minor border dispute. The proportionality principle in *jus ad bellum* must take into account the general balance of powers over a large geographical region. This means that even a country going to war in self-defense is not justified in annexing a country or becoming a long-term occupier if it wins. Finally, a justified war can only be fought as a last resort. This fits very easily alongside just cause—all other forms of solutions must be found to avert war before it is declared.

Humans have learned in our long history of conflict that war is a dirty business. Once war begins, lots of anticipated and unanticipated negative effects arise from it and political institutions can be slow to stop it. War can wreck economies. Distrust between competing societies can reside long after the shells stop flying, and ethnic tensions can percolate. Governmental/citizen relations can be strained by conflict. Leaders considering whether to go to war need to ask themselves "Do we really want to do this?" Going to war due to real or imagined effects of climate change is yet another challenge and will certainly put world leaders to the test.

Review Questions

1) Let's assume Salehyan is correct that conflict is not necessarily related to scarcity. Does this say anything about the possible connection between climate change and scarcity? Why or why not?
2) Let's suppose global climate change contributes to conditions of scarcity. Would the principles of just cause apply as a justification to go to war (think about *jus ad bellum* here) to garner resources? Should carbon emissions be reduced as part of a national defense strategy?

Case 16 – The Rapidly Sinking Town –

Study Questions

1) What are some of the issues faced by towns in the Arctic with respect to climate change?
2) What measures are being taken for those living in towns sinking into the permafrost to adapt to its melting? What might need to be done in the future?

It may be easy to dismiss some of the implications of climate change when its effects are only seen as being far to the north in the Arctic Circle where few are really affected. This is particularly true if your approach to environmental ethics is an *anthropocentric* (human-centered) one. Yet, an issue recently receiving more press is the destruction of small villages of native populations in

Alaska, Northern Canada, Greenland, Northern Sweden, and Northern Russia. The vast majority of these communities are built on permafrost, with some of them within close proximity to the ocean. With the loss of permafrost, many of these towns are sinking at an alarming rate, causing extensive damage to infrastructure, foundations, and to buildings themselves. Even getting around in these communities is increasingly difficult, with networks of wooden planks necessary to connect residences and businesses to each other.

Even without the melting of permafrost, building is tricky at the Arctic Circle. But its loss only makes construction and maintenance more difficult and brings with it costly infrastructural challenges to towns in the extreme north. Sara Brown, Director of Community Operational Support with the Northwest Territories Association of Communities, believes that the costs of upgrades and renovations to buildings in northern Canada could be anywhere from $250 to $420 million Canadian. Brown adds, "This doesn't just affect community infrastructure. It affects residents' homes and the amenities they enjoy as well. So, the better informed we all are about this, the better we'll be able to deal with it in a proactive way as opposed to a reactive way."[101]

Permafrost is soil or rock that remains below 32 degrees F not only throughout the year, but from year to year. However, a thinner, 'active' layer of soil lies just above the thicker layer of more solid permafrost. The active layer freezes and unfreezes throughout the year, but it is increasing in depth as temperatures in the Arctic region warm. With the active layer not freezing as it has in the past, the permafrost underneath is also becoming warmer, leaving the ground less stable for structures on the surface.

Initial signs of the problem came at the beginning of this century, appearing first with larger, heavier structures. In October of 2001, the floor of the Arctic Winter Games Arena in Iqaluit, Nunavut, Canada began sinking. Ice creation became impossible by 2003 and caused the arena's closing. After a $2.2 million dollar repair, the facility finally reopened in October of 2010. Even when facilities sitting on melting permafrost are renovated, it is only after substantial costs and there are no guarantees that more maintenance work or even rebuilding won't be necessary in the future.

Technology is available to help keep arctic buildings in place, but of course, it sometimes fails as well. A school in Ross River, in the Yukon Territory of Canada, had to be rebuilt after melting permafrost left it unstable. The school

---

101 Laura Wright, "Thawing Permafrost Sinks Building, Hikes Costs in North," *Canadian Broadcasting Corporation News*, (website), November 18, 2011. Available at http://www.cbc.ca/news/canada/north/thawing-permafrost-sinks-buildings-hikes-costs-in-north-1.1108686 (accessed June 1, 2015).

was built with thermosiphons, which are heat exchangers that draw energy out of the ground. This technology was first introduced with the building of the Trans-Alaska Pipeline in the 1970s. Thermosiphons come in two types. Passive thermosiphon systems use cold winter air to cool chemical fluids that circulate in pipes under the building to keep the permafrost frozen. More active systems use pumps to speed up the fluid flow, much like in a refrigerator. In the Ross River school, the pipes were not spaced properly and hence did not keep the permafrost from melting.

The problem in the north of Canada has become so persistent that the Canadian Standards Association has established guidelines incorporating climate change considerations into the infrastructural design of facilities in its permafrost regions. While many more towns are dealing with melting permafrost than in the past, the idea that a town's existence might be threatened with destruction by loss of permafrost is not new. In 1960, the town of Aklavik (in Canada's Northwest Territories) was built on a mixture of frozen silt and ice permafrost that was 1,000 ft. deep. However, the thin active layer of two to three feet that thawed each summer became more unsettled with each passing year, causing the town to be unofficially nicknamed "the 'mudtropolis' of the arctic."[102] With concerns over not only the melting permafrost, but also of floods and the erosion from the nearby Mackenzie River, the Canadian government investigated new sites to resettle the townspeople. A spot was found 70 miles along the same river and the new town of Inuvik was born. Even though Inuvik is built on permafrost, at the time it was thought to be a "safe spot" as it is at a higher elevation. Inuvik was expected to grow to 5,000 residents; as of 2011, the town had a population of 3,484.

But as the old saying goes, "The funny thing about change is how things stay the same." The so-called "safe spot" is now a "hot spot" of melting permafrost. The planned northern community is a contemporary poster child for the shift from climate change *mitigation* (with an approach to global warming involving calls for wholesale curbing of greenhouse gas emissions) to *adaptation* (with more urgent preparations and possible evacuations to protect people from negative conditions caused by climate change). Pilings must now be driven at least 20 meters (~65 and a half feet) into the ground just to build a new family residence in Inuvik. But soon, even that may not be deep enough to keep

102 "Small Town in Arctic is Sinking in the Slush," *The Milwaukee Journal*, August 11, 1960. Available at https://news.google.com/newspapers?nid=1499&dat=19600811&id=jBQoAAAAIBAJ&sjid=4yUEAAAAIBAJ&pg=7329,415706&hl=en (accessed June 1, 2015), p. 17.

houses from sinking into the mud. As Mayor Denny Rodgers reports, "There are areas in town... that are being washed away."[103]

By 2050, it is expected that about 20% of the world's permafrost will be melted away, at least as predicted by the IPCC in its 2007 Report. Yet, there is still some desire to spend money on infrastructure due, ironically enough, to melting ice. With the polar caps under threat, there are potentially new economic opportunities in the Arctic with the opening of the Northwest Passage. In 2013, a bulk cargo ship passed through the passage for the first time into Baffin Bay, along Greenland's southwestern coast. *The Nordic Orion*, a 740-foot ice-strengthened vessel, was carrying coal from Vancouver, Canada to Finland. All countries with access to the Arctic Ocean are pursuing opportunities for financial gains in the region with shipping lanes opening that could allow new major transportation routes. And then there is the oil. With the possibility of huge oil reserves becoming available in the Arctic with the retreat of the North Pole ice cap, governments are pushing to lay claim to these regions.[104] About a quarter of the world's remaining oil and gas lies in the Arctic, with 13% of it being conventional, undiscovered oil reserves and 30% of it undiscovered natural gas.[105] Large amounts of coal are also sequestered in Arctic soil.

This may be a reason why former Canadian Prime Minister Stephen Harper made an announcement in Inuvik that he envisions the first all-weather road that will link southern Canada to the Arctic coast. The first stage of the road is near completion as a 140 km gravel highway is being built from Inuvik to the north coastal village of Tuktoyaktuk (also known as "Tuk"). The scheduled completion date of the entire project is 2017. This transportation link will connect to the Dempster Highway, which runs through the Yukon. The stated purpose of the road is to gain economic benefits while saving northern Canadian residents hundreds of dollars a year in shipping costs.[106]

Canadians living in the north have sought a road linking them to the southern part of the country for decades, since at least the 1960s. Suitability studies

---

103 Niahm Scallan, "Melting Permafrost Called Ticking Time Bomb," *Toronto Star*, December 1, 2011. Available at http://www.thestar.com/news/canada/2011/12/01/melting_permafrost_called_ticking_time_bomb.html (accessed June 1, 2015).

104 See MacKenzie Funk, *Windfall: The Global Business of Global Warming* (New York: Penguin Group USA, 2014), especially the introduction and chapters 1 and 2.

105 United States Energy Information Administration, "Arctic Oil and Natural Gas Resources," (website), 2012. Available at http://www.eia.gov/todayinenergy/detail.cfm?id=4650 (accessed June 17, 2016).

106 The Canadian Press, "Prime Minister Harper Visits Inuvik," *Global News*, (website), January 8, 2014. Available at http://globalnews.ca/news/1068681/harper-visits-inuvik-to-make-announcement/ (accessed June 2, 2015).

for a route didn't occur until 1998, with territorial officials meeting with the federal government about the plan a few years later. In 2009, the federal economic development agency for northern Canada (CanNor) funded a project plan for the road. In 2012, likely seeing the possibilities of facilitating future oil industry development, Harper pledged $150 million of federal support for the proposed $300 million project. That federal financial commitment grew to $200 million by 2015 (all sums in Canadian dollars).[107]

There are many anticipated benefits of building the new road. The most obvious is reducing shipping costs to citizens in the north of Canada, including saving an average of $1.5 million on transporting supplies, which today must come via expensive air delivery services. There are visions of expanding a fledgling tourist industry, as there could be an annual boost of $2.7 million in tourist income as eco-visitors seek to have easier transportation access to the Arctic Sea. And of course, oil companies, such as Shell Canada, while having their own ice roads already, would embrace an all weather road to decrease their transportation costs.[108]

Needless to say, there are the same kinds of challenges to road building as there are for any structures constructed in the extreme north. Permafrost will have to be preserved to keep the ground underneath the road from sinking. Additionally, engineers must design an alternative to cutting into surface layers and building them back up with fill, which is standard practice in the transportation industry. Instead, they will blanket the whole roadway with a subsurface fabric.[109]

In addition to the current construction of the Inuvik to Tuk road, the Northwest Territories seek to build a separate, longer, and even likely harder to construct 800- kilometer (~500-mile) route from Wrigley to Inuvik to increase the capacity for expanding mineral and fossil fuel extraction. This transportation infrastructure may also reopen the possibility of a Mackenzie natural gas pipeline proposal. The hurdles to completing the road will be even higher when considering that construction crews will have to do the work in the winter to decrease the amount of damage done to the permafrost. This project is estimated to cost about $1.7 billion (in Canadian dollars).[110]

Some interesting moral questions come out of what looks now like economic gains to both northern and southern Canadians. Given the challenges of building and maintaining facilities in the Canadian Arctic, should this

107 The Canadian Press, "Prime Minister Harper."

108 The Canadian Press, "Prime Minister Harper."

109 The Canadian Press, "Prime Minister Harper."

110 The Canadian Press, "Prime Minister Harper."

investment of resources be made (especially with the use of public dollars)? Is this Arctic development strategy self-defeating? Even though it is enticing to harness the resources of the Arctic, does it make sense to build up an infrastructure that may likely sink into the muck if all of the permafrost melts? And won't most or all of it be melted away if we continue to use fossil fuels unabated—the very fuels that will be extracted, burned, and emitted as CO_2 and other greenhouse gases that lead to climate change?

Review Questions

1) The melting of the permafrost in the Arctic is on a long list of negative effects linked to anthropogenic climate change. Which do you think the Canadian government has a stronger obligation to do—protect northern communities from losing permafrost, or to develop their economy by oil, gas, and coal extraction while hoping that technological advances keep infrastructure and buildings in towns like Inuvik intact? As an alternative, what do you think of the possibility of the Canadian government aiding residents with migration out of the Arctic to other areas where they will not have these concerns?
2) Consider your answer to question 1. Would your answer change if you found out, as seems to be the case, that the majority of indigenous people (especially among the younger generation) favor energy exploration? Furthermore, let us say a different administration came to power in Canada and part of its agenda was to advance more environmentally sustainable policies. Do you think it would be morally permissible for this administration to override the local majority and place major restrictions on energy development on the grounds that such activities place northern communities at greater risk with loss of permafrost?

Case 17 – Climate Change May Stunt Your Growth –

Study Questions

1) In what ways might climate change stunt one's growth as shown in the case of West Timor?
2) In the recent history of Timor-Leste (East Timor), what has happened that has made obtaining good climate data difficult?

In the first decade of this century many children under the age of five in Indonesia's West Timor suffered a shared plight—severe malnutrition. A typical infant at 15 months old weighed approximately 24 pounds, but some children in the country weighed even less than half that amount, at 10 lbs. West Timor is the Indonesian part of the island of Timor, which includes 1.5 million residents. A severe drought in 2008 crippled farm production in an area heavily reliant upon agriculture. Until that extended dry spell, it had recently been relatively self-sufficient. According to estimates of the Helen Keller Society (HKS), CARE, and Church World Service (CWS), in excess of 50% of children under age five in West Timor suffered from malnutrition.[111] Of these children, it is believed nearly 1 in 10 endured acute malnutrition, which means they were near death. In some areas of West Timor, the estimated malnutrition rate was as high as 70%, which is greater than in some areas of Africa. 61% of children in West Timor also experienced stunted growth. As noted in the joint study from HKS, CARE, and CWS, we learn, "Stunting is the result of extended periods of inadequate food intake, poor dietary quality, increased morbidity or a combination of these factors. This finding indicates that the diet has been very poor quality for a very long time."[112]

Climate change is being blamed for the drought conditions, bouts of which have been intensifying in both West Timor and some northern areas of Timor-Leste (East Timor) during recent years. It is becoming increasingly difficult for people in this remote region to live off the land as they have for many generations. Aid workers note that it is hard to see this way of life existing for much longer.

In a report written in 2008 by the Stockholm Environment Institute as part of a project to support Timor-Leste's government in integrating climate change considerations into its assistance programs, the drought was a central focus. The report indicates that the main climate hazards in the country (which won its independence in 2000 from Indonesia) are floods, tropical cyclones, and landslides, with possible drought in the north. Timor-Leste has a hot climate with little variation in its temperature from season to season, with an average annual temperature of 24 degrees C (75 degrees Fahrenheit) at sea level, but with cooler temperatures in the mountainous regions. Precipitation increases from north to south, with the south receiving more rain from both the southeast and the northeast Indonesian monsoon seasons. However, El Nino effects

111 Arwa Damon, "Ruthless Drought in West Timor Puts Children in Crisis," *CNN*, (website), July 6, 2008. Available at http://www.cnn.com/2008/WORLD/asiapcf/07/06/westtimor.children/ (accessed May 7, 2015).

112 Damon, "Ruthless Drought in West Timor."

in the region caused the drought-like conditions of 2008, as the 2007 El Nino was a major factor in the loss of 30% of cereal yields in the country.[113]

Trying to pin down long-term meteorological patterns for Timor-Leste is difficult as records are sporadic, especially from the 25-year period of tumultuous Indonesian rule from 1975–2000. Farmers report that they have been experiencing more variability in the weather and the traditional planting cycles no long really fit with the seasons. Mountain communities have also noticed temperature increases. Researchers do have data from the rest of the Indonesian archipelago. These data confirm some of the experience of farmers, as the timing of the rainy season has changed. Since 1990, average temperatures in the archipelago have risen by 0.3 degree C (0.5 F). While there is some indication of a trend towards drying in the southern Indonesian archipelago, it is unclear if this is also happening in Timor-Leste.

Climate projections indicate that temperatures will rise worldwide over the remainder of this century and Timor-Leste will likely experience this phenomenon as well. By 2100, it is estimated that just as with the global average temperature, the temperature in Timor-Leste will increase within a range of 1.5–3.7 degrees C (2.7–6.7 F), with a mean increase across climate models of 2.5 degrees C (4.5 F). Despite a high degree of confidence in the temperature projections, estimates are less clear about the long-term trends in precipitation in both West Timor and Timor-Leste. Part of this is due to uncertainties over the future of both the Australian monsoon seasons and El Nino, which could affect regional rains. However, generally speaking there seems to be a trend toward increased rainfall intensity in the wet season, with longer dry seasons.[114]

There has been little research done to determine long-term social, political, and economic implications from climate change on Timor-Leste, especially if there are no actions taken to lessen greenhouse gas emissions. But again, researchers might have some sense of what will happen there by extrapolating from the climate data taken from bordering Indonesia. Food security will very likely be an issue. The World Food Program estimates that 20% of the population in Timor-Leste is food insecure. An additional 24% of the population is in danger of becoming so. This is mainly due to poverty, political instability, inefficient farming techniques, and poor soil. Certainly, variability in weather patterns will heighten these problems. Internally displaced groups project to be the most vulnerable to these increasingly anticipated food shortages. Any

113 Stockholm Environment Institute, "Climate Overview: Timor Leste (East Timor)," (website), 2008. Available at https://weadapt.org/knowledge-base/small-islands-and-climate-change/timor-leste-east-timor (accessed May 8, 2015).

114 Stockholm Environment Institute, "Climate Overview."

increases in El Nino activities will yield more frequent and longer droughts, particularly in the northern region of the country. This would obviously make food shortages more severe. With Timor-Leste's inefficient public water delivery system, leaks mean loss of water and this will be a more profound problem as water supplies dwindle. Extreme weather events, which may well be more frequent with climate change, would likely increase flooding. This would endanger seasonal crops and bring widespread food shortages to the area. These weather events will also leave mountain slopes unsuitable for growing and at increased risk of mudslides.[115]

In 2013, the Timor-Leste government and the United Nations' Development Program jointly organized a full-day conference on the impact of climate change on the country. The meeting was designed to survey the final draft of the Initial National Communication on Climate Change (INC) Report. H.E. Antonio da Conceição, Minister of Commerce, Industry and Environment, best summed up what the future likely holds for his country in the opening address to the conference, "Climate change, in addition to socioeconomic conditions, increases Timor-Leste's vulnerability to disaster. The INC report provides recommendations for the development and implementation of adaptive and mitigative measures, such as renewable energy, reforestation, to name just a few, to help reduce poverty, ensure sustainable development and contribute to the global effort to reduce GHG emissions."[116]

Mitigation strategies in the environmental context, which are ploys to lessen the adverse effects of climate change, might have some use in Timor-Leste though there will be limits on their effectiveness. After all, Timor-Leste produces relatively low amounts of carbon emissions, about 0.2 tons of carbon dioxide per year and 0.1 kg/GDP. These are negligible when compared to the global average, which is roughly 4.22 tons per capita and 0.75 kg/GDP. This is unsurprising given the country's history of civil unrest and lack of manufacturing. There is at least one sense in which this might be an opportunity—since infrastructure needs to be built for power (especially in rural regions), this is a chance to implement efficient lighting systems. This opens the door to widespread use of renewable forms of energy. On the other hand, since Timor-Leste is already not contributing much to global climate change, this country serves

115 Stockholm Environment Institute, "Climate Overview."

116 "UN Report Says Climate Change Will Strongly Effect Timor-Leste's Development and Growth Challenges and Opportunities," *LDC News Service*, (website), September 16, 2013. Available at http://ldcnews.com/un-report-says-climate-change-will-strongly-effect-timor-lestes-development-and-growth-challenges-and-opportunities/ (accessed June 17, 2016).

as a moral challenge. More affluent countries have much more room to reduce because they emit much more, including "luxury emissions."

Nonetheless, there seem to be some measures the country could take toward climate change mitigation while protecting itself at the same time. Timor-Leste is heavily forested, but is becoming less so. Slash and burn methods of clearing land for agricultural purposes have become increasingly popular over the past few decades. About 41% of Timor-Leste remains forested. Yet, the country has lost 30–50% of its forest cover since the early 1970's.[117] From 1990 to 2010, there was an average deforestation rate of 1.336 percent per year.[118]

Some obvious worries about the loss of forest cover involve losses of a number of ecological services which forests provide that can mitigate the effects of climate change, including protection of soil from erosion. Also, there is less forest to sequester carbon as slash and burn practices continue unabated.[119]

But given its low emissions, Timor-Leste is more focused on climate change adaptation than mitigation. Adaptation involves improving a country's resilience in the face of adverse effects of climate change. Increasing public awareness about the dangers of climate change and devising emergency strategies to escape natural disasters have been substantial components of this plan.

There are obvious questions about how resources will be available to help Timor-Leste pay for mitigation and adaptation strategies.[120] There are calls for more developed countries to pay for the extensive mitigation and adaptation costs of climate change in places such as West Timor (through funding Indonesian climate change mitigation and adaptation initiatives) and Timor-Leste. Justifications for this transfer of resources arise on both practical and moral grounds. With respect to the practical justifications, developed nations have simply more room to cut emissions and the wherewithal to subsidize mitigation and adaptation strategies in poorer countries. Additionally, more affluent developed countries have contributed, and continue to contribute, greater proportions of the GHGs linked to climate change than have Indonesia and Timor-Leste, and thus should pay more. Finally, developed countries have

---

117 Global Environment Facility, "Conserving Timor Leste's Rich Forest Land," (website), Available at https://www.thegef.org/gef/node/11123 (accessed May 11, 2015).

118 Timor-Leste Ministry of Economy and Development, National Biodiversity Working Group, "Timor-Leste's Fourth Annual Report to the UN Convention on Biological Diversity," (White paper), October 2011. Available at https://www.cbd.int/doc/world/tl/tl-nr-04-en.pdf (accessed June 17, 2016), p. ix.

119 See Case 29—"Not Seeing the Forests for the Deserts."

120 This issue also came up in the Maldives. See Case 12—"Climate Change Accords and Vanishing Islands."

derived more benefits from these activities while Indonesia and Timor-Leste have suffered more of the burdens.

Review Question

1) Relative to other nations (particularly those with highly productive capacity) Timor-Leste contributes very little to greenhouse gas emissions. Also, given its political instability and the poverty of the population, there is the question of whether other, more affluent nations, have an obligation to aid Timor-Leste in mitigation strategies. Do you think more affluent countries that emit higher concentrations of greenhouse gas have such an obligation? If not, why not? If so, would this obligation be less for affluent countries that contribute relatively low levels of greenhouse gases to the atmosphere?

CHAPTER 6

The Greening of Institutions

Case 18 – Food Ethics, Obesity, and Voting with Your Fork –

Study Questions

1) What does Pollan mean by saying that "eating is a political act"?
2) What are the major causes of the obesity "epidemic" in America?
3) What are the challenges to small farmers in delivering goods to consumers in the current food distribution network?
4) What is paternalism? How does paternalism appear in the debate about food consumption in the United States?
5) What ethical issues arise from the advertising of food products to children?

In his landmark book, *The Omnivore's Dilemma*,[1] Michael Pollan reminds readers that they can "vote with their forks" as "eating is a political act." Borrowing from an earlier adage from Wendell Berry,[2] these sayings exemplify the choice and thus the power consumers have to sway food policy (or at least to influence food producers). Reformers of how food is produced argue that markets ultimately follow what its participants want. If we demand that our food is fresh, nutritious, and local, then that is what we will get. Our consumption behaviors can drive *what* producers produce and *how* they produce it, even in the face of a regulatory structure that seems to privilege the production of certain types of food. But why should we alter the way food is currently produced?

Some say that one reason is to fight what has been called an "epidemic" in the United States right now—obesity. Obesity has been defined by the National Institutes of Health (NIH) as a body mass index (BMI) of 30 and above. A BMI of 30 is roughly being 30 pounds overweight. One's BMI is his weight in kilograms divided by his height in meters squared.[3] Two-thirds of all Americans

---

1 Michael Pollan, *Omnivore's Dilemma, A Natural History of Four Meals* (New York: Penguin Books, 2006).

2 Berry noted that "eating is an agricultural act." See his "The Pleasures of Eating," in Wendell Berry, *What Are People For?* (New York: North Point Press, 1990), pp. 123–125.

3 National Institutes of Health, "What are Overweight and Obesity?" (website), July 13, 2012. Available at http://www.nhlbi.nih.gov/health/health-topics/topics/obe (accessed July 24, 2015).

 | DOI 10.1163/9789004339996_008

are obese or overweight. Medically speaking, obesity is a primary driver of heart disease. In particular, it renders people insulin resistant over time—this in turn, causes hypertension and diabetes.

As further evidence that obesity is an epidemic in the United States, critics of the current state of food production point to many troubling trends in the weight of Americans over the past several decades. Obesity rates in the US from roughly 1960–1980 were relative moderate with 10.7–12.8% of adult males and 15.7–17.1% of adult females being classified as obese. But in the 1980s, we began seeing a rapid increase in obesity rates up to the 2008 figures of 33% in men and 36% in women. Perhaps the most alarming trend to health care officials is the recent rapid increase in the morbidly obese rate (BMI >= 40).

Nothing in the near future suggests these trends will be reversed, at least if we examine the health status of children. Around 18% of children in America are obese with childhood obesity rates tripling in just one generation. 40% of Texas fourth graders are obese or overweight, leaving the state with the sixth highest childhood obesity rate in the country. This is particularly worrying as children who are obese are much more likely to remain so as adults.

In the landmark Bogalusa Heart Study, 77% of its participants who were obese as children became or remained obese as adults. Other recent studies have shown that about half of females who were obese as teenagers became severely obese as adults. In contrast, in the Bogalusa study only 7% of participants who were not obese as children became obese as adults. Obese children are eight times more likely to develop hypertension. Additionally, over the past 25–30 years the level of physical activity of children has been steadily declining, a key contributor to the overall erosion of the health of American kids.

According to Jack Shonkoff, who serves as Director of Harvard University's Center on the Developing Child, "[t]he evidence is incontrovertible that children who are obese are at greater risk for diabetes, later hypertension, heart disease, asthma, and a variety of other health problems."[4] But astonishingly to the health care community, most Americans don't seem alarmed about this issue. As Francis Collins, Director of the National Institutes of Health adds, "If you were told that 'well your child was at risk for cancer' that would get your attention. If you were told your child were at risk for some brain disease—that would get your attention. Well, obesity should be on that list."[5]

We also cannot escape the wealth and racial disparity element of the obesity problem. The lowest income neighborhoods and towns in the US tend to have the highest obesity rates. Overall, a child born in America in the year 2000

---

4 See *Weight of the Nation*, Dir. Dan Chaykin, HBO documentary, 2012.

5 *Weight of the Nation*.

has a 1-in-3 chance of developing Type 2 diabetes. In African-American and Latino communities, this rate rises to 1-in-2.

Powerful commercial forces profiting from selling unhealthy food exacerbate the issue. Some who worry about the current food model in the US claim that parents need to provide an environment where their children will be protected from the forces of advertisers who constantly entice them to buy something "around every street corner." Dr. Elsie Taveras, a pediatrician and Co-Director of the Obesity Prevention Program at Harvard Medical School, predicts that given the current obesity rate and the risks of cardiometabolic ailments it fosters, this generation will very likely be the first to have a shorter life expectancy than their parents.[6]

The metabolic profiles of obese children are surprisingly similar to those of obese 25-year-old adults. They tend to have elevated insulin, increased risk of becoming diabetic, and disturbed blood lipids. Being obese poses significant negative health effects and parents should be aware that the current crisis doesn't merely concern their children being teased about their physical appearance. To simply focus on the social relations aspects of being overweight, as many parents do, is only the tip of the iceberg.

Physicians are trying to get parents to understand that childhood obesity is a much more significant problem mainly due to adverse health outcomes. Not addressing a child's obesity could mean chronic liver or heart disease for her by her mid-30s. But children do not need to wait until their adult years to suffer the consequences of obesity. Years ago, physicians would never have thought that type-2 diabetes could be found in 10–12 year-olds (consider the fact that type-2 diabetes used to be called "adult onset" diabetes). But now this is not an uncommon occurrence at all. All of this underlines how important it is to stop or at least curb obesity rates in the population at an early age. Yet despite these warning signs, only roughly 10% of American parents seek medical help for their obese children. Texas State Comptroller Pat Coombs warns that, "[O]besity will crush the United States. We will fade into the rearview mirror in oblivion. We could have done something different. We should have done something different. And we lacked the moral fiber and love for our children to do the right thing."[7]

The current school lunch program in the US doesn't help matters much. Consider that from the time children are aged 5–18, they will be served lunch

---

6 American Heart Association, "Overweight in Children," (website), August 2014. Available at http://www.heart.org/HEARTORG/GettingHealthy/Overweight-in-Children_UCM_304054_Article.jsp (accessed July 24, 2015).

7 *Weight of the Nation.*

outside of the direct supervision of their parents during the majority of those days. For many children it is closer to two meals per day. 99% of all US public schools participate in the National School Lunch Program, with 77% of them also providing breakfast. So, for most children, schools provide between 40–50% of their caloric intake per day. 20% of students eat breakfast, 40% of them have a snack, and 90% of them eat lunch at school.

Parental perceptions of school lunch are that it is (by and large) nutritious. In their most recent survey of parental opinions concerning school lunch, the US Department of Agriculture (USDA) found that 88.9% of parents believed school lunch was very to somewhat healthy.[8] However, in reality, the USDA showed that 94% of schools served lunches failing to meet that department's standards for a healthy meal. 80% of the schools served lunches exceeding the USDA recommended allowances for saturated and total fat.[9] A recent analysis revealed that 20% of the rise in the BMI of teens is associated with increased availability of junk food in schools.

Part of the reason for this is financial. Budget constraints in public schools have caused them to make significant cuts in a number of programs, including food service. Public school lunch programs must be self-sustaining, which in turn means that to survive they have to sell a sufficient amount of food. To attract students as customers, they tend to set up cafeterias food court style, with burger and pizza stations. This provides students with choices, but those already accustomed to consuming unhealthy foods are prone to buy the unhealthiest products on offer in the lunchroom. This arrangement also ultimately undermines the school lunch program, as unhealthy foods tend to win greater market share in the cafeteria, which provides incentives for stocking increasing amounts of processed foods.

Also, school lunch programs in America are designed to be as efficient, shuttling students through the cafeteria as quickly as possible. This is necessary, as many schools do not have the facilities to house large numbers of students all at once for meals. This means students need to eat in shifts. They don't have much time to eat and some teachers describe the situation as dismal. At some schools students only have 20–30 minutes to eat during lunch. Some educators think that students having little time to eat and being given low quality food leads to retention issues, behavioral problems, and exhaustion, all of which are not conducive to high quality learning. There is little money for bringing in fresh and nutritious food products to cafeterias. It is clear that given the choice, many school administrators and teachers do not eat lunches provided by

8 *Weight of the Nation.*

9 *Weight of the Nation.*

their schools due to health and weight concerns. In sum, many argue the American school lunch program is a pre-packaged system based on maximal efficiency where food quality receives short shrift. This raises the question of what the true function of a school is and what the role of food is within it.

When Harry Truman signed the National School Lunch Program Act in 1946, he said, "Nothing is more important to our national life than the welfare of our children." There has been some movement on attempts to change food policy with respect to the National School Lunch Program (NSLP) though progress is very slow. In 2012, President Obama signed a law that in effect (and for the first time) gave authority to the Secretary of Agriculture to set nutritional standards for all food sold in American public schools. In 2010, Congress passed the Healthy Hunger-Free Kids Act. The act dictates that the USDA must develop standards of higher nutrition recommended by the Institutes of Medicine (IOM). It also calls for investment of more money into the NSLP. The act, as written, requires doubling the amount of fruit served at breakfast, increases in the quantity of vegetables in lunches, including servings of dark green and orange varieties. It does permit potatoes (including French fries) to be listed as a vegetable, but limits them to being served twice a week. An immediate provision required in the act was that 50% of the grains served are whole grains. This was to be increased to 100% within two years. As written, the act originally called for a reduction in the sodium content of food in public schools by 53% over 10 years. Additionally, it initially mandated a reduction in saturated fat to less than 10% of all the calories served in a week and that any chocolate milk must be fat-free.

However, this law was altered in the final budget bill allowing for not only French fries, but also tomato sauce to be classified as vegetables. It also delayed implementation of the whole grains requirement by two years. This means that pizza can be still be served and meet the new vegetable standards as of this writing. These changes came after the frozen pizza and potato growers lobby pressured legislators to place limits on the new standards. Shortly thereafter, Courtney Rowe of the USDA issued the following statement: "While it is unfortunate that some in Congress chose to bow down to special interests, USDA remains committed to practical, science-based standards for school meals that improve the health of our children."[10] However, by the 2014–2015 school year, the requirement that all foods served to children in public schools need to follow nutritional standards took effect. This means, for instance, via the NSLP a

10 Rob Nixon, "Congress Blocks New Rules on School Lunches," *The New York Times*, November 15, 2011. Available at http://www.nytimes.com/2011/11/16/us/politics/congress-blocks-new-rules-on-school-lunches.html?_r=0 (accessed January 7, 2015).

mandate that all foods must be "whole-grain rich" was implemented.[11] Schools must also serve more fruits and vegetables at breakfast.

But it is unlikely that only institutional policies and programs are to blame for the increasing obesity rate in children, as cultural factors are also likely at play. For example, having televisions in children's bedrooms has been correlated with substantial weight gain and obesity. It used to be thought that the greatest peril for children watching television is that they will become "couch potatoes"—not getting enough outdoor activity and thus becoming heavier. But actually, the bigger worry now is that they are consistently bombarded by ad campaigns compelling them to think their happiness is wrapped up in buying the latest high sugar breakfast cereal or MacDonald's Happy Meal. According to Nielsen data from 2008–2010, preschool children were exposed to 54% more ads for energy drinks over that time span.[12] In 2008, children aged 2–5 saw almost as many ads for 5-Hour Energy Drink as they saw for CapriSun. Teens are the primary target audience for energy drinks, flavored water, and sports drinks. The marketing seems to be working. The energy drink market soared from 2004–2009, increasing in sales by 240% during that timeframe. The American Academy of Pediatrics finds this sort of advertising to be particularly worrisome, as they recommend children and adolescents should never consume sports drinks.

The most recent data available do show some of these trends reversing. In 2013, there was a notable decline in youth exposure to television advertising for sugary and energy drinks overall. Teens were exposed to 30% fewer of these ads when compared with 2010 and other children viewed 39% fewer. Nonetheless, advertising of both classes of beverages to teens and other children is still pervasive. Preschoolers (ages 2–5) saw 144 ads for unhealthy drinks, teens (ages 12–17) 287, while all other children (ages 6–11) 169 in 2013.[13] Recent data also suggest that overall children saw 8% more food-related advertisements on TV in 2013 than they did in 2007, and that food advertising to

---

11 United States Department of Agriculture, Food and Nutritional Services, "Healthier School Day: Tools for Schools Serving Whole-Grain Rich," (website), 2014. Available at http://www.fns.usda.gov/healthierschoolday/tools-schools-whole-grain (accessed November 12, 2014).

12 *Weight of the Nation*. Also see Cathryn Dembek, Jennifer L. Harris, and Marlene Schwartz, "Trends in Television Food Advertising to Young People: 2013 Update," (White paper), Yale Rudd Center, May 2014. Available at http://www.yaleruddcenter.org/resources/upload/docs/what/reports/RuddReport_TVFoodAdvertising_6.14.pdf (accessed January 7, 2015).

13 Rudd Center, "Analysis of Nielsen Data," (website), 2014. Available at http://www.sugarydrinkfacts.org/resources/SugaryDrinkFACTS_Report_Results.pdf (accessed January 7, 2015).

adolescents increased by 25%. The Rudd Center speculates that companies may have increased their efforts to advertise to an older, but still vulnerable population.[14]

Some are concerned that even if children and teens are not consuming energy drinks, sugar-sweetened beverages are enough of a threat to their health. Researchers have found that a fifth of teenagers in the US consume three or more sugar-sweetened beverages a day. This is the equivalent of having an extra meal. There is no respite from sugary drinks in the coffee shop either as a standard-sized Starbucks frappuccino contains 25 teaspoons (or 100 grams) of sugar. Sugar, of course, turns into fat.

Furthermore, food marketing to kids would not be so ubiquitous if it were not so effective. A substantial body of work confirms that heavy advertising translates not only into increased short-term sales, but also to long-term customer loyalty ensuring profits in the long run. It is estimated that $1.5 billion per year is spent in the US food industry on marketing its products to children. The foods marketed most aggressively to children are also the least nutritious. Studies have shown that once children are attracted by certain brands, they are likely to continue purchasing them later in life as brand loyalty is set at a young age.

This has led to a major debate about whether there should be limits placed on advertising to children, and not just on television but also in the medium primed to be the next desired domain for widespread food advertising—the Internet. As far back as 2006, the IOM recommended that advertising of unhealthy food to children should be limited. Currently, food companies determine for themselves what food is labeled "healthy" versus "unhealthy" as the industry has been allowed to self-regulate. Some legislators agree with stricter regulation of advertising to children. But others are skeptical that there is really any advantage to having governmental regulation of this matter, questioning whether health experts really will affect positive change through what is seen as bureaucratic inefficiency. These skeptics also tend to believe the need for curtailing advertising of food products to children is overblown.

In congressional hearings on the matter, the food industry has touted its own voluntary measures for improving the nutritional value of what they sell to children. But most health experts have suspicions about the soundness of such claims. Those arguing for greater regulation accuse companies selling food targeted to children with being disingenuous; of merely wanting to look as if they are reforming to make their food healthier, but in reality not following through. As of this writing, the report to Congress regarding the effects of

14 Dembek et al., "Trends in Television."

advertising on children is still being revised, as the previous recommendations of the IOM have been summarily rejected.

Obviously it isn't just the content of the message from advertisers that vexes those who argue for limiting ad campaigns targeting children; it is the content of the food itself. A major cause of the obesity crisis is how the American diet has changed over the past forty years. The simplest explanation is that we eat far more calories than we used to. A quarter of this caloric increase can be attributed to refined sugars from corn. Another quarter comes from added fats, which mainly comes from soy. The other half comes from refined grains such as cornstarch, wheat, etc. Much of the food sold in the grocery store is processed and ultimately made from corn and soybeans. These products tend to be high in both salt and starch, and correspondingly quite low in unadulterated nutrients.

As we have seen, the largest sources of sugar for children and adolescents are sweet beverages, such as soda, juices, and juice drinks. Researchers have found a high correlation between the consumption of sugar-sweetened beverages and obesity. These foods also tend to have few or no nutritional benefits. Sometimes, this seems rather obvious. Consider soda—for every 12 ounces of it, it has 10 teaspoons of sugar. However, while juices tend to be considered as healthy, close examination of them suggests differently. Some juices may contain the exact amount of sugar as equivalent quantities of soda. 12 ounces of grape juice has 10 teaspoons of sugar and 12 ounces of orange juice has 8 teaspoons of sugar. 48% of the sugar people consume comes from some sort of sugar-sweetened beverage. Of course, in the US "sugar" here generally means "high fructose corn syrup," which is a lower-cost alternative. When soft drink companies agreed to pull soda out of school cafeterias, they may have done so because they already had replacements in mind. They do not only own soda brands, but also juice and sports drink subsidiaries. These products are still permitted in schools.

In contrast, Americans tend not to eat those foods, such as fruits and vegetables, which studies show combat obesity and its attenuating negative health effects. Only one out of ten people in the US eat the recommended daily serving of vegetables and fruits each day, as found by the Centers for Disease Control (CDC).[15] Consumption of fruits and vegetables is associated with lower blood pressure and reduced cancer rates. Also, they tend to be rich in necessary

---

15 Centers for Disease Control and Prevention, "Strategies to Prevent Obesity and Other Chronic Diseases: The CDC Guide to Strategies to Increase the Consumption of Fruits and Vegetables," (Atlanta: U.S. Department of Health and Human Services, 2011). Available at http://www.cdc.gov/obesity/resources/recommendations.html (accessed January 7, 2015).

vitamins and minerals, including potassium. Fruits and vegetables are typically lower in fat content, especially unsaturated fat.

Even when they are available in our local grocery stores, fruits and vegetables have usually been shipped long distances before their arrival. 90% of fruits and vegetables consumed in Iowa (which has a robust agricultural sector) originate from outside of the state. Nationwide, the amount of produce imported from outside of the US has tripled since the 1990s. This can lead to some particularly strange scenarios, such as health care and other institutions in California receiving tons of produce from other countries such as Chile with 7,000 miles travelled. This occurs in a state with weather highly conducive to growing local produce roughly year around.

The vast majority of farms in the US produce monocultured commodity crops such as either soybeans or corn. Over 70% of the arable land in the US is committed to corn, soybeans, and wheat.[16] In comparison, less than 3% of American land is planted with fruits and vegetables.[17] Many farms in the US used to have diversified livestock operations on site growing oats, alfalfa, and hay. But economic incentives to grow monoculture commodity crops have proven too enticing for many farmers to resist.

There is a significant market for commodity crops such as corn. Also, corn farmers are the recipients of direct financial payments from the federal government, with government-backed crop insurance available in case there is a bad harvest year. The USDA spent $14 billion on insurance in case of crop or income loss in 2012, which was seven times the amount for the fiscal year of 2000.[18] Since 1995 up until the latest year for which we have figures (2013), the US spent $292.5 billion on farm subsidies overall, much of it for corn, soybeans, wheat, and rice.[19] In the 2008 Farm Bill, Congress directed the USDA to spend $45 billion on farms, but the lion's share of the funds were funneled to those specializing in producing large-scale commodity crops. With such a structure of incentives, it is not surprising why anyone would produce corn.

However, these incentives do not extend to fruit and vegetable farmers. They received less than 3% of funds allocated in the 2008 Farm Bill. This presents obvious challenges to Americans eating more fruits and vegetables, as such products tend to be more expensive than processed foods. The price of

16 United States Environmental Protection Agency, "Major Crops Grown in the United States," (website), April 11, 2013. Available at http://www.epa.gov/oecaagct/ag101/cropmajor.html (accessed July 24, 2015).

17 *Weight of the Nation.*

18 *Weight of the Nation.*

19 *Weight of the Nation.*

fruits and vegetables has risen precipitously since 1980, with a 117% increase. Comparatively, the cost of soda (which contains substantial levels of high fructose corn syrup as a sweeter) has only increased 20% during the same time frame. According to food economist Eric Finkelstein, the chief reason for this increase in fruit and vegetable costs is that more farmers have followed incentives and now produce corn, soybeans, rice, and wheat. According to Director of the Rudd Center for Food Policy and Obesity at Yale, Kelly Brownell, this heavy production (buoyed by artificial subsidies for commodity crops) drives down the cost of food manufactured from them—usually processed food.[20]

Where does all of the corn go? Only a tiny fraction of it directly feeds people. One reason is the use of corn in the ethanol industry, which in fact is seen as a worrisome competitor with the food market.[21] Most of the small fraction of corn directly feeding people (and some might balk at whether this use "directly" feeds humans) is diverted into sweetener manufacturing such as high fructose corn syrup. Corn is a very productive crop and is extremely versatile. Both of these factors are fueled by huge investments over a long period of time in its research, breeding, and promotion.[22] It also doesn't hurt that corn can be grown almost anywhere in the US. But some have argued that corn as a *crop* has become so dominant in America that it really can be thought of more as a *system* that has little to do with feeding people directly given its use in the production of ethanol, high fructose corn syrup, and as animal feed (where it indirectly feeds humans who eat the meat).[23] Any commodity so ubiquitous (you can get a visual of this just by driving across the corn belt of the US—starting in central Pennsylvania and heading toward western Nebraska) and so versatile can only become dominant.

In contrast, even where there is the will to increase the market share of fruits and vegetables, there is the problem of how to ramp up the scale of production. Most fruit and vegetable farms are small and their owners often deal with a razor's edge profit margin. This makes access to fruits and vegetables difficult at times even for those who can pay for them. There is a double-edged sword here. The prices are too high for people with limited income to purchase fruits and vegetables, but fruit and vegetable farmers could not stay in business

20 *Weight of the Nation.*

21 See Case 30—"Food Vs. Fuel: Unintended Consequences of Corn Ethanol."

22 Jonathan Foley, "It's Time to Rethink America's Corn System," *Scientific America*, March 5, 2013. Available at http://www.scientificamerican.com/article/time-to-rethink-corn/ (accessed July 24, 2015).

23 Jonathan Foley, "It's Time to Rethink."

without charging relatively high prices for their goods. It is hard to know how to expand access under these conditions.

There is currently little interest from state and federal governments in figuring out how fruits, vegetables, and other specialty crop markets can be developed, as only 15% of USDA research funding goes into these agricultural products. Considering that the US spends hundreds of millions of dollars on health care, the argument made by advocates for the expansion of access to fruits and vegetables is that a shift of some of that money to "small agriculture" would have an indelible, positive impact on health. Reallocating money to fruit and vegetable farming would create a level playing field, making its products more cost competitive. This could ultimately lead to a much healthier population with a much smaller investment of taxpayer money.

Beyond the incentives issue, there is the challenge that fruit and vegetable farmers do not possess a strong infrastructure for distributing their products. They don't possess the holding capacity—freezing or refrigeration facilities—to keep stockpiles of produce. They also don't likely have the machinery and labor to facilitate large-scale ramping up of production to meet consumer demand as it increases.

Despite the odds that smaller-scale locavore and organic fruit and vegetable farmers face, there are arguments for why locally grown food is better for consumers in terms of taste and nutrition. These farmers and their advocates believe that if consumers were to have even roughly equal access to locally grown and organic (both in terms of physical availability and cost) versus conventionally grown produce, consumers would choose the former. In conventional farming, taste has been eroded by an industrial food sector demanding easier to ship produce—often genetically modified to have tougher hides to endure miles of transport without spoilage. Compare this to locally sourced or organically produced crops, which advocates argue taste better and provide more nutritional value without genetic modification.

Of course, what we like to eat, other things like to eat as well, whether that means other animals or bacteria. Crops with good flavor and high nutritional value are going to be attractive to a variety of life forms making it hard to sustain the quality and quantity of that product. Thus, locally sourced fruits and vegetables rot and are susceptible to infestation by insects. This demonstrates the desirability of such food, but obviously also the importance of having it locally sourced to maximize taste and nutritional value, but near its consumers. More people would purchase fruits and vegetables if the emphasis for farmers were to prioritize their produce's taste as opposed to its transportability. This bears out through the emergence of the foodie and farm-to-table movements, as a growing segment of consumers are drawn to locally sourced fruits, vegetables, and meats. Admittedly, this is a niche market. The trick is to secure more mainstream market penetration of higher quality foods.

With respect to transportability, there is indeed a disconnect between small-scale fruit and vegetable farmers and the industrial distribution chain. Smaller farms have little access to large distribution networks designed to more easily accommodate larger industrial farms. But some larger distribution companies have become more interested in including locally sourced fruits and vegetables into their supply networks. This is because a growing number of consumers now demand such products. Smaller-scale farmers are increasingly savvy about entering larger distribution lines. From networking at farmers markets, vegetable farmers have learned to pool their resources to contract with aggregators who will collect their products for pickup by larger distribution chains.

The reasons why we should care about the issues raised by food production and distribution are legion. First, there are obviously health concerns. As mentioned earlier, many common adverse health effects are the result of obesity, including type 2 diabetes, high blood pressure, sleep apnea, and chronic fatigue, as well as muscle and joint pain. Many health care practitioners and policy makers alike worry that if we don't take this crisis seriously soon, Americans individually and collectively are going to suffer severe adverse consequences, both physically and economically. The projected health care costs of ignoring the obesity epidemic are staggering. In the case of an individual state, Texas, it is estimated that by 2025, private employers will incur an additional $30 billion per year in health care costs if the current trend of obesity continues. Some argue that an unhealthy workforce whose productivity is eroded will also likely impede the ability for companies and taxpayers alike to pay those bills.

Agricultural reformers in the US contend that the overwhelming costs of poor health on us individually and societally will eventually force a dramatic shift to the production of healthier food. As some evidence that this is already occurring, we can look to the expansion of direct-marketing agricultural models (such as farmers' markets and community supported agriculture)[24] and the farm-to-table movement. There are 7,100 more farmers' markets in the US than there were a decade ago—a 300% increase. Common-sense partnerships between small farmers and health care institutions are being developed, such as hospital-based farmers' markets.[25] This is at least one way to enhance access to farm-fresh vegetables and fruits for both patients and health care personnel, encouraging healthy diets. The farm-to-table movement is not merely the

24 See Case 40—"Community Supported Agriculture."

25 Ironically, the usual fare served in most hospitals is not particularly healthy, as it is heavily processed and high in sugars and fats.

domain of gourmet restaurants. It encourages people to foster arrangements between school lunch programs and local farms to bring in-season vegetables to children in America's schools. Some evidence suggests this model is expanding as well. The number of farm-to-school programs in the US has increased from 5 to 2,500 in the past fifteen years. Other signs that the agricultural system might be shifting is that despite the challenges small farmers face, the USDA reports the number of farmers under the age of 35 has doubled in the past ten years.

Food options have even changed in supermarkets. Some large chains, including Whole Foods and Trader Joe's, have hitched their wagons to the increased popularity of organic and local foods, thus leading these grocers to expand their inventories. Even supermarket chains that mainly sell processed foods are jumping on the bandwagon in response to an increasingly lucrative market in organic and local fruits and vegetables. Ball Food Stores, a chain of supermarkets in the Kansas City area, is one example. As David Ball, its CEO and President explains, "I sell a lot of processed foods in this store. But I also know there's a demand for good, healthy alternatives and I think it's our responsibility as retailers, wholesalers, and manufacturers to provide those alternatives to our customers."[26]

Ball has been working with an alliance of small farmers to source locally grown fruits and vegetables for his 29-store chain, selling them at prices competitive with those of other markets. The leadership of Ball Food Stores saw it was in their chain's interests to supply refrigeration trucks to stow produce from several local farmers to sell in their stores. Again, cooperation between small fruit and vegetable growers seems to be essential for their entry into the mainstream food distribution system. The independent, individual, small vegetable farmer going it alone will never be able to scale up production to a point viable enough to even supply an individual supermarket. So, small farmer groups have sprung up throughout the US using the hub method; first pooling resources to build infrastructure for storage, and then collectively pooling their crops to bring them to market.

Unquestionably, challenges remain. Even if all colleges, universities, schools, and hospitals were to express interest in making locally-sourced food available to their respective communities overnight, currently there would be no way to fully accommodate that need. What appears to be overshadowing all of the efforts of farmers to enter into the distribution system is that even with access, the current number of small farmers can't produce nearly enough to have much of an impact on overall health trends. The reason for this stems from farm policy. There is just not sufficient incentive for many farmers who

26 *Weight of the Nation.*

might otherwise raise fruits and vegetables to do so if they can instead produce high commodity crops heavily privileged by current policy. Yet, farm policy will not change unless there are voices loud enough to influence politicians, something that small-scale farmers have not been able to do thus far. Without "buy in" from several farmers, there will be little chance that healthy food will be easily accessible to many people. There is no way the obesity epidemic will be stemmed without an overhaul of the incentives structure for agriculture. Otherwise, overproduction of corn and soybeans will continue with the use of more processed foods as conduits for its sale and production.

Astoundingly, despite what many have seen as the critical importance of food and the sundry ethical issues surrounding it, historically philosophers have been relatively silent on the subject. Until Peter Singer's sustained arguments for the protection of animal interests chronicled in his groundbreaking work *Animal Liberation* came on the scene in 1975, there had been only sporadic reflections on what we eat and why we should (or shouldn't) eat certain things. It is likely not surprising to philosophers that the ancient Greek thinker Aristotle (384–322 BC) managed to say at least a few things on the topic, as he had something to say on almost every subject. Most of Aristotle's thoughts on food focused on vices of excess, often using food as an example. His virtue-based ethic[27] stressed the moral significance of temperance as a virtue in response to gluttony as a vice of excess with respect to food and beverage. This stance was adopted and exalted by early Christian thinkers and constituted a partial food ethic that has tendrils into our contemporary attitudes toward moderation and prudence.

The French philosopher Jean-Jacque Rousseau (1712–1778) was one of the very few Enlightenment philosophers who mentioned food at all. In the context of social justice, Rousseau focused attention on food scarcity issues and thus food was important insofar as it plays a crucial role in what he called a person's "right to life." Scarcity thus imposed a significant threat to the individual rights of man. Rousseau pointed to inequalities in possession of the basic necessities of life to demonstrate how the modern world, even as it flourishes with technological advances, ironically fails in preserving human dignity.

Though it may seem apocryphal to say, this was basically the extent of thinking about food philosophically except for a few random reflections on the treatment of animals from Kant and Bentham. Immanuel Kant infamously noted that the only reason why beating a farm animal is morally problematic is because such action against domesticated creatures may habituate abuse of those he really thought mattered—rational human beings. Though Bentham's

27 See the "Virtue Ethics" section in Chapter 1 on general moral theories.

assessment of the value of animals took their interests into account (and thus served as a touchstone for fellow utilitarian Singer's account), it was also relatively brief. Bentham noted that the main criterion for the moral interest of any being is not its capacity to reason, but in its ability to experience pleasure or pain—that is, its sentience. Since animals possess sentience, their interests should also count.

However, applying ethical thought to issues involving food consumption and production is now quite common, a fact to which the recent flurry of academic and popular books, congressional hearings, and college courses can attest. In light of this work, further scholarly attention to food ethics is being fostered. Given the vicissitudes of modern culture and what seems in Western countries to be something akin to an obsession with a consumerist lifestyle, some thinkers have reminded us that there are other ways to live "the good life." Consider Kate Soper's critique of the current popular conception of happiness that often invokes a heavy commercial element involving fast food, rapid transport, and generally speaking a veneration of made-to-order services designed to provide instant gratification. She argues such a lifestyle obliges us to work excessively to afford luxuries, but to ironically not really have the time to enjoy living a life filled with all of our gadgets. She proposes an alternative vision of a life of pleasure not so dominated by consuming things, but instead invites us to enjoy simpler pleasures of conversation, friendship, and general enjoyment of each other's company.[28]

Given Soper's analysis, one might link the current food system in America to this consumerist lifestyle. Beyond food policy that clearly advantages certain crops over others, overabundance of some select crops has fit well with a society that exalts quick and easy food that requires little preparation. It is part of a society that sees great value in things that are fast and cheap. The food industry has little incentive to change its modus operandi. Consider this—it would cost the industry about $36–40 billion if all people in the US reduced their food intake by even 100 calories.

Philosophical questions dealing with free will arise with ruminations about food. As one commentator put it, "What degree of free will do people really have if by just knowing someone's zip code you can pretty well predict their BMI?"[29] One could argue that while individual responsibility plays a role in obesity rates, 60–70% of the risk of becoming obese originates from genetic factors, which could surely not be anyone's fault. While there is no single "fat"

28 Kate Soper, Martin Ryle, and Lyn Thomas, *The Politics and Pleasures of Consuming Differently* (London: Palgrave Macmillan, 2008).

29 *Weight of the Nation.*

gene, there are likely several hundred that play their own individual roles, but collectively make some in the population much more likely to become obese. Most of these genes are associated with food intake, and the implication is that if one is placed in an environment where there is constant temptation to eat unhealthy foods, the likelihood of staying healthy is very small. Evidence suggests there is no "nature vs. nurture" in the possibility of becoming obese. More accurately, obesity is in the realm of "nature *and* nurture."

Paternalism also arises in this case study. Paternalism is the policy or practice on the part of people in authority of restricting the freedom and responsibilities of those subordinate to them in the subordinate's supposed best interest. Paternalistic considerations come out when we reflect on questions of what we should eat. To what extent should it be the function of the state to facilitate the health of its citizens? This is clearly a question of political philosophy considering what the role of the state and its regulatory agencies ought to be. Should people choose for themselves what the good life is or should the state play a proactive part in defining the good life for its citizens? What if people find it to be part of the good life to eat Twinkies on demand and prefer to trade off the healthiest food for the cheapest price and the highest convenience? What if given the choice between eating food more likely to make them healthy versus eating processed food that tastes good, but erodes their health, people choose the latter? I may really like easy, cheap, high fat food and would be willing to give up 10–20 years of my life and a healthy existence because, well, "YOLO" (You Only Live Once)! After all, part of the reason why the food system in the US has its current form is due to a reaction to inflation in food prices and worries about food scarcity in the 1970s.

Yet, what of relationships between vulnerable populations and the current state of food? What are the interests of children, versus their parents, versus the state? Those who advocate for the farm-to-table movement want to serve healthier meals to children and prepare them to make healthier choices as adults. Of course, some will see this not as a form of education, but indoctrination. On the other hand, food system reformers argue that paternalism is part and parcel of guiding youth responsibly. Anti-paternalistic arguments might apply readily to adults, who have full rational capacity to make decisions over their own lives. However, children are part of a vulnerable class and by virtue of not having had a chance to yet fully develop choice-making capacity, should not suffer from poor health in the meantime. Accordingly, they should be exposed to healthy food as children so that they can have the experiences necessary to make their own informed choices as adults. (There are few worries they won't be exposed to unhealthy foods!) For vulnerable populations, the state has a viable interest in making decisions for child welfare, even in the

US where parents are given a good deal of latitude in decision-making for their children without interference from the government.

Moreover, there is the existential fact that children are "thrown" into this world. The German philosopher Martin Heidegger (1889–1976) observed that by no choosing of our own, we come into this world (he called this concept "thrownness"). The French philosopher Jean-Paul Sartre added that we are left alone with a radically free will with which to make decisions. Since children are created without their choosing, this fact implies special moral consideration. As Anthony Iton, Senior Vice President of Healthy Communities and The California Endowment puts it,

> The child hasn't created the environment in which he or she is living. The child hasn't grown the foods or marketed the foods, or decided which aspect or the sorts of the food distribution system are going to be active in their community. The child doesn't decide where parks will be situated. The child doesn't decide how much education they're going to be able to get about nutrition and active living in their school system. The obesity crisis is a moral challenge for us as a society. But fundamentally the root of it is the question of what kind of society are we? And what kind of environments do we owe all of our children?[30]

Perhaps this leads to the question of who has responsibility for children while they are at school. If one argues schools take on the fiduciary obligation for the safety and well-being of children during the school day, then the conclusion would seem to be that they are responsible for serving fresh and nutritious food. It seems to follow from this that as laws govern institutions, governments are ultimately responsible for the health and welfare of children in those settings.

But this brings up another issue beyond the specter of paternalism and leads us back to the question of what are the core functions of our educational institutions. How is the education of students to be *balanced* with their health and safety? For example, when there are tight budgetary constraints, how should other competing interests fare when pitted against direct instructional needs? Currently, when school lunch program funding competes with other educational uses of that money, the latter is going to win out.

Even those who are not convinced the food system needs to change due to health concerns for Americans might be persuaded due to its effects on the bottom line. It is estimated that Texas businesses have experienced a $9.3 billion

30 *Weight of the Nation.*

annual increase in their cost of operation because of the poor health of their employees. These businesses expect to incur, by 2030, $32.5 billion in losses. These are real economic costs that can be primarily attributed to obesity and poor nutrition. These costs might increase pressure to do something about obesity, if not by funding fruit and vegetable growth, at least by decreasing incentives for corn and soybeans, thus leveling the market playing field. Right when we might think large-scale agribusiness interests are too great for fruit and vegetable farmers to overcome, we are reminded of the potential market for the latters' products. Consider the market clout of elementary, middle, and high schools when combined with colleges and universities as well as other institutions. These are potential coalitions of consumers who could serve as a new and robust market leading to reform of our current food system.

Review Questions

1) Some have argued the main obstacle to poor people (particularly those in the inner city) eating healthier is that they do not have easy access to inexpensive fruits and vegetables. The urban poor in America essentially live in "food deserts" whereby their neighborhoods are serviced mainly by convenience stores carrying few fruits and vegetables. They also may not be able to afford transportation to go to green grocers in other locations. But food system reformers seem to assume that given the opportunity, poor people would choose to buy healthier foods. This would need to be done by enticing green grocers to the inner city, reducing the cost of fruits and vegetables, or perhaps even less likely, by reducing income inequality. Do you think if fresh fruits and vegetables were on average as cheap as processed foods that poor people would choose to buy them even if accessible? Given your answer, would this say anything about people if they knowingly chose foods that are unhealthy for them despite having the opportunity to select healthier alternatives?
2) Given the choice, do you think school children would be more willing to eat healthier food if it were more easily available at their schools? What should the role of the state be with respect to the diet of children served by public educational institutions?
3) Some argue food advertisers shouldn't target children as consumers for unhealthy food because it is harmful. Also, such marketing is wrongful in that it is deceitful (i.e., it misleads children into thinking that sugary cereals are "part of a nutritious breakfast") and is not protected speech under the First Amendment in the United States Constitution. Do you agree? Why or why not? Should advertising to children be limited in any way? If so, to what extent? If not, why not?

4) On the one hand, getting fruits and vegetables from small farmers to market is challenging due to the modern distribution chain that makes access extremely difficult for small farmers who lack the infrastructure to connect to it. From this, one might reason that if consumers demand such products, the infrastructure will be built. However, even if there were dedicated and substantial interest in locally sourced fruits and vegetables tomorrow, there would be no way to fill demand since not enough of them are produced. What do you think might be a viable solution to what seems like a "catch-22" problem?

Case 19 – Chief Sustainability Officer? –

Study Questions

1) What does a chief sustainability officer (CSO) do in the context of this case study? Why did companies begin to create this executive position?
2) What are the "three stages of sustainability"?
3) What is corporate social responsibility (CSR)?
4) What are some of the worries about corporations having CSOs?

Since the beginning of the 2000s with corporations coming under increased pressure to do their part in protecting the environment, the shape of the corporate hierarchy began to transform. Initially, in response to this increased level of ecological accountability, companies hired a staff of lawyers and spin master advertising executives to keep regulators in check. But since the complexity of what it means to be "sustainable" has evolved, current executives may not specialize in environmental regulation or corporate best practices with respect to ecology. As a result companies in this new business climate began hiring their own "chief sustainability officers" (CSOs), which is an executive-level position within a corporation overseeing its environmental programs and services. Some in the business world still find this position a bit mystifying, but generally speaking CSOs champion and monitor the efforts their companies are making in being green.[31] As George Serafeim, Associate Professor of the Harvard Business School notes, "Companies are monitoring the impact they're having environmentally and on society, and the appointment of a CSO reflects

31 Dina Gerdeman, "What Do Chief Sustainability Officers Do?" *Forbes*, October 8, 2014. Available at http://www.forbes.com/sites/hbsworkingknowledge/2014/10/08/what-do-chief-sustainabilty-officers-do/ (accessed June 2, 2015).

an underlying need for companies to not only monitor but also improve their performance."[32]

While CSOs oversee environmental issues such as their company's impact on water, air quality, and waste management, their purview is more extensive in some corporations. Some CSOs closely review the working conditions of its employees, monitor the status of the supply chain, develop worker safety procedures, and encourage design of products that are both environmentally suitable and profitable to the company.

Since regulators, costumers, and even some employees expect companies to have a position on how they source their products or deal with the disposal of their waste streams, the number of corporations with CSOs has dramatically increased. From 1995–2003, two times as many companies hired fulltime CSOs, and this number doubled yet again by 2008. The first American publically-traded company with an official CSO was DuPont, with Linda Fisher accepting the role in 2004.[33] By 2015, there were three-dozen publically-traded companies with an executive with the CSO title and many others having someone in a similar sort of role.

Serafeim, who co-authored a paper with Kathleen Miller (CEO of Miller Consultants) on what the role of a CSO really entails, detects "three stages of sustainability" with which a company might be engaged.[34] Serafeim and Miller surveyed 66 CSOs with similar duties from 27 different industries in their analysis. The first stage of sustainability is likely not all that surprising since regulators demand that companies keep track of their sustainability efforts, thus *compliance* is of primary importance. Some companies engage in activities that go just barely beyond what is required, such as having a green team or employing recycling programs. But usually this level of engagement doesn't necessitate having a full-time or even formal CSO on board.

The second stage of sustainability is *efficiency*. At this stage, revising corporate policies and practices may involve enough complexity to require the expertise of a CSO. The primary goal in this stage is to save the company money and thus CSOs are charged with finding efficiencies in processes to save water and energy, reduce waste, and lessen carbon emissions. As Serafeim reports, "These things are an easy sell. They're good to do and the obvious thing

32 Gerdeman, "What Do Chief...?".

33 Gerdeman, "What Do Chief...?".

34 See George Serafeim and Kathleen Perkins Miller, "Chief Sustainability Officers: Who They Are and What They Do?" in Rebecca Henderson, Ranjay Gulati, and Michael Tushman, eds. *Leading Sustainable Change: An Organizational Perspective* (New York: Oxford University Press, 2014), ch. 8.

to do."[35] Note here that the CSO still has to make a business case to the CEO—sustainable changes do not only mean what is environmentally friendly, but they have to somehow be acceptable financially or somehow give the company have a better reputation that will concretely profit it in the long run. Miller and Serafeim note that many companies stay in this phase of sustainability.

The third and highest phase of sustainability that few companies reach is the *innovation* stage. At this stage, CSOs seek more thorough integration of sustainability into the core of its business' missions. Sometimes these efforts transform the company. Even in this stage, the impetus for sustainability has much to do with the bottom line. But companies can sometime see that their long-term viability is linked to ameliorating social problems such as climate change, water management, or obesity.[36] Serafeim and Miller found that at the innovation stage of sustainability, the CSO takes control of the sustainability functions of the company in an expanded role by delegating responsibilities to various departments, where these business units become accountable for successful (or unsuccessful) green initiatives. The conclusion of this stage is that companies have an outward perspective to see how their customers and suppliers can operate in a more sustainable way.[37]

Predictably, there is debate about whether CSOs really represent much of a sea change in the way corporations do business, or produce more of a show to try to appease the environmentally conscious. Perhaps companies have departments that deal with sustainability, but their function is *merely to look* as though it is being sustainable. For these companies, without being all that effective at changing their relationship to the environment, or by simply doing the bare minimum through legal compliance, can "check off" that they are giving due ecological diligence and appear to be more "sustainable" than they really are.

Given the possibility that companies may not be as green conscious as they seem to be, do CSOs really matter? Christine Bader poses this very question in her article, "What Do Chief Sustainability Officers Actually Do?"[38] As she points out, corporations that have been linked to large-scale environmental disasters, such as the 2010 Deepwater Horizon oil spill in the Gulf of Mexico and the Rana Plaza factory collapse, had CSOs. How can we know CSOs are

35 Serafeim and Miller, "Chief Sustainability Officers," ch. 8.

36 Gerdeman, "What Do Chief…?".

37 Gerdeman, "What Do Chief…?".

38 *The Atlantic*, May 5, 2015. Available at http://omnifeed.com/article/www.theatlantic.com/business/archive/2015/05/what-do-chief-sustainability-officers-actually-do/392315/ (accessed June 3, 2015).

driving genuine change in their own companies and whether they are changing business culture overall in regard to the environment?

Baden thinks one way to do this is by looking at how money is spent on sustainability issues. She points out that some CSOs have substantial budgets to implement numerous programs geared toward company environmental stewardship. For example, MGM Resorts CSO Cindy Ortega is leading a $30 million project to install more energy efficient lighting in all 17 MGM resorts. Another case Baden cites involves the executive vice president of sustainability and external relations, Elaine Dorward-King, who is a member of a four-person committee at Newport Mining, including the Chief Executive Officer, the Chief Financial Officer, and the head of exploration. Collectively, they decide both how the company allocates its financial resources and give the final approval of all major company projects going forward.[39] CSOs in this position have a good deal of clout. Yet, most CSOs do not have substantial budgets and much of their work is behind the scenes. Consider the software company EMC. All of its software and hardware products must undergo an energy efficiency analysis before reaching the market. However, this is simply part of the regular operations at the company and standard for the industry. Thus, it is not a function that is really "owned" by the sustainability department and would be done whether a CSO was on board or not.

One could also look for evidence of the effectiveness of CSOs by discerning if they help guide the long-term direction of the company. CSOs are commonly asked about short-term initiatives (e.g., recycling). Yet, a company serious about sustainability also expects their CSO to have a voice on what the risks, trends, and opportunities the company should consider over the long-term. This is a sign that a CSO is genuinely a driving force in a corporation. If CSOs are queried about such things, it is likely that they will be will be asked to closely investigate how the company's products are sourced, or how to design products in an environmentally friendly way. Owens Corning CSO Frank O/Brien-Bernini gives the following example: "[D]eveloping a strategic view on whether building codes in India, or China, or Brazil, mimic what's gone on in Western Europe and the United States, helps to frame our opportunities in those regions. Can we influence that trajectory? Where can we lead change, and where should we be responsive"?[40]

No matter how much influence CSOs actually have on corporate culture, another interesting trend emerging is who is hired for CSO positions. In a study of 36 CSOs (most with that title, but some who don't yet play an equivalent

39 Bader, "What Do Chief Sustainability Officers Actually…?".

40 Bader, "What Do Chief Sustainability Officers Actually…?".

role in their companies) at top organizations conducted by Ellen Weinreb, she found that 42% of them are women.[41] While the majority of CSOs are still male, she discovered that the number of female CSOs had risen dramatically in just three years, when only 28% of CSOs were female. This constitutes an increase in female CSOs of 52%. There is the question of whether, as the position is evolving and becoming more widespread in the business community, it is becoming seen as more of a women's role. As Weinreb puts the point, "I'm often struck by these photos of the leadership team where you see all men [ex]cept for the one woman—and she comes from Human Resources. There are some "C-level" functions that are seen as more feminine; human resources, for example. I wouldn't want the CSO to take on a gender stereotype."[42]

A number of interesting moral issues arise from this case study, especially if we see the CSO position arising out of the movement calling for *corporate social responsibility* (CSR). CSR is a term used in the business ethics literature referring to a management concept in which companies are expected to integrate social and environmental concerns into the business model and in their interactions with stakeholders.[43] One issue alluded to earlier was whether the move by companies to have CSOs is ultimately deceptive. Are they merely using the existence of the position to make them look as if they are really committed to sustainability, while not really changing the fundamental way in which the company does business? The CSO of a major oil company can say drilling is being made safer by using one platform with multiple wells branching out from it, instead of utilizing multiple vertical wells that expand the developmental footprint of the extraction facility. But, there seems to be a fundamental risk in drilling in general that is not being addressed. With a CSO as its public face, the company can claim an attempt at good environmental stewardship even while contributing heavily to greenhouse gas emissions by supplying fossil fuels to consumers. The company can win the public relations battle without really changing environmentally irresponsible behaviors.

Those who genuinely care about CSR want ethical values to become imbued into the fabric and culture of companies. But one can imagine a cynical take

---

41 Ellen Weinreb, "CSO Back Story: The Evolution of the Chief Sustainability Officer," (White paper), 2014. Available at http://weinrebgroup.com/wp-content/uploads/2014/10/CSO-Back-Story-II.pdf (accessed June 3, 2015).

42 Quoted in Joel Makower, "Ellen Weinreb Checks the Pulse of the CSO Profession," *GreenBiz*, (website), October 21, 2014. Available at http://www.greenbiz.com/blog/2014/10/21/ellen-weinreb-checks-pulse-cso-profession (accessed June 3, 2015).

43 United Nations Industrial Development Organization, "What is CSR?" (website), 2015. Available at http://www.unido.org/en/what-we-do/trade/csr/what-is-csr.html (accessed June 5, 2015).

on this point if we consider how CSR came to the business world. It seems like there was a critical *outside* perspective that called for companies to become more socially and environmentally responsible because they were shirking that responsibility. It was only under the duress of public sentiment and the demands of regulators that caused companies to create divisions of sustainability in the first place. The question is, can the culture of corporations really be changed in this way? Can those who purport to care about the environment expect that companies will "see the light" and care as well?

How might advocates of CSR respond without merely admitting that compliance is the fundamental point and ultimately businesses don't need to like supporting CSR measures as long as they execute them? Is it possible that they might take a clue from Aristotle? Aristotle was known as a virtue ethicist[44] who thought we only become morally excellent people when we constantly practice the virtues; we become virtuous people by acting in virtuous ways. Over time, one learns to be brave by doing what brave people do and by trying to hold the same set of dispositions and attitudes that they do. One goes through the process, even if for some time she doesn't really feel like displaying the virtue. She may not even understand why some virtues are considered to be virtuous. But slowly, yet surely, she becomes a courageous person by acting in the ways that the courageous person does. If one goes through actions and practices while having the correct reasons for doing the right thing, she will actually *become* a morally good character. Correspondingly, one might expect the reason why a parent incessantly asks a child to take care of her dishes after she finishes eating, to say "please" when making a request, and to say "thanks" when someone honors said request, is to habituate the practice. Not only does the behavior become a habit, genuine senses of the virtues of responsibility, gratitude, and thankfulness become part and parcel of the child's character as she grows into adulthood. In adulthood, she can effortlessly, without having to be told, without having to be persuaded or "bribed," or even without thinking, *be* honest, sincere, compassionate, and humble.

In applying this to the case of trying to instill corporate social responsibility, perhaps a similar sort of phenomenon happens. By habituating the process of incorporating green values into corporate practices, business people simply get used to having them and don't have to think about employing earth-friendly strategies—they just do. Yet, adopting new societal beliefs and practices takes a long time, likely with many missteps. It will be interesting to see what happens in the case of corporate sustainability measures and what place they ultimately take in business culture.

---

44 For more details, see the section on "Virtue Ethics" in the Chapter 1 on general moral theories.

Review Questions

1) A folk definition of ethics is that it is what you do (morally speaking) when no one is looking. Moreover, some have said the law alone will not be enough to develop the good society. Constant monitoring of human action is very difficult (if not impossible) and the threat of punishment is not always a successful deterrent from doing worng. Hence, members of a society need *internal* constraints on their actions—a motivation to be good even when they might not be caught and/or punished for morally inappropriate activities. How does this apply to corporations? Can we expect that corporations will not only police themselves, but also adopt a code of ethics going beyond legal requirements? Or is this effort merely a way for companies to compete in the "greening of the market"?
2) Should companies have CSOs? If so, why? What do you think the function of a CSO should be? If you don't think companies should have CSOs, why don't you? Do you have an alternative companies should use to deal with environmental issues? Ultimately, what do you think a company's role should be with respect to the environment?
3) Consider the initiative to install more energy efficient lights in MGM resorts. One might point out that changing the lighting is a paltry effort at being green given the existence of a large-scale luxury complex. Analogously, environmentalists commonly target golf courses, rejecting claims that they are made "green" simply by using some water conservation methods. Do you think the MGM case shows CSOs are effective? Is it too much to expect CSOs to change the core business practices of a corporation?
4) Notice the trend in the recent hiring of CSOs (or those in roughly equivalent roles) toward recruiting women. What are we to think of that? Should anyone be concerned about this trend?

Case 20 – The Demise (and Rebirth) of the Electric Car –

Study Questions

1) In what ways have state policies encouraged the development of zero emission vehicles? What factors in the automotive industry have pushed back against such policies?
2) Discuss the early history of electric vehicles and how they fared against gas-powered autos. After resurgence in the early 1990s, what caused the

decline of electric car development? What has sparked a more recent revival in the manufacturing of electric cars?

Who Killed the Electric Car? is a 2006 documentary chronicling the auto industry's creation of electric vehicles only to destroy them. In 1990, California passed the Zero Emission Vehicle (ZEV) Mandate. This law required that by 1998, two percent of all motor vehicles in the state must produce no emissions, increasing to ten percent of all vehicles by 2003. This meant that for a car manufacturer to sell internal combustion engine vehicles in California, it also had to sell some electric models to meet the mandate. In response to these regulations, major car manufacturers rolled out a few all-electric models (including General Motor's EV1, Chevrolet's S-10 EV, Nissan's Ultra EV, Ford's Ranger pickup EV, Toyota's RAV 4 EV, and Honda's EV Plus).

There are a number of benefits to driving electric vehicles. Engineers have noted that operating an electric car likely costs only a fourth of what it does a conventional gas vehicle. On balance, electric vehicles produce less of a carbon footprint than their gas-powered brethren. But this has a great deal to do with their energy efficiency. While electric vehicles themselves do not emit carbon dioxide, this doesn't mean they are emissions free. This is because over 80% of electricity in the United States is produced using coal, which emits a significant amount of CO2. So, one of the challenges electric vehicle manufacturers have to grapple with is finding a renewable and non-carbon emitting source to generate electricity, such as wind power.

While it may be easy to believe that the more recent turn to the electric car is unprecedented, a bit of history of the origin of motor vehicles will likely disabuse us of that notion. At the end of the 19th century when the automobile began to replace the horse and buggy as our primary means of transportation, the first practical electric vehicles were developed (mainly in Britain and France). At that time, they were locked in a head-to-head battle with gas-powered autos for market share.[45] Electric motor vehicles actually enjoyed an early advantage as they were thought to be quieter, easier to drive, and didn't emit the smelly exhaust fumes gas-powered machines did. The five-speed electric cars could go about 25 miles per hour, which made them ideal for city use. They became rather popular in urban areas, especially with women.

But with time, gas-powered cars won out because they could be fueled much more quickly and had more horsepower than electric cars. In 1908, Henry Ford's Model T became popular, dominating the market to the point where

45 Rebecca Matulka, "History of the Electric Car," *Energy.Gov*, (website), September 15, 2014. Available at http://energy.gov/articles/history-electric-car (accessed June 4, 2015).

we wouldn't hear from the electric car alternative for decades. His vehicle was affordable and widely available. By 1912, the Model T only cost $650, making it much more economical than electric vehicles that fetched some $1,750—a comparatively exorbitant sum at the time.[46] With gas-powered car domination of the market, automakers showed little interest in developing electric models. Gas was relatively cheap and the infrastructure for fueling cars spread quickly making petroleum widely available. Still, even Ford himself was interested in designing and producing electric vehicles for some time. He was friends with Thomas Edison, who was famous for his work with electricity and the invention of the incandescent light bulb. In 1914, they even discussed the possibility of co-developing an inexpensive electric car, but the plans never came to fruition.[47]

The first glimmer of interest in electric cars returned during the OPEC fuel embargo of the 1970s, when oil shortages caused long lines at the fuel pumps in the US as well as skyrocketing prices for gas. But for many preceding decades, there had been little change in the technology of electric batteries as there was no interest in financing their development. However, the crisis did spur interest in turning to some alternative energy sources to power automobiles. To exert more domestic control of fuel for cars by the enhancement of energy alternatives, Congress passed the Electric and Hybrid Vehicle Research, Development, and Demonstration Act in 1976. As the name of the Act suggests, it authorized the US Energy Department to support research and development of electric and hybrid vehicles.[48] Some car manufacturers toyed with developing the electric car in the 70s, with General Motors building a prototype for an urban electric vehicle in 1973. American Motor Company manufactured a series of electric jeeps that were part of a pilot program in 1975 for the United States Postal Service. NASA also helped raise the profile of electric vehicles in 1971 when its moon rover became the first manned vehicle to drive on the moon.[49]

However, electric carmakers couldn't take advantage of this renewed enthusiasm in America for electric cars, as battery technology was still underdeveloped. Electric cars of the 70s were plagued by a number of critical problems, such as not being able to operate at highway-grade speeds—they could only reach 45 mph. Their average range of travel without having to be recharged at that time was only about 40 miles. In the light of what petroleum powered

46 Matulka, "History of the Electric Car."

47 Matulka, "History of the Electric Car."

48 Matulka, "History of the Electric Car."

49 Matulka, "History of the Electric Car."

autos could achieve, electric cars couldn't gain much traction in the automotive market.

More recently interest in electric cars has had less to do with high gas prices and possible fuel shortages, and more to do with environmental concerns. As a firmer link has been established between climate change and its anthropogenic causes, such as the burning of fossil fuels in internal combustion engines, alternative fuel sources are sought. This is interesting, because the usual line is that no change will come in a market dominated by a type of product until there is an economic need, but this didn't seem to be the primary driver towards electric vehicles in this case. As mentioned earlier, new environmental policies in California (which has a huge consumer base for automobiles) as well as new federal laws, with the passage of the 1990 Clean Air Act Amendment and the Energy Policy Act of 1992, took effect. These measures imposed new transportation emissions regulations and interest in electric cars arose yet again.

Some manufacturers simply modified popular existing models. One consequence was that electric engines were built to be more powerful and could now achieve highway-grade speeds (or at least something close to it) and could also travel up 60 miles without recharging.[50] This offered electric vehicles a place in the showrooms of car dealerships as viable competitors with internal combustion engine cars.

The EV1 received the most press and was featured in *Who Killed the Electric Car?* But it was also unique as it was built from the ground up and not merely modified from an existing model. It was popular with its owners as it performed well for an electric car at the time and fared favorably versus gas-powered vehicles, with a range of 80 miles. The EV1 could also accelerate from 0 to 50 mph in only seven seconds. Despite having a loyal consumer base, the EV1 was discontinued in 2001 because General Motors executives claimed the vehicle's high production costs made it economically infeasible.

Of course, behind the scenes the automotive industry is a very powerful lobby, which advocated for rescinding, or at least scaling back, federal regulations on emissions. These efforts, along with lowering oil prices in the later 1990s, also led to a decline in the appeal of producing electric cars. But what was important according to advocates of electric vehicles was that even out of the spotlight of the press, scientists and engineers have continued their work on electric vehicle technology, including batteries in research supported by the US Energy Department and other governmental agencies.[51]

---

50 Matulka, "History of the Electric Car."

51 Matulka, "History of the Electric Car."

This leads us to the 21st century when once again there has been a call for automobiles running on alternative fuels, along with increased efficiencies—environmentalism was part of this charge. More active and well-funded players emerged in the automotive market leading to another wave of enthusiasm for electric vehicles. This has once again thrown down the gauntlet to large auto manufacturers to develop electric cars. But there are certainly hurdles to clear. Car owners cherish fuel efficiency without a major sacrifice in horsepower, as well as easily accessible refueling. Some think hybrids bring a number of those traits to the table. The Honda Insight's launch in 1999 introduced the first hybrid on the market in nearly a century. But a car that took the industry by storm and has actually become a viable player in the automotive market, the Toyota Prius, followed up the Insight.

There are basically four types of electric-powered vehicles available, including plug-in electrics, plug-in hybrids, hybrids, and hydrogen fuel cell cars. The market for electric cars today is still small, but it is growing. Over 3% of all new cars sold are electric vehicles and this could grow to 7% by 2020. Today, most major automakers are either building or planning to build some sort of electric vehicle. The company that has thrown all of its chips into the plug-in electric vehicle card game is Tesla, based in Palo Alto, California. But recently, it has been rumored that Apple will also likely enter into the electric plug-in fray, as it has been luring top engineers away from Tesla with signing bonuses thought to be around a quarter of a million dollars. Apple seems to be seeking out experts to build its own electric batteries division. While Apple is currently being sued in Massachusetts by electric car battery maker A123 Systems for illegally poaching talent away from other car manufacturers, there is a history here as Tesla hired away 150 Apple employees as the automaker was getting established.[52]

Even with the success of the Toyota Prius and other hybrid cars, they are beyond the means of most consumers. Prius owners have an advantage over owners of plug-in vehicles in that they don't have to hunt for special electric charge stations or have to wait long to fuel them. But Toyota still faces a challenge. Those driving a Prius may enjoy a (smug?) satisfaction of being an environmentally conscious consumer who drives a "smart and successful auto." Yet, this makes the vehicle seem more of a boutique item with a large cult following than a genuine competitor in the mainstream auto market.

---

52 Deepa Seetharaman and Edwin Clark, "Apple Poaching Auto Engineers to Build Battery Division: Lawsuit," *Reuters*, February 19, 2015. Available at http://www.reuters.com/article/2015/02/19/us-apple-autos-lawsuit-idUSKBN0LN04Y20150219 (accessed June 4. 2015).

Interesting moral issues arise out of this case. Owners of the EV1 who were forced to return their cars were often quite distraught doing so. They couldn't understand how a well-functioning vehicle that met environmental goals was controversial or why the auto industry was lobbying against regulations lending to electric vehicle development. After all, promoting electric vehicles would likely develop a viable long-term market for General Motors. It also afforded them a chance to innovate and become the leader in a profitable new market. EV1 Advocates were despaired when they realized where their cars were going—to be scrapped. In a car-loving society such as the US where people can develop a "bond" with their car, this end was disheartening to former EV1 owners. They used normative language in their complaints—"the auto industry shouldn't be doing this." Many EV1 owners felt betrayed by GM as they couldn't fathom how a company that had developed a highly desirable product could simply destroy it.

For some environmentalists, nothing short of an electric car will do. They doubt that even if hybrids dominated the market this would limit greenhouse gas emissions sufficiently. There is also the worry of the "curse of the half measure." A widespread acceptance of hybrids might again stall development of the more carbon-neutral solution—electric vehicles.

Review Questions

1) Representatives of "Big Auto" might claim the car industry shouldn't be regulated by governmental agencies with respect to carbon emissions—arguing that the cost of producing such vehicles will decrease profits and ultimately be passed on to consumers. They might add that free market solutions to any sort of greenhouse problem are superior. Citing free market economists such as Friedrich Hayek and Milton Friedman, they might argue that if enough consumers demand electric vehicles, entrepreneurs in the auto industry will develop them. Do you agree with all or any part of this line of argument? What potential fallacies (if any) might lurk in such reasoning?
2) Some complain the electric car is just another example of an "elitist" green product priced out of the buying capacity of the average consumer. Assuming that you would experience cost savings on fuel by driving an electric car and that it would emit much less carbon than conventional vehicles, which vehicle would you purchase? If you would purchase the electric vehicle, would your primary reason for doing so be due to having less of a negative impact on the environment, or for some other reason?

Do you think people have a moral obligation to buy an electronic car if it can be shown to lower emissions?

3) Some oppose electric vehicles on the grounds that they do not really reduce emissions given that dirty fuels such as coal currently produce most electricity. Do you think this is a justifiable reason to resist development of electric cars? If so, why? If not, why not?

Case 21 – A Natural Internment –

Study Questions

1) What makes a "green burial" green?
2) Why are green burials becoming more common? What features of traditional burial practices do some people find objectionable?

The Natural Burial Company of Portland, Oregon, has found a truly niche market in the already niche green market. There has been an increasing interest in "green" burials and the company's Ecopod, a biodegradable coffin made entirely out of recycled paper, was one of the first to fill this consumer preference. It is designed to be planted in the ground and gradually dissolves to nourish the soil with its nutrient-rich contents. Other biodegradable caskets are made out of formaldehyde-free plywood, hand-woven Somerset willow and fair-trade bamboo—they also come in a number of styles and sizes.[53] Some people forgo the casket entirely, electing to be buried in a cloth shroud with a board underneath.

The trend toward alternative burial practices is a response to modern cemeteries that resemble parking lots—with lots of land being appropriated by concrete vaults, caskets made of steel, and residual embalming fluid left over from the preservation process. Cynthia Beal, the founder of the Natural Burial Company, envisions that the products she designs will lead to the establishment of more natural burial parks.

There are some obstacles to green burial, such as legal and public health issues. Human remains hold toxins and simply "putting them out back" to "naturally" decompose can be a bit of misnomer. Not only can dead bodies be vehicles for dangerous pathogens, if bodily fluids leech into the earth they can cause problems for water supplies. Other obstacles are social and natural

53 April Dembosky, "Biodegradable Coffins," *The New York Times Magazine*, December 9, 2007, p. 57.

burial can be seen as unconventional. As some advocates of the green burial movement recognize, neighbors do not always appreciate having people buried next door. As Brandy Gallagher explained, "If you thought that there were going to be NIMBYs ("Not In My Back Yard") with an eco-village or a green building project... imagine if you know that your Aunt Sarah is going to be buried in your back yard—what would the neighbors say!"[54]

But traditional burial procedures involve the introduction of toxins into soils. For those concerned with environmental degradation this has driven alternative approaches to internment. In conventional burials, embalming fluids can contaminate ground and water. In fact, as much as 3–6 gallons of formalin may be used in a body during its preparation for burial. Embalming fluid is routinely used if there is a public showing of the deceased. Yet contrary to popular belief, its use is not required by law if there is an immediate burial. Also, formalin doesn't prevent the spread of diseases. In fact, outside of the United States embalming fluid is not commonly used. The materials used in caskets, beyond the steel already mentioned can include materials such as toxic adhesives that can leech into the soil. About 75% of the two million caskets made in the US each year are constructed out of metal sprayed with toxic finishes. Hardwood and metal caskets decompose slowly along with the remains within them. Concrete vaults are also commonly used in conventional burials, but the manufacturing process of concrete releases carbon dioxide into the atmosphere, contributing to climate change. The use of these materials may be part of the reason why casket manufacturers generally rank in the top 50 of the US Environmental Protection Agency's list of hazardous waste producers. Finally, consider what conventional cemeteries look like. They are pristine looking, well-manicured, lush green gardens with many headstones. But to maintain them requires heavy use of water, weed killers, and pesticides, not to mention frequent mowing.

We accumulate many toxins over time (bioaccumulation) through the animals we eat (mercury) and the pills we take (pharmaceuticals). Pharmaceutical components are a fast emerging problem in the waste management industry and the public health sector as treatment plants can't remove them from water systems. However, cremation may still work in these circumstances to remove or mitigate some toxins via burning (so cremation may still be a solution). But then again, if part of the worry of the mainstream approach to disposing of remains are the carbon footprints the deceased leave behind, it takes a good deal of energy to heat the kiln sufficiently to convert flesh and bone to dust.

54 See "Interview with Brandy Gallagher," *YouTube* (Online video), 2014. Available at https://www.youtube.com/watch?v=HcyIWpAhv6s (accessed June 21, 2016).

Along with being energy-intensive, cremation also contributes to air pollution. Roughly 330 pounds of CO2 is emitted on average from the burning of a single body, whereas only about 86 pounds are emitted in burial. This is not to mention that burning bodies with dental fillings can emit 200 milligrams of mercury into the air with every cubic foot of exhaust. In Britain, about 20% of the country's mercury emissions can be linked to cremations.[55]

Proponents of natural burials claim they have another virtue—they are relatively cheap. The primary reason why people elect cremation is on financial grounds. While prices vary, a traditional burial can cost around $7,000. Comparatively, green burials can cost as little as a few thousand dollars, with internment of cremations costing even less.[56]

Natural burials are popular in Canada, but sites are also found in several US states, including (as of 2016) California, Colorado, Connecticut, Florida, Georgia, Hawaii, Indiana, Illinois, Maine, Massachusetts, Michigan, Minnesota, New Jersey, New York, North Carolina, Ohio, Oklahoma, Oregon, Pennsylvania, South Carolina, South Dakota, Tennessee, Texas, Utah, Washington, Wisconsin, and the District of Columbia. According to the Funeral Consumers Alliance, only seven states require that a funeral director oversee a burial, including Connecticut, Indiana, Louisiana, Michigan, Nevada, New York, and Utah. Green burials have been common in the United Kingdom for many years to help alleviate land shortages for cemeteries.

Green burials can be done in a variety of ways. One might want a traditional funeral, but would rather have a biodegradable casket. Others may wish for no use of formaldehyde and to be plotted in a green burial center while picking out a specific type of tree and stone as their memorial markers. Some funeral homes offer both traditional and green burial options. Frye Funeral Home in Tipton, Iowa, provides green burial caskets, burial shrouds and blankets, and memorial markers made to be "suitable for green burial sites."[57] While they do not have embalming free burial options, they do have non-formaldehyde embalming fluids.

---

55 See "Exit Strategies." *Economist* (2010) 396 (8700): 74–5. Available at http://ezproxy.lvc.edu:2048/login?url=http://search.ebscohost.com/login.aspx?direct=true&db=aph&AN=53783708&site=ehost-live (accessed June 7, 2015).

56 Cheryl Corley, "Burials and Cemeteries Go Green," *National Public Radio* (Radio transcript), December, 16, 2007. Available at http://www.npr.org/templates/story/story.php?storyId=17232879 (accessed July 22, 2016).

57 Fry Funeral Home, "Green & Natural Burials," (website), 2011. Available at http://www.fryfuneralhome.com/services/green-burial/ (accessed June 4, 2015).

Some who choose green burials have shared moral values underlying their choice. Integrity means having one's remains disposed of in a manner that is meaningful to him; leaving the Earth is as environmentally friendly way as possible. Another value is simplicity; that in the light of a consumerist lifestyle involving a sometimes harrowing and expensive labyrinth of environmentally taxing materials and practices, green burials are seen as straightforward, economical, and sustainable. Others choose green burials in part as a protest to what they see is an overly commercialized American funeral industry that rakes in $21 billion per year. Some think funeral parlors have too much control over an important ritual—how to leave this world. One complaint is that the funeral industry is unethical; generating huge profits from families of the deceased, who are grieving and vulnerable. Also, some argue that individuals should have more control over how their bodies are disposed of—perhaps wishing to be buried on their own property.

Beyond embalming techniques and caskets with finishes that leave contaminants in the soil, modern cemeteries are criticized for their inefficient use of land. The natural burial model is thought by its advocates to preserve open space and incorporate cultural landscapes into cities and regions.[58] Green burial could also be seen as a consciousness raising exercise—by using burial practices that are more ecologically sensitive we can raise awareness of environmental issues and also see death as a natural process.[59]

With the baby boomer generation aging, there will be a huge increase in the number of plots needed in cemeteries already running out of space. Simply building more graveyards won't solve the excessive demand for space to bury America's dead. One looming problem is where are you going to build them? As urban planning professor Chris Coutts observes, "In the US context we thought that space was infinite, but there are limits to the frontier."[60] Coutts has teamed with Carlton Basmajian to study future trends in land use as they relate to the funeral industry in the US. Seventy-six million Americans are estimated to reach the average age of life expectancy of 78 between 2024 and 2042. Assuming that they will someday occupy standard-sized burial plots, this would require roughly 130 square miles of grave space—and this doesn't

58 Alexandra Harker, "Landscapes of the Dead: An Argument for Conservation Burial," *Berkeley Planning Journal* (2012) 25(1): 150–159.

59 Harker, "Landscapes of the Dead."

60 Quoted in Amy Biegelsen, "America's Looming Burial Crisis," *The Atlantic*, October 31, 2012. Available at http://www.citylab.com/housing/2012/10/americas-looming-burial-crisis/3752/ (accessed June 5, 2015).

even count trees, pathways, and roads. This would be an area about the size of Las Vegas.[61]

Ultimately, we are left with moral questions about individual autonomy and what obligations are owed to the living, the dead, and the yet unborn. Do we owe the same degree of obligation to the dead as we do to the living? Do we owe the same degree of obligation to the dead that we do to future generations? Do we owe it to others who remain when we pass to leave a cleaner environment? Do we owe that and more to the yet unborn? How do we begin to resolve conflicts in these obligations when they arise?

Review Questions

1) One might argue that humans have obligations (whether to other living humans, future generations of humans, animals, or the environment) while they are living. But usually we don't think of these duties lasting after our deaths. Do you think we have moral obligations to take environmental protection into account as we make arrangements for how our remains are to be interred?
2) Imagine there is a conflict between family members over how a relative's body is to be buried. How should a decision be adjudicated if some family members insist that the remains of their loved one be placed in a more conventional steel casket designed to preserve the body as long as possible, while others insist on a natural burial?
3) Even assuming one is environmentally conscious, do you think the Ecopod is the best device for leaving the smallest impact on the earth? Are there other alternatives such as cremation that are more environmentally friendly?

Case 22 – Green Building Designs –

Study Questions

1) What does LEED certification mean? What factors are considered when judging whether a building should be LEED certified?
2) What are some "green" building principles?

61 Biegelsen, "America's Looming Burial Crisis."

Officials at the Aldo Leopold Nature Center (ALNC) based in Monona, Wisconsin—an organization devoted to preserving and enhancing the legacy of its namesake, the famous naturalist—wanted to expand the Center, but in a way reflecting green design principles consistent with Leopold's "land ethic." The Center's administrators and designers thus collaborated to create a new space combining sustainable and recycled materials. By all accounts, they succeeded as evidenced by achieving a LEED (Leadership in Energy and Environmental Design) "silver" designation for the expansion—one of the highest levels of "greenness" in the building industry. The goals of the project were to protect vital ecosystems, enhance biodiversity, reduce solid waste, improve air quality, and conserve resources with the use of passive solar energy, wind power, and water efficiency systems. These specific goals roll into promoting two primary values—minimizing impact on the land and creating a community embodying principles of sustainability. For its efforts in green building, the ALNC was also chosen for the 2014 Sustainable Small Business Award from *In Business* magazine. This award is given annually to Madison area businesses that make significant eco-conscious decisions.

Leopold is known for his "land ethic," which is an approach to nature based on viewing the environment as a set of organic ecosystems that can be healthy or unhealthy.[62] The ALNC building expansion is a testament to green building principles in accordance with this tradition. Its "silver" LEED rating can be attributed to scoring highly in categories such as sustainable site development, water savings, energy efficiency, materials selection, and indoor environmental quality.

The ALNC is situated for maximal exposure to the sun and has large side windows to capture light during daytime work hours. The system of exterior and interior windows is designed to allow for greater light flow into individual offices and other workspaces. This reduces the need for electric lighting for a substantial number of hours of daily building operation. The east-west orientation of the new addition to the Center also enhances the use of the photovoltaic (PV) and solar thermal panels placed on its roof. This array of solar panels allows energy consumption within the building to be just 50% of what one would find in a similarly-sized conventional building. The 17.172 kW solar array captures the sun's power with 54 modules in 3 rows generating 20,918 kW hours of power per year. The PV system is connected to the local Madison Gas & Electric (MG&E) grid and sends its surplus energy out to other customers in the area.

---

62 See "Leopold's Land Ethic" in Section 1.

A solar thermal system is the main source of heat for the new portion of the building, warming not only the air, but also the water. Hot water is stowed in large storage tanks in the Center's lower level. The building also has thick concrete floors that absorb solar heat as it comes through an extensive set of south-facing windows. As the concrete warms it develops a "thermal mass" that releases heat and warms the interior of the building for several hours.

Outside the building one can find an electronic vehicle (EV) re-charging station installed by MG&E. MG&E is known for being a leader in reusable energy, helping to build the infrastructure and capacity for energy conservation throughout the Madison area. The charging station not only provides present users of EVs with a place to juice up their cars—it also serves as a reminder of what could be the future of transportation.

With respect to efficiency measures, water used on site is sourced by aquifers deep below the soil's surface. There are several building features suited to conserving water. The building is retrofitted with water efficient toilets and washers. A WaterSense line from Kohler is also installed within the Center. Kohler is a Wisconsin-based company recognized as an industry leader in water efficiency for bathroom and kitchen installations, having won the US Environmental Protection Agency's prestigious Water Sense award. Another local company (The Brothers Main) provided a Maytag EPIC z high-efficiency washer that saves a great deal more water than conventional models.

Energy and atmospheric elements are also important in designing a building that meets LEED certification standards. When electric lighting is used the Center strives to only have it sourced by LED (Light Emitting Diodes) bulbs. LED bulbs are one of the most energy-efficient lighting options available as they are approximately 10 times more efficient than traditional incandescent light bulbs and roughly two times more efficient than compact fluorescent lights. To further increase energy efficiency, the ALNC uses motion sensor switches for the lights.

To create a comfortable environment in a cost effective and energy-efficient manner, the ALNC is fitted with state-of-the-art insulation. The roof, floor, and ground walls of the Nature Center have more insulation than what is required in the local building codes. This enhances the capacity of the building to retain cool or warm air depending upon the seasonal needs of its users. Energy efficient appliances beyond the washers and dryers already mentioned include a refrigerator from Sub-Zero, a local company.

With respect to the materials and resources utilized in building the Nature Center expansion, the emphasis was on collecting and refurbishing recycled items. Salvaging and reconstituting used items keeps them out of landfills and reduces the need to construct new materials for building. Donors supplied several items incorporated into the main structure. The post-and-beam framing in the new addition was supplied by Glenville Timberwrights from the town of Baraboo, just an hour's drive from the ALNC. Installation of this previously used and reconditioned framing not only keeps these materials from being disposed of, but also preserves some area architectural history.

The carpeting and tile for the site were castoffs from other local construction projects and were thus saved from the dump. Sergenian's Floor Coverings, a company known for developing and retailing "green" flooring, donated carpet, ceramic, and vinyl tile to the expansion project. Additionally, Epic Systems, another company acknowledged for its sustainable building practices, donated carpet for the office areas in the ALNC as well as lighting fixtures, which were retrofitted to utilize LED bulbs. Dyson Airblade hand dryers were installed in the bathrooms to reduce paper waste. Twenty rooms worth of furniture was donated, including office chairs, tables, and the reception desk of the Center. Other furniture came from Smith and Gesteland, LLP. Other chairs and tables were donated by American Girl, which is a locally-based, but now nationally renowned, doll company. Target Interiors also contributed tables and other office furniture to the project.

Poorly designed or faulty HVAC systems can cause an assortment of health issues and drive up heating and cooling costs extensively. After all, buildings account for 72% of electricity consumption.[63] Though these are likely not "sexy" expenditures, HVAC system efficiency features can go a long way toward creating a truly sustainable building over the long term. As alluded to earlier, the ALNC uses an integrated heating and cooling system with efficiency controls to best conserve energy. There is in-floor radiant or hydronic heating in the building. Fluid is heated in the solar thermal system and pumped through a series of tubes running throughout the building's concrete floors. This heats rooms from the ground up. Most of the HVAC equipment in the Center is composed of two-stage, variable speed, 95% efficient furnaces and 20 SEER air conditioning units. Consultation with the engineering department at MG&E resulted in a highly efficient system at an economical cost. Both computer equipment and people give off heat in the work environment and this needs to be taken into account when planning an HVAC system. This often requires cooling for a comfortable workplace. But in the spring and fall, economizers

63 United States Department of Energy, "Energy Efficiency Trends in Residential and Commercial Buildings," (website), October 2008. Available at http://apps1.eere.energy.gov/buildings/publications/pdfs/corporate/bt_stateindustry.pdf (accessed June 22, 2016), p. 9.

use the ambient air for cooling instead of running air conditioners, which in turn saves on electricity use.

The Center has more decorative features that not only lend a "homey" feel to the place, but are also environmentally friendly. The Valcourt Versailles high-efficiency wood-burning fireplace heating the lobby while providing an aesthetically pleasing touch (and using a renewable resource), was donated by Top Hat Heating. Granite stones originally used as ballasts for ships sailing from Belgium to Chicago in the late 1800s, were repurposed as fireplace stones in the Center (salvaged by Glenville Timberwrights). And this isn't merely the second use of this granite, as the stones were previously reused as pavers for Chicago's early streets. Wooden flooring throughout the original Center was procured from a now demolished warehouse in New Orleans. The yellow heart pine boards (about 100 years old) were re-planed by Resplendent Mill Shop and then shipped (at a greatly reduced rate) by W & A Shipping of Fort Atkinson, Wisconsin, to the ALNC. This allowed for fewer trees to be harvested for the Nature Center expansion. Artists at the Scottsdale Art Factory crafted the antler chandelier that lends to the rustic look of the main lobby in the ALNC. It was constructed from naturally–shed elk antlers and weighs some 1,400 lbs. Smaller deer antler chandeliers adorning breakout rooms throughout the Center consist of naturally-shed deer antlers gleaned from northern Wisconsin.

An art glass wall in the new addition created by the glass artist, Renee Miller Knight, enhances the indoor environmental quality of the facility. Along with being aesthetically pleasing, it also provides daylight to the main working rooms of the Center. The wall is meant to evoke the northern lights and is framed by Leopold Logs (more on those later). It is imprinted with boughs from pines harvested from the woods surrounding the Leopold family's "shack"—a small dwelling just outside of nearby Baraboo where the Leopold's lived and worked in the 1930s. Only minimal material containing VOCs (volatile organic compounds) were used in the facility.[64]

But the attention to eco-friendly building principles was not just adopted for the interior of the Nature Center, there was also a great deal of time and care taken on the landscaping and functional layout of the property beyond its walls. The outdoor deck is a prime example of how green building concepts are used on the grounds. Trex Company, the world's largest manufacturer of

64 VOCs are emitted as gas from liquid and solid forms and can have short-term and long-term negative health effects on those exposed. They are commonly emitted from construction and building materials, including paint and adhesives.

wood-alternative decking built the deck from recycled plastic products. Not only is the reuse of plastic helpful to the environment, it again reduces tree harvesting and lasts longer without the worry of splintering and rotting of wood that often results from harsh Wisconsin winters.

Even the driveway leading up to the Center employs green design concepts, such as the reuse of asphalt. DRS Paving of Madison recycles old pavement as material for a number of local projects. This means fewer components that go into asphalt had to be mined and processed for new pavement. The Center installed a rain garden to alleviate erosion issues caused by a wooded drumlin near the building that previously washed out trails and the front entrance area during heavy rainfalls or spring runoff from melting snow and ice. The new siting of the building due to the addition allowed for the construction of rain gardens collecting runoff. Rain gardens absorb water that otherwise disrupts other ecosystems in the area. Native plants that flourish in wet environments are also beneficiaries of these gardens.[65]

The entrance to the Center is flanked by four pine logs (Leopold Logs) culled from trees planted by Aldo Leopold and his family at the shack. These trees were harvested in 1996 by volunteers using draft horses, then stripped, and left outside to weather naturally. They are now structural supports for the old lobby serving as a symbol of Leopold's land ethic and as a tribute to his work in environmental education.

While there are obviously the "nuts and bolts" of green building, principles of architectural design serve as its guidelines. While not all advocates of green building necessarily agree on all of them, there is significant overlap of the principles they endorse. Ken Yeang observes what he calls the "five propositions of green planning and design." One is that the approach to green design should be the "seamless and benign biointegration of four sets of infrastructures: The 'grey' (ecoengineering systems and utilities), 'blue' (water management and sustainable drainage), 'green' (green infrastructure), and 'red' (human systems, spaces, society, and regulations), as a method for resolving green design issues by blending of all these four sets of ecological and artificial armatures into masterplan form and built form."[66]

The second principle takes green design to be the seamless biointegration of human-made things with the natural world. This is to be done at three

65 See Case 46—"A Garden for Rain."

66 "What is Green Design and Planning? Five Propositions," in Andy van den Dobbelsteen, Machiel van Dorst, and Arjan van Timmeren, eds. *Smart Building in a Changing Climate* (Amsterdam: Techne Press, 2009), p. 16.

levels; physically, systematically, and temporally. To help explain what biointegration means in this context, Yeang adds that the term should be thought of as one would consider medical and surgical integration as part of successful prosthetic design.

A third proposition of green building is that design should mimic ecosystems as much as possible. Here the architect needs to pay close attention the structure, processes, features, and functions of ecosystems.[67] For example, he/she needs to consider recycling, energy efficient systems, and a holistic balance of biotic and abiotic parts of the ecosystem.

A fourth proposition of green design and planning is that ecodesign is not only considered as constructing new artificial ecosystems that are "living and urban" or even as rehabilitating already developed ecosystems and cities. Ecodesign must also be thought of as restorative—reclaiming impaired and devastated existent ecosystems back into a recognizable local landscape.[68]

The fifth proposition set forth by Yeang is the most comprehensive and most complicated. However, the basic idea is that green design is in large part a theoretical model as well as a collection of technical and design components.[69] To understand this better, ecodesign involves global monitoring of the environment and destruction caused by humans, as well as an understanding of natural disasters and impacts on our built environment. Human activities and industries are not divorced from nature, but instead are married as sets of environmental interactions.

While there is a good deal of support for green building, the main question that arises is how much will it cost. Does economic reality poke holes in the case for sustainable design? Will green building be sabotaged by questions of cost and whether it is worth it for mainstream building projects?

Review Questions

1) The ALNC is a privately funded, non-profit institution and thus one might argue they can call for any "green" building features they wish. Yet, how much of a role should environmental sustainability play in a *publicly funded* institution, such as a college campus? Any role at all? Is there a difference in what role it should play at a private university versus a public one?

---

67 Yeang, "What is Green Design," p. 18.

68 Yeang, "What is Green Design," p. 19.

69 Yeang, "What is Green Design," p. 20.

2) Do you think green building can be justified even if the short-term costs are higher than using conventional building materials and techniques? Is there any sort of obligation to employ green building design in new construction?
3) Should the state use its coercive power to establish policies mandating green design, or short of that to endorse any sort of incentives to encouraging it? Why or why not? On what grounds?

CHAPTER 7

Recreation and Environmental Ethics

Case 23 – The Walleye War –

Study Questions

1) What are usufructuary rights? How are they related to the land treaties referred to in this case study?
2) What role has spearfishing played in the traditions of the Anishinaabeg (Ojibwe) people? Try to find multiple ways.
3) Describe the boat launch protests? What moral issues arise during these events?
4) At the end of this case study it is noted that, "While spearfishers feel much safer going on the lakes now that the Walleye War is over, the controversy evolves in new ways." What are some of the "new ways" into which the spearfishing controversy has "evolved"?

One day in February 1974, brothers Mike and Fred Tribble, two members of the Lac Courte Oreilles Ojibwe band in Wisconsin decided to fish off the reservation. It would be hard to imagine that such a normal seeming fishing trip would be of much consequence today. But within a decade of their action, a simple fishing expedition was an important part of a wave of controversy that would help lead to the reassertion of Native American treaty rights, a federal court decision affirming those rights, and subsequent violent protests at fish landings throughout Wisconsin.[1] These events would further attract national media exposure and foster connections between indigenous peoples throughout the world.

To set the context for the Tribble brothers' action, we need to replay some American history. As settlers pressured for westward expansion, they demanded that the United States government make land available for exploration and the extraction of a host of resources. In Wisconsin, the primary desire

1 Wisconsin's Attorney General at the time, Don Hanaway, referred to the Tribble brothers' incident as "the most important fishing citations ever issued in Wisconsin." See Charles F. Wilkinson, "Treaty and Fishing Rights of the Wisconsin Chippewa," The Oliver Randall Lecture, Occasional Paper from the University of Wisconsin Law School, April 19, 1990.

 | DOI 10.1163/9789004339996_009

was to open what were at that time territorial lands to timber harvesting and mining. Valuable tracts of white pine could be used for building materials and the extensive waterway system of the territory could transport lumber from the forests of the Northwoods to mills downstream for processing. Southwestern Wisconsin possessed rich lead deposits that could be mined for use in a number of industries. Iron ore was plentiful in the northern reaches of the territory.

In a move demonstrating that it recognized tribes as sovereign nations, the US government sought to negotiate a land deal allowing for these activities to feed a growing industrial revolution throughout the expanding young country. This is an important acknowledgement, as we shall see later, for the treaty rights controversy. In 1837, the first of the treaties that opened land to non-tribal timber and mineral extraction was signed, with the second following in 1842. Together these treaties ceded all Ojibwe lands in Wisconsin to the US government. While this opened land for industrial and settlement purposes, the Ojibwe successfully negotiated a series of usufructuary (or "use") rights they would retain in the land in perpetuity. As hunting and fishing were obviously crucial to their survival, the Ojibwe in Wisconsin kept the right to harvest wild rice and other gathering rights in the ceded lands. Importantly, some of the gathering rights were for purposes beyond physical subsistence. For example, Ojibwe gathering rights include the ability to collect berries, sage, and other plants crucial for tribal religious ceremonies. The Ojibwe regard some plants, including sage and tobacco as sacred and access to them is necessary for medicinal, spiritual, and cultural reasons.

Fishing takes a central place in the cultural identity of the Ojibwe and within that communal narrative walleye are vital. As the anthropologist, Larry Nesper, observes:

> When the Ojibwe reconstructed and reimagined their history during the spearfishing dispute in the late twentieth century, they gave primacy to fishing, emphasizing cultural practices that had previously been in the background. A contemporary traditional story of the origin of spearing walleye at night stresses its importance and close connection to the origin and identity of the Anishinaabeg who live in the area of Lac du Flambeau. "How Winaboozhoo Learned to Firehunt" was written by Flambeau band and Crane clan member Ernie St. Germaine in 1990 and appears in *Winaboozhoo Adisokan* used as the children's curriculum in Family Circles. The story is set in the present, at the site of the Old Village where the Bear River flows into Flambeau Lake. A boy asks his grandfather how the old man learned to spear. Grandfather says his

> own grandfather taught him and perhaps, Winaboozhoo, taught him. The grandson's question "Who taught Winaboozhoo?" sparks the tale.[2]

The story goes on to explain how spearfishing came to the Ojibwe. It is a paradigmatic case of the significance of spearfishing to the Ojibwe in general and to the Lac du Flambeau band in particular. Walleye (*oka*) are privileged in the culture and stand out from other fish during the spawning season. There is little doubt that walleye have had an important status historically among Native Americans throughout the Great Lakes region and have come to be the object of a practice central to the identity of contemporary Ojibwe.[3]

The treaty of 1854 was of particular importance since the Ojibwe agreed to live on reservations in Wisconsin and to cede all of their lands in Minnesota. All of this means that while the Ojibwe sold millions of acres of land to the US government, they were careful to retain the ability to use it (which of course is *non-reservation* property) in important ways.

Yet, in the years after these treaties, what became the state of Wisconsin declared it had the right to manage hunting and fishing within its borders. Hence, Ojibwe who attempted to exercise what they understood as being unlimited access to fish and game on non-reservation land in the state were arrested. The threat of punitive action likely deterred Ojibwe hunters and fishers over the decades to the point that by the 1970s many of them did not venture off the reservations to engage in these activities. If they did so at all, they spearfished in secret and in avoidance of game wardens, engaging in a practice called "violating" within the Ojibwe community. It is also possible that the interest in "violating" was on the decline by the early 1970s.

But then, along came the Tribbles. Mike and Fred Tribble left the Lac Courte Oreilles reservation to spearfish through the ice of Chief Lake. They were arrested by two game wardens on charges of illegal spearfishing. What the wardens likely did not know was that the Tribbles had recently taken a history course on treaty rights at the College of St. Scholastica in Duluth, Minnesota. Despite showing the wardens a copy of the original 1837 treaty as proof that they could spearfish on ceded lakes off the reservation, the brothers were issued citations.

In their subsequent case (heard by the federal district court), the attorney for the Tribble brothers argued that they were merely exercising their rights as granted by the treaties of 1837, 1842, and 1854, which taken together granted the rights to hunt, fish, and gather on ceded lands in Wisconsin, Minnesota,

2 Larry Nesper, *The Walleye War* (Lincoln, NE: University of Nebraska Press, 2002), p. 56.

3 Nesper, *The Walleye War*, pp. 57–58.

and Michigan to the Ojibwe. In what is now the landmark *Lac Courte Oreilles vs. Voigt* court decision, federal judge James Doyle of the Ninth Circuit upheld the tribal rights of the Ojibwe in the ceded territory.

With this court decision, more Ojibwe were inspired to spearfish on non-reservation lakes in Wisconsin to the dismay of mainly white sport fishermen. As the former head of Wisconsin's Department of Natural Resources (WDNR) Division of Enforcement George Meyer described it, the reactions to the *Voigt* decision by non-Indians in northern Wisconsin was "like a train hitting a wall—that's the social dynamite."[4] In the 1980s, northern Wisconsin faced stiff economic pressures. The economy of the area had little industry and was heavily dependent on farming, logging, paper manufacturing, and some secondary industries that grew along side them. Northern Wisconsin is also very rural, set far from the urban influences of Milwaukee and Madison to the south. Over the years, there has been a much heavier reliance on the tourist industry as vacationers are lured to the area for both summer and winter recreational opportunities. More specifically, lakes providing some of the best walleye and bass fishing in the country draw many fishers to northern Wisconsin.

Soon after the court decision, tensions were likely exacerbated by newspaper headlines reporting that the federal court decision had granted "special rights" in the form of "unlimited [fishing] rights given to the Chippewa [Ojibwe]." The fires of protest were further flamed when Judge Barbara Crabb ruled in *Lac Courte Oreilles IV* (*LCO IV*) that "effective tribal regulation of the harvest precluded state regulation except to conserve a particular species or if there were a risk to public health or safety. This became known as a '100 percent ruling'—that because 'a modest standard of living' could not be gained by harvesting all the ceded territory resources, the [Ojibwe] had the right to 100 percent of the harvestable quota of fish and deer."[5]

Conflict between white anti-treaty protesters and the Ojibwe escalated by 1987 when an organization of white northwoods anglers (called Protect America's Rights and Resources—or PARR) began protesting against Ojibwe spearfishers at boat landings. There were a number of factors that precipitated these acrimonious encounters. First of all, the courts recognized that only those who are members of tribes that ceded lands to the US government are legally permitted to spearfish walleyed pike. All other fishers must use hook and line to harvest walleyes. Most fishers agree that spearfishing is a more efficient way to harvest walleyed pike. Also, as stipulated by the original treaties

4 Rick Whatley and Walter Bresette, *Walleye Warriors: An Effective Alliance Against Racism and for the Earth* (Philadelphia: New Society Publishers, 1994), p. 48.

5 Whatley and Bresette, *Walleye Warriors*, p. 35.

and reaffirmed in some subsequent federal court decisions, tribal members are legally permitted to spearfish any time of the year. However, in practice the Ojibwe walleye harvest season usually begins just after the ice has completely left the lakes until mid-May (unless the tribal spearfishers reach their total fish quota before that time). The angler season is basically from the 1st Saturday in May to March 1st.[6] Spearfishers venture into lakes at night using motorboats with flashlights designed to attract the attention of walleyes, drawing them to the surface of the lake to be speared. White anglers pointed to these different fishing regulations as discriminatory against non-tribal fishers and on this basis complained that the Ojibwe were being given special privileges.

While there were a number of protests against spearfishing throughout the late 80s and early 90s, it was the Butternut Lake protest of 1987 that set the stage for some of the most tense and violent clashes between spearfishers and protesters in Wisconsin. Large crowds of protesters, sometimes intoxicated, threw rocks at spearfishers and their families, while berating them with racial and sexual insults. Gunshots from protesters were reported at the boat launches in attempts to threaten spearfishers off from the lake. Protesters were also accused of running spearfishers and their families off the roads leading to the boat launches and for lodging assassination threats against tribal judges. The protesters also employed dangerous tactics such as boat ramming and boat swamping as means of intimidating the Ojibwe spearfishers.

Some anticipated trouble before the 1987 spearfishing season at Butternut Lake since it lies very close to where PARR had been organizing protests. The incident occurred for two hours, despite the presence of police officers and WDNR wardens. Two female Ojibwe elders were pulled off from a truck and pushed to the ground by some in the crowd. The anti-treaty protesters declared the Butternut Lake incident as a victory for their cause, inspiring more frequent protests at other boat landings in subsequent years.

The Butternut Lake protest was just the first example of vitriol and violence cast on spearfishers, their families, and "witnesses" (including a number of whites who came to observe and construct reports of the violence at the boat landings). Witnesses and family members of spearfishers also gathered with the hope of lending moral support and a counter presence to the PARR protesters. The witnesses served as a way of gathering evidence of unlawful activities by the protesters that could lead to building a legal case. To launch their boats, spearfishers often had to walk through areas where they were subject to vicious

6 Steven Hewett and Timothy Simonton, "Walleye Management Plan: Moving Management into the 21st Century," Administrative Report #43, (Madison, WI: Wisconsin Department of Natural Resources, January 1998), p. 45.

verbal abuse, including racial epithets such as being called "timber niggers" and "red niggers." Female Ojibwe were the targets of graphic sexual threats as spearfishers and their families endured catcalls of "I want to rape your squaw." Placards were meant to send messages of intimidation, including one that became a popular bumper sticker in northern Wisconsin in the late '80s and early '90s that read, "Kill an Indian, Save a Fish." Other signs, such as a picture of a revolver pointing toward the viewer declared "Spear This" exacerbated the threat of violence at the boat landings.[7]

Anti-treaty protesters leveled a number of complaints on Ojibwe spearfishing beyond the notion that they believed Native Americans are given reverse-discriminatory preferential treatment over other citizens in regard to hunting and fishing. First of all, there was the economic argument that spearfishers would take too many walleyes that are a highly desired species of game fish for tourists, who flock to northern Wisconsin for the chance to catch them. Anti-treaty protesters claimed that since Ojibwe spearfishers are able to fish the lakes earlier in the season than other fishers, they would get to have "first dibs" on trophy fish. Accordingly, spearfishers have opportunities unavailable to white sportfishers as the stocks in the lakes would be depleted by the time the angler season opens. We have already noted that northern Wisconsin is heavily reliant on the tourist trade for its economic survival. In sum, the worry was that heavy Ojibwe fish harvests would deter a significant number of tourists drawn to northern Wisconsin lakes not so much for the solitude and beauty of the region, but by the promise of trophy fish (usually considered to be over 28 inches long by sportfishers in Wisconsin). Hence, anti-treaty protesters argued that the cumulative depletion of walleye from years of spearfishing would become a major drain on the tourism sector of the local economy.

For fishers within the state, walleyed pike are the most sought after fish, with 31% of respondents choosing it as their top fish to harvest in a survey conducted by the WDNR, with largemouth bass second at 15%.[8] Even when fishers chose another game fish for the top spot, walleye still scored highly in the same survey as 19% of participants ranked it as the second most preferred fish to catch.

The timing of the catch was also a concern for anti-treaty protesters. Walleye spawn in early to mid-spring, coming in close to shore where they are relatively easy to catch on a barb in five feet or so of water. Females are filled with eggs,

---

7 To locate the poster and more context on anti-treaty placards used at the boat landing protests, see Whatley and Bresette, *Walleye Warriors*, p. 39.

8 Hewett and Simonton, "Walleye Management Plan," p. 72.

haven't had the chance to drop, and thus when harvested it is hard to restock the lake.

A second argument made by the anti-treaty protestors was that while one of the major claims the Ojibwe make for why spearfishing is important to them is on traditional cultural grounds, spearfishers undercut their own position by using modern technology to fish. To understand this counter-argument, we need to put it in context. The Ojibwe assert that spearfishing for walleye is not merely a means of harvesting meat to feed their families. They also argue that spearfishing is a traditional cultural practice conducted by the tribe for generations. This significant cultural dimension of fishing and hunting is a reason why the Ojibwe insisted on retaining rights to spearfish on ceded lands in the first place.

However, anti-treaty protesters were quick to point out what they saw as a contradiction in the Ojibwe argument that spearfishing preserves traditional tribal customs and is thus a culturally protected practice. On the one hand, one might be sympathetic with the notion of tradition as a way of arguing for fishing and hunting no matter one's cultural background. After all, anti-treaty protesters might seek to protect fishing privileges on similar grounds, wishing to continue a practice in which past generation's engaged. However, this sympathy is based on not only seeing something as being a tradition (whether it is the general practice of fishing and hunting), but also to see it *as being done in a traditional way*. Anti-treaty protesters contended that the Ojibwe argument for spearfishing rights based on tradition is eroded when spearfishers do not use traditional methods of fishing such as paddling birch-bark canoes with torches and hand-held spears. Instead, anti-treaty protesters chaffed at the notion that there is anything "traditional" about Ojibwe spearfishing at all since contemporary tribal spearfishers have replaced birch-bark canoes with motorboats, hand-held spears with mechanical ones, and torches with incandescent headlamps.

A third argument made by anti-treaty protesters against Ojibwe spearfishers was much more general and not related to spearfishing per se. It was that the Ojibwe have become welfare dependents and a drain on non-tribal taxpayers because they insist on staying on reservations. Native American "separateness" and lack of cultural assimilation into mainstream American culture cause them to be seen as a "special" cultural group granted "special" rights and status not offered to Caucasian citizens. Anti-treaty protesters believed that one problem with being physically and culturally separate is that being beholden to minority cultural values does not prepare Ojibwe for life in a modern, democratic, capitalist society where they might prosper if not for holding on to the "old ways." Furthermore, some anti-treaty protesters argued that life on the reservation, with its high unemployment, lack of economic opportunities, and other attenuating hardships (such as rampant alcoholism) keep Native Americans in a culture of dependence with heavy reliance on the federal government.

Anti-treaty protestors believed the federal courts (and perhaps activists judges with a pro-Native American agenda) overstepped institutional bounds and made decisions on matters that should be left to states to decide. This complaint highlighted the issue of states rights versus federalism. Beyond the legal question of whether federal courts are meddling in matters belonging to state legal authorities is the normative question of whether there *ought* to be more local control over fish and game regulation.

How did pro-spearfishers and their sympathizers respond to these anti-treaty arguments? First, they insisted the impression that the Ojibwe were granted "special" fishing rights was mistaken and in part the media was to blame for perpetuating this false perception. As the Tribble brothers tried to explain when they were arrested on Chief Lake, the collective agreements in the land transfer treaties allowed the Ojibwe to *exercise rights they already had*. All the courts did was recognize and reassert those treaty rights. Moreover, these rights were not negotiated through a special procedure privileging Ojibwe views on land and property ownership, but instead via a judicial system favoring a Lockean conception of property handed down through British jurisprudence to the US. The Ojibwe retained these use rights despite the fact that they were forced to negotiate contracts in the context of a political philosophy they didn't necessarily share, "played by the rules," and hammered out terms consistent with Anglo-American legal tradition. Disallowing Ojibwe these rights was a blatant violation of the terms of the contract. White anglers couldn't simply say, because the Ojibwe negotiated well under trying conditions "that this happened a long time ago" and that the treaty stipulations were "special" rights that should be stripped away because they were now inconvenient.

To the charge that the Ojibwe were imperiling fish stocks and receiving preferential treatment to harvest the most coveted fish, pro-spearfishing advocates used a couple of responses. First of all, the tribal governmental organization that handles game and fish resource regulation (Great Lakes Indian Fish and Wildlife Commission or GLIFWC) attempts to work in concert with the WDNR, as both groups are mandated to take stewardship of natural resources seriously. Because the Ojibwe use walleyes and other natural resources not just for subsistence or entertainment purposes, they recognize a deeper value in their stewardship. How this is accomplished on a practical level is that the Ojibwe have negotiated a quota of fish with the WDNR that they can take from lakes in the ceded lands off the reservation. Part of the reason for this agreement is to ensure that there are fish for future generations, but also there is recognition of the inherent value of nature. While technically under the treaties mentioned above the Ojibwe could argue they are under no legal obligation to curb the numbers of fish they take off reservation, they have voluntarily accepted limits consistent with sound wildlife conservation.

Soon after the *Voigt* decision, Ojibwe leadership called upon its members to voluntarily refrain from hunting off reservation until an agreement could be made with WDNR and GLIFWC officials to manage the natural resources in question. Also, no lake with less than 500 acres could be speared, nor could one that was being used as a wildlife refuge or for research. Fish longer than 20 inches were initially designated as off limits to Ojibwe spearers, but later some provision was made for limited trophy-sized take. This was an attempt to protect trophy fish for sportfishers, as well as to allow for more mature fish to reproduce. Another major provision in the first spearing agreement between the Ojibwe, GLIFWC, and the WDNR was that each of the six Ojibwe bands in Wisconsin would be able to name six lakes to fish for the season. Prior to the 1986 spearing season, more regulations were instituted, including a twenty-five fish bag limit per night, with spearing ending at 1:00am on weeknights and 2:00am on weekend nights. Initial quotas were also set, with Ojibwe bands being allowed to take 10% of what was called the "total allowable catch" (or TAC). The TAC is the total number of members of a fish species that can be sustainably taken from a body of water each year, a figure estimated to be at 35% of the total estimated population.

Additionally, Ojibwe spearfishing advocates argued that the size of the Ojibwe spearfishing harvest paled in comparison with other sport fishing. The Lac du Flambeau spearers only harvested roughly 2 percent of the 670,000 annual take of walleyed pike by sportfishers in the northern Wisconsin ceded lands during the main years of the boat landing protests. This low number was achieved despite the fact that early on spearfishers could take fish for other tribal members (even those who had no intent of participating). This meant that even if some spearfishers were able to take advantage of spearing many more fish than individual anglers, few enough Ojibwe speared so as to ensure the tribes would not exceed quotas.[9] Through most of the '80s and '90s, tribal spearers took around 25,000 to 30,000 walleyes annually on about 150 lakes in the ceded territory.[10] Angler harvests were projected to around 125,000 on those same lakes each year.

9 Nesper, *The Walleye War*, p. 77.

10 Hewett and Simonton, "Walleye Management Plan," p. 28.

During every year from when the first legal spearfishing for walleye began in the winter of 1983 until 2009, Ojibwe spearers also took a lower number of walleye than their allowable quotas (sometimes by a substantial amount). According to GLIFWC data, the lowest percentage of walleyes taken in any one year from 1995–2009 was 50% of the Total Quota in 2001 (Tribal Quota = 45,321; Actual Harvest = 22,743). The highest percentage of walleyes taken in a single season over the period from 1995–2009 was in 2000, when 74% of the walleye quota was taken (Tribal Quota = 40,762; Actual Harvest = 30,367).[11] While both the walleye quota and percentage of walleyes taken started to climb in the 'oos, the percentages of walleyes taken versus the quota in those years still tended to be in the 60–69% range.

Tribal harvests have thus accounted for a range from 20–30% of total harvest on lakes with a 15-inch size limit. These harvests have amounted to only about 10% of total harvests for lakes where there is no size limit for anglers. To use muskie harvests for the sake of comparison, according to GLIFWC, the muskie take from Wisconsin lakes by Ojibwe spearfishers from 1985–1991 was as follows: 86, 55, 196, 158, 118, 303, and 185. During the same time period, the average estimated number of muskies taken by non-tribal sport fishers was 9,454.[12]

It is also interesting that all told very few Ojibwe spearfish in the ceded lands off from the reservation, with only 15 percent of tribal members fishing. Of those who fish, half of them did so only a few nights per week and took a total of a few dozen walleyed pike.[13] In Wisconsin, on lakes not in the ceded territories, the bag limit for walleyes is 5. In 1989, a sliding bag scale was introduced that required a reduction in the daily bag limit in lakes within the ceded territory where the Ojibwe wished to exercise their right to harvest walleye at more than 1% of the safe harvest total. That is, bag limits can be reduced to 4, 3, 2, 1, or 0 per day depending on the proportion of safe harvest that is declared.[14] For most anglers in Wisconsin fishing on lakes that are a part of the ceded territory, this amounts to a 2 or 3 fish daily bag limit.[15] Depending on the number of walleyes actually taken by spearfishers from when the ice is clear to the beginning of the angler season, the bag limits for anglers can be adjusted. More recently in 2009, 492 tribal spearfishers fished on 181 lakes with

---

11 Jennifer Krueger, "Open Water Spearing in Northern Wisconsin by Chippewa Indians During 2009," (Odanah, WI, GLIFWC, January 2010), p. 11.

12 Krueger, "Open Water Spearing," p. 218.

13 Nesper, *The Walleye War*, p. 4.

14 Hewett and Simonton, "Walleye Management Plan," p. 44.

15 Hewett and Simonton, "Walleye Management Plan," p. 71.

walleyes being taken in all but two. They harvested some 32,947 fish, with 98% of them being walleyes, having an average length of 15.3 inches.[16] The tribal quota was exceeded on only three lakes and only by a total of 16 fish. Over the past two decades, the total number of spearers fishing on lakes in the ceded territories has ranged from 400–500. Tribal quotas were selected for 481 lakes in the ceded area for 2009, with spearers allowed to take 53,706. Needless to say, the Ojibwe usually take substantially fewer fish than the amount allotted to them in most years. According to GLIFWC, from 1989–2011, spearfishers harvested some 585,000 walleye. In the same time period, non-tribal anglers hooked tens of millions.[17]

With respect to Ojibwe to assimilation into mainstream American culture, there is not much evidence to support the claim that even if they agreed to move off from the reservation they would necessarily share the values anti-treaty protesters tout. Even after the boat landing protests, what is startling is that despite extreme pressures from the modern, dominant culture (both explicit and implicit) is the extent to which Ojibwe have held onto beliefs about nature that tend not to be shared by many in mainstream American society. Fish harvesting, hunting, and gathering practices are not merely for subsistence or recreational, but bound up in a much deeper cultural meaning for the Ojibwe.

The "Walleye Wars" effectively ended in June of 1991 when the State of Wisconsin ceased its legal challenge to the Ojibwe's spearfishing rights. For the most part, the range of attitudes of anglers toward tribal spearfishers today is reported to be from "grudging acceptance to placid indifference.[18] This, of course, doesn't mean anglers don't hold some lingering resentment. They face lowered bag limits when the DNR imposes them to conserve the overall walleye population and these limits are stricter whenever Ojibwe spearfishers announce they wish to take more walleye than in the previous year.

More recently there has been an upswing in the number of walleye taken by spearfishers. While this hasn't reignited protests at boat launch sites, it has caught the attention of the WDNR. In 2009, spearfishers harvested 32,201 walleyes and this number increased to 34,156 in 2010. The Lac du Flambeau band proposed spearing 20,970 walleyes during the 2011 season, which would have been triple the number actually speared (6,188) in 2010 (65% of number

---

16 Krueger, "Open Water Spearing," p. i.

17 Jeff Starck, "Walleye Wars Evolve into Peace on the Lakes," *Wausau Daily Herald*, June 21, 2011. Available at http://archive.wausaudailyherald.com/article/20110522/WOF16/105220714/Walleye-Wars-evolve-into-peace-lakes (accessed August 24, 2015).

18 Starck, "Walleye Wars Evolve."

allotted to the tribe). The main reason given for the rise in spearing has been due to a poor economy and the need to feed families with fish. Spearfishers realize that by not taking their limit, they are giving up opportunities they can't afford to, especially in families where many of its members are disabled, too old, or too young to fish.

As of 2015, the walleye take requested by the Ojibwe on non-reservation lakes was 68,226—a record number—up 7% (or 4,738 walleyes) from the previous year. In recent years, spearfishers have harvested about 50% of the proposed number, taking 27,725 in 2014. This has caused some tension between the tribes and the WDNR, which believes the requests run counter to evidence of declining walleye populations. The WDNR claims that when the Ojibwe increase their tribal quota, stricter bag limits have to be imposed on anglers to compensate and thus sustain walleye populations. However, some Ojibwe representatives accuse the WDNR of misleadingly implying that the bag limit (which is unpopular with anglers) is the only thing standing between sustainable walleye populations and their extinction. Tensions died down when spearfishers only took 28,400 fish and the WDNR raised its bag limits on anglers accordingly. But each time the Ojibwe declare that they will take more fish, tensions rise again as the WDNR warns that angler bag limits need to be lowered.

While spearfishers feel much safer going on the lakes now that the Walleye War is over, the controversy evolves in new ways. One is left to wonder if litigation over fishing rights and catch limitations will occur again in the coming years.

Review Questions

1) What does it mean to be culturally authentic? Does an argument for the importance of access to a resource on the grounds of traditional practice get undercut if new techniques or technologies are used?
2) Some have argued (including a member of the St. Sault Marie Ojibwe band in Michigan) that Indian treaty rights ought to be abrogated on the basis that the agreements were originally made by "full-bloods," but that now on some reservations, such as the Lac du Flambeau, there are many more "mixed bloods." Should a racial or cultural group lose rights on the basis that over generations they become more genetically diverse? Think about your answer in relation to the answer you gave to question 1 just above.
3) In this case study there were not merely tensions between Ojibwe spearfishers and white anti-treaty anglers, but also between the Ojibwe and state regulatory agencies such as the WDNR. a) What do you think the

balance (if any) should be between the rights of the Ojibwe to use a natural resource on the one hand, and the responsibilities of regulatory agencies to protect resources on the other? b) What do you think the balance should be between permitting fishing privileges to non-tribal anglers and the responsibilities of regulatory agencies to protect resources? Were your answers to parts a and b of this question different? What do you think accounts for your answers being the same or different?

Case 24 – Foul Ball: Ballpark's Effluent out of Play? –

Study Questions

1) What was the environmental problem discovered at Miller Park?
2) Explain what a combined storm water/sanitary system is. What is a CSO and why might such an event be worrisome?

Miller Park, the baseball facility serving as the home field for Major League Baseball's Milwaukee Brewers, apparently had a design flaw when it was first built in 2001 that only came to light in 2007. An inspection confirmed that some of the bathrooms at the park flushed directly into the Menominee River, which eventually feeds into Lake Michigan. Of course, wastewater from the park is supposed to go to a treatment plant and not into storm drains or directly into the river. What caused speculation that there was a sewage discharge problem at the stadium, however, is chronicled in a longer, wider-reaching, and more tangled tale of waste management. This is not a uniquely Milwaukee story; it is also an issue related back to an aging and crumbling American public utility infrastructure plaguing many cities.

In the Milwaukee metropolitan area, as in most US cities, there is a system of storm water drains and a system of sanitary drains. Storm water drains draw water from the surface urban landscape that collects from rains falling on streets, sidewalks, and parking lots. It originates from surface watering activities, delivering it into a larger storm water system that eventually empties into streams, rivers, or directly into lakes. In Milwaukee, this water is not treated since it is not expected to carry microbes from human fecal matter. Sewer pipes carry human waste from homes and businesses through sanitary system conduits that lead to wastewater treatment plants. After this water is treated, it is then discharged into nearby waterways.

There are two items to note here. First of all, storm water drains can obviously still carry contaminants, such as oils, metals from leaching, salts, pesticides, fertilizers, fecal matter from animals, and so forth. Second, while there are two separate systems in many cities, including Milwaukee, they are part of an overarching, *connected* system called a combined sewage/water system. Yet, water from one system is only expected to mix with the other during severe snow melt or rain events, when overflow is allowed to avoid (or at least lessen) the chances of flooding basements.

Combined sewer overflows (CSOs) are a problem in such systems and some newer urban areas have built completely separate systems with each having adequate capacity to handle extreme water inputs to eliminate the possibility of CSOs.[19] Some older cities have completely separated the systems but at great expense, hence most continue to use combined sewage/storm water systems. In the US alone, there are 772 cities using combined sewage/storm water systems. In Milwaukee (as is also the case in other cities) sanitary system tunnels are buried deeper than conduits for the storm water system. As a result, leaking sanitary pipes are not likely the cause of human waste leaking into the storm water system. Some cities with combined sewage/storm systems treat all water in waste facilities recognizing both systems contain contaminants. But again, treating rainwater is an expense that many taxpayers don't want to shoulder. It is possible that contaminants entering storm water systems cannot be properly treated in waste removal systems. Under US federal law, cities with CSOs must monitor them, do what they can to avoid them, report out to the public when they occur, and specify if affected rivers, streams, lakes, and beaches should be closed to the public if water contamination levels render them too harmful for human exposure.

In 2007, the *Milwaukee Journal Sentinel* reported that tests revealed something that should not be found at the city's storm water discharge pipes—genetic markers for human feces. Some of the earliest findings came in January of 2007, at locations not directly downstream from Miller Park, such as Honey Creek, yet leading to widespread testing of other discharge sites in the city. Honey Creek, which runs through Milwaukee and its suburbs of West Allis and Wauwatosa, was reported to be "more sewer than natural stream."[20]

While still a problem, some levels of fecal matter are expected as rains wash some animal waste into the stream—likely as much as one would find in a

19 United States Environmental Protection Agency, "Combined Sewage Overflows," (website), December 15, 2014. Available at http://water.epa.gov/polwaste/npdes/cso/ (accessed May 13, 2015).

20 Don Behm, "Honey Creek Tainted with Human Waste," *Milwaukee Journal Sentinel*, January 21, 2007. Available at http://www.jsonline.com/news/milwaukee/29416799.html (accessed May 13, 2015).

sanitary sewer. What was unexpected in the Milwaukee Metropolitan Sewage District (MMSD) study was that along with the fecal matter of geese, gulls, and most likely livestock (found at the State Fairgrounds near which there are storm water discharge pipes), there were also microbes from human waste. While Milwaukee has a combined sewer/storm water system, assuming a well-functioning wastewater system, there would only be the possibility of finding human fecal matter from the storm water discharge pipe if there had been a recent CSO. However, there hadn't been such an event when the storm water discharge was tested by MMSD.[21]

This left only a few other plausible possible causes. Either sewage tunnels were somehow leaking waste into storm tunnels or there were misconnections of sanitary pipes into the storm drain system from private homes or businesses. The first possibility would be more unlikely given that sanitary sewer mains lie *beneath* storm water mains. However, there was a slim chance that leaks in the laterals (pipes that are supposed to connect from private homes to the sanitary system) and lying closer to storm water conduits could cause contamination. So, the most likely option was that some laterals from homes and businesses were misconnected, either intentionally or by mistake. But in any case, the fix would be expensive. Repairing either misconnections or leakage of laterals from private homes into storm drains can cost from around $3,000 to $12,000 and considerably more for businesses depending on their size.

Samples taken from other sites showed similar results as those found at Honey Creek. A preliminary analysis of storm pipes discharging into Underwood Creek, Bradford Beach, and importantly for our case study, the Kinnickinnic and Menominee Rivers, revealed that samples from at least a third of the pipes had traces of human feces. Yet, if there is a city that has the ability to find out just how big of a waste management issue it faces, it is Milwaukee. This is because it is home to the University of Wisconsin-Milwaukee's Great Lakes WATER Research Institute, which in 2007 was the only laboratory in the United States where testing for human pathogens from storm systems could be done. Based on tests of those early samples, the Director of the Institute, Sandra McLellan reported that, "Perhaps 10 percent to 20 percent of bacteria in storm water are from human sources." Her lab tested the samples for the *Bacteroides* genetic markers that come directly from human feces. About half of human feces is made up of this kind of bacteria.[22]

In the Honey Creek study, *Bacteroides* were found in 19 of the 20 samples taken, about 95%. The samples were taken from a storm water discharge pipe

21 Behm, "Honey Creek Tainted."

22 Behm, "Honey Creek Tainted."

at the junction of N. 79th St. and Mt. Vernon Ave in Milwaukee, through which water flows into a concrete lined stream running just south of W. Blue Mound Rd. The bacteria were also found in 90% of the samples taken from a storm discharge site at N. 80th St. and W. Wisconsin in Wauwatosa and in 80% of the samples at N. 84th St. (or Glenview Ave) in Milwaukee, just downstream from the State Fairgrounds. They were present in samples taken during both rainy and dry days, and during the time period when no CSOs occurred. With the presence of *Bacteroides* in these sites, MMSD knew that others would need to be tested. Since Miller Park was not far away and preliminary tests showed some contamination of the Menominee River, into which some of Miller Park's effluents enter, a more direct set of tests at the ballpark were warranted.

Finding *Bacteroides* in bodies of water isn't merely gross, but also alarming, as they are indicators of the presence of such viruses as *Cryptosporidium* and other pathogens. Milwaukee has a history with *Cryptosporidium*. The city suffered an outbreak in 1993, when one of its two water treatment facilities failed. Though it is debated what the exact origins of the outbreak were, it has been speculated that farm runoff of manure into Lake Michigan was likely the cause.[23] The outbreak killed 104 people (most of them elderly and/or immunocompromised).[24] Short of fatalities, the health of many people drinking the city's water was also adversely affected. From March 3 to April 28 of 1993, an estimated 403,000 of the city's 1.61 million people (of which 880,000 were served by the malfunctioning treatment plant at the time) fell ill with fever, stomach cramps, vomiting, diarrhea, and dehydration. It was the largest waterborne disease outbreak recorded in US history. While wastewater treatment failures are always possible whether storm water is contaminated with human pathogens or not, there is still an obvious worry about adding pathways for disease into areas where there are good chances for human contact with *Cryptosporidium* or any other waterborne diseases.

More specific concerns about what Miller Park might be contributing to storm water effluent contamination came when genetic markers for bacteria from human waste were found in 12 samples of storm water from a pipe discharging into the Menominee River from the stadium. It was revealed that, as was suspected, workers mistakenly connected sewer pipes into the storm

23 William R. MacKenzie et al., "A Massive Outbreak in Milwaukee of *Cryptosporidium* Infection Transmitted Through the Public Water Supply," *New England Journal of Medicine* (1994) 331: 161–7.

24 Neil J. Hoxie et al., "Cryptosporidiosis-Associated Mortality Following a Massive Waterborne Outbreak in Milwaukee, Wisconsin," *American Journal of Public Health* (1997) 87 (12): 2032–2035.

water system during Miller Park's construction. Given that the park opened in April of 2001 and the plumbing issue was not discovered until spring of 2007, raw sewage flowed from executive suites directly into the Menominee River via storm drainage conduits for six years (at least while the park was open during the baseball season that lasts from April to October). All of this has been particularly troubling to environmentalists since there has been an ongoing effort to clean the river through the Menominee Valley Restoration Project. The cost for correcting the effluent problem was over $10,000, but this doesn't include the cost of the environmental impact years of raw sewage spillage likely caused.

There are, of course, significant questions about whether there should be stricter standards on combined sewage and storm water runoff systems, or whether they should all be replaced with separate systems. Moreover, should cities be allowed, no matter how they control their sanitary and storm water systems, to allow any effluent go into waterways without being treated? Finally, are there ways of curtailing the amount of grease, oil, salts, metals, and other contaminants that seep into bodies of water? Who should pay for this cost? Is this a general responsibility of all taxpayers, or should this ecological service be regarded as a consumer-based responsibility—where those who discharge waste are directly charged a fee for its treatment? Given the crumbling infrastructure of water utilities in the US, it should not come as a surprise if these sorts of questions become more pressing in the future.

Review Questions

1) Do you think the owners of Miller Park (the Southeast Wisconsin Professional Baseball Park District and the Milwaukee Brewers Baseball Club Limited Partnership) were morally culpable for discharging raw sewage into the Menominee River, since they claim they *did not intend* to do so and didn't know about the problem until testing occurred in 2007? Or do you think the plumbing contractors during construction are morally responsible? Or are both parties responsible? Please explain your answer.
2) If the owners and/or contractors are culpable, what would be the morally appropriate response? To simply fix the flushing problem, or to also compensate others for environmental harms? How should we calculate the cost such harms? To whom should compensation be paid?
3) Should cities be forced to treat all water flowing through both its sanitary and storm water drainage systems? If not, why not? If so, why and who should pick up the costs of these regulations?

Case 25 – Ticks and Deer –

Study Questions

1) What has been the history of the deer population in Maine? When did a decline occur? What conservation strategies were put in place to help deer populations recover? Has the number of white tail deer increased more recently?
2) What are some of the worries about high deer populations?
3) What is Lyme disease and how is it transmitted? What factors make it a difficult disease to manage? What factors go into its rapid spread and what might the Maine Department of Inland Fisheries and Wildlife do to reduce it?

Deer hunting in Maine has been restricted to the point where if hunters wish to have the option to harvest either sex of deer, they have to apply for a special permit. Permits can only be won by lottery and they are allocated according to Wildlife Management Districts, of which there are thirty in the state. During the 1980's and 1990's, wildlife researchers in Maine were concerned about the low fecundity rate of the deer herd generally and the lack of females in particular, drawing the conclusion that the population could only reverse its steady decline via strong conservation measures.

Beginning in 1983, the state implemented strategies such as establishing designated hunting districts and restricting some take on antlerless deer, using either-sex hunting days until 1985. In 1984, Maine also imposed a uniform 4-week firearms hunting season for deer that continues to this day. In 1986, The Maine Department of Inland Fisheries and Wildlife (MDIFW) further curbed the hunting of female deer by allowing most hunters "bucks only" licenses while issuing a more limited number of "Any-Deer" permits (more locally known by hunters as "doe permits"). Animal rights groups applauded these efforts as they read them as a signal that hunting privileges would gradually decline in the state. Many of these advocates worried about the status of the species locally, and some also opposed the hunting of deer on a moral and/or aesthetic basis.

Apparently, the strategy of limiting the taking of does has worked, perhaps too well in the opinion of some. Deer populations have bounced back dramatically as record herds have been detected this past decade, which has led to a number of inconveniences for Mainers. Vehicle collisions and damage to agriculture have increased as the size of the state herd has climbed. Even in far northern Maine, where deer have struggled to rebuild their numbers, there were 203 bucks harvested in the fall of 2012, the largest number in that region since 1963. The total deer harvest in Maine in 2012 was 21,365, up from 18,839 taken in 2011—a 13% increase.[25]

Additionally, deer carry ticks, which leads to perhaps a more alarming problem for Mainers—the increased threat of Lyme disease. Deer ticks (also called black-legged ticks) are the primary carriers of the disease. There were 918 cases of Lyme disease reported in Maine from 1993–2003, with estimates of an average of 200 cases per year for the rest of the decade. But this figure has risen dramatically relative to the human population of the state since then. In 2013, the number of confirmed cases of Lyme went over 1,000 for the first time, with another 246 probable cases of the disease. This comes to 84.6 cases per 100,000 people, one of the highest per capita rates of occurrence of Lyme in the United States (only nearby Vermont and New Hampshire were higher in 2013).[26] The first documented case of Lyme disease contracted in Maine came in 1986.[27] It is now estimated that 1 in 4 deer ticks are infected.

Anyone familiar with Lyme disease, either as one its victims or as a health care professional diagnosing and treating it, can attest that it is very hard to detect and manage. Diagnosing Lyme (especially given some of its wide-ranging symptoms similar to those of other diseases) is particularly tricky. But figuring out and practicing useful courses of treatment is even harder. A number of the symptoms are similar to those suffered by patients with rheumatoid arthritis, as Lyme disease manifests itself in muscle joint pains, typically in the arms and legs, but also in the neck. Lyme disease patients present with a crick in their necks and often move as if they are trying to "crack" their necks to alleviate the pain. Other symptoms can be initial presentation of rashes at the site of the tick bite and later, more worrying secondary symptoms such as memory loss and neurological problems.

Lyme disease is transferred to humans via tick bites. The ticks are infected with a bacterium called *Borrelia burgdorferi*. Ticks get the infection from biting other infected animals such as deer and mice. Deer and mice are

25 John Holyoke, "Biologists Says Maine Deer Herd Rebounding Nicely," *Bangor Daily News*, March 29, 2013. Available at http://bangordailynews.com/2013/03/29/outdoors/biologist-says-maine-deer-herd-rebounding-nicely/ (accessed June 9, 2015).

26 Centers for Disease Control, "Reported Cases of Lyme Disease by State or Locality, 2004–2013," (website), March 19, 2015. Available at http://www.cdc.gov/lyme/stats/chartstables/reportedcases_statelocality.html (accessed June 9, 2015).

27 Aislinn Sarnacki, "Maine Tick Myths, Revealed," *Bangor Daily News*, November 7, 2012. Available at http://bangordailynews.com/2012/11/07/outdoors/maine-tick-myths-revealed/ (accessed June 10, 2015).

asymptomatic but carry the bacteria. And of course, as mice or deer usually don't bite humans, ticks become a powerful vector for the disease in us. While the risk of infection in humans, even when bitten by ticks is rare, there are some common misconceptions of risk factors. For instance, it was once generally thought that victims of tick bites were rarely susceptible to Lyme if the tick (which will try to stay attached and "burrow" into its host after biting) is removed within 36 hours of the bite. However, some leading researchers have found that ticks only attached for about five hours and removed improperly could infect its human host. While the circular ring, target-like patterns around the bite area are standard ways of diagnosing the presence of Lyme disease in humans 1–4 weeks post-exposure, this assessment is not completely reliable. While the target pattern rash can emerge after bites in the abdomen, those positive for Lyme sometimes exhibit triangular patterns around bites in other areas of the body. How Lyme manifests itself really depends on differences in their hosts. Unfortunately, this makes diagnosis even more difficult and whether someone definitively has Lyme is practically impossible to know.

Deer ticks in Maine remain in the woods in the fall, prowling for warm hosts to carry them through the winter—usually deer or moose. The problem is that hunters are also in the woods during the autumn deer season. Beyond the fact that ticks are still present in Maine's forests when there are a goodly number of potential victims for them to prey upon, there are some popular myths that need to be dispelled.[28] One folk remedy espoused for removing ticks is to hold a lit cigarette or a match to a tick's backside, allegedly allowing them to retreat on their own. While this method might work for the common dog tick, it usually doesn't for deer ticks. When deer ticks bite, they hold on tenaciously. They insert their mouthparts (called hypostomes) deeper into the dermis of their hosts than do other ticks. Also, their hypostomes come equipped with more backward-pointing denticles (barbs) than other ticks, while secreting a glue-like substance under the skin that allows them to adhere more securely. To make matters worse, they have enzymes allowing them to detach from the cement-like secretion when they are engorged. Many young ticks that could carry Lyme are also too small to treat with a lit cigarette or match without the risk of burning the host's skin. Besides, there is a more important reason not to take these measures—the tick might just vomit or increase salivation, thus injecting even more pathogens into the wound. (Instead, you should carefully use tweezers, clasping the tick as close to the skin as possible, and using

28 Sarnacki, "Maine Tick Myths."

gradual pressure, slowly and steadily pull straight up and out).[29] Other myths include the idea that ticks die in the winter and that if you don't have a rash, then you don't have Lyme disease—both of these beliefs have proven to be unfounded.

There are the ticks, but what about their most popular non-human animal hosts? Maine has one of the largest recognized subspecies of the white tail deer. White tail bucks in the state reach the age of maturity by about the age of five and can grow to 400 lbs. Most are between 200–300 lbs. live weight and stand between 36" to 40" at the shoulder. Does are considerably smaller, usually weighing 120 to 175 lbs. live weight. Fawns generally grow rapidly going from 4–10 lbs. at birth to 85 lbs. live weight by the time they reach six months of age. Importantly to those who are travelling through woods and fields by car, they can dart out quickly and reach speeds of 40 mph at full gait, carried by long, strong, graceful legs.[30]

As in other areas of the US, the number of hunters for white tails in the Maine woods and fields has declined over the past few decades. In the 1950s, about 150,000 hunters regularly sought deer. This figure rose to a record 215,000 in 1982; in the early 1980's hunters outnumbered deer in the state. But since then, the number of deer hunters has steadily decreased, with anywhere from 170,000–175,000 today. This decline in hunters and the increase in deer population led to the state legislature allowing the MDIFW to use proscribed hunts to cull deer in some population or disease problem locations beginning in 1993.

In the background of all of this are arguments against hunting put forth by animal advocacy groups such as People for the Ethical Treatment of Animals (PETA). PETA points out that while hunting may have been necessary for survival in prehistoric times, it is not today. Instead, most people hunt for the "thrill of the chase." PETA adds that killing deer leaves their families fragmented with fawns sometimes orphaned. Furthermore, not all hunters are skilled and can wound deer without finding the carcass to use, which is seen by many animal rights advocates as causing unnecessary suffering and waste of life. Quick kills, claims PETA, are exceedingly rare and deer suffer when they are wounded. Using figures taken from the Maine Bow Hunters Alliance, it is

---

29 See University of Manitoba tick expert Kateryn Rochon describe this method in "How to Properly Remove a Tick," *YouTube* (online video), June 25, 2013. Available at https://www.youtube.com/watch?v=27McsguL2Og (accessed June 10, 2015).

30 Maine Department of Inland Fisheries and Wildlife, "Deer: Whitetails in the Maine Woods," (website), 2013. Available at http://www.maine.gov/ifw/wildlife/species/mammals/deer.html (accessed June 9, 2015).

estimated that 50 percent of animals that are shot with crossbows (which are now allowed for use in hunting deer during the firearms season in the state) are wounded but not killed. Moreover, PETA claims that a study conducted of 80 radio-tagged white-tailed deer reveals that 11 of 22 shot with "traditional archery equipment" were wounded but not recovered by hunters.[31] A British study concluded that 11% of deer taken by hunters with firearms took two or more shots to be killed and some of the wounded deer suffered for more than 15 minutes before dying.[32]

Along with direct suffering, hunting opponents also charge that hunting disrupts the migratory patterns of deer. Gunfire causes stress to old and young animals alike, causing adults to abandon dens, leaving their young vulnerable to natural predators. Hunting also disrupts the mating rituals of deer, resulting in lower birth rates. Gunfire alters their eating patterns and does not allow them to build up sufficient food stores to provide energy during the lean times in winter.

PETA members see hunting not as a wholesome sport, but as a needless and cruel exercise in killing. They point out that a true sport involves willing participants who compete with each other on a "level playing field." Hunters retort that hunting is a form of "fair chase" as long as hunting laws are followed and that their low take rates indicate hunters don't have "improper advantage." However, PETA members are not convinced by these counter arguments. They stress that deer do not have much of a chance to outrun bullets from rifles and shotguns. While arrows may not kill deer as often, they certainly cause damage to their targets. Finally, deer clearly cannot fight back.

PETA also accuses deer hunting advocates of devising sham arguments to try to justify their practices, such as that hunting is necessary as a method of deer population control. PETA contends that if nature is left to "take care of its own" there won't be as many problems with population density as when humans interfere with deer. For example, natural predators seek out only the weakest and sickest prey, which curbs the population and leaves its fittest members. Instead, hunters undercut this rationale by taking only the strongest members because they want the biggest thrill, the most meat, and the largest trophies. Even if hunting may help with population control of some species, it hasn't been particularly helpful in managing sustainable populations of

31 People for the Ethical Treatment of Animals, "Hunting," (website), 2015. Available at http://www.peta.org/issues/animals-in-entertainment/cruel-sports/hunting/ (accessed August 21, 2016).

32 People for the Ethical Treatment of Animals, "Hunting."

white-tail deer, as there are still a substantial number of them despite decades of hunting.

Some philosophers undergird these anti-hunting arguments by detailing moral principles they believe even their opponents would accept. Gary Francione and Anna Charlton contend that two simple, widely supported principles fortify their case for a vegan diet.[33] The first principle is that we have a moral obligation not to impose unnecessary suffering on animals. Francione and Charlton claim that no one doubts this principle. The second principle is that while animals matter morally, humans matter more. With respect to the first principle, Francione and Charlton are mindful that there may be long debates as to what qualifies as "necessary" versus "unnecessary" suffering, or when suffering is necessary. Nonetheless, they insist this debate is immaterial for their argument. All that matters is that there is widespread agreement on suffering that is *not* necessary—that is, to inflict pain on animals simply for one's pleasure, amusement, or entertainment. With respect to the second principle, Francione and Charlton note it doesn't matter why we think humans are more important than non-human animals. For the sake of their argument, they don't even think they need to defend the principle.

The point is that *many who eat meat* accept these principles as being true. But even given that they do, Francione and Charlton argue that if animals (including fish) have any moral standing at all, humans cannot justifiably eat them or consume products made from them, such as eggs, milk, cheese, and so forth. Thus, humans would be obligated to follow a vegan diet.[34] Francione and Charlton add that humans can acquire all of their necessary nutritional requirements with much less negative impact on the environment without consuming animals and their products. Animals not only experience great pain when they are raised and killed for food, but the development of their products also leads to their suffering. The only reason to eat meat or other animal products then is to gain pleasure, but this is wrongful given our acceptance of the first principle above.

But others argue that the ethics of hunting deer is more complex than it would seem on its face. Interestingly, some hunting advocates also appeal (at least in part) to suffering to make their arguments. There is the question of whether there is a moral obligation to hunt if one cares about the following four factors of a species such as white-tail deer; their health, their suffering,

---

33 See their *Eat Like You Care: An Examination of the Morality of Eating Animals* (Louisville, KY: Exempla Press, 2015), pp. 4–5.

34 Francione and Charlton, *Eat Like You Care*, p. 4.

human health, and human suffering. One possible argument for an ethical obligation to a hunt a species (such as white-tail deer) is that without doing so, their population will likely increase. Furthermore, while the number of white-tail deer is high, their quality of life is low. In high densities, they denude vegetation, not leaving enough food for all deer. This causes some to starve, which is clearly one way for any organism to suffer. Overpopulation of deer also allows for diseases to spread more quickly throughout the population and causes them to die from ailments that they otherwise would not likely have died from. On these grounds, measures to cull members of the deer herd causes more of them less harm than they otherwise would have endured. Hence, if one seeks to reduce white-tail deer suffering, she should hunt. Of course, if one were to accept such an argument, it would be contingent on whether it could be demonstrated that enough deer could be killed quickly.

Finally, there is the tricky business of defining what one means by the "overpopulation" of deer. The deer population is not equally distributed throughout Maine. For example, in northern, eastern, and western portions of the state there are fewer deer than local wildlife ecologists (and of course, hunters) would like. This is bad news for the hunting industry in these areas of the state, which rely on high harvests to not only keep local sportsmen and sportswomen happy, but also to draw in much needed revenue from hunters outside of the region and particularly from out-of-state. But the situation is much different in southern Maine where there are more deer, as well as more people to hit them in their vehicles or to be infected by ticks deer may be carrying. Notice that even though deer populations in Maine are 60% higher than they were in the early 1980s when hunting restrictions went into effect, the tone of the reports on the status of the deer population is somewhat subdued. They indicate that while the white-tail population is improving, it still is not what it "should be" despite some of the highest buck harvests in the past 50 years.

What should we conclude from that? That the number of females is still critically low? That hunters are simply dissatisfied because they don't detect the presence of the sex of deer that they desire to harvest? What if Maine had, by the lights of its hunters, enough deer allocated throughout the state, but this population was so large it denuded the forest landscape and caused substantial harm to the ecosystem? In fact, this is currently playing out in other states, such as Pennsylvania, where an excessive deer population causes massive degradation of forest ecosystems for all species. Nevertheless, local hunters are upset with state officials because they still think it is too hard to harvest deer. Once more, we can ask if the increase in Lyme disease is really impacted so much by having more deer around or by other factors? What about the increase in the number of car accidents caused by deer? Does the rise in the number of these instances really suggest anything other than there are more cars? Does it tell us that there are "too many deer"? And from whose perspective does it matter?

Review Questions

1) What do you think it means for there to be "too many (or too few)" deer? Do you think there are legitimate worries about a high deer population in your state? Why or why not? If you think it is an issue, are there alternative measures to alleviate the problems of having a high deer population?
2) No matter what else you might think about the ethics of hunting, do you believe the public health and safety concerns illustrated in this case study should have any bearing on the moral and legal permissibility of hunting?
3) Aside from any public health concerns, is the nuisance of a high deer population enough of a justification for hunting them?
4) Hunting enthusiasts claim that restrictions on hunting by local municipal governments and/or by private owners who post their land with no hunting signs contribute to deer overpopulation and deer tick problems. Do private property owners (especially those who own large acreages) have an ethical obligation to allow hunting on their land? Should they be forced to allow hunting on their land by the state?
5) Much of this case concerns the health and well-being of humans and deer. But what rights to life (if any) do ticks have in this case? Regardless of whether ticks possess rights, do humans have any obligations toward them? If one cares about protecting deer, why doesn't that concern extend to ticks? Assuming that neither ticks nor deer are suffering species decline, is there any reason for finding deer to be more morally valuable than ticks?

Case 26 – Green Stadiums –

Study Questions

1) Give some examples of green stadiums. What are some of the features of a "green" stadium?
2) What are some of the motivations for building green sporting facilities?

With the proliferation of professional sports teams recently looking to secure (and often finding) new places to play, there has been an interesting twist on the building of their arenas and stadiums. One of the questions now raised when a proposal for a new sporting facility is presented, given an uptick in environmental awareness, is what sort of impact will it have on the local ecosystem? Additionally, there is a wider background concern about how arenas and stadiums fit with values of sustainability more generally.

Some of the examples of ballparks built to achieve "green" principles use cutting edge innovations and extreme engineering feats. For example, at Chase Field in Phoenix, Arizona, home of the Diamondbacks Major League Baseball (MLB) team, an alternative approach is used for cooling the facility when its retractable roof is closed. Instead of the usual air conditioning systems of most arenas, the Northwinds cooling system at Chase vacuums in air over ice made in an adjacent building. The system brings the temperature down by at least 20 degrees F within a few hours, a remarkable feat given that it is battling the 100-plus degree F heat outside on most days during the summer months when the Diamondbacks play. This system allows for the roof to be open longer (which is necessary to maintain the natural turf) because it is much more efficient in cooling down the facility quickly than the original air conditioning system was. However, the system is expensive as it still costs ~$2 million to cool Chase Field for Arizona's 81 home games per season.

In 2011, the Diamondbacks collaborated with Arizona Power Service (APS) and contracted with HKS Architects and Renewable Energy Contractors to retrofit the park with a solar pavilion with a 20-year life span. The 17,280 square foot solar array not only provides shade for fans when the roof is open, but also generates about 75 kw of solar power—enough to provide for the electrical needs of 8–10 homes. APS plans to use the solar array as a demonstration site, storing the power in batteries to be used to charge electric vehicles just outside of the stadium.[35]

MLB has also made a more concerted effort to educate fans about green design and sustainable living practices at the ballpark. At Chase Field, the solar array provides a visual segue to exhibits that promote recycling, renewable energy, electric vehicles, and energy efficiency. Environmental awareness campaigns beyond MLB encourage fans to make green choices at the stadium. Baseball fans can discover which ballparks have the best vegan options by

35 Major League Baseball Advanced Media, "Chase Field History," (website), 2015. Available at http://arizona.diamondbacks.mlb.com/ari/ballpark/information/index.jsp?content=chase_history (accessed July 14, 2015).

visiting PETA.org, which has complied a ranked list (the top five include, in order, San Francisco's AT&T Park (Giants), Philadelphia's Citizen's Bank Ballpark (Phillies), Detroit's Comerica Park (Tigers), Atlanta's Turner Field (Braves), and Denver's Coors Field (Colorado Rockies)).[36]

Large stadiums that house major sporting events throughout the world can use copious amounts of resources, some of which are scarce and/or can contribute to climate change. Large video scoreboards and Jumbotrons consume lots of electricity, frequently supplied by coal-fired power plants often cited as the worst emitters of CO_2. Vast refrigeration systems are needed to keep the beers, sodas, and ice cream cold on game days. Breaks in the action between innings or especially during halftime involve the use of hundreds of gallons of water and sewage systems have to handle all of the human waste flushed from rows of urinals and toilets (hopefully to waste treatment facilities).[37] Even just outside of the stadium there are energy use and emissions concerns attached to large sporting events, such as vehicles idling in parking lots and entryways.

To combat these environmental worries, sports teams have had to improve their image as good environmental stewards and their stadiums need to reflect that commitment. The National Football League's (NFL) Philadelphia Eagles play their home games at Lincoln Financial Field (known locally as "The Linc"), which is now retrofitted with 11,000 solar panels since their installation in 2013. These panels, along with 14 wind turbines erected around the facility generate power for the stadium. The Eagles are in the midst of what they call their "Go Green" campaign (the team's colors are green, silver, and white), launched in 2003. The Eagles also reduced their use of water and electricity. Recycled paper products are used for all tissues in the stadium. Leftover grease and oil from the extensive network of grills and fryers throughout "The Linc" is collected and processed for biodiesel production. Even the cheerleader calendars can't be found in paper form anymore, but are available digitally on the team's website. Since their deployment, all of these conservation measures have combined to cut the Eagle's energy consumption in half.

Other NFL franchises have been lauded for implementing green features in their stadiums and revising their environmental policies. The San Francisco 49ers moved into Levi's Stadium, a 68,500-seat facility they built in Santa Clara, California. The $1.2 billion facility will be the first with a "living roof"—a garden of green and flowering plants covering the top of an eight-story tower of

36 People for the Ethical Treatment of Animals, "Top Ten Veg-Friendly Major League Ballparks," (website), May 18, 2011. Available at http://www.peta.org/blog/top-10-vegetarian-friendly-major-league-ballparks/ (accessed August 20, 2014).

37 See Case 24—"Foul Ball: Ballpark's Effluent out of Play"?

luxury boxes along the edge of the stadium.[38] This 18,000 square foot design feature helps reduce the stadium's energy use by forming a natural layer of insulation. Pedestrian bridges leading to and from the stadium are affixed with solar panels to collect rays from the California sun. This energy is used in car charging stations in the parking lot to fuel up plug-in electric vehicles in the parking lots. Levi's Stadium, as does a growing number of NFL venues, also includes wind turbines and other low-carbon alternatives to save energy and reduce its footprint.

Part of the motivation for more teams to join in the green stadium effort began with the launch of the Green Sports Alliance (GSA) in 2011. Not only are professional teams involved, but college ones as well since facilities (especially in the case of National Collegiate Athletic Association (NCAA) Division I programs) may be even larger than those of the pros with respect to seating capacity and land use footprints. The GSA has both environmental bona fides and big money behind it as the Natural Resources Defense Council and Seattle Seahawks owner (and former Microsoft co-owner) Paul Allen's Vulcan Inc. co-founded the organization. Its list of members includes almost half of the teams from the NFL, two-thirds from MLB, and half of the National Hockey League (NHL), in addition to 21 NCAA Division I universities.

The GSA was formed to harness the power of sports to raise environmental awareness. The organization's officials recognize that when sports teams go green they provide powerful examples for supporting environmentally friendly practices in the public sphere. These measures also aid teams by lowering their long-term energy costs, even though heavy initial green infrastructural investment is required. The idea is that everyone gains when these sustainability initiatives are taken. As Scott Jenkins, the Chairman and President of GSA notes, "Everywhere you turn, it's win-win." Jenkins is also the General Manager of Mercedes-Benz Stadium, which will serve as home to the NFL's Atlanta Falcons beginning in 2017. This $1.2 billion facility will include a rainwater collection system that irrigates the grass surface and cools the stadium.

Nationals Park, the home venue of MLB's Washington franchise, started the green stadium trend in American professional sports by becoming the first major sporting stadium to achieve the U.S. Green Building's Council LEED Certification in 2008. It is a LEED Silver rated facility for New Construction based primarily on its energy efficient lighting and innovative cooling system. While Nationals Park is still lauded for being one of the "greenest" stadiums in

38 Jim Carlton, "Some NFL Teams are Going Green," *Wall Street Journal*, May 18, 2014. Available at http://www.wsj.com/articles/SB10001424052702304677904579537882691550494 (accessed June 8, 2015).

the MLB, newer baseball parks are achieving even higher standards of sustainable construction and operation. The Miami Marlins opened Marlins Park in the spring of 2013 and won the Gold LEED certification—a designation earned mainly by virtue of its energy efficiency. Its mechanical, lighting, cooling, heating, and electrical systems were all state of the art when it was built and the ability to score highly on so many conservation fronts make it considered by many as the most environmentally friendly MLB stadium. Another park built since Nationals Park opening, Target Field (home of MLB's Minnesota Twins), is praised because it has won not only one, but two LEED certifications—one for New Construction and another Operations LEED for sustainable practices in the park's operation.

It may be expected that new stadiums would incorporate green building principles. But, even older MLB baseball facilities have been retrofitted to meet sustainability goals. For example, venerable Fenway Park, built over a century ago for the Boston Red Sox, underwent a number of green updates in 2010, including new LED lighting, waterless urinals, and a number of other energy saving features. In 2015, a rooftop garden (dubbed "Fenway Farms") was added on the third base side of the ballpark, supplying about 4,000 lbs. of produce annually and absorbing storm water.

While we see the motivations team owners and stadium authorities might have for utilizing green strategies for sporting facilities, interesting moral questions arise for fans. For example, if my team fails to support green measures and I believe that they should, do I have an obligation to try to persuade them to do so? Furthermore, imagine that my team doesn't support green measures even after prodding. Should I then disavow my loyalty to the team and support another that does make contributions to environmental sustainability? On the other hand, imagine that one denies the science of climate change and remains extremely skeptical that few or any of the dire consequences of business as usual practices will come to fruition. Should that fan criticize her team if it promotes measures on grounds of combatting global warming?

Review Questions

1) What obligations do you think those who build stadiums should have with respect to their design? To what extent should they strive to be "ecologically friendly" if at all? Would your answer to the previous parts of this question change if you were reminded that a high proportion of professional sports stadiums are publically funded, as is the case with many facilities for NCAA Division I athletic programs at public universities?

2) Let's say we agree that stadiums should be built or renovated using green design concepts. Consider the costs of new stadiums. New Yankee Stadium, which opened in 2009 cost $1.5 billion, and Levi's Stadium (2014) and Mercedes-Benz Stadium (2017) each cost $2.1 billion. Presumably with each new stadium comes a new environmental footprint. Given these factors, should there be a moratorium on the construction of new sports facilities? Additionally, should there be a call to update existing stadiums and ballparks to either meet, or at least strive to meet, green certification guidelines?

Case 27 – Cockfighting and Culture –

Study Questions

1) How pervasive is cockfighting in the United States and where does it occur? What legal measures have been taken in the US to ban cockfighting? Do some states have more relaxed restrictions on cockfighting than others and in what ways are they less constraining? What have the consequences of more lenient laws been?
2) Why has cockfighting been banned in so many states? What are some objections to cockfighting?
3) What have been some of the consequences of cracking down on cockfighting? Has this aggressive approach eliminated the practice in places where civil authorities have tried to enforce laws against it?

Cockfighting was banned in the United States by law in 2007 (the last two states to impose bans were New Mexico and Louisiana (enacted in 2008)). In 2014, as part of the Farm Bill and its attached Animal Fighting Spectator Act, it became a federal deferral misdemeanor to watch any organized animal fight, with a fine of $100,000 imposed on anyone convicted of doing so. But since the 2007 ban, a reprieve from it was given to Louisiana and New Mexico. Oklahoma was allowed to repeal its ban in 2014. Even in some states that are not exempt from the ban, such as Kentucky, laws preventing cockfighting are fairly weak and enforcement of them is lax. The practice still exists legally in many places (such as Nicaragua, Cuba, Mexico, France, Colombia, Venezuela, Ecuador, Panama, Philippines, Peru, Saipan, and the Canary Islands). Cockfighting is even the national sport of some countries or is at least a close second (such as in the baseball-obsessed Dominican Republic).

Opponents of cockfighting regard it as a deplorable practice amounting to staged animal cruelty for the cheap amusement of humans. With its association to gambling (cockfights usually involve betting on which rooster will win), opponents see another morally unsavory element—that the lives of animals are traded for profit. The losing cock is usually killed in the fight. With that said, most states have updated and strengthened their laws, with forty of them charging those who operate or participate in cockfighting with a felony. As mentioned earlier, with a 2014 update of the Farm Bill, it became a federal crime to attend or bring a minor to a cockfight.[39]

When the ban was enacted, the state of New Mexico committed vast resources in an attempt to end cockfighting with raids of suspected rings, but has had little success.[40] Many of the cockfighters and spectators simply moved underground. Admittedly, New Mexico's penalties upon conviction were relatively light (a first offense is a minor misdemeanor) and the state still attracted cockfighters from four of its five neighboring states where participation in the sport is a felony.[41]

Ed Lowry breeds fighting rooster lines and is director of the New Mexico Gamefowl Association, a group advocating cockfighting. He sued the state and was part of the movement that helped New Mexico gain legal exemption from the ban. Lowry and his attorneys made their case on the grounds of tribal, religious, and cultural sovereignty.[42] However, it will be interesting to see if states that currently have exemptions will be able to keep them. Public opinion seems to be turning against any form of staged animal fighting with the spotlight thrown on the topic since Michael Vick's dogfighting indictment. Heather Ferguson, the legislative director for Animal Protection of New Mexico, an animal rights advocacy group, stated that a hotline set up to report animal cruelty receives about 90 calls every two weeks. The local cockfighting industry (which once generated $80 million a year) appears to be in decline, as feed stores and hotels in areas known for cockfighting have seen a 70% loss in profits.[43]

Cockfighting involves two roosters coaxed to battle each other to the death, typically as spectators wager on which cock will emerge victorious. The roosters

---

39 Wayne Pacelle, "Weak Laws Bring Cockfighters, Mafia Flocking into Kentucky," *Huffington Post*, (website), May 8, 2014. Available at http://www.huffingtonpost.com/wayne-pacelle/weak-laws-bring-cockfight_b_5290848.html (accessed June 10, 2015).

40 Adam B. Ellick, "A Ban on Cockfighting, but Tradition Lives On," *The New York Times*, July 6, 2008. Available at http://www.nytimes.com/2008/07/06/us/06fight.html?pagewanted=all&_r=0 (accessed August 21, 2016).

41 Ellick, "A Ban on Cockfighting."

42 Ellick, "A Ban on Cockfighting."

43 Ellick, "A Ban on Cockfighting."

are reportedly pumped up with stimulants and usually outfitted with razor-sharp blades (called gaffs) designed to cause injury to its opponents. This often results in punctured lungs, broken bones, gouged eyes, and lacerations that critics claim are left untreated in winning birds. The fights themselves can last from several minutes to about an hour with both birds sometimes dying. Critics say that even when a bird is not killed from the fight it suffers tremendously. The roosters cannot escape from the pit, which means they essentially have to "win or die." There have been cases where handlers have been killed by accidently after being slashed with the razor-sharp gaffs attached to their own birds.

Critics of cockfighting have cited other concerns beyond cruelty to animals and potential dangers to their handlers. As mentioned, gambling is normally an integral part of these events. Both cock owners and spectators wager large sums of money (sometimes in the thousands) on the results of the fights. Owners whose birds win a derby by virtue of being victorious in a series of fights can expect a large windfall. This income (usually tens of thousands of dollars) is unreported and thus not taxed. With lots of money at stake and given that the means by which it is gained are illegal, many of owners and spectators come to cockfights armed and cockfight-related homicide is not uncommon.[44]

Opponents of cockfighting also note a correlation between cockfighting and the distribution of illegal drugs. Authorities cite cases of only learning about cockfighting operations through their narcotics investigations. Worst of all, children are sometimes brought to watch the fights. Some cockfighting venues have special "family nights" where children under the age of 12 are admitted for free. In this way, a tradition of cruelty and illegal behavior is reinforced and passed down to the next generation.

Yet, even some police officers question aggressive enforcement approaches to cockfighting like those taken in New Mexico. While some raids have been successful (including one that netted a sum in excess of $20,000 dollars), many high-ranking officials, even when opposed to cockfighting, worry that the crime is being disproportionately emphasized. As one officer noted, "We don't even investigate misdemeanors of other crimes, we laugh at those investigations." He added that vital police resources are squandered on raids of cockfighting operations: "We wasted $10,000 on a recent misdemeanor. I'd rather use that on a D.U.I. checkpoint and take 20 people off the road in three hours and save lives over chickens, but whoop-de-do, a misdemeanor."[45]

---

44 The Humane Society of the United States, "Cockfighting Fact Sheet," (website), March 9, 2011. Available at http://www.humanesociety.org/issues/cockfighting/facts/cockfighting_fact_sheet.html?credit=web_id80597821 (accessed June 10, 2015).

45 Ellick, "A Ban on Cockfighting."

In 2014, federal agents serving undercover made busts at two major cockfighting venues, the Big Blue Sportsman's Club and the Red Bush, both in Kentucky. In the affidavit supporting several indictments, Special Agent Stan Wojtkonski of the United States Department of Agriculture reported that the perpetrators boasted about selling roosters to members of Mexican drug cartels. Members of the Mexican Mafia were said to have attended cockfights in the pit of one of the defendants.[46] The cousin of one of the cockfight pit's owners was the House Speaker in Kentucky, who has been a vocal opponent of legislation that would make cockfighting a felony in the state. Kentucky and neighboring Ohio have some of the weakest anti-cockfighting laws in the US and are a part of what is sometimes called the "cockfighting corridor" that stretches south from these two states through Tennessee and Alabama to Mississippi. Many of the fighting roosters come from farms in Mississippi.[47]

Anti-cockfighting groups charge that despite the most recent policy efforts, laws in "cockfighting corridor" states have not really changed in decades. Politicians there have not been eager to match their own state's laws with stricter legislation found in other states or the federal government. Opponents insist that this neglect has allowed free access to cockfighting opportunities that draw in criminal elements, as not only is holding fights illegal, but also one can be charged on separate counts of betting on fights, thus facing gambling charges. Some accuse local law enforcement of turning a blind eye to known cockfighting establishments. The Humane Society of the United States (HSUS) smuggled out undercover footage from a cockfighting pit in Manchester, Kentucky, where uniformed police officers stood by watching fights without intervening. Cockfighters from the neighboring states of Virginia and North Carolina have crossed the border into eastern Kentucky to engage in the practice, again primarily to take advantage of weak laws and lax enforcement. Others have been attracted to Kentucky from South Carolina, Ohio, Michigan, West Virginia, Maryland, and Georgia. Cockfights also seem to have local support as hundreds of spectators attend.[48]

During the trial of those suspected and then later convicted of animal cruelty in the Big Blue Sporting Club case, United States Attorney Timothy J. Heaphy charged that,

---

46 Pacelle, "Weak Laws Bring Cockfighters."

47 Brian Ross, "Washington Keeps Cockfighting Legal in Three States," *ABC News*, (website), March 30, 2015. Available at http://abcnews.go.com/WNT/story?id=131142 (accessed July 14, 2015).

48 Ross, "Washington Keeps Cockfighting Legal."

> The Big Blue Sporting Club was a sophisticated criminal operation that collected substantial profits by abusing animals. The fighting birds were horribly mistreated by these defendants. Cases like this help bring illegal private gambling and animal cruelty out of the shadows. Thanks to the persistence and creativity of law enforcement, this operation has been exposed, its profits disgorged, and its operators held accountable.[49]

One defendant was convicted to 18 months in jail and two others were sentenced to 10 months each. The defendants were also forced to forfeit over $905,000. This money was generated mainly from fight proceeds collected from spectators and participants who were only allowed entry with a valid membership card, costing a one-time $20.00 fee. Participants were charged $2,500 per entrant and at times 400 people attended the fights while bulldozers blocked a roadway to prevent entry or exit from the venue until the fights concluded. Some take issue with cockfighting on the international stage as well, with worries that it facilitates the spread of bird flu and is the primary reason for the disease's pervasiveness in Southeast Asia.

Despite its critics, cockfighting has its supporters, and others simply don't think it's a significant issue. Utah State Representative Earl Tanner took a unique tact in trying to defend cockfighting. He first noted that Americans eat 7 billion chickens a year, after being raised in pens, slaughtered, and then boiled. He wondered how this compares with cockfighting. Tanner argued, "We eat a lot of chicken nuggets here, guys, and the thought that what these cockfighters are doing is worthy of being a felony is a little bit absurd."[50] He added that fighting is what comes naturally to roosters and when they die they do so from another rooster: "[w]hen the end finally comes they're not being strapped upside down inside a huge factory. By animal standards, this isn't such a bad life that these fighting roosters have."[51] Others share these sentiments, while painting cockfighting participants not as criminals, but as "regular folk" who happen to enjoy the practice. As the director of the American Animal Husbandry Coalition puts it, "They're a cross-section of America. You know,

49 "New Details Released in 'Big Blue' Cockfighting Operation," *WSAZ News 3*, (website), October 10, 2014. Available at http://www.wsaz.com/home/headlines/Floyd-County-Ky-Couple-to-be-Sentenced-in-Cockfighting-Case-278682911.html (accessed July 8, 2016).

50 Robert Gehrke, "Utah May Toughen Cockfighting Laws After Years of Trying," *The Salt Lake Tribune*, March 10, 2015. Available at http://www.sltrib.com/news/2276551-155/utah-may-toughen-cockfighting-laws-after (accessed August 21, 2016).

51 Gehrke, "Utah May Toughen Cockfighting Laws."

we're God-fearing people who pay our taxes. [Cockfighting] is brutal, but it's not cruel."[52]

Lowry, who advocates cockfighting on cultural grounds, cites symbolic and nationalist reasons to promote the practice, "A gamecock shows me what an American should be like. You defend to the death."[53] Some New Mexican legislators add that many of the state's Hispanic citizens regard cockfighting as an important part of their culture. While fighting two proposed cockfighting bans (both voted down in the New Mexico Senate's Conservation Committee in 2005), State Senator Phil Griego of San Jose endorsed leaving the decision on banning cockfighting to local jurisdictions: "To do away with cockfighting is to do away with a major part of our culture, of our Hispanic heritage. We have to take a stand here as rural representatives and to make absolutely clear that the rural ways of life continue to exist."[54] Lowry was relieved by the results of the vote, thinking that it sent a clear message, "'Leave us alone,' that's what [the committee] was saying."[55]

Cockfighting supporters packing the New Mexico Senate floor during a special hearing thought the proposed bans would abolish a generations-old tradition, as well as decease jobs and community networking that come with the practice. There is some resentment from those involved with either breeding or fighting gamecocks such as Juan Gonzalez to what they see as undue meddling from outsiders. As Gonzalez says, "Whether you like it or not, we don't appreciate you coming to our state and saying it's wrong. We've been doing it for hundreds of years. It's not wrong to us."[56]

Even some celebrities attended the hearing. Actresses Ali McGraw and Rue McClanahan were enlisted by animal rights groups to help advocate against cockfighting. But others, such as Wilfred Brimley, were more skeptical and spoke out against the proposed ban.

Opponents of cockfighting believe we have to examine claims about the supposed cultural importance of cockfighting more closely, asking what kind of character such culture fosters. While some practices are deeply entrenched, they may also be deeply immoral. This sentiment certainly came out in a

52 Quoted in Ross, "Washington Keeps Cockfighting Legal."

53 Quoted in Ellick, "A Ban on Cockfighting."

54 Kate Nash, "2 Cockfighting Bans May Die," *Albuquerque Journal*, February 23, 2005. Available at http://www.abqjournal.com/xgr/308380xgr02-23-05.htm (accessed July 14, 2015).

55 Nash, "2 Cockfighting Bans."

56 Nash, "2 Cockfighting Bans."

statement from United States District Judge James P. Jones when he admonished the defendants in the Big Blue case while conveying their sentence,

> It does not enhance the human being to inflict pain on animals. It simply doesn't. It's something that ought to stop. There is no good purpose for it, and, as the government points out, bad things happen around these types of events—gambling obviously occurred, people spending money that they have no business spending, and we can all imagine the grief that may occur in families and individuals who become addicted to something like this. It diminishes us as human beings to treat animals in this fashion, and I think Congress has enacted a law that deserves enforcement.[57]

Also, there are certainly self-identifying Hispanics who don't endorse cockfighting on cultural grounds. State Senator Mary Jane Garcia of Dona Ana, who was a sponsor of one of the two proposed cockfighting bans in 2005, said she was disappointed in losing the vote. She added that she doesn't consider cockfighting to be part of her Hispanic culture. "We'll just keep persisting. I think the culture (in the Capitol) will begin to change."[58]

Review Questions

1) Do you think cockfighting is a sport? Does this matter with respect to showing that it is a traditional practice that should be protected on this basis?
2) Communitarian political philosophers think local communities should have a strong role in deciding what are to be considered morally acceptable practices for themselves. They also argue that ultimately cultural preservation is so critical that many traditions other societies might find to be morally problematic should be protected. Advocates of cockfighting might thus contend that the practice has a long-standing tradition in New Mexico and Louisiana, and should be protected in these places on cultural grounds. Do you think this could be a strong enough argument to allow for legal cockfighting in New Mexico or Louisiana? Do you think special exemptions should be allowed to local groups desiring to continue cockfighting?
3) Some think the case for cockfighting has suffered a major blow due to what are now other notorious animal fighting cases (such as the Michael

---

57 "New Details Released."

58 Nash, "2 Cockfighting Bans."

Vick dogfighting case) that galvanized public opinion largely against these practices. But what if someone were to say dogfighting is morally problematic and cockfighting isn't on the grounds that dogs are more aware of their plight than roosters? Would you agree? Why or why not? Do you think cockfighting is morally wrong, though not as morally reprehensible as dogfighting? On what grounds do you defend your position?

4) Consider the point Representative Tanner of Utah makes about cockfighting. Millions of chickens are killed in the food industry every day (often suffering great pain and deprivation in the process) with no legal prohibition, yet cockfighting is banned in most of the US. Do you think this is morally consistent? What, if anything is the moral distinction between cockfighting and the killing of roosters in food production (especially since several recent undercover operations have exposed cruelty to chickens in poultry plants)? Overall, do you agree with Tanner's position? Why or why not?

CHAPTER 8

Environmental Reform and Unintended Consequences

Case 28 – Acids and Bases (... of Geoengineering) –

Study Questions

1) What is ocean acidification? Explain the process.
2) What happens to coral due to ocean acidification? Explain the process.
3) What is geoengineering and what are some examples of it related to this case study? What have been pointed out as some moral benefits and moral difficulties with geoengineering?

When most people think about a carbon problem relating to the environment, they are likely to assume it just *is* the problem of destructive climate change from carbon dioxide's role as a greenhouse gas. Ever-increasing carbon emissions are leading to an atmosphere oversaturated by gases that contribute to the greenhouse effect, which is warming the Earth and causing adverse climatic conditions for all life. However, what is sometimes called "the other carbon problem" is quickly drawing the attention of climate scientists and oceanographers. Just five years ago, very little was said about *ocean acidification*, which is defined as chemical changes in the ocean due to carbon emissions. The problem has been less studied than the more widely known (and heavily debated) notion of climate change. In fact, not that long ago those worried about anthropogenic contributions to the warming of the Earth were grateful that the ocean could serve as a carbon sink, as it was regarded as a counter to the rapidly growing concentration of CO2 in the atmosphere.[1] However, due to greater degradation of marine environments such as coral reefs due to the world's oceans becoming pH neutral from their absorption of CO2, this problem is receiving a spotlight.

Actually, the term "ocean acidification" is a bit of a misnomer. The issue isn't so much that the world's seawaters are acidic or becoming more so. Seawater

1 National Oceanic and Atmospheric Administration, PMEL Carbon Program. "Ocean Acidification: The Other Carbon Dioxide Problem." (website), 2015. Available at http://www.pmel.noaa.gov/co2/story/Ocean+Acidification (accessed March 11, 2015).

 | DOI 10.1163/9789004339996_010

is slightly basic (with a pH > 7) and the term refers to the process whereby the world's oceans are moving toward pH neutral conditions. According to a 2014 International Panel on Climate Change Report, there is "high confidence" that marine organisms will endure increasingly lower oxygen levels with more ill effects from ocean acidification.[2] These combined factors will leave both polar ecosystems and coral reefs highly vulnerable.

Oceans absorb about 25% of the CO2 emitted into the Earth's atmosphere each year. This means that as increases in CO2 concentration level occur in the atmosphere, they do so in the ocean as well. When carbon dioxide is combined with water and a carbonate ion, this produces 2 bicarbonate ions. As CO2 is absorbed by seawater, chemical reactions occur that lower its pH, carbonate ion concentration, and saturation states of calcium carbonate minerals, making the water less alkaline. Currently oceans are slightly alkaline with a pH of 8.1. Yet, dissolved CO2 results in increased [H+] thus lowering the pH of seawater.[3] Also, the hydrogen ions produced by carbonic acid bind to carbonate (CO32-) to form bicarbonate (HCO3), thus decreasing the amount of carbonate in water. Since the industrial revolution the surface carbonate concentration of the tropical and southern oceans has decreased by 10%.[4]

The problem is that the consumption of bicarbonate ions by marine animals impedes the calcification process that allows them to build bones and shells. Carbonate calcium minerals are the building blocks for the skeletons and shells of several marine organisms. However, the undersaturation of carbonate calcium minerals, which occurs in seawater with lower pH, does not allow for marine organisms to quickly and properly build shells. Furthermore, when carbonate concentrations fall too low, even existing CaC3 begins to dissolve. This means that not only do corals, mollusks, echinoderms, calcareous algae, and other shelled marine organisms have difficulty building new shells,

---

2 Intergovernmental Panel on Climate Change, "Summary for Policymakers." in C.B. Field, V.R. Barros, D.J. Dokken, K.J. Mach, M.D. Mastrandrea, T.E. Bilir, M. Chatterjee, K.L. Ebi, Y.O. Estrada, R.C. Genova, B. Girma, E.S. Kissel, A.N. Levy, S. MacCracken, P.R. Mastrandrea, and L.L. White, eds. *Climate Change 2014: Impacts, Adaptation, and Vulnerability. Part A: Global and Sectoral Aspects. Contribution of Working Group II to the Fifth Assessment Report of the Intergovernmental Panel on Climate Change* (Cambridge, UK and New York, NY, USA: Cambridge University Press, 2014), pp. 1–32 at 13.

3 Marah Mardt and Carl Safina, "Covering Ocean Acidification," *Yale Climate Connections*, (website), June 24, 2008. Available at http://www.yaleclimateconnections.org/2008/06/covering-ocean-acidification-chemistry-and-considerations/ (accessed April 6, 2015).

4 Mardt and Safina, "Covering Ocean Acidification."

but they also have a hard time maintaining them.[5] Predictably, the skeletal building capacity of fish is also decreased due to ocean acidification.

The cycle for anthropogenic causes of ocean acidification begins with the burning of fossil fuels, cement manufacturing, and land use changes.[6] These drivers lead to atmospheric changes as CO_2 is pumped into the stratosphere.[7] When precipitation occurs, CO_2 is washed into oceans, which further causes increased CO_2 bicarbonate ions to exist. At the same time, CO_2 carbonate and pH decrease.[8] When combined with the warming and de-oxygenation of oceans we're witnessing, there are adverse effects on marine ecosystem.

Ocean acidification is occurring more rapidly than it has over the past 300 million years. In 1850, the global mean surface pH of the ocean was around 8.2, 0.1 higher than it is today. Note that the pH scale, like the Richter scale, is logarithmic, so what may be initially seen as a negligible change actually represents a 30% increase in ocean acidity over roughly the last century and a half. Projecting into the future, several climate models forecast continued lowering of pH in the Earth's oceans. The RCP 2.6 model[9] (where carbon emissions would peak between 2010 and 2020 and rapidly decline after that) predicts that by 2050, ocean pH will be 8.05, though this model indicates this rate of change would flatten out between 2050–2100. However, the RCP 8.5 climate model (sometimes called the "business as usual model" where emissions continue to rise during the 21st century) indicates that pH will fall quite precipitously to between 7.7–7.8 by 2100. This means that by the end of the century, the global mean of surface ocean waters could well become 150 times less alkaline than they are today. This would result in ocean pH that has not occurred on Earth for more than 20 million years.[10]

There are already deleterious changes to food webs, ecosystems, and assemblages of marine life attributable to ocean acidification and this would only

---

5 Mardt and Safina, "Covering Ocean Acidification."

6 Intergovernmental Panel on Climate Change, Working Group I, "Climate Change 2013: The Physical Science Basis," *Fifth Assessment Report of the Intergovernmental Panel on Climate Change* (Cambridge, UK and New York, NY, USA: Cambridge University Press, 2013), 6.3.2.

7 Intergovernmental Panel on Climate Change, Working Group I, "Climate Change 2013," 2.2.1.

8 Intergovernmental Panel on Climate Change, Working Group I, "Climate Change 2013," 3.8.2 and 30.2.2.

9 RCP stands for "Representational Concentration Pathways."

10 National Oceanic and Atmospheric Administration, PMEL Carbon Program, "What is Ocean Acidification? The Chemistry," 2015. Available at http://www.pmel.noaa.gov/co2/story/What+is+Ocean+Acidification%3F (accessed March 11, 2015).

be expected to worsen into the future without substantial reduction in carbon emissions. Additionally, it isn't merely that more marine animals are dying overall that concerns scientists; it is instead the loss of whole species, thus eroding the biodiversity of the sea. And of course, changes in water chemistry from increased carbon absorption is not an isolated process, as changes in biogas production are linked to feedback loops in the climate; sometimes in ways we don't fully understand.

One of the most pressing worries for oceanographers and marine ecologists is the anticipated adverse effects of ocean acidification on coral reefs. One reason for these concerns is that healthy corals harbor some of the most robust ecosystems of marine flora and fauna on Earth. Thus, corals are considered to be keystone species, serving as a foundation for biodiversity in the seas. Corals are marine invertebrates of the class *Anthozoa* and the phylum *Cnidaria*. Other animals living under the *cnidarian* phylum include jellyfish and sea anemones. Corals can come in a wide array of different colors, shapes, and sizes, but they share the same basic structure; with a mouth, simple stomach, and tentacles bristling with thousands of stinging cells called cnidoblasts that serve as protection and a way to stun prey.[11]

Each individual animal is called a polyp, but corals tend to live in compact colonies made up of thousands of individual polyps. The colonies are constructed through a process called "budding" whereby individual polyps clone themselves. Corals can exist in both deep, cold waters and shallow, tropical waters. However, temperate and tropical corals can only be found in waters between 30 degrees latitude North and 30 degrees latitude South of the equator.

There are soft and hard corals. About 800 species of hard corals exist in the world and these are typically called "reef-building" corals. Reefs composed of these hard corals are usually part of a larger ecosystem of mangroves and seagrass beds. On the other hand, soft corals, such as sea feathers, sea whips, and sea fans do not secrete the calcium necessary for building hard calcareous exoskeletons known in the hard reef-building corals. Soft corals instead build wood-like structures for support and have fleshy rinds for protection. These corals, like hard corals, live in colonies that resemble trees or plants. But you can identify soft corals by their tentacles, which occur in numerals of eight and tend to look feathery. Soft corals also attach themselves to sea caves and ledges, often hanging upside down to catch their prey, which are floating by along currents. These corals tend to live in oceans from the equator to the North and South Poles.

11 International Coral Reef Initiative, "What are Corals"? (White paper), 2015. Available at http://www.icriforum.org/about-coral-reefs/what-are-corals (accessed March 19, 2015).

Hard corals construct reefs first by extracting copious amounts of carbon from seawater. In turn, they build a hardened structure to use as protection and accommodate their growth. Thousands of individual polyps thus create massive carbonate structures. But corals also provide protection and habitat for thousands, if not millions, of other species. In fact, coral reefs are the largest living structures on Earth and the only ones that can be seen from space. Reef-building hard corals can be found in swallower waters that are about 100 feet deep with temperatures between ~60–90 degrees Fahrenheit, where lots of light filters through the water.

Corals have adapted in interesting ways during their 200–300 million years of evolutionary history. Reef-building corals are intriguing due to their symbiotic relationship with microscopic, single-celled algae called zooxanthellae. Zooxanthellae live in the tissues of coral polyps. The coral provides shelter for the zooxanthellae, while in turn the algae provide the polyps with food.[12] They share space, nutrients, and gas exchange. More specifically, the corals provide compounds necessary for the zooxanthellae's photosynthesis through their metabolic waste products. In return, the algae provide oxygen and remove the corals' wastes.[13] Most importantly for the corals, they gain nutrients from the algae; as a byproduct of photosynthesis, the zooxanthellae generate carbohydrates that are instantly absorbed internally by the polyps. The corals use these to build fats and synthesize their carbonate calcium skeletons. About 90% of organic material produced by zooxanthellae end up entering the corals, making them a major contributor to the health of reefs. The pigment within the millions of zooxanthellae living inside the translucent polyps give healthy corals their vibrant colors.[14]

Light drives corals to compete with a variety of species for space on the seafloor, but this makes them very susceptible to environmental stresses. Also, due to the symbiotic relationship corals have with zooxanthellae, any changes in the environment that impact the latter will also dramatically affect the former. Corals reproduce slowly and their growth rate is sluggish. The more massive the coral colony is, the slower growing it is. Massive corals typically only

---

12 Smithsonian Institute, "What is Coral? A Coral Polyp and Zooxanthellae," (website), 2015, Available at http://ocean.si.edu/ocean-photos/what-coral-coral-polyp-and-zooxanthellae (accessed July 3, 2016).

13 United States National Atmospheric and Oceanic Administration Coral Reef Conservation Program, "Symbiotic Algae," (website), 2015. Available at http://coralreef.noaa.gov/about-corals/coral101/symbioticalgae/ (accessed March 19, 2015).

14 United States National Atmospheric and Oceanic Administration Coral Reef Conservation Program, "Symbiotic Algae."

grow 0.5 to 2.0 centimeters (cm) per year. Even under ideal conditions, with lots of food, abundant light filtration, consistent temperature, and moderate wave action, corals can only grow up to 4.5 cm per year.[15]

An emerging consensus thinks that ocean acidification is having a devastating effect on coral, but just how much of a problem is this for marine life? After all, one might think corals cover a relatively small area (when compared to the global ocean floor). An estimated 284,000 to 514,000 square kilometers of the Earth's surface is covered by shallow coral reefs, with even more covered by deep, cold-water reefs. This means that cumulatively shallow coral reefs worldwide cover an area somewhere from the size of the country of Ecuador (low estimate) to the country of Spain (high estimate). But to put this into context, this area of roughly 198,000 square miles is only 0.015 of the 140,000,000 square miles of the Earth's ocean areas. Yet, this fact is actually cause for grave concern, as recall that corals play a crucial role in sustaining sea life. What is amazing to consider is that in this fraction of the world's oceans, coral reefs are home to about a quarter of the world's aquatic species. In short, no other ecosystem supports so many life forms in such a small area.

There are socio-economic consequences associated with ocean acidification. Fisheries aquaculture is imperiled and countries that depend heavily on fishing for subsistence face obvious food security issues. Protection of coastlines that is usually provided by healthy corals along reef coasts will likely lessen as dead coral reefs disintegrate. The tourist industry will be adversely affected in shoreline areas as prime ocean real estate is lost to rising seas and more species disappear that attract recreational tourists. Worldwide, there is a $30 billion coral reef tourism industry (driven mainly by diving) that is under threat.[16] All of this could lead to climate regulation as well as investigating carbon storage possibilities in the future.[17]

Many groups (such as the Intergovernmental Panel on Climate Change (IPCC)) are working on policy options and providing forums to discuss how to combat ocean acidification, including the UN Framework Convention on Climate Change: Conference of the Parties and the Conference on Sustainable Development (Rio+20). Additionally, work conducted by the Convention on Biological Diversity could produce vital input on policies for dealing with the issue. Of course, several organizations have already deliberated on possible ways to mitigate climate change via the reduction of the emission of greenhouse gases through international cooperation.

15 International Coral Reef Initiative, "What are Corals?".

16 Mardt and Safina, "Covering Ocean Acidification."

17 IPCC, Working Group II, 2014, Ch. 6.

A more radical approach for action on climate change is *geoengineering*. Geoengineering is the large-scale manipulation of environmental processes that affect the Earth's climate, mainly in an attempt to counter climate change. Some geoengineering options might include ocean fertilization, methods for pulling carbon out of the atmosphere, artificial trees, and solar radiation management (i.e., controlling sunlight before it reaches the Earth). A final policy option is the use of local or regional acts, laws, and policies aimed at reducing some of the stresses caused by ocean acidification.[18]

Many moral issues arise with geoengineering despite the intent to use it to reverse destructive anthropogenic changes in our climate. As Duncan McLaren of *The Guardian* has pointed out, with great power comes great responsibility.[19] Geoengineering holds great potential mainly because it may have powerful tools at work to try to fix climate problems. Consider *solar radiation management* (SRM). This technology is poised to allow for quicker, more flexible, and greater control over climate change. Generally speaking, SRM is an umbrella term for various techniques whose purpose is to reflect or redirect radiation back out into space. By reflecting a small percentage of sunlight out into space using aerosols, we can cool the Earth's atmosphere, thus reducing global warming. Limited research on SRM indicates some promise in this approach. SRM could reduce global temperatures in an extremely short time span, perhaps within a few months of initial deployment. Though it couldn't eliminate regional temperature and precipitation changes brought on by climate change, SRM could mitigate them. SRM could be quite cheap when compared with the relatively high costs of placing strict restrictions on greenhouse gas (GHG) emissions.

But as with most new techniques, there are potential limitations and costs associated with SRM. First of all, critics of SRM point out that these techniques only mask some of the negative effects of high GHG concentration levels in the Earth's atmosphere. SRM obviously doesn't reduce emissions and isn't a substitute for reductions in carbon output. There are other problems caused by GHG beyond merely the greenhouse effect, including of course, ocean acidification in the case of CO_2 and it seems unlikely that these issues will be addressed by SRM. Second, there will likely be unanticipated side effects of the use of SRM. Some of these will be geopolitical. Who will decide whether SRM will be used? SRM techniques could be initiated unilaterally despite the opposition of

18 IPCC, Working Group II, 2014, 30.6.7.

19 Duncan McLaren, "Where's the Justice in Geoengineering?" *The Guardian*, March 14, 2015. Available at https://www.theguardian.com/science/political-science/2015/mar/14/wheres-the-justice-in-geoengineering (accessed July 5, 2016).

several countries. Third, it is likely that once SRM is implemented, it will need to be used as long as the atmosphere has high concentrations of GHG. This could mean multiple uses of the technique over time without knowing its cumulative effects. You can't simply stop SRM once it is started, as its sudden termination would likely lead to widespread and substantial climate change. One problem that could result from the use of SRM techniques is that cooling over large landmasses can greatly reduce rains that are usually generated during the monsoon season. In areas of the world that greatly depend on heavy seasonal rains for water, this could have a dramatically adverse effect on agriculture.[20] Given all these uncertainties, there is understandably a high level of anxiety about the use of SRM.

However, it appears that the IPCC prefers SRM to *carbon dioxide removal* (CDR) strategies as a possible climate fix because of the Earth's extremely complex carbon feedback loops (as we'll see below). Also, spraying albedo (reflective) particles into the atmosphere is much quicker and cheaper. But again, for some other environmental problems, such as ocean acidification, SRM is not very helpful at all, and may actually worsen the condition of the world's oceans. Recall that SRM merely reflects solar radiation to cool the Earth and fails to reduce CO_2. Besides, part of the attraction of SRM is that it is seen as a way to lessen global warming *without* having to take aggressive steps in reducing the world's emissions. SRM is billed as a cost effective way to lower global temperatures without placing a collar on economies that currently rely greatly on industries linked to heavy emissions of GHGs. But all of this completely ignores the problem of decreasing pH in our seas. In sum, many environmentalists complain that SRM treats the symptoms, but does nothing to cure the causes of both climate change and ocean acidification.

This returns us to the alternative geoengineering approach that was *not* taken seriously by the IPCC as a response to climate change and ocean acidification—CDR. This approach is designed to take CO_2 out of the atmosphere in various ways and sequester it in some form.[21] CDR typically involves adding a liquid to CO_2 to create a chemical reaction that extracts it from the air, a technique that is already used on submarines and spacecraft.[22] Yet, the problem is that once

20 Joel Achenbach, "Short-term Fixes for Long-Term Climate Problems? Not So Fast, Experts Say," *Washington Post*, February 8, 2015. Available at http://www.washingtonpost.com/national/health-science/upcoming-report-on-technological-fixes-for-climate-change-adds-to-debate/2015/02/07/07bbe150-ac6e-11e4-9c91-e9d2f9fde644_story.html (accessed March 26, 2015).

21 Achenbach, "Short-term Fixes."

22 Achenbach, "Short-term Fixes."

CO_2 is extracted, you have to put it somewhere. To remove copious amounts of CO_2 out of the atmosphere and water to reduce the greenhouse effect and ocean acidification would necessitate creation of a massive repository. Likely the best places to permanently stow excess CO_2 would be in porous rock formations, such as abandoned oil and gas extraction sites. This has only been done to a limited degree so far. CDR is expensive and could potentially be environmentally disruptive.

Carbon scrubbing is another CDR technique that removes carbon from the atmosphere in a few different ways. But doing so, in general, isn't the most efficacious way to reduce CO_2 emissions because little carbon can be extracted from the process as it is extremely diluted. There may be more promising ways of utilizing carbon scrubbing, as some point pollution sources such as smokestacks have been installed with scrubbers where CO_2 is more concentrated for removal. This is done in some coal-fired power plants, whose stacks can be retrofitted to remove CO_2 as they emit their effluent. Yet again, the process is expensive and regarded as cost prohibitive by most manufacturers. Thus retrofitting coal-fired plants with scrubbers hasn't be widely practiced.

Other "low-tech" solutions for pulling carbon dioxide out of the atmosphere, such as *reforestation*, may be more palatable. But considering the worldwide trend seems to be toward *deforestation* and acknowledging the scale of the problem, these techniques individually appear to be futile.[23] There are, as always, economic and political barriers to coordinated solutions since few governmental entities are interested in paying two types of costs—the direct financial costs of developing technologies to use CDR and the opportunity costs of reducing production for the sake of lowering CO_2 emissions.

The Royal Society produced a report in 2009 reviewing several ideas for deliberately intervening in the Earth's climate to reduce global warming. The report essentially concluded that geoengineering strategies such as SRM and CDR warrant further study. The Society thus partnered with the Environmental Defense Fund (EDF) and TWAS (which is the academy of sciences for developing countries) to further explore geoengineering's acceptability and feasibility. They formed the SRM Governance Initiative (SRMGI) with the goal of using the best science to inform future plans for whether to use (or to what degree to use) SRM. Furthermore, the group would develop the framework governing SRM in case it is used. But the report was not meant to make specific recommendations about whether SRM should be deployed.

In the Fifth Assessment Report of the IPCC, it was recommended that geoengineering techniques, including SRM, should continue to be pursued. This

23 See Case 29—"Not Seeing the Forests for the Deserts."

is despite criticism that geoengineering generally may cause adverse global side effects in the long-term. As noted in the 2007 report, "If SMR were terminated for any reason, there is *high confidence* that global surface temperatures would rise very rapidly to values consistent with the greenhouse gas forcing" [Emphasis in original].[24] This 2007 report was superseded by a follow-up report entitled "Climate Change 2013: The Physical Science Basis" which is also called "AR5." Contained in its body is a recommendation of SRM use over CDR methods, as the later are limited in their effectiveness on a global scale. However, the report recognizes that neither approach is ideal due to potential long-term negative consequences. Still, some believe geoengineering is coming given the dire predictions of catastrophic climate change. "While the entire community of academia still pretends not to know about the ongoing reality of global geoengineering," comments Dane Wigington at Geoengineering Watch, "the simple fact that they are now discussing geoengineering in the last IPCC report indicates that the veil is beginning to lift."[25]

We already noted the important question of who gets to decide whether to deploy SRM. Other moral questions arise with its possible use. Will the temptation to use SRM as a way for some countries to control others be too great? The worry is that the potential harm of some governments holding others hostage is not worth the risk of advocating use of the technology. Of course, some say SRM is already being developed and consequently it is better to focus on how to use it responsibly. With that said, it's not surprising that the joint Royal Society/EDF/TASC Report centered on the possible implications of different governance structures adjudicating the use of SRM.

Since CDR strategies are generally deemed less radical than SRM techniques and are regarded to possibly lessen impacts of ocean acidification, they have not been scrutinized as closely. However, it is still important to investigate the possible ethical implications of CDR. Up front, there would likely be resistance to using the procedure as it would be seen as "unnatural." Of course, this is a common response to a number of other environmental interventions. For example, genetically modified organisms have been objected to in just this way for a long period of time.

Some ethicists are dissatisfied with the "unnatural" objection to interventions of any type on the grounds that they reduce to "the yuck factor" or some

---

24 As quoted in Rady Ananda, "Solar Radiation Management, Geoengineering, and Chemtrails: IPCC Warns Policymakers Not to Stop 'Solar Radiation Management,'" *Global Research*, November 5, 2013. Available at http://www.globalresearch.ca/solar-radiation-management-geoengineering-and-chemtrails/5356632 (accessed July 5, 2016).

25 Rady Ananda, "Solar Radiation Management."

other vague objection based solely on sentiment that new technologies challenge, "the way things are." As applied to CDR, these ethicists would say it isn't enough to merely shun it because it "doesn't feel right." Even if CDR is ethically unacceptable, it has to be shown to be so by revealing some potential or real violation of rights, undue harm and suffering, or dereliction of moral duty.

Additionally, some forms of CDR are not completely immune from potentially causing some environmental risk. In the case of ocean ion fertilization, some changes could occur in sea ecology that scientists would have to study before the strategy is used. Other CDR techniques carry lower or negligible risks to both humans and the environment. That said, many of the moral issues about whether we should use CDR to combat the effects of ocean acidification and climate change reside in whether we should use scarce resources to develop these techniques and who should pay for them. The cost and our relative lack of technological knowhow could limit the efficacy of CDR. CDR is slow and it would take a long period of time to reduce CO2 levels in the oceans. In their 2015 report, the National Research Council (NRC) cites that the cost of CDR procedures (outside of the low-tech strategy of reforestation) will likely be prohibitively expensive, further noting that some believe CDR could well have costs equivalent to or exceeding those of developing wind and solar energy.[26] No matter what, it would seem that work would have to be done soon to lessen or remove technical barriers to developing affordable and efficient CDR techniques. With regard to policy, the NRC was clear that some advantages of CDR methods include that they don't really introduce global risks and raise fewer and less difficult questions with respect to global governance than SRM.

Some call for mandatory sequestering of carbon as a condition of fossil fuel extraction.[27] Carbon sequestration advocates point out that climate change will not likely be properly dealt with by focusing on annual emission quotas. This is because it is the cumulative effects of anthropogenic contributions to greenhouse gases that genuinely alter the atmosphere. With this in mind, they call for a general policy scheme linking any fossil fuel extraction to carbon emissions. Companies would be required to sequester an "adequate fraction" of their carbon output to be permitted to extract fossil fuels. Proponents of

---

26 The National Academy of Sciences, "Climate Intervention: Carbon Dioxide Removal and Reliable Sequestration—Reflecting Sunlight to Remove Earth," *Report in Brief*, (website), 2015. Available at http://dels.nas.edu/resources/static-assets/materials-based-on-reports/reports-in-brief/climate-intervention-brief-final.pdf (accessed March 26, 2015).

27 Myles R. Allen, David J. Frame, and Charles F. Mason, "The Case for Mandatory Sequestration," *Nature Geoscience* (2009) 2: 813–814.

this policy think the effect would be the same as having a carbon tax without having to use such a direct measure, which has faced stiff political resistance. Companies would react to the regulation of having to sequester carbon by treating it as an operational cost and raise consumer rates to maintain their profit margins. Thus consumers of fossil fuels would still be paying for the costs of carbon sequestration even without a formal tax.

The proposed advantages of a mandatory sequestration plan would be to split off the direct link between consumption and carbon policy. Advocates see this as a less economically disruptive approach to reducing overall carbon emissions, including those into seawater. The mandate for carbon sequestration would develop a market for permanently locking up carbon in abandoned oil reservoirs, as mentioned earlier. A mandate has the added attraction of being a simple, market-driven mechanism. There would be a single policy for a single energy industry, which would tweak (instead of completely reengineer) the fossil fuel market. Also, this strategy would be seen as much simpler than having to negotiate emission quotas with individual carbon emitters on a continuous basis. An adequate fraction would be determined by a ratio between the cumulative emissions at the time that the policy is established to total outstanding allowable emissions at that time (currently this comes to around half a trillion tons). As cumulative emissions reach the allowable total, the sequestered adequate fraction of extraction (or "SAFE") would be 100%. This would ensure that the total carbon concentration budget is not exceeded.[28]

As is likely clear from the policy discussion above, there are very complex issues involved with geoengineering. Ultimately a number of them have to do with more than merely assessing what is technologically possible or prudentially acceptable. Some moral philosophers have begun, no matter how tentatively, to look at moral questions involved with technological interventions designed to mitigate the effects of climate change.

Stephen Gardiner lays out an array of moral arguments that have been put in play both in favor of and against geoengineering.[29] Some proponents of geoengineering think a variety of techniques to reduce the greenhouse effect and ocean acidification should be on the table for consideration. Gardiner calls this the pluralist "let many flowers bloom" or broad "portfolio" approach.[30] On this approach, both controversial and uncontroversial ways of handling

28 Allen et al., "The Case for Mandatory Sequestration,".

29 "Some Early Ethics of Geoengineering the Climate: A Commentary on the Values of the Royal Society Report," *Environmental Values* (2011) 20: 163–181.

30 Gardiner, "Some Early Ethics," p. 176.

climate change should at least be investigated. This would likely be a necessary step to sorting out more and less viable strategies.

Others have argued that geoengineering might well buy time for future mitigation, as "not ready for prime time" alternative energy technologies can be developed while SRM and CDR ward off the worst effects of climate change. Still others argue from a more prudential perspective, saying geoengineering should definitely be developed and deployed "just in case" there is an imminent environmental catastrophe.

Turning back to the Royal Society/EDF/TASC Report cited earlier, it endorses the pluralist approach in two ways. First of all, the authors appear to be critical of those who more tepidly advocate geoengineering only on last resort grounds, instead thinking that more "upfront" intervention strategies are justified. Second, the authors of the Report also indicate that the focus on SRM, with relative indifference shown to CDR, is unwarranted and further research into the latter method deserves a closer look. As Gardiner points out, there is initially a very strong motivation to accept this pluralist approach. It promotes a diversity of strategies in an open-minded approach conducive to innovation.

However, Gardiner identifies some potential limitations for the pluralist approach. He notes that we have to bear in mind that government-sponsored research is competitive and funds are limited. That said, it would make more sense to prioritize which sort of research projects warrant funding. Also due to concerns with economies of scale, promoting a select group of higher priority projects is likely to have better outcomes than funding many proposals, especially if this avoids lower levels of funding over a wider array of projects.

Gardiner also worries that a generally open attitude to geoengineering proposals might be questioned. After all, some projects might be more deserving than others and on those grounds alone we might want to accept more viable proposals. On top of that, some projects may be quite controversial because they are risky and consequentially have a large burden of proof against them. These factors will likely work against the pluralist approach.

Another argument for further investigation of geoengineering and its possible use has to do with 'buying time.' The Royal Society position on this type of position is to say that geoengineering might hold promise to "augment"[31] or "complement"[32] more conventional methods. From this perspective, such use may be "a route for buying back some of the time lost in the international

31 The Royal Society, "Geoengineering the Climate: Science, Governance, and Uncertainty," Policy document 10/09, September 2009. Available at https://royalsociety.org/~/media/Royal_Society_Content/policy/publications/2009/8693.pdf (accessed July 5, 2016), p. ix.

32 The Royal Society, "Geoengineering the Climate," pgs. xii, 57, 61.

mitigations negotiations."[33] Gardiner notes that it is for this reason that the Royal Society favors CDR. The thrust of this argument is that since there has been undue political inertia in addressing carbon emissions in the past, humanity might be able to "make up" for this by directly removing excess greenhouse gases later this century. But even this argument is not a full-throated endorsement of geoengineering and treats it more as a second-best option to conventional mitigation and adaptation strategies. It also seems that geoengineering is to be utilized only after conventional measures to combat climate change fail.

As mentioned earlier, the "geoengineering as last resort" argument treats the practice as inherently much riskier than convention mitigation and adaptation strategies. As presented, this argument suggests that geoengineering should only be used in an emergency situation when dire climatic conditions are imminent. It is, as Stephen Gardiner puts it, "a 'lesser evil' argument that ought to be accepted only very reluctantly, when things are so bad that drastic measures are only justified to avoid catastrophe,[34] and where this might have further implications."[35]

We have already examined the "unnatural procedure" objection to both SRM and CDR. But there are also *moral hazard* concerns with the use of geoengineering. Moral hazard is a term that comes to us from economics. The basic idea is that when people have some sort of insurance, they will tend to take greater risks than they otherwise would without it. This is mainly due to them being protected from the worst results of their actions. As a result, insurers will be barraged with claims. To have a better understanding of these issues, it may be best to look outside the scope of the environment and focus on health care. Health economists have long worried about comprehensive health care coverage schemes (such as a single payer system) on the grounds that when people have comprehensive coverage of their health care, they have a perverse incentive to overuse available resources. After all, if there is no or little individual cost (say a low premium in the form of taxes paid in and low or non-existent

33 The Royal Society, "Geoengineering the Climate," p. 45.

34 Stephen Schneider, "Geoengineering: Could—or Should—We Do It?" *Climatic Change* (1996) 33: 291–302; Paul Crutzen, "Albedo Enhancement by Stratospheric Sulphur Injections: A Contribution to Resolve a Policy Dilemma?" *Climatic Change* (2006) 77: 211–219.

35 Quoted in Gardiner, "Some Early Ethics of Geoengineering," p. 177. On this second point, see also Stephen Gardiner, "Is 'Arming the Future' With Geoengineering Really the Lesser Evil?" in Gardiner et al. *Climate Ethics: Essential Readings* (Oxford, U.K.: Oxford University Press, 2010).

co-payments) there is an incentive to request high levels of care without regard to its cost to the system.

Some think that if geoengineering were used a similar structure of perverse incentives would be in place for both consumers and energy providers who wish to supply demand. Geoengineering (or perhaps just the promise of its future use) might be treated as an insurance policy. If consumers believe that technological fixes are available (especially if they are thought to be relatively inexpensive) for reducing the worst effects of climate change, they will lose any interest in reducing their consumption of fossil fuels. Therefore, energy producers will continue to emit greenhouse gases (particularly CO_2) as the greenhouse effect would be lessened.

The implication of geoengineering being seen as a form of insurance will be to syphon off support for both conventional mitigation and adaptation strategies, as well as for alternative energies, locking us in to a high carbon economy. After all, if one can keep using lots of fossil fuel energy without having to pay that much more because CO_2 emissions needn't be reduced, he would likely consider this a win-win result. If simply spraying sulfates, thus redirecting the sun's radiation back into space can cool the atmosphere, why go through the trouble of painful mitigation and adaptation? But of course, this would be to completely disregard the possibility that the cure may be worse than the disease. The effects of SRM might be adverse and other climate issues would not be addressed. This might be more of a worry for SRM than CDR, but even using CDR without reducing carbon emissions may have similar moral hazard effects.

No matter what, with major carbon-emitting nations reluctant to reduce emissions to slow climate change and ocean acidification, pressures to pursue some form of geoengineering will likely only increase.

Review Questions

1) Why are there ecological concerns about ocean acidification and what social implications are anticipated when large amounts of carbon are absorbed into the world's seas?
2) What is the moral hazard objection to geoengineering? Do you think moral hazard concerns should give us pause about geoengineering? Why or why not?
3) Do you think any kind of geoengineering is morally acceptable? Why or why not? Do you find solar radiation management to be more or less morally acceptable than carbon dioxide removal? Why or why not?

Case 29 – Not Seeing the Forests for the Deserts –

Study Questions

1) At what rate is deforestation occurring? What are the purposes of clearing forestland?
2) What is the connection between deforestation and the removal of carbon from the atmosphere?
3) What is the link between deforestation and desertification?
4) Describe Costa Rica's PES program and its intended purpose. In what sense is it a regulatory plan? In what way is it an incentives-based program? What are some of the results of the program?

Deforestation is the clearing and destruction of forests to obtain wood for consumption, open land for agricultural purposes, and to make way for expanding settlement frontiers. It is thus driven by economic and urban development pressures, caused by the expansion of facilities in which humans work, live, and play, or through expansion of land converted to agricultural use for crops, dairy production, and/or ranching. The Earth is still 30% forested, but trees are vanishing quickly, with losses of rainforest each year equivalent to an area the size of Panama.

In the last decade of the 20th century, carbon loss and removal have been intricately connected to tropical and boreal forest clearance and forest regrowth respectively. How much forest and other accumulated biomass regrowth in temperate and boreal forests have been able to offset carbon losses due to tropical rainforest deforestation is a contested point. Some evidence suggests that emissions from deforestation have generally been about 5.8 GtCO2/yr.[36]

Of course, deforestation is not just a modern phenomenon. Throughout history humans have cleared land for both crops and livestock. There is evidence that in many parts of the Americas, including the eastern part of the United States, Central America, Mexico, Peru, and the coastal regions of Venezuela, native cultures systemically used fire to clear forest, thus managing game and creating openings for growing crops. Archeologists have hypothesized that prehistoric construction and charcoal remains in the one time plurinational state of Bolivia and Brazil indicate very large swaths of the Amazon Basin may have

---

36 G.J. Nabuurs et al., "Forestry," in B. Metz, O.R. Davidson, P.R. Bosch, R. Dave, L.A. Meyer eds. *Climate Change 2007: Mitigation. Contribution of Working Group III to the Fourth Assessment Report of the Intergovernmental Panel on Climate Change* (Cambridge, U.K. and New York: Cambridge University Press, 2007).

been cleared for farming. The extensive forest area of the western Amazon might have been significantly smaller than it is in the present day.[37] While deforestation has occurred over time, the grave concern is with its scale and rapidity in our contemporary era. If deforestation (albeit slowed over recent years) continues at its current pace unabated, the Earth's forest cover will be lost by 2100.

A recent estimate is that roughly 100,000 km2 of tropical rainforests are deforested each year and another 100,000 km2 are degraded (perhaps in part by inefficient technologies in forest harvesting). The accuracy of such estimates is improving dramatically with better quality satellite imagery. But certainty about how much land area is deforested is limited, as different techniques are needed to cobble together estimates. For example, the NASA Landsat system cannot detect the clearing of understory below the top canopy.[38] For these measurements, radar remote sensing can plug part of this gap as this technology can detect forest clearances in the understory (such as a coffee plantation). Regardless of these limitations, it appears that at one time tropical forests covered 16 million km2 worldwide, but now the Earth has about 8–9 million km^2. Also, estimates are that Asia and Latin America have lost 40% of their original forest cover, and Africa has lost a little over a half.[39]

While the rate of worldwide deforestation is shown to be declining, it is still high. There is wide variance of deforestation rates from country to country. The Food and Agricultural Organization of the United Nations (FAO) reported that approximately 13 million hectares of forestland were either lost to natural calamities or converted to other land uses each year from 2000–2010 compared with 16 million hectares in the 1990s.[40] The countries with significant reductions in their rate of forest loss in recent years have been Brazil and Indonesia. In contrast, both countries had the highest rates of forestland loss in the 1990s. On the other side of the spectrum, Australia has reported the highest rate of forest loss since 2000, mainly attributable to severe drought and subsequent

---

37 Michael Williams. *Deforesting the Earth: From Prehistory to Global Crisis* (Chicago: University of Chicago Press, 2002).

38 Regents of the University of Michigan, "Global Deforestation," (website), 2015. Available at http://www.globalchange.umich.edu/globalchange2/current/lectures/deforest/deforest.html (accessed April 15, 2015).

39 Regents of the University of Michigan, "Global Deforestation."

40 Food and Agriculture Organization of the United Nations, "Deforestation and Net Forest Area Change: Facts and Figures," (website), November 4, 2011. Available at http://www.fao.org/forestry/30515/en/ (accessed April 10, 2015). Also see FAOUN, "Global Forest Resources Assessment," (website), 2010. Available at http://www.fao.org/docrep/013/i1757e/i1757e.pdf (accessed April 15, 2015).

heavily destructive forest fires. Large-scale planting of trees is having a substantial impact on reducing the net loss of global forests, including afforestation and the natural expansion of forests in some specific countries and regions. Thus, the net loss in worldwide forestland in the first decade of the 21st century was about 5.2 million hectares per year (equivalent to a land area about the size of Costa Rica), as compared with the 1990's when it was about 8.3 million hectares.

Tropical rainforests are largely concentrated in developing countries. Yet, the economic interest in them is global; to feed consumer demand, there is greater pressure to exploit their vast natural resources.[41] The necessity of economic development combined with available global markets has meant that deforestation has been able to proceed at alarming rates. It bears repeating that the direct, primary cause of deforestation is clearing land for crops and pasture for ranching (as is particularly true in the Amazon). Thus, the main purpose of forest clearing is subsistence farming. Removing trees can be done quickly and cheaply through slash and burn methods, exacerbating their disappearance.

Another reason for rapid forest loss in general and the rainforest in particular is the practice of commercial logging, whether practiced legally or illegally. Timber extraction yields many different wood and paper products that enter the global market. Logging also supplies wood for domestic use in building and for fuel, such as charcoal. Roads are conduits for further harvesting of timber, leaving more remote areas increasingly accessible. Roads aren't necessarily built for logging, but their construction has an especially dynamic interplay with the tree harvesting industry. For example, countries often build roads to improve transportation infrastructure.[42] While the roads themselves directly cause limited deforestation, they also open previously inaccessible (and a key here) previously unclaimed land. After a road is built to gain access to timber, a network of additional roads follows, offering access to further development.

Another related reason for deforestation is settlement. After the timber runs out, cleared land is attractive to settlers. In this way, whatever may remain of the forest in a recently logged area is slashed and burned for cropland and pasture. This cycle happens repeatedly, beginning with an enterprise as seemingly innocuous and beneficial as transportation. Heavily logged areas become

---

41 National Atmospheric and Space Administration, "Tropical Deforestation," *Earth Observatory*, (website), 2015. Available at http://earthobservatory.nasa.gov/Features/Deforestation/deforestation_update3.php (accessed July 6, 2016).

42 National Atmospheric and Space Administration, "Tropical Deforestation."

more fire-prone and eventually deforested by repeated exposure to accidental fires on pastures or adjacent farms.[43]

Large-scale commercial enterprises beyond logging also play an increasingly significant role in deforestation.[44] In the Amazon, industrial-scale cattle ranching and soybean production for global markets is causing the clearing of forests. Likewise in Indonesia, rainforest lands are cleared to continue development of commercial palm tree plantations used for biofuel export. This has become a major cause of deforestation in Borneo and Sumatra.[45] There is obviously an interesting environmental tradeoff here as biofuels are exalted as ecologically friendly alternatives to other heavier polluting fuel choices. Yet, biofuel production puts an additional strain on vital forest areas due to the range of ecological services they provide, such as water filtration and carbon sequestration.

A final reason for the decrease in forestland worldwide is the expansion of urban areas. Constructing new real estate developments involves the clearing of trees from lands adjacent to population centers. This isn't merely a manifestation of urban sprawl, as rural and suburban people also desire to have quick, easy access to consumer goods and services. Consequently, malls and commercial properties are built to meet this demand. It is readily apparent that large-scale deforestation does not have a single cause.

In the exploration of the causes for deforestation, it is likely unsatisfying to simply report out on its direct causes. There are underlying factors at root to explain human motives and behaviors toward forest areas. For example, poverty is often cited as the chief underlying cause of deforestation. But many scholarly works suggest this conclusion is too simple. Human migration to forest areas is driven in part by poverty. Poor people move into areas where slash and burn techniques, given that they are relatively low cost, allow them opportunities to convert land to productive capacities. However, one underlying factor alone is not sufficient to fully explain a global phenomenon such as deforestation.

We also need to consider national policies encouraging activities leading to deforestation. As we have already seen, for the sake of economic development, infrastructure is constructed, such as road and railway expansion projects. These policies have had the unintended adverse effect of significant deforestation (especially in the Amazon and wide areas of Central America). Other ways governments have incentivized activities promoting deforestation include

43 National Atmospheric and Space Administration, "Tropical Deforestation."

44 National Atmospheric and Space Administration, "Tropical Deforestation."

45 National Atmospheric and Space Administration, "Tropical Deforestation."

agricultural subsidies, tax breaks, and timber concessions. Sometimes the economic issues are truly global as a country's level of foreign debt and expanding global markets for wood and other products from rainforests play a role in forest clearing. Likewise, factors that are even largely outside of the control of individual nations, such as low domestic costs for labor, land, and fuel can encourage deforestation as opposed to more sustainable land use practices.[46]

Finally, other indirect factors contribute to deforestation, such as access to technology. Open availability of technology can work either way—to diminish or to enhance deforestation. Technological innovation provides tools for large-scale agriculture that can lead to widespread and rapid forest clearing. On the other hand, older inefficient technology in the logging industry can cause long-term "collateral damage" of adjacent forest areas. This has a further consequence of making logging expansion in such areas much more likely.[47]

Deforestation can lead to a number of ecological problems, one of which is desertification. The soils of forests are moist. However, without a forest canopy to block out the sun's rays, these soils dry out quickly. When trees are cleared away, significant hydrological changes in the land can occur underground. As root systems disappear with the harvest of trees, moisture is lost, causing the soil's desiccation. In arid areas this can eventually result in deserts. Trees also play a crucial role in continuing the water cycle; they return water vapor back into the atmosphere. If there were no trees to perform this function, many additional former forestlands could quickly become barren deserts, again especially in dry places. The Sahara desert in northern Africa continues to extend southward by 48 kilometers each year in just this way.[48]

Deforestation leads to other hydrological changes that adversely affect the land. Forest areas play a key role in mitigating soil erosion. The loss of trees, on the other hand, causes widespread erosion throughout the tropics. The root systems of trees are extensive and by virtue of their position as the largest parts of plants are essential in retaining nutrient-rich topsoil. In the tropics, only a small number of locations possess good soils, but these can be quickly washed away during heavy rains. Erosion not only has an adverse effect on forests themselves, but also means topsoil losses for farmers and ranchers as crop yields decline and grasslands for grazing are denuded. Thus, the pressure to rejuvenate the soil for agricultural purposes translates into expenditures on

---

46 National Atmospheric and Space Administration, "Tropical Deforestation."

47 National Atmospheric and Space Administration, "Tropical Deforestation."

48 National Geographic, "Deforestation," (website), 2015. Available at http://environment.nationalgeographic.com/environment/global-warming/deforestation-overview/ (accessed July 6, 2016).

imported fertilizers from abroad. Yet, even newly placed fertilizers may not stay for longer than a growing season or two as with the loss of adjacent forest lands, erosion strips away any of this fertilizer as well.

In the 1980's, soil erosion was a major problem in Costa Rica where about 860 million tons of valuable topsoil was lost each year (note that major efforts in limiting deforestation has reduced the erosion problem there a great deal, as we'll see later in this case study).[49] Great Red Island in Madagascar loses so much soil to erosion (approximately 400 tons per hectare) that its rivers turn a dark red color. This runoff stains the surrounding Indian Ocean and astronauts viewing this phenomenon from space have remarked that it looks like Madagascar is "bleeding to death."[50] In the Ivory Coast, we have more evidence that the *rate of increase* in soil loss is very substantial when forest is cleared. There, a forested slope lost only 0.03 tons of soil per year per hectare. On unforested, but cultivated slopes, the annual loss of soil was 90 tons per hectare per year. However, the greatest losses of soil in the Ivory Coast came on bare slopes, with 138 tons of topsoil lost per year per hectare.[51]

Soil erosion obviously causes a host of problems. Runoff from heavy tropical rains on cleared forest floors carries soils into local watersheds. Hydroelectric projects and irrigation infrastructure can lose efficiency due to siltation. Industrial facilities may suspend operations due to, ironically enough, a lack of water. Collections of silt carried downstream can raise riverbeds, which in turn cause the increased severity of floods. Displaced soils can also result in shoals and sandbars, which raise barriers to river navigation. Increased sediment loads in rivers negatively affect aquatic life as fish eggs are smothered, resulting in lower hatch rates. Suspended particles in rivers leave water cloudy. As this murky water reaches the ocean, it can lead to regional loss of coral reefs as sunlight is blocked, thus adversely impacting local fisheries. Siltation also destroys mangrove forests, which are important to not only the structural integrity of coastlines that protect communities from sea level rise and storm surges, but also provide critical habitat to wildlife.[52]

---

49 Joanne Omang, "In the Tropics, Still Rolling Back the Forest Primeval," *Smithsonian* (1987) 17(12): 56.

50 Quote from photographer Frans Lanting in his *Madagascar: A World Out of Time* (New York: Aperture Foundation, Inc., 1990).

51 United Nations Educational, Scientific, and Cultural Organization/United Nations Environmental Programme/Food and Agriculture Organization, *Tropical Forest Ecosystems* (Paris: UNESCO, 1978).

52 United Nations Educational, Scientific, and Cultural Organization/United Nations Environmental Programme/Food and Agriculture Organization, *Tropical Forest Ecosystems*.

Deforestation also makes significant contributions to climate change. Removing trees obviously reduces forest canopy, which otherwise would block out sunlight during the day and hold in heat at night. Disrupting this process on a local scale will likely lead to extreme temperature swings that can place undue stress on plants and animals alike in these severe microclimate conditions. Globally, forests play a critical role in absorbing greenhouse gases contributing to climate change. Decreases in forest areas (particularly substantial rainforests such as the Amazon) correspondingly mean larger amounts of carbon dioxide entering the atmosphere and the increased pace and severity of global warming.[53]

Another consequence of deforestation is loss of biological diversity. Often forest areas provide critical habitat to a host of plant and animal species. About 70% of the world's land flora and fauna use forests as their habitat.[54] Of these, many will not survive the destruction of their habitats caused by deforestation. With the disappearance of forest habitat and no clear substitute, whole species tend to vanish as well.

Preventing deforestation would simply mean to stop cutting down trees (or less radically, to have a one-to-one matching scheme that for every tree removed, another is grown in its place). More realistically, the response might be to more carefully manage forest areas by eliminating clear-cutting. Eliminating clear-cutting will keep large tracts of forestland intact. This strategy would likely have to be buttressed by insuring a balance between planting enough young trees to replace older ones harvested in any given forest. While the number of tree plantations has increased in recent years worldwide, this accounts for only a tiny fraction of the Earth's forested land.

Some countries, such as Costa Rica, have tried a mix of economic and regulatory policies to protect forests. Costa Rica is a Central American country that at one time held vast rainforest cover. But in much the same way as has already been described, the nation lost a great deal of it. With increased concern about both the net amount of rainforest lost and the alarming rate at which it was being lost, the Costa Rican government decided to take aggressive means to at least slow, if not reverse the trend of deforestation.

Costa Rica used two main strategies to combat deforestation. The first was to prevent landowners from changing the way they used their land. The second was to pay landowners in return for protecting forests. This second strategy

---

53 United Nations Educational, Scientific, and Cultural Organization/United Nations Environmental Programme/Food and Agriculture Organization, *Tropical Forest Ecosystems*.

54 United Nations Educational, Scientific, and Cultural Organization/United Nations Environmental Programme/Food and Agriculture Organization, *Tropical Forest Ecosystems*.

was couched in the context of the types of services ecosystems provide for the benefit of all Costa Ricans. So, landowners are compensated for the part they play in helping ecosystems provide these services. Among the ecological services rainforests offer are habitat to species, thus contributing to the biodiversity that keeps ecosystems healthy. Promoting tree growth is vital, as they store carbon that would otherwise simply go into the atmosphere thus contributing to the greenhouse effect, or into the oceans contributing to their acidification. In turn, landowners can protect rainforests by regulating river flows to decrease soil erosion that has been a major problem in Costa Rica.

Since 1997, payment-for-ecosystem schemes (PES) have been applied to over 1 million hectares of forest in Costa Rica. The result has been to return the country's forest cover to over 50% of its land area. This is a dramatic increase from just over 20% in the 1980's. Three core principles, best described as a cross between moral and ecological ideals undergird the program: equality, effectiveness, and low implementation cost. The hope was also to "fit" this reforestation program into the framework and goals of international conservation. Also, PES needed to reflect the realities faced by Costa Rica's people and landscape, as well as the fact that several different policies came into play simultaneously.

There has been popular support for the program as some landowner participants have seen benefits, but that doesn't mean PES is ideal. As one beneficiary of the scheme living in the Osa Forest Reserve put it,

> PES has helped us a lot and has given me the opportunity to ensure my economic future, not one hundred percent, but I'm not complaining. The paperwork to obtain PES is getting more difficult… I don't know whether I will be eligible in the next period… if I am not eligible, I will probably start cutting, and SINAC [the conservation area authority] does not have the capacity to stop us. You should understand that during the last 25 years, under the banner of 'protecting the environment' the authorities have progressively taken our rights, which IDA [the Land Reform Institute] previously said were ours… so it's natural that us locals are angry.[55]

One interesting characteristic of the PES scheme is that it is designed to facilitate conservation by appealing to several different ideological views about forest conservation in general, and about economics specifically. The 1996

55 David N. Barton, "Payments for Ecosystem Services: Costa Rica's Recipe," *International Institute for Environment and Development*, (website), November 29, 2013. Available at http://www.iied.org/payments-for-ecosystem-services-costa-rica-s-recipe (accessed July 6, 2016).

Forest Law codified the PES incentive plan at the same time as it banned land use change in forests. With the ban in place, the Forest Laws may appear to set up a paradox. To some, they are really just another forest owner subsidy scheme that Costa Rica had actually abandoned in the mid-1990's as part of the International Monetary Fund (IMF) Structural Adjustment Program. Others think features of the PES seem more like a neoliberal forest policy, despite the ban on changing land use. That is, some have called PES a national *quid pro quo*—a sort of compromise between regulators and landowners. Forest stakeholders are given some concessions to accept a ban on forestland use changes in the Forest Law.

One's perspective on whether PES is a neoliberal conservation policy will likely depend on one's views of incentives (carrots) vs. regulations (sticks) with respect to land use. One might think PES is an attempt to shift from general tax financing of forest conservation to user financing—particularly via water use fees. On this line of reasoning, PES is a move toward reducing monitoring costs of forest conservation and using competitive contracting, thus favoring larger over smaller forest landowners by having a fixed cost for monitoring. This system thereby privileges individual contracts over group contracts. Finally, some believe PES is increasingly captured by anonymous legal entities, including companies.

But on the other hand, Costa Rica now utilizes a national-level policy where priorities are set according to criteria that match PES conservation policy goals. This has replaced the original "first-come, first-served" approach to enrollment in PES. What the new policy seems to favor are applications from indigenous territories, areas with low social development, and properties under 50 hectares. The PES program has had little success in obtaining support from the private sector. Correspondingly, most of the funding for the program still comes from public sources.

Though the incentives as put in policy are monetary, the motives of those joining PES don't appear to be merely financial. Administrators determine the level of payments and thus unlike "pure" market mechanisms, PES is not driven by supply and demand. The rates of payment, in fact, have stayed at roughly the same level they were in 1996. Thus, in "real value" terms they are roughly only half the value they were two decades ago. Despite this depreciation of compensation, the number of contracts signed annually is actually higher now than when payments were at their highest level in 1998.

There are other values to landowners in signing on to PES.[56] PES and the Forest Law do not only allow, but also lend support to landowners in evicting

56 Barton, "Payments for Ecosystem Services."

squatters through the use of civil authorities. Before the implementation of these measures, landless farmers (many immigrants to Costa Rica) would cultivate previously unused farmland. Some poorer landowners who used to colonize farmland, now borrow against future PES payments to pay for the paperwork to legalize their tenure. In many areas of the country, protection against land tenure difficulties, high transaction costs, and weak conservation enforcement outweigh the size of the payment.

Public authorities are also given certain benefits in ease of administration due to the PES program. Landowners must have their paperwork in order to join the PES program, which makes them easier to track and ensures that the National Forest Fund can receive monies to pay for social security benefits of farm employees in a timely fashion. Updated information also makes it easier for the Minister of Finance and local municipalities to collect taxes. Finally, the PES program has legitimated conservation efforts in national protected areas, as landowners who had land "expropriated" within national parks are now compensated until the government can buy their land.

In any case, PES is a unique policy mix. As David N. Barton describes,

> PES in Costa Rica is a useful example of a balancing act between the 'commodification' of nature and the 'fair development' [of it]. There is a certain tension between the voluntary nature of conservation and the desire for private financing of PES, and the public targeting, regulation of enrollment and still much needed public funding. PES has changed along with changing tastes in conservation. Its flexibility has made it a politically viable policy for almost 20 years in Costa Rica. The PES Programme has remained quite inexpensive and there is some evidence to show PES has been as effective, or ineffective as the country's national parks in conserving forest cover—depending on which conservation cuisine your prefer. PES is an eclectic and successful policy mix in its own right, as well as a key ingredient in Costa Rica's conservation recipe.[57]

It can be difficult for people who are very concerned with forest loss and live in relatively affluent societies to imagine there could be much moral controversy over deforestation. After all, from this perspective, negative ecological changes should be sufficient to convince us that clearing trees must be stopped, even if this means some economic losses occur. However, it is not that easy. As mentioned above, significant clearing has come from non-corporate, non-industrial sources—that is, the global poor. It is hard to imagine from a perspective of

57 Barton, "Payments for Ecosystem Services."

not having to worry about having easy access to potable water, enough food to eat, and adequate basic shelter what measures people have to take when they don't have such necessities. Cleared forests render fields open for farms to feed families. We are, in essence, forced to face up to some of the hardest of moral problems. We have to pit grave concerns about the environment against the survival of some of the world's most vulnerable and least well-off populations. With the issue of deforestation then, we have to ask who or what gets sacrificed.

What is to be done when logging and/or agriculture resulting in deforestation are the primary ways through which impoverished people survive? Poverty in itself is a too easily identifiable underlying cause of deforestation. But that doesn't mean one can just as easily dismiss it. We are forced to ask the question of whether the detrimental effects of clearing the world's forests are so great that it is morally permissible to sacrifice the livelihoods (and perhaps lives) of thousands of people.

Despite the regrettable results in the short-term for some of the poor, one might argue that deforestation is a greater long-term concern as it might ultimately destroy a society (including societies to which the impoverished of the Amazon belong). There are only so many trees to sustain the poor who are using them—and at this rate nothing will be left for their future generations. Droughts, when combined with greatly denuded areas from deforestation, will not allow for the plant growth upon which the Amazon's poor also rely. This is not to mention the temperature extremes that would occur on the ground without forests to modulate microclimates. Nor does this really take into account the worldwide struggles of dealing with global climate change. From this perspective, wouldn't strong measures for combatting deforestation be easily defended?

The counter to this, of course, is that it is hard to deny that policies preventing deforestation would likely involve abandoning people who rely on resources exploited via deforestation *right now*. Does this unduly sacrifice their well-being for the consciences of much economically better off people in places shielded from the immediate adverse effects of curbing deforestation? Would such a policy merely show that relatively affluent people value their own lives more than those of poor people currently facing life or death choices in the Amazon? Is assuaging "green guilt" worth it?

Other moral issues arising from deforestation take root in debates about which specific policies should be enacted in response to it. Should conservation laws and incentives for reforestation be used in countries besides Costa Rica (perhaps in Brazil, which holds the vast majority of the world's remaining rainforest)? Should governments around the world try to accept an accord not only focusing on limiting carbon emissions (mainly from the burning of fossil fuels), but also on possible remediation strategies such as reforestation? And wouldn't this require that countries signing on to such agreements attain or retain a certain amount of forest cover themselves? Perhaps reforestation could be rolled into a country's plan to reduce CO_2 emissions, with the idea that this might be a cheaper way of reaching such goals than by employing carbon scrubbers and other carbon emission mitigation technologies.

There are also the obvious questions about rights to land use and how much control individual landowners ought to relinquish to the state. Some landowners in Costa Rica may complain it is bad enough that the Forest Law places severe restrictions on their personal use of land, but more of an insult when the state asks them to "pay for the privilege" by spending time and treasure to push forward a conservation agenda they may not fully embrace and/or may not be in their economic self-interest. From this perspective, private landowners may think that if arguments for conservation entail that forest cover services are public goods, then it should be paid for by the public through tax revenues. Otherwise, a few individuals, just by virtue of property ownership, are unjustly required to shoulder too many regulatory burdens and asked to take on further economic ones in pursuit of the public good. Their argument might be that a more laissez-faire market approach should be developed where the state, non-profit conservation groups, or concerned private citizens pay market prices for land to set aside for national parks. This could be done in concert with a market for *conservation easements*. That is, short of land transfers, any agent could negotiate a contract to entice landowners to manage their property in accordance with reforestation and other conservation measures. But in any case, this should be a voluntary market exercise available to all.

Of course, the opposing argument from the regulatory side is likely that *everyone* is ultimately implicated in the possible environmental collapse caused by extreme deforestation. One might add that those who are deriving the most benefit from an environmentally degrading activity are justly asked to take on more regulatory and economic burdens than those who don't. Besides, landowners are receiving some compensation for their troubles in the PES program. The question then seems to boil down to what is ultimately the fairest way of parsing out responsibilities for a shared environmental issue.

Review Questions

1) What might be a utilitarian response to the moral conflict between the needs of poor people versus the deep environmental costs of deforestation?

Is this the best sort of moral approach to land use with respect to deforestation?

2) Let's say we use a Kantian deontological approach to the problem of deforestation. Would the emphasis on the respect for persons as ends in themselves always trump concerns about the environment?
3) What "green virtues" (if any) might underpin calls that we should combat deforestation?
4) Ultimately, do you think adoption and expansion of PES is a just solution to provide for the needs of poor people in the Amazon and encourage economic development at the same time? Why or why not?

Case 30 – Food vs. Fuel: Unintended Consequences of Corn Ethanol –

Study Questions

1) What are some of the plants from which ethanol can be manufactured?
2) What are some of the unforeseen, unintended, consequences of heavy ethanol use in gasoline or as a fuel in its own right?
3) Elaborate on some of the history of motor vehicles fuels in the United States.

Sometimes what seems to be a sound solution to one crisis causes other problems that are unintended, unforeseen, but decidedly negative. In the search for renewable energy sources in the 1970s to replace non-renewable ones such as fossil fuels, a promising candidate appeared to be ethanol. During this time, corn appeared to be the best source for ethanol in the United States. Ethanol has been touted as a clean-burning alternative to what has been seen as a diminishing crude oil and coal supply. It can be manufactured from a number of different organic materials with different levels of energy efficiency. For example, ethanol is processed from sugarcane in Brazil and from wood in Sweden. Research using cellulose from plants as an ethanol source suggest it may be the cleanest burning, but its time to market is still considerable.

Corn was initially a popular candidate feed stock for ethanol because it was already being mass-produced and heavily subsidized by the US government. The use of corn ethanol was seen as not only a viable way of decreasing American dependence on foreign oil, but also a way to offer farmers a profitable new market that could effectively end the need for government subsidies for corn. US farmers are heavily subsidized to produce corn, so new ways to use it have always been sought. There has been a major shift from using cane sugar (or no sweetener at all) in many American foods to high fructose corn syrup. Advocates of corn ethanol argue its use as a fuel can reduce carbon emissions *and* support farmers by diversifying the corn market from merely the food sector into the energy sector as well. Since 1978, the US government has given tax incentives designed to make ethanol competitive with gasoline in the energy market. Consequently, corn ethanol advocates believed they were promoting a "win-win" proposition.

As the corn ethanol industry ramped up, at least farmers were happy, as they had steady demand from the energy sector for their yields and prospects looked even better for the future. With demand so heavy from ethanol producers, farmers could barely keep up. However, an unintended consequence came from the corn ethanol solution to the fossil fuel energy problem—so many food products use high-fructose corn syrup and corn is so heavily used for feeding beef cattle, it became a scarcer and expensive commodity. With so much corn going into ethanol production, food prices rose worldwide. For every bushel of corn entering the ethanol plant, they were thus unavailable for both high fructose corn syrup production and cattle feed. Prices skyrocketed for corn-based food products. For a while, the beef industry had a "solution" to this problem, they simply killed as many bulls for beef as they could and sent their meat to market. But of course, this market soon crashed and the price of beef also increased dramatically.

Ethanol is also known as ethyl alcohol and has the chemical formula C2H5OH—the same alcohol that is found in alcoholic beverages. While only recently has it been used widely, ethanol was first burned in motor vehicles decades ago. In fact, Henry Ford built what he called a "quadcycle" in 1896 that ran on pure ethanol. Ethanol is currently mixed with regular gasoline, usually consisting of 5 to 10% of the total volume. In some states, such as California, corn ethanol has taken the place of methyl butyl tertiary ether (MBTE) as a component of gasoline.[58] Roughly 90 percent of all gasoline sold in California contains at least 6 percent ethanol.

Studies suggest that when ethanol is blended with gasoline for use as a vehicle fuel, it has some emissions benefits over burning gasoline alone.[59] This is especially true when gasoline is compared with high ethanol concentration

---

58 California Energy Commission, "Ethanol as a Transportation Fuel," (website), 2016. Available at http://www.consumerenergycenter.org/transportation/afvs/ethanol.html (accessed July 6, 2016).

59 United States Department of Energy, "Ethanol Vehicle Emissions," (website), December 16, 2014. Available at http://www.afdc.energy.gov/vehicles/flexible_fuel_emissions.html (accessed June 22, 2015).

fuels such as E85.[60] Since ethanol burns cleaner than regular gasoline, its use aids in meeting air quality and emissions standards in the US. Some vehicles are equipped to run on much higher concentrations of ethanol, including flexible-fuel vehicles (FFVs) that burn E85, composed of 85% ethanol and 15% gasoline. Major car manufacturers have manufactured several million of these vehicles in the US.

When comparing fuels, life-cycle analysis is sometimes used because it is important to know how a particular fuel performs while taking into account not only what is emitted when it is burned, but also how it is extracted, processed, and disposed of—charting its complete lifecycle. The costs and benefits for all fuels become clearer when considered in this light. As the geoscientist Alan Carroll reminds us, there is no free lunch. With respect to greenhouse gases, using ethanol mixed with gasoline as a fuel has lower carbon emissions than simply using gasoline. With respect to carbon emissions specifically, much of this difference can be attributed to the fact that the carbon dioxide released while burning ethanol is offset by the CO_2 that is captured when the corn used to make ethanol is grown. (This is obviously also the case no matter what organic matter is used in its production.) So FFVs emit less CO_2 than do conventional gas powered vehicles per miles travelled.[61]

Additionally, in studies conducted at the Argonne National Laboratory in 2007, researchers found that when using life-cycle analyses, corn-based ethanol vehicles specifically performed better than conventional vehicles with respect to carbon emissions—with a 19–52% reduction.[62] The differences depended on the source of energy used during ethanol production—a 28% reduction when using natural gas and a 52% reduction when using biomass. However, this same study served as a reminder that there may be better alternative sources of ethanol on offer than corn, as sugarcane ethanol produced using energy from biomass could achieve a 78% reduction in CO_2 emissions when compared to gasoline. Cellulose ethanol did even better, with an 86% reduction in CO_2 emissions when it was produced using energy from biomass.[63]

There are some limitations to keep in mind for any ethanol. Adding small amounts of ethanol to gasoline can cause gasoline to evaporate more quickly (this generally happens while the vehicle is sitting or refueling) and this

60 United States Department of Energy, "Ethanol Vehicle Emissions."

61 United States Department of Energy, "Ethanol Vehicle Emissions."

62 See Michael Wang, May Wu, and Hong Huo, "Life Cycle Energy and Greenhouse Gas Emission Impacts of Different Corn Ethanol Plant Types," *Environmental Research Letters* (2007) 2: 1–13.

63 Wang et al., "Life Cycle Energy," pp. 1–13.

phenomenon must be calculated into total emissions. The use and storage of ethanol doesn't completely eliminate emissions of regulated pollutants, greenhouse gases, or toxic chemicals. Also, the level of emission benefits is contingent on several factors such as the type of vehicle being used, engine calibration, and blend level.

Yet, ethanol remains attractive as a way of developing fuels holding the promise of wide applications to the transportation sector. This progression toward biofuels will likely involve a relatively lengthy process of research and development, but there are already examples of the biofuel industry entering mainstream society. Alternatives to conventional gasoline are permeating practices in the public sector; the US Department of Energy touts such cases as California's push to invest in expanding infrastructure for the production of biofuels of various types to produce ethanol. The agency also points to one Idaho county's conversion to a fleet of FFVs as a way in which corn ethanol is being put to good use as an alternative to conventional gasoline.[64]

Its advocates note that in addition to environmental advantages, corn ethanol use expands America's agricultural economy by putting more money in farmers' pockets. This provides a significant financial stimulus to America's rural areas that have suffered from joblessness and economic stagnation. Iowa corn farmers have been big winners in the push for corn ethanol, with a large influx of cash that has spilled over into their local communities—on top of the $10,000-a-year farm subsidy checks that they still receive from the federal government. Other states also provide ethanol production and use incentives, usually in the form of tax credits, to encourage the industry.

Proponents of corn ethanol for energy use believe its domestic production diverts a great deal of money away from foreign governments linked to aiding and abetting terrorist organizations and with whom the US has poor to worse diplomatic relations. To these advocates, investing in corn ethanol simultaneously enhances national security and economic growth.

Even though the US is the world's largest producer of ethanol by total volume, perhaps the country most heartily embracing its production as an alternative to conventional gasoline is Brazil. Overall, Brazil is the second largest producer of ethanol in the world, but is the greatest producer per capita with the largest fleet of FFVs. In 1977, the Brazilian government mandated that all vehicles had to burn E20 (also called "gasohol") that is composed of 20% ethanol and 80% gasoline. Since that time, the ethanol sector in the Brazilian automotive market has surged, with all light vehicles having left gasoline completely behind—100% of such vehicles run entirely on ethanol. According to

64 Wang et al., "Life Cycle Energy," pp. 1–13.

2010 figures, Brazil produces about half of the amount of ethanol that the US does, with some 6.5 million gallons per year (compared with 13.2 million gallons in the US). The US only surpassed Brazil in ethanol manufacturing in 2005. As noted earlier, one important difference between ethanol production in the US and Brazil is that the latter derives its ethanol from sugarcane as opposed to corn.

Some ethanol advocates promote its production from non-food plants such as switchgrass or wood. Others believe ethanol should be produced from invasive species such as kudzu (see Case 34—"Filling up on Kudzu"). These are attempts to avoid the issue that even if corn ethanol helps to reduce CO2 emissions, it causes rising food prices that contribute to the starvation of the poor worldwide. Countries around the world are heavily dependent on grains for substantial portions of their residents' diets. The US, however, is a dominant player in the world's grain market, even though we produce more corn than wheat. As production of corn ethanol ramped up in 2007–2008, pork and beef production suffered significantly from rising feed costs and made it much less profitable. So along with stiff food prices for US consumers, corn ethanol production also "put the squeeze" on pork and beef farmers. These effects paled in comparison with the food price spikes faced in countries with little to no domestic wheat production.

In the wake of rising food prices between April 2007 and April 2008, the US Congressional Budget Office was prompted to conduct a study on the impact of ethanol production on food prices and greenhouse gas emissions.[65] At the time of the study, about a quarter of US corn went into ethanol production, roughly 3 billion bushels. At the height of the food price spike, ethanol production in the US stood at 9 billion barrels, which was a record yield at the time. This all had an effect not only on the spending of individual consumers, but also on the federal school lunch program. It was found that about 10–15% of the rise in domestic food prices from April 2007-April 2008 was caused by the diversion of corn to ethanol production. This may seem like a small percentage, but in actual dollars it had a major impact, increasing federal spending for food programs (specifically the Supplemental Nutrition Assistance Program (SNAP) formerly the Food Stamp program and child nutrition programs) by an estimated $600 to $900 million dollars.[66]

---

65 See United States Congressional Budget Office, *The Impact of Ethanol Production on Food Prices and Greenhouse Gas Emissions*, (website), April 2009. Available at https://www.cbo.gov/sites/default/files/111th-congress-2009-2010/reports/04-08-ethanol.pdf (accessed July 6, 2016).

66 United States Congressional Budget Office, *The Impact of Ethanol Production.*

Along with effects of corn ethanol on food prices, there is also the question of whether burning corn ethanol lessens contributions to climate change. In 2009, the state of California used an "all-things-considered" approach to estimate the effects of burning corn ethanol during deliberations about whether to continue championing it as an effective way to reduce carbon emissions. An assessment of corn ethanol as a cleaner energy alternative was presented to the California Air Resources Board (CARB), the regulators in charge of reducing the state's greenhouse gas emissions. The report stated that corn ethanol, no matter how much it directly reduces tailpipe emissions from motor vehicles, raised corn prices to the point where there were perverse incentives to expand its production in rainforests and conservation lands. Rapid deforestation caused by corn production offsets the carbon emissions benefits of burning corn ethanol instead of conventional fuels. According to the CARB, the associated loss of these heavily forested land areas that provide valuable carbon sinks has to be calculated into the total emissions of the all biofuels. Additionally, the CARB insisted that the indirect effects of replacing cropland that would otherwise grow food to using it for energy production, had to be included in a holistic fuel assessment. When these factors were taken into account, the CARB concluded that burning corn ethanol would not reduce greenhouse gas emissions overall and may even indirectly contribute to higher emissions. Therefore, they decided to promote the use of other energy alternatives seen as cleaner, such as electricity, hydrogen, and cellulosic ethanol to replace *both* gasoline and corn-based ethanol to power vehicles in the state.[67]

Other critics note that any benefits of corn ethanol are offset by the use of petroleum and nitrogen in the very process of growing and harvesting corn. The use of heavy farm equipment and machinery in the processing of corn certainly involves some use of petroleum. Corn plants also absorb copious amounts of nitrates and often this leads to farmers increasing their yields by overapplication of nitrogen, which can be washed away if it is not fully absorbed by crops, leading to erosion into creeks, rivers, and lakes. These nutrient-rich sediments eventually contribute to dead zones in river deltas and oceans, where hypoxia kills off almost all marine animals. The most obvious example of this can be seen the Gulf of Mexico, into which feeds many industrial and agricultural pollutants from farms in the Mississippi River valley.

While several issues arise here, it is interesting to note a sort of cost-benefit analysis conducted by various actors in this debate. First of all, unlike many

---

67 Matthew Cimitile, "Corn Ethanol Will Not Cut Greenhouse Gas Emissions," *Scientific American*, April 20, 2009. Available at http://www.scientificamerican.com/article/ethanol-not-cut-emissions/ (accessed June 22, 2015).

other cases of cost-benefit analysis, this time what is being measured is not quantified in terms of money. Instead, money is replaced by carbon dioxide. The CARB, as well as others skeptical that corn ethanol use really aids in the reduction of greenhouse gases, don't only include factors that go directly into the lifecycles of the fuels being compared. They also examine the actual and anticipated effects of land use domestically and internationally to see if, on balance, when both emissions *and* sequestration of carbon are taken into account, which competing fuel offers the greatest reduction in total CO_2 levels. Other critics of corn ethanol go even further as they worry about pitting fuel versus food as a moral concern. They contend it is greedy and harmful to cause undue suffering and put lives at stake in other countries just so Americans can transport themselves relatively cheaply and generate higher agricultural profits.

Corn ethanol enthusiasts, on the other hand, challenge this perspective by arguing that the reach of CARB's assessment is far too extensive and thus, less accurate. They worry that corn ethanol is given short shrift by taking farmland loss into consideration and that the calculations are at best speculative and imprecise. US policymakers also cannot be expected to be responsible for what might happen internationally with rainforest loss. It is not in the US's purview to regulate wheat production in Russia or rice prices in China. Even domestically, corn ethanol proponents argue it is uncertain whether cropland used for producing ethanol would necessarily have otherwise been used for food production.

This debate boils down to whether there are national or international obligations with respect to food production. Should there be a concern that too much American corn is being produced for ethanol when food prices spike in other countries? Under such conditions, is there an obligation on the US government to encourage its farmers to produce corn for food instead of for ethanol? Furthermore, do farmers themselves have ethical obligations to produce corn for food instead of for biofuels, even when they can make a higher profit on corn sold for ethanol production? The only certainty in this debate is that it serves as another example of the often-lurching development of attempted technological fixes of a deep environmental problem and the unanticipated negative consequences of trying to reach this goal.

Review Questions

1) What moral problems arise from the conflict between the need for cheap food versus the need for cheap fuel? What do you think are some possible solutions to the conflict that pass moral muster?

2) This case raises interesting questions regarding national versus international obligations. What do you think some of these duties might be? Which obligations do you think are more important and why?
3) This case also introduces the concept of the lifecycle assessment of alternative fuels to try to determine their true level of carbon emissions and accompanying contributions to climate change. The factors taken into account in the Argonne study included what other fuels are used in the extraction, production, and transportation contributed to total emissions. However, while the Argonne study stopped there, the CARB also took into account the effects of deforestation worldwide would have in total greenhouse emissions. Which factors do you think should go into these determinations and why?

Case 31 – An Internal Animal Interests Battle: Birders vs. Cat Lovers[68] –

Study Questions

1) What are some of the conflicts that can occur between the advocates of different species of animals as reflected in this case study?
2) Why did Stevenson shoot the cat? What argument does he give for doing so? What was the response to his killing of the cat?

We all know it is common for people to prefer some species of animals to others—we've certainly heard of "cat people" versus "dog people." To the chagrin of ecologists, the public will sometimes show great passion and willingness to marshal copious resources for the furry, cuddly and/or otherwise charismatic fauna, while ignoring other less photogenic animals and failing to come to their aid when necessary. This difference in attitudes towards different species of animals bears out not only in the amount of media attention they receive, but also in what levels of protection they're granted. In a sense there is competition between species for the affection and attention for humans—the continued health and sustainability of non-human animals depend upon it. In listing species as endangered, politics is at play and popular creatures stand a

68 Bruce Barcott, "Kill the Cat that Kills the Bird?" *The New York Times*, December 2, 2007. Available at http://www.nytimes.com/2007/12/02/magazine/02cats-v--birds-t.html?pagewanted=all (accessed June 11, 2015).

better chance for protection. Sometimes though, these competitions lead to some impassioned responses when there is conflict between beloved species.

According to news accounts, Jim Stevenson, an avid birder and director of the Galveston Ornithological Society, pulled himself away from watching television one evening in 2007 to do one of his favorite things—observe birds at San Luis Pass, just outside of his home in Galveston, Texas. He parked his van near the end of a bridge connecting the city with Follett's Island and found a spot in the grass-speckled dunes. He saw some piping plovers, which are listed as a federally endangered species in the United States. But he also noticed something else; a feral cat was stalking the birds, likely emerging from a group living under the bridge. Stevenson said that he tried to protect the birds by securing the cat, but was unable to do so. The next morning he returned with a .22 caliber rifle and took matters into his own hands, shooting the cat dead.[69]

This is how he explained his actions: "Piping plovers are tame, abiding creatures. They roost in the dunes and can't hear or see a cat creep up on them. The American taxpayers spend millions of dollars to protect birds like piping plovers and yet here are these cats killing the birds, and nobody's doing anything to stop it."[70]

John Newland, a tollbooth attendant on the bridge heard Stevenson's shot and quickly came out to see what happened. He often fed the cats under the bridge and had become fond of them, calling them his "babies." When he discovered that Stevenson had killed the cat, he yelled, "I got you! You quit shooting my cats!" Stevenson tried to flee the scene, but the police arrived quickly to arrest him. Stevenson soon became a sworn enemy of the feline-loving world.

But there is a history behind this tale as bird lovers and cat lovers have battled for over thirty years, hurling insults at each other. Organizations advocating birds and cats have gone to the mat as well. The National Audubon Society (NAS) claims, "Worldwide, cats may have been involved in the extinction of more bird species than any other cause, except habitat destruction."[71] The American Bird Conservancy (ABC) calls for direct action to protect birds, trying to persuade cat owners to lock up their cats. In what is called their Cats Indoor campaign, the ABC asserts that by keeping cats inside, not only will the environment be better off, but so will the cats themselves.[72]

---

69 Barcott, "Kill the Cat...?".

70 Barcott, "Kill the Cat...?".

71 Barcott, "Kill the Cat...?".

72 Mel White, "North American Birds Declining as Threats Mount," *National Geographic*, June 21, 2013. Available at http://news.nationalgeographic.com/news/2013/06/130621-threats-against-birds-cats-wind-turbines-climate-change-habitat-loss-science-united-states/ (accessed June 11, 2015).

Cat lovers think it's senseless to try to prevent cats from doing what comes naturally to them—hunting birds. Even if owners can keep their cats away from birds, the real problem occurs with feral cats, of which there are 50–90 million in the US (obtaining accurate numbers is obviously hard because they are not registered and don't have reliable migratory routes or nesting grounds). Feral cats live independently with their offspring. Several were former pets until their owners decided they didn't want or couldn't keep them—and thus "released them into the wild." This doesn't mean that feral cats don't have champions. Alley Cat Allies is an organization that attracts celebrities to speak on behalf of stray cats and advocate on their behalf.

Cat advocates were quick to call Stevenson's act cruel and criminal, with some going so far as dubbing him "Evil Galveston Bird Lover" on various social media sites. The president of the Houston Audubon Society condemned what he called Stevenson's "illegal" actions. Yet, some birders praised Stevenson for what he'd done. For example, a Texas fourth-grade science teacher and birder suggested that Stevenson be given a medal for protecting plovers.[73]

Bird populations are declining worldwide with mounting worry that more species will soon go extinct. As of 2014, 1,373 (or one in eight) of the world's bird species were threatened globally. 213 critically endangered species of birds are so perilously low in numbers that they are at risk of imminent extinction.[74] Some species have not been observed in many years and may already be extinct. These critically imperiled species generally have tiny populations, with fragmented or limited ranges, or are declining rapidly in members. The overall picture is not much rosier, with 80% of the globe's threatened bird species suffering population decline. There have also been more recent instances of precipitous population collapses.[75]

Over the past five hundred years, invasive species have been either partly or wholly to blame for the extinction of at least 65 bird species. Invasive species are also the main threat for at least 50% of all currently threatened bird species. This is the most common contributing factor for the more recent vanishing of birds. Bird species on islands are especially in peril, as three-quarters of them are adversely affected by invasive species. These invasive species have included rats and cats, both of which have had by far the most significant negative impact on birds globally, though other species such as snakes have also contributed.[76]

---

73 White, "North American Birds Declining."

74 BirdLife International, "Spotlight on Threatened Birds," (website), 2014. Available at http://www.birdlife.org/datazone/sowb/spotthreatbirds (accessed June 11, 2015).

75 BirdLife International, "Spotlight on Threatened Birds."

76 BirdLife International, "Spotlight on Threatened Birds."

In the US, bird loss (particularly song birds much loved by birders) has been rapid. The NAS, in a report called "Common Birds in Decline," presents data suggesting that some common species once thought to have healthy populations have fallen in number by as much as 80% since 1967. Nineteen other species chronicled in the report have seen 50% reductions.[77] Smithsonian scientists report that feral cats kill many more birds than was previously thought, with between 1.4 and 3.7 billion birds estimated to be taken each year in the lower forty-eight states. Their report is based on 21 studies of cats and birds in the US and Europe, concluding that domestic cats preying on birds may be the single greatest human-related reason for declining bird populations.[78]

On the other hand, couldn't one say that while cats may kill many birds, they are not in any way blameworthy (particularly feral cats)? Feral cats have already suffered being abandoned by their former owners and left to survive on their own, with their offspring likely never to experience the comforts of domestication. Feral cats are left to hunt or scavenge for their food—what else are they supposed to do? Surely, one might claim, feral cats are the wrong targets for either blameworthiness or elimination. Instead of killing them, could they be managed by sterilization? Or maybe there should be a concerted education program for cat owners spotlighting why abandoning their felines has negative outcomes. Perhaps stricter measures against those abandoning cats should be enacted, involving heavier penalties.

We may need much more robust solutions to this issue, but we have to consider what we think the most important moral priorities are when dealing with conflicts between species. We have to dive into thorny issues of whether cats have a morally higher status than birds and if so why. Is it because more people identify and prefer cats to birds and on that anthropocentric basis cats win out? Or might we simply say that since domesticated cat populations are high relative to those of birds, we should take action to try to lessen cat predation on birds? After all, if a bird species really is endangered, shouldn't it be protected on that basis?

Review Questions

1) Consider the case Stevenson tries to make for why he shot the cat. In part he tries to justify his actions by noting that even though cats don't know plovers are endangered, *he* knows they are and no one was acting to protect the birds. Also, cats are not an endangered species. Do you

77 Cited in White, "North American Birds Declining."

78 White, "North American Birds Declining."

think Stevenson was morally justified in killing the cat to protect members of an endangered species? If so, why? If not, why not? What should Stevenson have done instead?

2) Should feral cats be considered an invasive species? Why or why not? Should domesticated cats be considered an invasive species? Why or why not?
3) What do you think it means for cats or birds to have rights? Do you think cats or birds (or both) should have rights? If so, why? How strong are these rights? If not, why not? If cats or birds don't have rights, what should keep humans from treating them however they wish to do so? If cats and birds both have rights, how ought we to adjudicate between those rights when they come into conflict?

Case 32 – A New Way of Handling Invasive Species? –

Study Questions

1) What is an invasive species?
2) What does it mean for a rate of "take" (or killing) of members of an animal species to be compensatory? What implications does this notion have for the prospects of controlling invasive species by consuming them?
3) What are some potential advantages of promoting invasive species as food sources? What are some possible drawbacks?

For conservationists, one major challenge to reclaiming territory for native species has been in finding effective ways of eradicating invasive plant and animals from targeted areas.[79] Invasive species are singled out for elimination because they endanger native plants and animals. They commandeer space and resources from native species and do not fit with the conservation plans to return, or maintain, local ecosystems to their "normal" or "natural" state.

Many methods are employed to combat the infiltration of non-native species of plants and animals, but a relatively new one should be highlighted for its innovation. J.M. Franke's *The Invasive Species Cookbook: Conservation through Gastronomy*[80] is designed to provide recipes to those who wish to make tasty meals while helping the native environment. Among the delicacies is a recipe for nutria, a rodent regarded as a pest in the state of Louisiana (i.e., it is referred

79 See Case 42—"Beware the Invaders!".
80 Wauwatosa, WI: Bradford Street Press, 2007.

on a website selling Franke's book as the "scourge of the southern wetlands"). Lovers of coastlines loathe nutria because they destroy such habitats. But the cookbook includes invasive flora as well. For example, Franke recommends that nutria should be served with pasta topped with pesto made of "ecologically noxious but highly edible" garlic mustard.[81]

This isn't the first time the gauntlet has been thrown down to chefs to make something palatable from the ecological hazard of nutria. In 1998, the state of Louisiana thought if anyone could make nutria taste good, it would be the famous culinary stewards of the food metropolis that is New Orleans. Chefs from that city were invited to develop a series of recipes highlighting nutria in part as a way of raising public awareness of the environmental damage they do, but also to think creatively about some strategies to eradicate (or at least to curb) them. Even further back, in 1993, noted New Orleans Cajun chef Paul Prudhomme placed a number of nutria food options on his menus to combat their takeover of southern Louisiana. Prudhomme invented such dishes as nutria etouffee, nutria sausage, and nutria gumbo. He also contributed to Nutriafest, a local festival promoting food made from the rodent.[82] The hope from state officials at the time was that these dishes would become popular enough for local hunters to provide them for the nation's markets. That dream has not come to fruition.

One of the main reasons given for why invasive species have been responsible for widespread ecological destruction is that they provide stiff competition for resources with native species in which the latter often lose. Invasive species have proven to be an economic drain, with estimates that some $1.4 trillion is spent globally to try to control them. They have contributed to high maintenance costs for some public utilities, such as water treatment plants along lakes (e.g., zebra mussels in the Great Lakes). In some cases electrical utilities infrastructure is damaged by invasive species. There are so many brown snakes on the island of Guam that they climb electric utility poles and sometimes cause shorts and subsequent blackouts.[83]

Given the constant battle against invasive species, conservationists seek many tools for the fight. Some hope that creating a market in invasive species will add to the arsenal. Considering that domestication of a number of non-native species is one of the reasons for their spread (e.g., unwanted pet

---

81 J.M. Franke's *The Invasive Species Cookbook*.

82 Cornelia Dean, "New Orleans Journal: From Gator Fodder to Haute Cuisine," *The New York Times*, July 27, 1993. Available at http://www.nytimes.com/1993/07/27/us/new-orleans-journal-from-gator-fodder-to-haute-cuisine.html (accessed July 15, 2015).

83 For more on these invasive species, see Case 42—"Beware the Invaders!".

boa constrictors let loose in the Florida keys), any measure to kill off invasives (including consumption) would seem to be morally appropriate. Reflecting on the crashing populations of a plethora of aquatic species due in part to invasives, any means available to mitigate this phenomenon would appear to be welcome. In this light, it would seem uncontroversial to consume invasive species, putting aside for a moment the idea that people may find them unpalatable. After all, if they were tasty they might be thought of as a "welcome new neighbor species" as opposed to an invader.

Yet, some scholars believe we should think twice before advocating the consumption of invasive species. For example, Daniel Simberloff, Sara Kuebbing, Martin A. Nunez, and Romina D. DiMarco present an interesting case for why this trend is not positive.[84] Their main message is that while gastronomy has been presented as a "silver bullet" for dealing with invasive species, there are several reasons why we should refrain from marketing them as consumer products.

The first two reasons are more prudential and legal than moral. First, the authors point out that these ploys are simply ineffective—nutria eating away the coastline habitat of Louisiana (not to mention Maryland and Delaware) are doing so unabated despite efforts to add them to restaurant menus. No significant dent has been put in the advance of the reviled kudzu plant either, as it continues to "eat the South" after already "consuming" 7.5 million acres and will likely add 150,000 more each year (see Case 34—"Filling up on Kudzu"). To genuinely reduce the population of any species, obviously its mortality rate must be higher than its birth rate. Yet with invasive species, harvest for consumption only leads to what wildlife biologists call *"compensatory mortality"* (i.e., the number of individual members that would likely have died by other means, including old age). Simberloff et al. note that even if the culling of a species is beyond compensatory, take must be systematic to overcome its birth rate. This is especially true of invasive species as one of their chief characteristics is a high *fecundity rate* (i.e., the rate at which a species reproduces).

The second reason not to encourage the eating of invasive species is that it is illegal in the United States to ship them. For instance, if a plant is federally listed as an invasive species it cannot be legally transported, marketed, bought, sold, or possessed. Laws also prevent the selling of meat that is not properly inspected, which means that wild caught or hunted nutria from Louisiana could not legally be on offer, in say, Illinois. In effect, even if there is a market

84 "Why Eating Invasive Species is a Bad Idea," *Ensia*, (website), September 9, 2014. Available at http://ensia.com/voices/why-eating-invasive-species-is-a-bad-idea/ (accessed June 23, 2015).

for an invasive species, it will have to be local. Thus, a consumer market for invasive species faces impediments to expansion as creating enough demand would likely prove difficult to sustain. Unless some changes occur in current legal codes, this market would remain fairly restricted. There are also likely independent reasons why these laws should remain in place.

The third and fourth reasons for not championing the eating and marketing of invasive species are normative. The third worry Simberloff et al. note is decidedly different from the first two in that it gives a reason to be hesitant about encouraging the consumption of invasive species even in if they *do* somehow become desirable. By promoting consumption of an invasive species, this may create a substantial market for it. This creates a perverse incentive to *retain* the invasive species to further supply demand and thus find *new* habitat for it to take advantage of in a new and flourishing market. Illegal introduction of non-native species of sport fish occurs quite frequently and provides examples of the kind of market expansion encouraged by consumption of an invasive species.[85]

The fourth and final reason why the promotion of invasive species consumption could be harmful is that again the species could not only become popular, but eventually regionally iconic. Though this reason may initially seem a bit farfetched, Simberloff et al. cite the case of wild boar in Hawaii. Wild boars were first introduced to the state as domesticated animals in the lowlands, but escaped captivity and now have bred generations of feral boar. They have proven to be an ecological menace, running roughshod over the islands, uprooting and eating endangered plants. Invasive species can also serve as vectors for disease with pathogens typically accompanying them. Sometimes they don't even need to be hosts to proliferate diseases. Wild boars are a case in point, as they cause massive erosion through their destruction of vegetation that in turn creates breeding pools for non-native mosquitos carrying malaria and pox that threaten bird species in the state. Despite these problems, efforts to eradicate wild boars in Hawaii are extremely controversial. Native Hawaiians have hunted them for generations and wild boar feasts are now deeply woven into their cultural fabric.[86]

Perhaps the most noteworthy invasive species on offer in culinary circles is lionfish. Lionfish were introduced from Indo-Asian waters to the Caribbean in the 1980s and they have become a voracious predator there ever since. Some have been observed to eat 30 fish in just a half-hour's time. They have not only become comfortable in coral reef habitat, but also near mangroves

85 Simberloff et al., "Why Eating Invasive Species."

86 Simberloff et al., "Why Eating Invasive Species."

along coastlines and in some estuaries. Lionfish have virtually no predators in the Caribbean or elsewhere for that matter. Spines containing venom provide excellent protection and they possess a configuration of unusually colored fins that confuse potential predators.

Yet, lionfish are considered to be quite palatable, with firm white meat and a mild taste that has been described as a cross between a hog snapper and a grouper.[87] They can grow to over a foot in length and weigh over two pounds, which makes them a substantial fish to eat. Governmental organizations have joined the effort to promote the consumption of lionfish as a means of diminishing their population. For example, the United States National Oceanic and Atmospheric Administration (NOAA) launched a campaign via several media outlets called "Eat Lionfish" in partnership with the Reef Environmental Educational Foundation (REEF) and other organizations to promote the species as a sustainable seafood option.[88]

But, even hunting lionfish can cause ecological damage. The usual take method is spearfishing, but missed shots can lead to coral damage or the destruction of non-targeted species. This point, as well as several other unintended, adverse consequences presented in this case underlines the notion that even when there seem to be "no-brainer" methods for administering environmental protection, we sometimes have to think again.

Review Questions

1) While eradicating invasive species may make sense ecologically, there is the question of whether these species *should* be sacrificed for this end. What do you think is the moral justification for eliminating invasive species (if there is one)? Are there any moral problems with the attempted removal of invasive species? Is there any justification for the removal of a non-native species if it is not having a major detrimental impact on its new environment? Could a deep ecologist possibly justify removal of such a species?
2) Imagine that over time, from a process of hunting and climate change, an endangered species migrates into an area where it has not previously inhabited. It becomes a predator of species native to its new ecosystem.

---

87 "Frequently Asked Questions About Lionfish," *SafeSpear.com*, (website), 2015. Available at http://www.safespear.com/v.php?pg=59 (accessed July 7, 2016).

88 National Oceanic and Atmospheric Administration, "Invasive Lionfish Threaten Coral Reefs and Fisheries," (website), December 15, 2014. Available at http://www.nmfs.noaa.gov/stories/2014/12/12_01_14impacts_of_invasive_lionfish.html (accessed June 24, 2015).

However, it is not hunted as much in this new location and has no non-human predators. How should this species be treated? As an invasive species to be eradicated? Should its members be re-settled back in its native environment? What if the species declines again in its native environment?

3) Recall the case of wild boar in Hawaii. On the one hand, it makes sense to consider them as an invasive species as they meet some of the usual characteristics. Wild boars were introduced from another region of the world and they have undesirable effects, such as uprooting and thus destroying native vegetation. On the other hand, hunting them has become a tradition among native Hawaiians who not only enjoy boar meat, but also the opportunity to gather at community festivals celebrating the hunt. In sum, do you think wild boar should qualify as an invasive species in Hawaii?

Case 33 – PETA vs. Itself and Others –

Study Questions

1) In what ways is Newkirk's approach to environmentalism controversial, both to those within PETA and to those outside of it? What are some examples of controversial tactics Newkirk and her organization have used to try to raise awareness of animal interests?
2) What are some of the moral principles PETA accepts? Is any use of animals acceptable to its membership?
3) What are some initiatives PETA cites as being successful? Why do some within the animal advocacy movement ultimately think the organization is doing a disservice to animal welfare?

Ingrid Newkirk may not be a household name, unless you happen to be an animal rights activist or part of the animal food and clothing industry that she loathes (and likely hates her back). Newkirk is the President and co-founder of the organization known as People for the Ethical Treatment of Animals (PETA). PETA is the world's largest animal advocacy group, with some 3 million members and an annual operating budget of $30 million—huge by animal rights advocacy standards. To some, Newkirk is certainly considered a pariah. But to others, her ability and willingness to orchestrate radical actions (from naked demonstrations to throwing fake blood on fur retail stores) is considered vital to raising awareness of cruelty to non-human animals.

PETA began when Alex Pacheco, while enrolled at George Washington University, founded it as a student group. At that time, Pacheco had mainly organized efforts to combat the practice of castrating pigs and cattle without anesthetic and the use of leghold traps. He decided to leave George Washington without graduating in 1979 to develop PETA despite having no staff and no money. Newkirk joined him, bringing with her experience with investigating and exposing controversial animal experimentation, while Pacheco had worked on ships protesting the harvesting of whales around the world. Together, they developed a powerful organization that has done over 80 investigations of everything from labs conducting experiments on monkeys to turkey factory farms. In these efforts, undercover members of PETA have complied notes and video documenting ill treatment of animals leading to shocking exposes of universities, medical facilities, and private businesses.

Newkirk claims that the core of her activities come from the basic principle that we, as human animals, are the same as any other sort of animal. PETA announces itself as an organization existing to combat "the undeclared war on animals since the beginning of time, because they are vulnerable." Some have declared war on Newkirk, at least as suggested by the hate mail she receives. In the opening scene of the HBO documentary *I Am An Animal: The Story of Ingrid Newkirk and PETA*, Newkirk reads from letters and emails that include passages like the following: "May I suggest that you have [your anus] stretched and used for the threshold for your front door at PETA," and "I wish you all [members of PETA] meeting up with Jeffrey Dahmer so he can rid the world of scumbags like you."[89] But others with legitimate animal right bona fides also have misgivings with Newkirk's strategies for protecting animals, such Friends for Animals President, Priscilla Feral, who thinks PETA actually trivializes animal rights with its activities and calls Newkirk "a media slut."

The premise behind PETA's work was clearly and succinctly put by Newkirk in her keynote address to the attendees of the 2014 Animal Rights National Conference: "Animals are not ours to eat, animals are not ours to wear, animals are not ours to experiment on, and animals are not ours to use for entertainment."[90] She adds that the term "humane meat" is oxymoronic. Even though Newkirk acknowledges free-range chickens do have larger crates and some dairy cows on small organic farms may be spared shots of BST and

89 *I Am An Animal: The Story of Ingrid Newkirk and PETA*, Dir. Matthew Galkin, HBO documentary, 2007.

90 Ingrid Newkirk, "PETA and Co-Founder Ingrid Newkirk at Animal Rights National Conference 2014," *YouTube* (online video), August 26, 2014. Available at https://www.youtube.com/watch?v=EusaXzYThZ4 (accessed July 7, 2016).

antibiotics, she doesn't think these creatures are likely "thankful" for their treatment. Chickens are still eventually going to the slaughterhouses and cows are going to be milked and otherwise manipulated against their desires.[91] As Newkirk puts it,

> [I]t is being suggested that we actually find it acceptable to eat the flesh of animals that were very much alive, had friends and family—or more likely, were deprived of them—and went through enormous trauma despite some small courtesies, such as perhaps 2 inches of additional space in their jam-packed prison cells. Yes, kicking the dogs six times a week instead of seven *is* marginally better, but that doesn't mean that we should go around suggesting that people kick the dog, just not as often, does it? Calling this sad flesh "humane" is like calling Britney Spears an opera singer. Yes, "Baby One More Time" may be easier on the ear than fingers on a blackboard, but it's hardly Wagner's "Ring Cycle" is it? I could go along with SLCBSU, or "slightly less cruel but still unacceptable" meat, but it's definitely still not *humane* by a long shot.[92]

Newkirk also chronicled a rather lengthy list of victories for PETA throughout 2013 and other news of general public awareness of the use of animals. Some examples cited include; PETA pressure on India forcing it to ban foie gras, Colorado's becoming the thirty-ninth state to ban greyhound racing, Mexico City's ban on animal circuses, and that all airlines have ceased shipping animals to laboratories for experimentation with the exception of Air France. Additionally, PETA's expose and the documentary film *Blackfish* revealed whale suffering at SeaWorld and led to divestment of the company that holds its major stock shares. Newkirk also briefly recounted the story of how film director Darren Aronofsky approached her just before filming *Noah*, a movie in which he would need to show pairs depicting all species of animals. He wanted to receive advice on how to properly treat animals while filming, but PETA's response was to say that there was no way to guarantee animal safety on the set. And so, Aronofsky improvised—with the exception of using two live doves, all of the "animals" that appear in *Noah* are computer-generated images.

The strident position of the organization has attracted the support of celebrities such as Pamela Anderson, Alec Baldwin, Bono, Ellen DeGeneres,

---

91 Ingrid Newkirk, "Let's Face It: There is No Such Thing as Humane Meat," *Huffington Post*, February 26, 2013. Available at http://www.huffingtonpost.com/ingrid-newkirk/humane-meat_b_2765996.html (accessed June 24, 2015).

92 Newkirk, "Let's Face It."

Angelina Jolie, Aishwarya Rai, Kristen Wiig, Sting, Elton John, Bill Clinton, Paul McCartney, Bill Maher, and many more. But to still others, including co-founder Alex Pacheco (who is no longer with PETA), Newkirk's activities represent a single-minded passion to protect animals that sometimes disregards respect for humans and damages the reputation of animal rights organizations. Newkirk has drawn the ire of many due to her high-risk, high-publicity approach to animal abuse investigations. Perhaps the most controversial of all of PETA's actions under Newkirk's direction was an ad campaign comparing the conditions of caged chickens and pigs to those of Jews during the Holocaust. In defense, Newkirk says, "Sometimes, the way you get discussion on the table is to do something jarring... I still hold that the [Holocaust] campaign is absolutely true."[93]

And then there is the use of scantily clad celebrities to convey messages concerning animal cruelty and the promotion of an animal-free diet. Actress Alicia Silverstone, who in 2004 was voted "Sexiest Female Vegetarian" by PETA, taped a public service announcement and appeared in a print advertisement for the organization. The commercial was a 30-second spot where Silverstone emerges nude from a pool. In the ad she offers a brief description of why she is a vegetarian, amounting to a testimonial about health benefits and how the dietary practice has transformed her life. Pamela Anderson appeared in a revealing ad in PETAs' "All Animals Have the Same Parts" campaign. Olivia Munn posed naked in an ad that proclaims, "Who Needs Fur to Feel Beautiful? I'd Rather Go Naked Than Wear Fur." While some have applauded these efforts as innovative and attention-grabbing in a way that escapes most animal rights organizations, these advertisements have been controversial to say the least, even with those who might be otherwise sympathetic to the animal rights movement. The main contention is that while "sex sells," we shouldn't use it even for what may be a good cause. Men have also appeared in sexually provocative ads for PETA, as evidenced by a series of them shot with Wayne Parnell (a cricket player from South Africa) that encourage viewers to boycott zoos and circuses.[94]

Some intriguing issues appear in deep relief with respect to this case study. The most obvious question is whether the ends always justify the means? Are there horrific historical events, such as the Holocaust, that shouldn't be used as analogues to the contemporary plight of animals in factory farms? Even if

---

93 See *I Am An Animal*.

94 People for the Ethical Treatment of Animals, "Wayne Parnell Strips Down to Stop Circus Cruelty," (website), May 27, 2011. Available at http://www.petaindia.com/features/wayne-parnell-new-peta-ad/ (accessed June 24, 2015).

one puts aside the question of whether the comparison is hyperbolic, is there something morally wrongful with using a human tragedy to try to persuade people to stop animal slaughter?

Also assuming that one finds using attractive bodies, whether they are male or female, to persuade others to buy things is at least tacky and at most exploitative, is there anything morally wrong with using sexy advertisements to promote normative ideals? If we think it degrades women to use images of them in other contexts, such as pornography in print and digital media, even when they consent to having their images used this way, how would the case be different (if at all) for advocating a vegan lifestyle?

Review Questions

1) Do you think any of Newkirk's more controversial actions are morally justifiable? Why or why not? What sorts of action by animal rights advocates would be morally permissible and on what grounds? Do you agree with Pacheco and others that the more controversial demonstrations by Newkirk and her colleagues actually harm the animal rights movement more than help it? Is there a moral justification more mainstream animal rights advocates have at their disposal to condemn radical animal rights activities?
2) The utilitarian moral philosopher Peter Singer thinks that what matters morally to show non-human animals have interests is sentience (the ability to feel pleasure and pain). He observes that animals, though in different degrees, can feel pleasure and pain. We should not then, according to Singer, kill and eat animals because it causes them distress and pain while only providing a bit of pleasure for humans. Now consider the following quote from Newkirk: "When it comes to pain, joy, loneliness, and fear, a rat is a pig is a dog is a boy." Do you think Newkirk is offering Singer's justification for the protection of animal interests? What similarities or dissimilarities do you see in their moral outlooks towards animals? Do you agree with either Singer or Newkirk concerning the moral value of animals?
3) We saw that some celebrities endorse PETA's mission. What should the role of celebrities be in advocating environmental causes? Should people listen more or less to the pleas of celebrities when it comes to animal rights advocacy? Some point out that while one male has appeared in PETA's racy ads, the vast majority of celebrity participants are female. They add that such tactics are exploitative and perpetuate views of women as sexual objects. What do you think of these advertising campaigns? If

you find them morally objectionable, on what grounds? If not, why not? What might Kant say about them?

Case 34 – Filling up on Kudzu –

Study Questions

1) How has kudzu been traditionally regarded in the United States, especially in the American South?
2) What are biofuels? What are some of the possible sources for biofuels?
3) What are some of the environmental advantages biofuels are supposed to have over other competing fuel sources (especially fossil fuels)?

In the hunt for alternative fuels, much has been made of the unsuitability of corn ethanol [CH3—CH2—OH] for this purpose. Since corn is extremely valuable in the food market its use as a fossil fuel substitute has exacerbated shortages and driven up world food prices.[95] It also takes many resources to grow and harvest corn grain for fuel. For those researching biofuels, the desire is to find plant matter that already grows well without lots resource inputs and simultaneously is not a food commodity.

In his research on how rising temperatures on Earth affects plants, Lewis Ziska has struck upon an unintended potential benefit of climate change.[96] Because of their ability to live and proliferate in environments with higher temperatures and greatly increased CO2, weeds could be a viable fuel source in a changing climate. Kudzu roots (*Pueraria lobata*) are composed of as much as 50% starch by weight and thus are excellent candidates for ethanol production. Also, their vines, which can grow by a foot a day, would be a very powerful source of alternative energy. Potentially, we could use market incentives to reduce the presence of what is called "the weed that ate the South" and develop an alternative to fossil fuels at the same time.[97]

"Garage scientists" have been cooking up kudzu to test its promise as an alternative fuel source.[98] Tennessee resident Doug Mizell has been making

---

95 See Case 30—"Food vs. Fuel: Unintended Negative Consequences of Corn Ethanol."

96 See Case 5—"To Be A Weed."

97 Christopher, "Can Weeds Help Solve the Climate Crisis?" p. 47.

98 "De-Vine Biofuel: Ethanol from Kudzu," *NBC Nightly News* (Television series episode), December 16, 2008. Available at https://www.youtube.com/watch?v=MhYux3QIJgo (accessed May 1, 2015).

"kudzunol" in his improvised laboratory since the late 1990's. A kitchen food grinder, a garden hose, and an old milk jug serve as his lab equipment. Mizell uses these common household items to ferment ground Kudzu into what is essentially moonshine.[99] The alcohol is ethanol, just as is found in corn. But one big difference between corn and kudzu is that kudzu is a non-food plant, and thus would put to use what former NBC news anchor Brian Williams called a "practically indestructible scourge."[100] Moreover, kudzu may be just as good a source of ethanol as corn. During Mizell's experiments, he found that an acre's worth of kudzu would produce about 10 gallons of kudzunol costing only $1 dollar per gallon. He ran it in his tractor as well as his truck. He built a small demonstration plant and his fuel was quite popular with some locals and even attracted the attention of a NBC news program, which aired a story on Mizell's invention in late 2008. He then founded Agro Gas Industries, but Kudzunol has not really caught on. However, it may be a step in the evolution of non-corn biofuels becoming more viable—first by gaining nascent interest, and then perhaps luring venture capital investment to develop the industry.

Kudzu is just one example of an organism now considered as an invasive species that was intentionally introduced somewhat ceremoniously years ago. Kudzu landscaped a garden in the Japanese Pavilion at the Philadelphia Centennial Exhibition of 1876. In the early 1900s, Kudzu was thought to be excellent forage for cows, pigs, and goats in the American South, especially in acidic and dry soils. As a tough, deeply rooted vine, it was introduced widely in the South as a tool against soil erosion and planted in gullies during the 1930's. Today, you can find it as far south and west as Texas, stretching south and east to Florida, northwest to Missouri, and northeast to Connecticut.[101] There are roughly seven million acres of kudzu in the southern US, with roughly 2 million of that on forestland. It has been spreading at a rate of about 150,000 acres a year.

Kudzu has been vilified not only because it grows over everything and is a gardener's pest. It is also destructive when it grows in heavy concentrations and is known to pull down electrical poles, break power lines, collapse buildings, and kill trees.[102] Highway departments across kudzu-infested areas spend

99 "De-Vine Biofuel."

100 "De-Vine Biofuel."

101 Missouri Department of Conservation, "Kudzu Control," (website), 2015. Available at http://mdc.mo.gov/your-property/problem-plants-and-animals/invasive-plants/kudzu-control (accessed July 7. 2016).

102 Missouri Department of Conservation, "Kudzu Control."

millions of dollars trying to control the hard-to-kill plant in what seems an epic losing battle despite the estimated $6 million a year spent on the effort. As Ray Rucker of the Tennessee Department of Transportation describes the progression of kudzu near area highways, "If you're not careful, it'll get into the road—right over the guardrail and into the road."[103]

Back in 2008, it wasn't just Mizell who thought there was promise in using kudzu as a feedstock for cellulosic ethanol. The US Department of Agriculture and the University of Toronto were also investigating this possibility. However, within just a few years, the idea was abandoned, most likely due to the economic crash. Their joint 2008 study indicates that kudzu would produce about 400 liters (~105 gallons) of ethanol per acre—about what corn produces. Rowan Sage, the UT investigator who led the research team, concluded that kudzu could thus be a viable alternative fuel. His team collected samples of kudzu from several different sites throughout the southern US during different seasons of the year. They measured the carbohydrates present in the leaves, roots, and vines, finding that the roots had the highest concentration (up to 68% by dry weight). This greatly exceeded the amount found in the leaves and vines, which contained just a few percent. Sage and his team estimated that kudzu could produce about 2.2 to 5.3 tons of carbohydrates per acre in a substantial portion of the South, or about 270 gallons of ethanol.[104] However, a large-scale, follow-up experiment planned to further study the idea was never conducted, as alternative fuel investments dried up due to the poor economy and reductions in fuel prices.

Biofuels are of interest for a number of reasons. Obviously, one of the major issues with fossil fuels is that they involve pumping oil that has sequestered carbon in it out of the ground to be burned and then emitted as greenhouse gases. While biofuels emit CO_2 during the combustion process just as fossil fuels do, it is from plant material that will be regrown and thus is carbon neutral. Using biofuels has the potential to be part of a "closed-loop" process, where the waste stream of CO_2 is simply fed back into the manufacturing process as an input. This leads us to a second promise of biofuels, according to their proponents, which is that they are renewable, whereas fossil fuels are limited in supply.

Of course, the trick for manufacturing a viable biofuel is finding the right biomass. Yet, there are some prospects, especially if cellulose from plant matter can be used instead of corn grain to make ethanol. There is much more

103 "De-Vine Biofuel."

104 Nasir El Bassam, *Handbook of Bioenergy Crops: A Complete Reference to Species, Development and Applications* (Washington, D.C.: Earthscan LLC, 2010), p. 229.

abundance of cellulose than corn grain—in fact, it is the most abundant organic molecule on Earth. It takes less energy to produce cellulose than it does for corn grain. Using cellulose-based ethanol may even remove carbon dioxide from the atmosphere.

There are limitations on the use of cellulose for ethanol. Cellulose is composed of a long chain of glucose (or sugar) that is protected by a tough molecular sheath. This makes cellulose difficult to break apart. So, to produce cellulose-based ethanol means first unraveling the outer layer and slicing the glucose beneath into its individual molecules. Only when all glucose molecules are unpacked can they be converted into ethanol by fermentation. Some types of fungi and bacteria break down cellulose. When dairy cows eat alfalfa or hay, microbes living in their stomachs break down these cellulose-laden foods. They do so by producing enzymes known as cellulase that dissolve the cellulose so that glucose can be absorbed. Researchers are thus studying fungi and bacteria to try to prepare cellulose for biofuel production. Part of this endeavor will likely mean finding microbes that already exist in nature with cellulase in them.

Jonathan Trent is just one of a host of scientists researching this possibility in his work with algae. Using waste treatment water in ponds, he introduces microbes into the mixture. With CO_2 and water, the algae produce oxygen as they grow. Heat is distributed to the surrounding water and once harvested, the algae can be used to manufacture a wide range of products—from animal food, to fertilizer, to cosmetics, and of course, biofuel. Trent estimates that microalgae could produce 5,000 gallons of fuel per acre per year. His plan is to float algae in large plastic bags along small portions of city coastlines. This is in large part because the medium for the microbes would use waste treatment water that would otherwise go out to sea. A large challenge to his research has been removing oxygen from the bags produced by the microbes so they won't suffocate.

Many questions surround the financial viability of biofuels. First of all, several technical problems and financial challenges to non-corn biofuel development have already been mentioned throughout this case study. Second, sometimes even the best ideas don't come to fruition as investment timing and even luck can advance some lesser products ahead of other better ones. Henry Ford's Model T first ran on ethanol. If Prohibition hadn't come to the US, perhaps the fuel industry would never have been dominated by petroleum. Yet given that many different avenues are being taken to find renewable fuels that won't compete with food resources, we shouldn't be surprised if our gas tanks are someday filled by something fermented instead of pumped or mined.

Review Questions

1) Suppose kudzu, or another biofuel alternative to corn such as switchgrass, became a viable source of ethanol. Suppose further that one or many of these alternatives is also shown to be at least superior (emissions wise) to the use of fossil fuels for automotive vehicles. Some might think this sounds promising, yet still worry that using alternative sources impedes adoption of the cleanest mode of powering vehicles with respect to emissions—electricity. In devising what sort of policy decisions we ought to adopt, we have to consider that developing one technology in the short-term might create delays in what might be a better technology down the road. Taking this into consideration, do you think the proper tactic should be to skip the development of switchgrass, kudzu, and algae to instead spend time and effort developing lithium batteries and the like?
2) Some other candidates for use in making cellulosic ethanol beyond those mentioned in this case study include giant reed. Some have wanted to use it as a feedstock for biofuel, believing it will have greater energy output than competing sources such as kudzu, switchgrass, and algae. But many scientists worry about this strategy, as giant reed can grow prolifically and promoting its cultivation might well lead to another invasive species disaster, such as with kudzu (See Case 32—"A New Way of Handling Invasive Species"?). Do you think plants that may become invasive species should be avoided for fuel production? Or do you think the risk of widespread use of an introduced species is worth it to develop the most efficient fuel source?

CHAPTER 9

Relations between Human and Non-Human Animals

Case 35 – The Loneliest Catch –

Study Questions

1) What is the "loneliest catch"?
2) What is overfishing? What are the factors that contribute to the fish featured in this case study being overfished?
3) What is the "tragedy of the commons"? Why do some think international fishing provides an ideal case study for the tragedy of the commons?

Cable television has popularized and sensationalized the commercial fishing industry by airing shows such as *The Deadliest Catch* and *The World's Most Dangerous Jobs*. But the element of the industry that goes relatively unnoticed is that along with the perils of fisherman fighting fatigue and rough seas is the possible unsustainability of the industry itself. There is the very real question of whether there will be any fish left to catch if they continue to vanish at the current rate.

Worldwide consumption of fish (including fish meal for animal feed and fertilizer) has tripled since 1960. Roughly a quarter of the world's fish population is either already depleted or slowly recovering from past exploitation.[1] The ocean is the largest source of food on Earth, with some 1.2 billion people consuming fish as their main means of protein. Some researchers believe that as compared with pre-industrial levels, stocks of large fish specifically have declined by an astounding 90%.[2] In another 50 years, we might see a complete collapse of most populations of fish species.

On large deep-sea fishing trawlers (some are converted super tankers), immense nets with openings of almost 250,000 square feet (or the size of four

---

1 Peter Singer and Jim Mason, *The Way We Eat: Why Out Food Choices Matter* (New York: Rodale, 2006), p. 111.

2 Ransom Myers and Boris Worm, "Rapid Worldwide Depletion of Predatory Fish Communities," *Nature* (2003) 423: 280–283 at 280.

 | DOI 10.1163/9789004339996_011

soccer fields and large enough to hold thirteen jumbo jets) can trap tremendous yields of fish (up to 500 tons). It is estimated that collectively, long-line fishing vessels use some 1.4 billion hooks a year to catch fish. Not only are commercially valuable fish in peril. Due to by-catch (i.e., marine creatures inadvertently caught in nets), many members of non-targeted species are hauled up and cast back into the ocean, often dead or with various degrees of physical damage. It is not uncommon for shrimp trawlers to throw 80–90% of what they haul back into the ocean.

It now seems all too common to hear a story of how some species of aquatic animal, once so prolific that only hyperbolic sounding analogies seem adequate to describe their immense populations, is now struggling for survival. For instance, at one time there were so many cod that fisherman joked one could walk across New England's coastal waters on their backs. While less poetic, the lobster population in Maine was so robust in Colonial times that Native Americans used them to fertilize fields and to bait their hooks for fishing. Likewise, the Pacific bluefin tuna (Thunnus Orientalis) held such a mantle in Northern Pacific waters, which once teemed with them mid-20th century. But heavy fishing has left the species "endangered" for several years now. It is estimated that its population is only 4% of what it once was.[3] Pacific bluefin tuna are among a group of greatly endangered species that make up the "loneliest catch."

The Pacific bluefin tuna has a muscle-bound, streamlined body allowing it to propel itself at high speeds. It is one of the fastest fish in the Pacific Ocean, sometimes reaching speeds of sixty miles per hour.[4] To gain information about this fast, mobile species, Pacific bluefin tuna populations are tracked via two programs. One is the "Tagging of Pacific Predators" program (TOPP) and the other is "Tag-a-Giant" (TAG). The TOPP program is designed to show the efficacy of electronic tagging to follow the migratory paths and behaviors of large predators in the Pacific. Fifty researchers in TOPP have tagged over 4,000 fish, some of which include Pacific bluefin tuna. TAG's mission is to gather the scientific data necessary to preserve the future of bluefin tuna in both the Atlantic

3 The Pew Charitable Trusts, "Rebuilding Pacific Bluefin Tuna: A Science-Based Blueprint," (website), July 11, 2014. Available at http://www.pewtrusts.org/en/about/news-room/news/2014/07/14/rebuilding-pacific-bluefin-tuna (accessed May 20, 2015).

4 National Oceanic and Atmospheric Administration, "Pacific Bluefin Tuna: Overview," *Fishwatch US Seafood Facts*, (website), February 24, 2015. Available at http://www.fishwatch.gov/seafood_profiles/species/tuna/species_pages/pac_bluefin_tuna.html (accessed May 18, 2015).

and Pacific Oceans. Correspondingly, TAG has also been heavily involved in tracking bluefins, having tagged at least 1,300 around the world.[5]

Tracking information indicates that Pacific bluefin tuna achieve impressive migratory feats, sometimes crossing whole ocean basins. They are a single-stock fish and are found mainly in the North Pacific, though they span a vast east-west range of the ocean, all the way from East Asia to the North American West Coast. Being a single stock fish means that Pacific bluefins only have one spawning ground, which has been located with electronic tagging data as being off the southern coast of Japan.[6] What this extensive migratory activity means is that Pacific bluefin tuna are often in international waters where there is no systematic regulatory protection of them from overharvest.

Even when their population decline began, fisheries management paid little attention to Pacific bluefins, lumping them in with other, more prolific species of fish.[7] But recently, their status has changed significantly. In November of 2014, the International Union for Conservation of Nature's Red List changed the Pacific bluefin's designation from "least concern" to "vulnerable"—which means the species is threatened with extinction.[8] With this change, Pacific bluefins now join both the Southern bluefin (ranked "critically endangered") and Atlantic bluefin (designated "endangered") as imperiled species.

The primary threat to all of these bluefin tuna populations is common—overfishing. In 2013, the US National Oceanic and Atmospheric Administration declared that Pacific Bluefins were overfished. A 2014 stock assessment conducted by the International Scientific Committee for Tuna and Tuna-like species in the Northern Pacific (ISC) corroborated NOAA's conclusions, adding specific concerns that large numbers of juveniles who had never reproduced were being harvested. In fact, the number of juveniles caught (90% of which had never reproduced) lead some fish scientists to worry that this might be the last generation of the species, as not enough would reach reproductive maturity to avoid the tipping point for extinction.[9] This is also known

5 Census of Marine Life, "Bluefin Tuna—Past and Present," (website), 2009. Available at http://www.coml.org/comlfiles/press/CoML_tuna_release_public.pdf (accessed July 22, 2016).

6 Tagging of Pelagic Predators, "Pacific Bluefin Tuna," (website), May 18, 2015. Available at http://www.topp.org (accessed July 8, 2016).

7 Fiona Harvey, "Overfishing Causes Pacific Bluefin Tuna Numbers to Drop 96%," *The Guardian*, January 9, 2013. Available at http://www.theguardian.com/environment/2013/jan/09/overfishing-pacific-bluefin-tuna (accessed July 8, 2016).

8 Eliza Barclay, "Why Some Chefs Just Can't Quit Serving Bluefin Tuna," *National Public Radio*, (website), January 7, 2015. Available at http://www.npr.org/sections/thesalt/2015/01/07/375366742/why-some-chefs-just-cant-quit-serving-bluefin-tuna (accessed July 8, 2016).

9 Barclay, "Why Some Chefs."

in the marine ecology world as *recruitment overfishing*. In marine ecology, "recruitment" means the number of juveniles that survive to the point of being reproductive. According to Michael P. Sissenwine and J.G. Shepard, recruitment overfishing occurs "when the spawning biomass is so low that recruitment decreases significantly and perhaps dramatically."[10] Though now this phenomenon is more often called *recruitment collapse*.[11]

The ISC study reveals that historically the Pacific bluefin tuna catch was from 9,000 to 40,000 tons per year.[12] In 2009, 20,000 tons were caught and the recent 5-year and historical average catches of Pacific bluefins were each 23,000 tons, with records going back to 1952. Japan is responsible for taking about half of the total catch. Until the 1980's, the US fishing fleet caught a significant number of Pacific bluefins. But Korea and Mexico have increased the size of their catches in recent years. There are many different ways to catch Pacific bluefins, but it is unsurprising that the Japanese fishing fleet, given their harvest yields, deploys all of them, including purse-seine nets, set nets, longline, pole-and-line, and trolling. Purse-seine net fishing is the most popular means of catching them.

Current spawning stock biomass for Pacific bluefins is near its lowest level ever (3.6%) and has been in decline for over a decade.[13] Fishery impact analyses from the years 1970–2010 indicate that Japanese coastal fishing groups have had the most significant impact on the population (i.e., in terms of expected spawning stock biomass).[14] Since about 1990 the impact of purse-seine fishing has increased, with it currently being the highest impact method of fishing on Pacific bluefins. Most importantly, catches of juvenile Pacific bluefin tuna have increased substantially.

---

10 Michael P. Sissenwine and J.G. Shepherd, "An Alternative Perspective on Recruitment Overfishing and Biological Reference Points," *Canadian Journal of Fisheries and Aquatic Sciences* (1987), 44: 913–918.

11 See Keith Sainsbury and Tom Polacheck, "The Use of Biological Reference Points for Defining Recruitment Overfishing, with an Application to Southern Bluefin Tuna," in *Population Dynamics for Fisheries Management* (Perth, WA: Australian Society for Fish Biology, 1994), pp. 265–274.

12 International Scientific Committee for Tuna and Tuna-like species in the Northern Pacific, "Pacific Bluefin Tuna Working Group," (website), July 1, 2014. Available at http://isc.ac.affrc.go.jp/working_groups/pacific_bluefin_tuna.html (accessed May 19, 2015).

13 International Scientific Committee for Tuna and Tuna-like species in the Northern Pacific, "Pacific Bluefin Tuna Working Group."

14 International Scientific Committee for Tuna and Tuna-like species in the Northern Pacific, "Pacific Bluefin Tuna Working Group."

In June 2013, citing the species rapid decline, the Inter-America Tropical Tuna Commission (IATTC) proposed a 2014 commercial catch limit set at 5,000 metric tons.[15] This recommendation went unheeded. Even when there is initial agreement on some fish take curtailment, there can be grave differences of opinion about what those limits ought to be. Take the case of Atlantic bluefin tuna, which is also highly prized by the commercial fishing industry. In 2008, scientists recommended that only 10,000 tons of Atlantic bluefin be taken in an effort to rebuild a greatly depleted stock.[16] However, the European Union and other fishing nations decided on a limit of 29,000 tons, far exceeding the suggested limit. Despite this inflated number, 61,000 tons of Atlantic bluefin tuna were actually caught. This case indicates (and serves as a warning for conservationists) that even when there are limits placed by governmental bodies, enforcement of these regulations is by no means guaranteed.

Of course, commercial and recreational fishing for Pacific bluefins have existed for thousands of years. But since the 1970s and 1980s there has there been a systemic pressure on the species from the fishing industry due to the sushi and sashimi markets, which have soared. Heavy demand and correspondingly high prices have led to intensive fishing, putting a major dent in the Pacific bluefin population. Fisheries worldwide have rapidly expanded which means commercial ships travel further from port and find more locations where the species has not been traditionally fished. This is true with other species of tuna besides bluefins. While two-thirds of tuna caught worldwide comes from the Pacific Ocean, only 1% of this yield consists of Pacific bluefin tuna.[17]

Several factors make rapid and intensive harvesting a substantial problem for Pacific bluefin in particular. It is a slow-to-mature species (with reproductive capacity beginning at around three to eight years of age or at 100 cm). They are slow-growing and (at least without misadventure) long-lived (more than 20 years). All of this leaves Pacific bluefins much more vulnerable to population collapse when compared with shorter-lived species of fish that reproduce more frequently. Also, since they have become highly prized and have such a long migratory route, conservation goals are difficult to achieve as multiple nations fish for them. Consequently, many parties need to come to agreement about whether and how to protect and restore the population.

---

15 NOAA, "Pacific Bluefin Tuna: Overview," 2015.

16 Asha de Vos, "Ending Overfishing," *TED Ed* (online video), May 21, 2012. Available at https://www.youtube.com/watch?v=F6nwZUkBeas (accessed July 8, 2016).

17 Harvey, "Overfishing Caused."

Amanda Nickson of the Pew Environmental Group, which has conducted studies on the dwindling Pacific bluefin population, notes the uphill climb the species faces:

> There is no logical way a fishery can have such a high level of fishing on juveniles and continue. The population of Pacific bluefin is a small fraction of what it used to be and is in danger of all but disappearing. It's a highly valuable natural commodity and people naturally want to fish something that gives them such a high return. This assessment shows just how bad the situation really is for this top predator. This highly valuable fish is being exploited at almost every stage of its life cycle. Fishing continues on the spawning grounds of this heavily overfished tuna species.[18]

Pacific bluefin tuna are a "top of the food chain" fish with few predators (barring humans). They are the largest of all tuna species, with the largest specimens at 680 kg. Marine biologists take the overfishing of large predators very seriously for many reasons. Despite a long history of being a successful species, Pacific bluefins are victims of their own virtues. Several nations can fish for them given their long migratory routes. They are highly prized to the extent where rich sushi connoisseurs are willing to pay exorbitant sums of money for their flesh. In 2013, one fish sold for more than $1.5 million at an auction in Japan.[19] In 2015, a 380-pound Pacific bluefin fetched "only" $37,000 at the Tsukiji fish market in Tokyo, which was a bargain given previous record sales. Throughout Asia and in high-end Western restaurants, Pacific bluefins are sold as a delicacy and, of course, for very high prices. Obviously, this is a scenario that makes it worth it for fisheries to hunt for remaining tuna given the heavy economic incentive. A single fish could make someone catching it rich for life.

Sushi and Sashimi lovers prize the tender, fatty, pink meat of the Pacific bluefin. The $37,000 "bargain" fish went to a Japanese restaurant chain called Sushi Zanmai. About 80% of the world's bluefin tuna are consumed in Japan. International conservation groups blame the Asian sushi and sashimi markets for exerting great pressure on all bluefin populations.[20] Both Asian and American chefs can't seem to avoid the allure of bluefin flesh, no matter the attention paid to the species' plight. Bruce Mattel, associate dean of food production at the Culinary Institute of America asserts, "Bluefin tuna belly is one of the most delicious things in the world." While chefs are certainly aware of

18 Quoted in Harvey, "Overfishing Causes."

19 Harvey, "Overfishing Causes."

20 Barclay, "Why Some Chefs."

how few bluefins are left, they can make a lot of money by keeping them on their menus. As Mattel adds, "I think most of the chefs are aware of the conservation issues, I don't know how you cannot be if you have a passion for fish."[21]

But chefs may not only be motivated to keep bluefins on their menus just because of the money. At least one celebrity chef claims to serve bluefin because he wants to acknowledge a way of life that is vanishing in front of our eyes. Jose Andreas owns and oversees several Spanish-themed restaurants throughout the US. He was planning to hold a special "Tuna Celebration" at one of his restaurants in Washington, D.C. focusing on a *ranqueo*, an event where a whole tuna is prepared in front of an audience, a coastal Spanish tradition. In small Spanish fishing villages, there is a long-standing fishing method based on the *almadraba*, an elaborate system of fishing nets used for centuries to catch fish, including Atlantic bluefins. Torres argues that local fisherman in small boats using traditional methods whose families have fished for generations aren't to blame for the plight of bluefins. The real culprits who have decimated their populations around the world are huge industrial fishing vessels. In defending his position, Andreas explains, "What I wanted to do was celebrate the way of life of the people in the Mediterranean, show the disappearing way of life, the ranqueo, the almadraba…I believe something radical should be done. We should be stopping those [large fishing] fleets following those tunas. If that also means stopping those almadrabas in Spain, Turkey, Italy I will say so…but I don't believe those almadraba catchers are the problem."[22]

Catherine Kilduff, who is staff attorney for the Center for Biological Diversity, doesn't think this kind of argument holds up well to critical scrutiny. While she agrees that the immense catches from super-trawlers account for most of the damage done to the bluefin population, she doesn't think there is any more justification for cooks using endangered fish from independent fishers as opposed to purchasing it from a large Japanese fish market. "The argument that there is a way to catch them that makes it sustainable is kind of a red herring."[23]

It should be mentioned that Pacific bluefins are not merely prized for their taste, nor merely valued by the commercial market. Anglers especially appreciate their fight when they are hooked. As noted earlier, they are very powerful fish and when brought on board after being caught fishers are amazed by their liveliness. All of this makes for a greatly entertaining sport fishing experience. Paul Hoofe, California's representative to the International Game Fish Association describes catching Pacific bluefins from a recreational fisherman's

---

21 Quoted in Barclay, "Why Some Chefs."

22 Barclay, "Why Some Chefs."

23 Barclay, "Why Some Chefs."

perspective, "Everyone knows that they're the hardest tuna of all the tunas to catch. They're very finicky, they taste great, they fight pound per pound incredibly, and they're just a beautiful fish."[24]

Interestingly, some think that sportfishing may be the place to begin advocating for the species. On this line of argument, if people who enjoy having the opportunity to catch fish and their passion for the species could be rallied to support Pacific bluefin conservation, perhaps a more widely supported movement could begin to place some sort of limits on commercial fishing. It may be useful here to consider the conservation movement for animals under threat of extinction during the era of big game *market hunting*. Teddy Roosevelt and a number of other politically and economically influential game hunters are credited with (along with other members of their hunting clubs) advocating on behalf of a number of imperiled big game species. They worked to set aside lands that could be used as wildlife refuges and in the case of Roosevelt during his presidency, for National Parks. Groups such as Ducks Unlimited have contributed significant time, energy, political advocacy, and money to save marshland serving as critical habitat for ducks. Perhaps sportfishers could have a similar effect for tuna.

Inevitably, there are powerful voices in the food and environmental movement (such as Michael Pollan) who think the best way to save species that are overhunted or overfished is to "vote with your fork." Abstinence from Pacific bluefin consumption and boycotts of purveyors of their meat would be the most effective strategies. As Cardiff explains, "Unfortunately, this is an environmental issue that's been very responsive in the past to economics. There are still people who want to buy it; that's why the price is so high. The only way to break that loop is to have people say that they are valuable in the oceans... Public demand that these fish do not go extinct is the only thing that is going to save them. Having people go into restaurants [serving it and telling them not to] is probably the most direct way to vote on these issues."[25]

The fishing industry fears that catch limits will decrease job opportunities and slow the economy. Globally fishing is a $250 billion industry and significant job losses would likely follow from curtailing it, especially if limits imposed were suddenly implemented. Many in the fishing industry recall when the cod markets crashed in the 1980s and 1990s causing grave economic despair not

---

24 Deborah Sullivan Brennan, "Bluefin Fishery On the Hook, New Limits Looming for Popular Gamefish," *San Diego Union Tribune*, September 12, 2014. Available at http://www.utsandiego.com/news/2014/sep/12/bluefin-tuna-limits-fishing/ (accessed September 3, 2016).

25 Barclay, "Why Chefs Can't," 2015.

only in US coastal communities, but especially in Western Europe and Canada. In 1992, when the crisis came to a head, Canadian Fisheries Minister John Crosbie closed the northern cod fishery.[26] This resulted in 40,000 Canadians losing their jobs as a $500 million a year industry came to a halt.[27] At the time, no one expected the moratorium. Several out-of-work commercial fishers went to the Ottawa TAGS (The Atlantic Groundfish Strategy) for aid. However, this support only lasted for four years after the collapse.

While the issue of what to do to support unemployed fishers is significant and there is no easy answer to that question, some think that continuing to fish will only lead to an eventual market crash anyway. The reality is that shrinking fish biomass doesn't care about the impact on fisheries. This is all the more reason why both the fishing industry and its regulators must pay attention to declines in fish stocks as they were happening. Some conservationists blame the Canadian government in the years leading up to the crash for over-projecting fishing yields and encouraging the industry to take even more deep sea cod. This was despite independent cod stock assessments showing the species in decline.[28] This is why it is so important to heed species decline warnings, since there will be no jobs in the fishing industry if there are no fish to catch. It is better to endorse gradual limits during times of population decline of marine resources than to resign them to decimation.

As was mentioned, several organizations have called for policies designed to limit fishing of Pacific bluefins. The IATCC only covers the Eastern Pacific Ocean but does include all of the nations responsible for the harvest of Pacific bluefins, including Japan, Korea, Mexico, and the US. The Western Pacific Fisheries Commission has also developed conservation and management strategies. This multi-nation organization that monitors bluefin tuna over most of the Pacific has agreed to reduce the catch of juvenile bluefin tuna to half of the average taken from 2002–2004 levels.[29] Before 2013, IATCC had no fishing limits on Pacific bluefin tuna in Eastern Pacific waters. But in 2012, IATCC nations agreed to limit the cumulative catch for all members to 10,000 metric tons for 2012 and 2013 combined. There was also a provision for non-member, but cooperating nations with a long-standing history of harvesting Pacific bluefins, allowing each of them to

26 Greenpeace, "Canadian Atlantic Fisheries Collapse," (website), 2007. Available at http://archive.greenpeace.org/comms/cbio/cancod.html (accessed May 21, 2015).

27 Erin Anderssen. "The Cod Collapse," *The UNESCO Courier: Research Library* (1998) 51(7&8): 44–46 at 45.

28 For a more detailed account of the cod fisheries collapse in Canada and its cause, see the introduction of Monbiot's *Feral*.

29 Brennan, "Bluefin Fishery."

catch 500 metric tons annually for 2012–2013. The proposal for 2014 reduced the permitted cumulative catch of all member nations to 5,000 metric ton. No US commercial vessels specialize in catching Pacific bluefins. Thus, if they are caught, it is usually in purse-seine nets or sometimes small gill nets. During the time period from 1999–2010, the annual average of US catch of Pacific bluefins has been 113 metric tons.

On the west coast of the US in 2014, recreational anglers were still permitted to take 10 Pacific bluefins per day, as there have been strong California coast fish runs in the first half of the 2010s. But in 2015, deliberations began on a proposal to only allow 1–5 fish to be taken per day, with some advocating a complete moratorium on recreational fishing of bluefins. California's representative to the International Game Fish Association, Paul Hoofe (a bluefin fisherman himself) thinks most anglers support limits as they see that despite the recent runs the species is in trouble, "I think the global tuna fishery worldwide is overfished." Yet, he also acknowledges that any limit of under three bluefins per day would be a blow to sportfishers in Southern California. As Buzz Brizendine, a representative of the Pacific Fishery Management Council (and captain of a sport fishing boat) adds, "I think it would hurt sportfishing in Southern California, and being as the recreational industry is such an economic driver in the area, that it would also hurt the local economy."[30]

Many conservationists believe that both the commercial and recreational fishing sectors will need to make a concerted effort to help the species to rebound. But, it will not be enough to simply rely on coastal fishing ordinances of small bluefin fisheries to overcome heavy bluefin losses from commercial overfishing. A significant international problem looms and bluefin advocates face a significant human theoretical problem, which is known in the philosophical, ecological, economic, and political theory literature as the *tragedy of the commons*.[31] This is not just an abstract concern; it also has deep practical consequences in environmental matters.

The tragedy of the commons is, at its heart, a sort of prisoner's dilemma. Here's how it works. Tragedy of the commons scenarios often occur when there is no plan or management system in place to govern a resource, eventually causing its depletion.[32] This might mean there are no clear property rights to the resource in question. When the ecologist Garret Hardin wrote his

30 Brennan, "Bluefin Fishery."

31 See also a separate discussion on the tragedy of the commons in Case 10—"The Lakes and Phosphorus Twins: Monona and Mendota."

32 Baylor L. Johnson, "Ethical Obligations in a Tragedy of the Commons," *Environmental Values* (2003) 12(3): 271–287 at 271.

groundbreaking article that gave widespread scholarly attention to this issue in 1968 (titled simply "The Tragedy of the Commons"), he used the example of cattle owners, whose cattle grazed on a common—in this case a field. What Hardin pointed out was that there is an inevitable trend towards overgrazing in the case of an open field that is owned by no one. Each cattle owner has an incentive to add cattle to the field, especially since they don't have to contribute to its upkeep. But for each animal added, there is a small bit of degradation of the resource. There is also no short-term penalty for adding additional cattle to the common. Each cattle owner, if he had to predict the behaviors of the other cattle farmers, would assume that they also have a strong incentive to add more cattle to the field. After all, each would think that if he didn't do so, someone else would get that benefit. In this light, it seems eminently *reasonable* to add more cattle.

But the tragedy of the commons does not only apply to fields where no one has exclusive ownership. We can think of the ocean as a common in the case of Pacific bluefins in that the ocean is not owned by anyone. The fish aren't really owned by anyone either until they are captured. This is particularly true in international waters, where fish are generally considered fair game since they do not lie in any nation's territorial waters and thus are not under any one country's jurisdiction. Each fisher wants to harvest fish from the ocean, but to do so without having responsibility for its upkeep or regulation. Again, as a common, no one owns it, but since there is no ownership, no one has the responsibility to clean it, fail to overfish it, or do anything that will maintain it. There is also no penalty in deploying more than one vessel to fish.

But, as in the case of the cattle in the field, we know that if some threshold is crossed, there will be no more fish to catch. Everyone could even realize there is benefit to all fishers in sustaining fish stocks. But there will be a cost involved in abstaining and some sufficient number of fishers will have to pay that cost to achieve sustainability. For fishers, this most straightforwardly means the cost of *forgoing* fish. However, fishers, perhaps just like everyone else, do not prefer to pay that cost if they can help it. So, what would seem to be the rational solution for me as a fisher? I would seek to reap the benefit of fishing without paying the cost. That is, fishers will have a huge incentive to take the opportunity to *free ride* on *others* who forgo fishing opportunities. Each fisher will wish to gain as much benefit as possible while accepting the lowest degree of risk and making the least possible effort to secure the benefits of sustainability. This temptation to free ride is what drives the tragedy of the commons.

Notice a couple of reasons why the tragedy of the commons is so intractable and insidious. First of all, unlike many problems, it does not require some evil character to make it occur. Also, it may be (or at least initially seem to be) completely rational and innocent to act in the way that ultimately causes the problem. Finally, the problem comes from lots of individuals acting intentionally in the same way for what seems to be in their own interests, and yet collectively causes an outcome that none of them desire.

Notice that Hardin is not using the term *tragedy* in its usual sense. Hardin does make it sound as if this kind of destruction and overuse of a common is *inevitable*. This is a depressing scenario and makes it seem as if nothing can be done. There is, then, the important question of whether human nature prevents us from falling into the logic of selfish destruction.

Many environmentalists believe that the way out of the tragedy of the commons is by enlisting individuals to reduce their take of resources in the commons. Or at least, in the international analogue that pertains to our case, each country should reduce fish take to a sustainable level. But some philosophers do not agree with this stance. Baylor Johnson argues that in a tragedy of the commons situation, it is *not* rational for individuals (or individual nations) to "go it alone" and voluntarily reduce using a resource. Instead of "fruitlessly reducing our individual use," we "should support a collective agreement to reduce everyone's use to the sustainable level."[33] He also thinks the grounding principle for why we have obligations to seek collective agreements in tragedy of the commons scenarios is out of respect for the common good.

Other philosophers, such as Elizabeth Kahn, identify the tragedy of the commons as an "essentially aggregative harm,"[34] which means that overusing a resource prevents others from being able to use it, which constitutes harm. Presumably this is the case when the resource contributes significantly to one's survival and is not available for current or future use by those in need. One might argue that this aggregative harm could be extended to the case of Pacific bluefin. Kahn seems to suggest using something like the *precautionary principle* in determining what obligations people owe to each other. She believes people in tragedy of the commons cases at least owe each other efforts to establish collective solutions to avoid the tragedy, because they should be expected to take reasonable precaution against actions that harm others.

While Kahn agrees with Johnson's analysis of the tragedy of the commons and his approach to avoiding it, she also thinks there are not only *positive duties*, or obligations to act in some way (usually to the benefit of others) that arise in these scenarios. She also thinks there are appropriate *negative duties* (i.e., obligations to refrain from engaging in some activity or acting some way).

33 Baylor L. Johnson, "Ethical Obligations."

34 Elizabeth Kahn, "The Tragedy of the Commons as an Essentially Aggregative Harm," *Journal of Applied Philosophy* (2014) 31(3): 223–236.

Kahn argues these negative duties provide additional moral reasons why commons users should collectively prevent overusing its resources. More specifically, these kinds of duties may meld together commons users to realize that they have obligations to make more robust efforts than do bystanders to avert losses of commons' resources. Moreover, negative duties may require developing longstanding institutions to have the authority to regulate use of the commons.[35]

Review Questions

1) What do you think of the argument that groups using traditional methods of fishing (e.g., the Spanish practice of the *almadraba*) should be exempt from any regulations curtailing Bluefin fishing? Is this a convincing case?
2) Let's make the (safe) assumption that curtailing Pacific bluefin tuna fishing would cause job losses. What do you think should be done to help those losing their jobs, if anything? Are they owed compensation? Should they be provided with state aid in finding new jobs? Explain your answer (you might also reflect on the case where Atlantic bluefin fishing grounds were formally closed by the Canadian government).
3) Consider Johnson's and Kahn's approaches to the tragedy of the commons and apply them to the Pacific bluefin case. What do you think Johnson and Kahn would say about this case study? Do you agree with Johnson and/or Kahn that there are certain moral obligations to others and that we should seek collective action in response to tragedy of the commons scenarios?
4) Do you see any way out the tragedy of the commons? If so, do you think your solution would apply to the case of Pacific bluefin tuna overfishing?

Case 36 – George, Tex, and a Sand County Crane Tale –

Study Questions

1) What were some of the main causes for whooping cranes coming close to extinction in the 1940's? What are some of the main challenges to whooping crane conservation today?

35 Elizabeth Kahn, "The Tragedy of the Commons," pp. 223–236.

2) What were some of the techniques used by Dr. Archibald to help whooping cranes successfully lay eggs?
3) What are some of the ways the histories of sandhill and whooping cranes intersect?
4) Why are some worried about opening a hunting season on sandhill cranes in Wisconsin with respect to whooping crane conservation? How are hunting and poaching different (especially since hunting and poaching involve the killing of individual members of species)?

Sometimes, the fates of two separate, but closely related species are deeply intertwined. This is the case for the sandhill crane, which is no longer endangered, and the whooping crane (otherwise known as a "whooper"), which is. Sandhill cranes can now be legally hunted in a number of states and are sometimes confused with whooping cranes. This fact obviously makes life more difficult for whooping cranes, which have been illegally shot in some recent cases.

Sandhills were immortalized in Aldo Leopold's *A Sand County Almanac*. After waxing poetically about the grace and charm of cranes in the Baraboo Hills of Wisconsin, Leopold turns to how habitat destroyed by the conversion of marshes into cropland, and how sometimes the spillover effects of such land use transition can further degrade other prime crane habitat. These factors hasten the decline of both whooping and sandhill crane populations. Leopold also notices that early conservation efforts, while yielding some benefits, led to other changes to the nesting areas of cranes that would continue their decline. As he writes in *A Marshland Elegy*, a chapter in his famous *A Sand County Almanac*:

> For a decade or two, crops grew poorer, fires deeper, woodfields larger, and cranes scarcer, year by year. Only reflooding, it appeared, could keep the peat from burning. Meanwhile cranberry growers had, by plugging drainage ditches, reflooded a few spots and obtained good yields. Distant politicians bugled about marginal land, over-production, unemployment relief, conservation. Economists and planners came to look at the marsh. Surveyors, technicians, CCCers, buzzed about. A counter epidemic of reflooding set in. Government bought land, resettled farmers, plugged ditches wholesale. Slowly the bogs are re-wetting. The firepocks become ponds. Grass fires still burn, but they can no longer burn the wetted soil. All this, once the CCC camps were gone, was good for the cranes, but not so the thickets of scrub popple that spread inexorably over the old burns, and still less the maze of new roads that inevitably follow government conservation. To build a road is so much simpler than to think of

> what the country really needs. A roadless marsh is seemingly as worthless to the alphabet conservationist as an undrained one was to the empire-builders. Solitude, the one natural resource still undowered of alphabets, is so far recognized as valuable only to ornithologists and cranes. Thus always does history, whether of marsh or marketplace, end in paradox. The ultimate value in these marshes is wildness, and the crane is wildness incarnate. But all conservation of wildness is self-defeating, for to cherish we must see and fondle, and when enough have seen and fondled, there is no wilderness left to cherish.[36]

The fires that were used to created hayfields needed by farmers spread out of control in dry conditions, lit peat in the bogs, and continued to burn. Public works projects such as reflooding (implemented by the Civilian Conservation Corps or "CCC") during the Great Depression aided in reducing fires and restoring some wetlands, but also had a separate negative effect. What Leopold is noting here is that both sandhill and whooping cranes are shy and secretive birds (especially during nesting season). Disturbing their mating rituals causes reduction in their reproductive success. The noise of vehicle traffic and the all too ready access of humans to the nesting sites of cranes worried Leopold greatly. During his time, the number of sandhill cranes had been greatly diminished. But as of 2013, the species had recovered remarkably, having a population of some 650,000 worldwide and about 72,000 in the migratory path that runs from Wisconsin to Tennessee, with numbers likely increasing.[37] Whooping cranes, however, though with a higher species population than they had during Leopold's lifetime, remain endangered and face several challenges not so unlike those encountered almost a century ago.

Ornithologists are not often invited as guests on *The Tonight Show*, which usually reserves such spots for famous musicians, movie stars, and sometimes politicians. However, one was in 1982 in the person of Dr. George Archibald. At first, Archibald was a bit hesitant to appear on the show. For one, he doesn't really like being in the media. But he was also afraid the point of his important work in whooping crane conservation would be lost as host Johnny Carson instead focused on (and poked fun at) the unusual techniques designed to aid

36 Aldo Leopold, *A Sand County Almanac and Other Writings on Ecology and Conservation*, Curt Meine, ed. (New York: Literary Classics of the United States, 2013), p, 89.

37 Operation Migration, "Proposed Sandhill Crane Hunt," (website), 2013, Available at http://operationmigration.org/InTheField/2013/07/24/proposed-sandhill-crane-hunt/ (accessed February 7, 2015).

the species overcome an important obstacle to its survival. Archibald admitted that even as he waited in the green room shortly before taping the show he felt quite nervous. Yet, he was able to get though the interview, having kept the integrity of his work intact and leaving twenty-two million audience members (and Carson himself) with some new awareness of an endangered species. To understand what sort of techniques a talk show host might make fun of and the problem the techniques were designed to solve, we need to look more closely at the plight of whooping cranes and the story of George, Tex, and the International Crane Foundation.

The whooping crane is the tallest North American bird, with males (which are larger than females) sometimes standing 1.5 meters (almost five feet) in height.[38] The average wingspan of a whooping crane is seven to eight feet across making them a majestic sight to behold. But despite their size, they are light (on average around 15 lbs.) as their bones are hollow to adapt them to flying.[39] Whooping cranes inhabit marshlands, grasslands, and some croplands, following rivers as part of their migratory route.

Archibald began working with cranes in the 1960s. At the time, whooping cranes, as well as fourteen other species of cranes worldwide, were greatly imperiled and their numbers had been steadily dropping over the past several decades. Wildlife biologists feared that most if not all crane species would soon go extinct. The low point for the whooping crane population came in the 1940s with only 22 birds left in the wild. The historic range of the whooping crane extended from the Arctic south coast to central Mexico, and from Utah east to New Jersey and south to South Carolina, Georgia, and Florida. Historically, the breeding range for the species was across the north central United States and Canada.

The main reasons for whooping crane decline were hunting, egg snatching, and most importantly, loss of habitat. It is estimated there were likely 700 to 1,500 whooping cranes in the 1850s. Particularly rapid decline of the species population came in the late 1800s and early 1900s during a time period when market hunting reigned. Overhunting, when combined with egg collection

---

38 Tom Stehm. "Species Status and Fact Sheet: Whooping Crane," US Fish and Wildlife Service, (website), 2001. Available at http://www.fws.gov/northflorida/whoopingcrane/whoopingcrane-fact-2001.htm (accessed February 2, 2015) and Natural Resources Foundation of Wisconsin, "Welcome Whoopers," Fact sheet, 2001.

39 Karen Sonnenblick, Sherry Klosiewski, and Beth B. Kienbaum, *A Closer Look at Whooping Cranes: Whooping Crane Education in Wisconsin and Eastern North America* (Madison, WI: Wisconsin Department of Natural Resources and Whooping Crane Eastern Partnership, 2008), p. 6.

and the conversion of marshland to agricultural land, caused whooping crane numbers to crash. As settlers moved further west, there was increased pressure on whooping crane nesting habitat as wetlands vital to their survival were drained and prairies plowed under for farming. Recognition of the dwindling number of the birds exacerbated their decline as hunters and museum curators scrambled to procure the few remaining specimens and eggs.[40]

Though it is still the rarest of the fifteen species of cranes that exist worldwide, the whooping crane has made a slow but steady recovery, with 106 members in Wisconsin during breeding season and 556 members overall. There are currently three sub-populations of whooping cranes in the wild. One is the Wood Buffalo/Aransas population (~300 members), which uses Wood Buffalo National Park in Canada as its breeding location and the Aransas National Wildlife Refuge on the Texas Gulf Coast as its wintering spot. A second sub-population consists of non-migratory cranes recently released in central Florida. The third is an eastern migratory group whose population has been rising gradually. Of the fifteen species of cranes worldwide, only the whooping crane and the sandhill crane inhabit the US.[41]

The whooping crane is an interesting case study for a number of reasons. First of all, it involves one of the most researched and well-known restoration efforts in the history of species conservation. Second, there is an intriguing comparison between whooping and sandhill cranes since while both have overlapping habitat, sandhills are flourishing whereas whoopers are not. Third, the two species are similar enough that they have played complementary roles in each other's conservation histories. Fourth, whooping crane conservation is an interesting effort at international cooperation between the United States and Canada, involving arduous navigation of thorny issues in international law and different conservation strategies in varying jurisdictions.

The gains the whooping crane has been able to achieve are attributable not only to international cooperation, but also to strict legal guidelines on their take and careful habitat preservation. But as was mentioned earlier, many of the chief obstacles to whooping crane survival persist. Shrinking habitat is obviously one issue as wetlands continue to be filled in for development projects. Reduced fresh water availability in the species' wintering grounds in the wetlands along the Texas Gulf Coast presents a persistent challenge. Additionally, pollution (chiefly from the oil and chemical industries) in these

---

40 Sonnenblick et al., *A Closer Look at Whooping* Cranes, p. 9.

41 United States Geological Survey, "Status Survey and Conservation Action Plan: Whooping Crane," *The Cranes*, Fact sheet, 2013. Available at http://www.npwrc.usgs.gov/resource/birds/cranes/grusamer.htm (accessed February 3, 2015).

wintering grounds is a serious barrier to the species' propagation. Utility wires (with which the birds sometimes collide), human disturbance, loss of genetic diversity within the population, predation, disease, and vulnerability to either human-made or natural disasters all pose significant threats to the species.[42] There have also been some recent cases of illegal shootings of whooping cranes, sometimes precipitated by the general public's inability to identify the species.

And, of course, there are fits and starts with any species recovery effort, as new initiatives commence and take hold while others fail. For instance, the Whopping Crane Eastern Partnership (WCEP) migratory sub-population is steadily increasing in numbers (as was mentioned earlier) since the reintroduction program began in 2001, while an experimental cross-fostered population of birds in the Rocky Mountains (introduced in 1975) was lost in 2002.[43]

There is an important lesson to be learned from the failed Rocky Mountain reintroduction experiment. This project involved taking eggs to locations where there were no whooping cranes to teach the chicks how to migrate. The hope was that sandhill cranes would be a migratory surrogate, leading whooping crane chicks south each winter and back north each spring. The premise behind the strategy was reasonable—to foster new populations and with few adult whooping cranes available, surrogates would be necessary. However, whooping cranes are *imprinted* birds—they associate with objects with which they first interact in their early development. So while sandhill cranes successfully hatched and taught whooping cranes how to migrate, when the whooping crane chicks grew to breeding age, they had no interest in breeding with other whooping cranes as they had imprinted on sandhill cranes and pursued them instead.

This brings us to the early conservation efforts for whooping cranes and some of the unorthodox practices used for their conservation alluded to earlier. It also introduces us to Tex, a female whooping crane who became a symbol for her species' survival. As Archibald puts it, "The story of Tex means much, much more than just dancing with this imprinted bird."[44] The breeding program for

42 United States Geological Survey, "Status Survey."

43 International Crane Foundation, "Historic Whooping Crane Numbers," (website), 2011. Available at https://www.savingcranes.org/images/stories/site_images/conservation/whooping_crane/pdfs/historic_wc_numbers.pdf (accessed February 3, 2015).

44 Much of the information for the story of George and Tex comes from a filmed interview with Archibald. See International Crane Foundation. "George and Tex," *YouTube* (online video), October 28, 2011. Available at https://www.youtube.com/watch?v=fsDYhdaJL98 (accessed July 9, 2016).

whooping cranes began in 1965 at the Patuxent Wildlife Research Center in Laurel, Maryland. The researchers at the Center imported eggs from Canada and from these efforts they hatched about a dozen whooping cranes. Tex was hatched from this group, produced by the only pair of whoopers in captivity at the time. Tex was an important chick. Her parents were at the San Antonio Zoo in the 1950's when there were fewer than 30 whooping cranes in existence—so just genetically speaking this crane family was significant. The pair produced fifty eggs, some of which hatched, but all of the chicks had died in the pen.

One major problem at the time was that researchers didn't know how to care for whooping crane young. The Zoo Director took Tex from her parents and raised her for the first six weeks in a box in his living room. But then, governmental authorities demanded that Tex be sent to Patuxent to live with whoopers in captivity there that had been procured from Canada. Archibald did research at the Center from 1968–1971 and agreed to work with Tex to see if he could induce her to lay eggs. Researchers at Patuxent had tried unsuccessfully to get Tex to breed. After obtaining the appropriate permits, Tex arrived at the International Crane Foundation in 1976, which had been co-founded by Archibald just three years earlier in Baraboo, Wisconsin.

Archibald and Tex bonded over the first summer and from then on he worked with her each spring. She was, in a sense, a bird who thought she was human. Since Tex had been exposed to humans for too long early in life, she became imprinted on Archibald and his colleagues. This meant that she wouldn't breed with other whooping cranes, and thus demanded a different breeding strategy—artificial insemination. This would require Archibald simulating whooping crane courtship rituals. His job was to do as best he could to mimic the mating behaviors of whooping cranes, including the deep knee bends that are part of elaborate dances with Tex. Other mating techniques involve researchers wearing crane suits so that the cranes wouldn't become imprinted on humans.

Archibald described part of his daily routine with Tex in an interview:

> I would arrive at her enclosure each morning at about 5am, which was before daylight, open the door and we would walk together up to the top of the hill. I would leap around doing deep knee bends and running around with her. It was actually quite exhausting. But it was sort of fun to wake up and dance with a crane. And then she would solicit for mating and the two people who would be doing the insemination would appear at the end of the field. I would turn her away from the door, stroke her back, she would elevate her tail and they would run out the door of the shack and inseminate her. And they would let her chase them and then

> she would come back and unison call and she would be successfully inseminated—including the egg that would become Gee Whiz."[45]

Gee Whiz was a chick that came from one of Tex's deformed eggs that was flattened on the ends. Archibald placed the egg with a reliable pair of sandhill cranes to serve as surrogates. One of the first questions Archibald and his research team considered was whether the egg was fertile. On inspection of the egg in the incubator with a bright light from a candling machine, they found an embryo that was very much underweight and quick action was needed to keep it viable. The egg was kept in the incubator and regularly monitored. Soon it was readily apparent that the egg was dehydrating very quickly, likely due to a shell that was too thin and thus losing moisture. Without some intervention, Gee Whiz would very likely have died. Upon consultation with a physiologist (Dr. Bernie Wentworth) with the Poultry Science Department at the University of Wisconsin-Madison, Archibald learned about an experimental procedure for adding antibiotics to duck eggs. If an egg slightly warmed in an incubator was then submerged in very cold water, the liquid content of the egg contracts, and the atmospheric pressure will drive water surrounding the egg into the vacuum created by the constriction of the liquid content.

But to try this procedure was risky—putting the egg in ice water could kill the embryo. However, the procedure worked—the egg was re-hydrated and started to gain weight. Gee Whiz hatched on day twenty-eight of the gestation period. Yet, the chick was not out of the woods yet. He was very small and wouldn't eat or drink at first after hatching, spitting out his meals. Archibald and his research team hypothesized that due to suffering dehydration in the egg, Gee Whiz's esophagus was stuck together and wouldn't let water and food pass to the stomach. Normally, if a chick doesn't eat or drink, it is tube fed, but there was hesitance on the part of the research team to do this because of the risk of rupturing the esophageal wall and killing the chick instantaneously.

Yet, as Gee Whiz continued to lose weight, the research team felt as though they had no choice and gently fed a lubricated catheter tube down his throat and administered nutrient fluid very slowly in small amounts (1cc at a time). Gee Whiz quickly recovered and was raised with a group of red crown cranes so that he wouldn't become imprinted on humans. Nonetheless, due to them being hand-raised by humans, cranes like Gee Whiz are not afraid of us. Despite his early life close calls with death, he went on to father several chicks.

---

45 International Crane Foundation. "George and Tex."

This case study is interesting as an example of early international wildlife conservation. Just as they are in the US, whooping cranes are listed as endangered in Canada. Protecting whooping cranes was the goal of the US conservation movement as early as the 1930s with the establishment of the Aransas National Wildlife Refuge by President Franklin Delano Roosevelt along the Texas Gulf Coast (though drought and the Great Depression initially hindered these efforts). But conservation efforts on only one end of a migratory route have little chance of success ultimately. It was the discovery of a significant wintering location of whoopers in the Wood Buffalo National Park that sparked Canadian involvement in conservation and habitat restoration efforts for the species.[46]

The first formal joint conservation agreement for whooping cranes between the US and Canada came in 1985 with a Memorandum of Understanding. This agreement to improve the coordination of conservation efforts for whooping cranes between the two countries was renewed in 1990 and 1995.[47] In the 1995 renewal of this agreement, a goal of increasing the size of the Wood-Aransas sub-population to 1,000 members was confirmed (this goal was first set in the 1994 US whooping crane recovery plan). In 1996, the Whooping Crane Recovery Team was established by the two nations to meet either annually or bi-annually, composed of five members from each country.

A more general framework for a partnership between the US and Canada for the identification and conservation of a wide variety of shared species (those that cross the border between the two countries) came in 1997. The document was entitled the "Framework for Cooperation Between the United States Department of the Interior and Environment Canada in the Protection and Recovery of Wild Species at Risk." The goal of the agreement is fairly straightforward—to prevent the extinction of species caused by human activities that share land on both sides of the border as habitat. The goal of the "Framework" is to be executed via the conservation of wildlife populations and the ecosystems upon which they rely.[48]

In these efforts, American and Canadian biologists share research findings, coordinate habitat conservation, collaborate on conservation techniques, and

---

46 Paul A. Johnsgard, *Sandhill and Whooping Cranes: Ancient Voices Over America's Wetlands* (Lincoln, NE: University of Nebraska Press, 2011), p. 49.

47 United States Department of the Interior, "Whooping Crane," *Conserving Borderline Species: A Partnership between the United States and Canada* (Washington, DC: Government Printing Office, 2001), p. 11. Available at https://www.fws.gov/endangered/esa-library/pdf/borderline_species01.pdf (accessed February 9, 2015).

48 United States Department of the Interior, "Whooping Crane," p. i.

sometimes conduct joint reintroduction measures. Whooping cranes provide an example of these cooperative enterprises, especially with respect to reintroduction of the species. As early as the late 1940s researchers in the US and Canada began sharing data pertinent to the conservation of whooping cranes.[49] Since the 1960s, ornithologists in both countries have bred whooping cranes in captivity to reintroduce them on both sides of the border.[50] More specifically, in 1967, the US Fish and Wildlife Service teamed up with the Canadian Wildlife Service to launch a captive breeding program for whooping cranes that would ultimately release whoopers back into the wild. Eggs were transported some 2,145 miles from the Wood Buffalo National Park in Canada to the Patuxent Wildlife Research Center. By 1989, the International Crane Foundation had established a facility for breeding whooping cranes in Wisconsin. On the Canadian side of the border, a breeding facility was begun in the Calgary Zoo in 1993. This was the same year that, using cranes from both US and Canadian breeding facilities, US officials began reintroducing the non-migratory flock of whooping cranes into Florida. In 1996, egg collection to be used for captive rearing was ended after 250 eggs had been collected.

Researchers in the US have assisted those in Canada since the 1990s with conducting aerial surveys to determine the success rate of crane reproduction in the nesting grounds of the Wood Buffalo National Park. In 1993, tracking efforts were made with sandhill cranes to locate their wintering migratory routes. These observations were made with the hope of figuring out whether sandhills might be used as migratory guides for whoopers. An essential part of the strategy to help guide the migration of whooping cranes since 1993 has been the use of ultralight aircraft, a project called Operation Migration. These aircraft help teach a specific migratory route to surrogate sandhill cranes. As populations of whoopers have increased along the eastern migratory route, Operation Migration now utilizes ultralights to directly guide them without the use of sandhill migratory surrogates. American and Canadian biologists also initiated studies into exploring reintroduction sites in the provinces of Manitoba and Saskatchewan in Canada, and Wisconsin in the US. These programs resulted in establishing the steadily increasing, reintroduced population of whooping cranes travelling between Wisconsin and Florida.

Of course, whooping cranes face hurdles to rebuilding their population on both sides of the border. In Canada, the chief concerns are that a bad storm or a drought could destroy eggs or early hatched chicks as the country's Wood Buffalo National Park serves as the species' last wild breeding ground. Along

49 United States Department of the Interior, "Whooping Crane," p. 11.

50 United States Department of the Interior, "Whooping Crane," p. 1.

with increased industrial buildup occurring along the migratory corridor used by whooping cranes, there are worries that since their wintering grounds lie in a densely developed area of oil refineries and chemical plants, habitat will be lost to the pollution from these industries.[51] An additional concern is that since whooping cranes migrate to the Texas Gulf Coast during the beginning of hurricane season, the entire flock could be eradicated in such a storm. These are legitimate issues for any species with a majority of its population so concentrated for significant portions of the year.

What are the moral issues that arise from this case? One issue shows the relationship between sandhill and whooping cranes. As mentioned earlier, sandhill cranes are relatively abundant, to the point where in some states there is a sustainable hunting season for them. But as was also noted, people mistake whopping cranes for sandhill cranes, which obviously causes more peril for the endangered whoopers. A case in point comes to us from an illegal shooting of a whooping crane in Waupaca County, Wisconsin, in 2013.[52] A 28-year-old hunter named Matthew Kent Larsen used a .22 caliber rifle to shoot what he claimed he thought was "a white sandhill"—an albino sandhill crane. The chief problem in this case is that there is no such thing as an albino sandhill crane, though the myth has persisted in popular culture in Wisconsin for some time. The crane was radio tagged, which helped researchers from the International Crane Foundation locate the body. Larsen pled guilty and was sentenced at the United States Magistrate Court in Green Bay.

Larsen agreed to be interviewed and explained how he came to shoot the whooper. He was on his family's property when he saw the crane in a wheat field located on an adjacent property. He left, but came back with a borrowed rifle and shot it. Larsen then texted a friend saying that he had killed an albino sandhill crane, but his friend replied that there are no white sandhill cranes and that he had likely just killed an endangered whooping crane. Larsen left the area and added that he wouldn't have shot the bird if he'd known it was a whooping crane and regretted killing it. As Wisconsin Conservation Warden Theodore Dremel reports, "He [Larsen] was remorseful about his actions and realized too late that he had made a grave mistake."[53] However, US Fish and Wildlife Service Resident Agent in Charge Pat Lund pointed out that, "Regardless of whether Larsen thought he was shooting a sandhill crane or a whooping crane, they

51 United States Department of the Interior, "Whooping Crane," p. 10.

52 United States Fish and Wildlife Service, "Foster a Land Ethic that Would Make Aldo Leopold Proud: Large Water Birds, An Identification Guide," *Newsroom*, (website), June 23, 2014. Available at http://www.fws.gov/midwest/news/730.html (accessed July 12, 2016).

53 United States Fish and Wildlife Service, "Foster a Land Ethic."

are both federally protected and neither can be legally hunted in Wisconsin. Incidents like this undermine the work of a huge network of conservationists who have worked for decades to bring whooping cranes back from the brink of extinction."[54] Larsen was required by the court to pay a $500 Migratory Bird Treaty Act fine as well as $1,500 in restitution to the ICF. He also lost his hunting and fishing privileges nationwide for two years.

This shooting is not simply an isolated incident, as there have been at least 20 whooping cranes shot from 2011–2015 in the US. Two whooping crane shootings occurred within just a few weeks of one another—including one in Louisiana in the fall of 2014 and another in Texas in early 2015. The shooting in Louisiana happened in Vermillion Parish, where state Department of Wildlife and Fisheries officials found a whooper with leg injuries caused by a rifle bullet on November 2, 2014. The agency announced a $10,000 reward for information about who might have committed the crime. The crane had been hatched from the Patuxent Research Center and released in Louisiana in January of 2014. In 2011, an experimental population of whooping cranes had been released in the state.[55] The late 2014 incident was the sixth whooping crane shot in Louisiana in recent years.

The other case occurred in Sand Lake at Aransas Bay near the National Wildlife Refuge. On January 4, 2015, the USFWS and the Texas Parks and Wildlife Department announced that they found a dead whooping crane, likely due to a shooting. The ICF offered a $27,500 reward for information that leads to a conviction in the crime.[56] While it is not clear how many of these cases were intentional shootings versus mistaken identity, there is still the concern that opening sandhill hunting seasons in additional states will only exacerbate the problem of illegal shootings of whoopers. Especially when considering how low the population of the species is, the loss of even a small number of whooping cranes places terrific strain on their ability to prosper.

Due to cases like these of misinformation and mistaken identity, opening hunting seasons on sandhill cranes are controversial to say the least. This raises an important moral question. Should the possibility of misidentification of one species be a strong enough reason to ban or otherwise curtail the hunting

54 United States Fish and Wildlife Service, "Foster a Land Ethic."

55 International Crane Foundation, "Endangered Whooping Crane Shot in Louisiana," (website), January 22, 2015. Available at http://www.savingcranes.org/whats-new/2015/01/endangered-whooping-crane-shot-in-louisiana/ (accessed February 15, 2015).

56 International Crane Foundation, "Suspected Shooting of a Whooping Crane in Aransas Bay, Texas," (website), 2015. Available at http://www.savingcranes.org/whats-new/2015/02/5307/ (accessed February 13, 2015).

of a similar species? This has certainly been an issue in Wisconsin, where there have been recently proposed bills for establishing a hunting season on sandhill cranes. Sandhills are already legally hunted in fifteen states, many of them in the West. Only two states east of the Mississippi River, Kentucky and Tennessee, allow sandhill crane hunting.[57]

In Wisconsin, there is some impetus to open a hunting season on sandhill cranes because the birds like to eat corn seed from the ground early in the growing season, to the displeasure of farmers. Even for farmers who endorse protection of wildlife, there are some who think there are too many sandhill cranes in the state.[58] For some hunters, the sentiment is that since the bird is not endangered and is purported to taste good (labeled by some connoisseurs as the "rib eye of the sky"), why shouldn't the option be available to hunt them?[59]

There is also the issue with whether, on balance, hunting helps or hinders wildlife conservation. With respect to funding conservation efforts in the US, dedicated revenue streams go to federal and state agencies responsible for these efforts. When someone pays for a hunting or fishing license, a duck stamp, or fees for using land for hunting, these monies directly fund the Services and state wildlife management agencies.

For the sake of comparison, note that the American funding scheme for conservation is significantly different from Canada's. Since the Canadian government is a parliamentary system, the underlying political ideology is not that the government exists to execute the will of the citizens of that country. Instead, the mission of Canadian elected officials is to do what they think is best for the interests of the country, even if that is counter to the political will of the public. Consequently, Canadian wildlife managers have had much less funding than their American counterparts. For revenues entering Canadian government coffers from licenses and fees for the take of wildlife flow into a general pool of funds that includes monies collected from other taxes. The government then deliberates over what they think are the best uses of such

---

57 "Hunting Sandhill Cranes Isn't for Everyone, but It Isn't Unethical," *Chicago Sun-Times*, April 6, 2014. Available at http://chicago.suntimes.com/uncategorized/7/71/195127/hunting-sandhill-cranes-isnt-for-everybody-but-it-isnt-unethical/ (accessed February 10, 2015).

58 Monica Davey, "Wisconsin Bill Would Allowing Hunting of Once-Rare Crane," *The New York Times*, February 23, 2012. Available at http://www.nytimes.com/2012/02/24/us/wisconsin-consider-hunting-of-sandhill-crane.html (accessed February 10, 2015).

59 Richard Simms, "'Bizarre Foods" TV Host Andrew Zimmern Samples Area Sandhill Cranes," *Nooga.com*, (website), December 12, 2013. Available at http://www.nooga.com/164635/bizarre-foods-tv-host-andrew-zimmern-samples-area-sandhill-cranes/ (accessed February 10, 2015).

funds. As it turns out, environmental protection in general has lost out to other agencies in funding allocation battles. This is bound to happen when there are admittedly other worthy sectors to fund such as health care, education, and economic development.

In the US, while having dedicated revenue streams have left wildlife protection agencies in better shape financially than in Canada, there is still a beguiling problem for American wildlife managers. There is an element of "whoever pays the piper calls the tune" in the dynamic of the American conservation system. More to the point, since hunters and fishers largely fund the wildlife management agencies, they represent a seriously powerful lobby. If, for example, hunters wish to have much more convenient deer hunting and thus more deer, they will petition their representatives to make that happen, even if regulators worry there already too many deer causing too great of an adverse effect on their habitat. This situation becomes problematic whenever there is a disagreement over conservation strategies and policy, causing a "planners versus hunters" mentality to persist that is counterproductive to conservation.

Some hunters also think they are put in a bad light by environmentalists who are anti-hunting and counter that they promote conservation in two respects. One is obviously that hunters actively fund measures through paying for licenses and other fees. But second, they participate voluntarily through advocacy groups for specific game species to help them flourish and these efforts sometimes aid other species as well. For instance, Ducks Unlimited has spent a great deal of money on reclaiming and restoring habitat that benefits not only ducks but also other waterfowl.

Additionally, hunters might argue that they have greater appreciation of animals than non-hunters who are omnivores. They might claim that hunting, killing, dressing, cooking, and eating animals shows them more respect than do non-hunters who don't even know where their meat comes from. Those who don't hunt, argue these hunters, are complicit in an industrial food system that includes CAFOs and factory farms well-documented for their inhumane treatment of animals.

Finally, while Leopold points to the interesting paradox that conservation efforts, no matter how well-intentioned, may go wrong if wildness is in some sense tamed by those efforts, flipping this situation and seeing it from another perspective is also telling. For what is the alternative to conservation efforts of regulatory organizations such as the CCC and its modern equivalents? Will public support for endangered species be mustered if it has little access to the wild? Instead, do we allow for "nature to take its course" by not interfering with wild things at all?

This makes us very briefly consider the controversial role of zoos in conservation. Some may consider zoos to be morally offensive due to the captivity of animals, not only from the standpoint of animal suffering (from not being able to freely move about) but also due to captivity by definition robbing animals of their wildness and conversely, robbing the wilderness of one of its chief values—its wildness. Yet, one might wonder if having people see rare species (especially as children) might actually encourage zoogoers to hold higher regard for those species, and perhaps inspire the next generation of wildlife biologists, conservation ecologists, or even environmental philosophers. Moreover, donations to zoos (even if they are small) sometimes go to the protection and restoration efforts of habitat for endangered or threatened species. These are all difficult issues that we as citizens in a democracy will have to weigh carefully.

Review Questions

1) What do you think Leopold means when he says, "The ultimate value in these marshes is wildness, and the crane is wildness incarnate. But all conservation of wildness is self-defeating, for to cherish we must see and fondle, and when enough have seen and fondled, there is no wilderness left to cherish"?
2) Argue for or against the hunting of sandhill cranes. Be sure to anticipate some possible objections to your position and address them as best as you can.
3) Do you agree or disagree with the argument that those who pay the lion's share of a program should derive the results they prefer? What reasons can you give for your position? If you agree with the argument, how do you address the concerns that the best conservation science may call for hunters to suffer some inconvenience or even restrict their hunting practices? If you don't agree with the argument, should you promote policies calling for a wider base of funding wildlife conservation if wild animals are really in the public trust?

Case 37 – A Plan for Karner Blues –

Study Questions

1) What is a habitat conservation plan?
2) What perverse incentives were caused by the original Endangered Species Act and how did this lead to the establishment of habitat conservation plans such as the one for the Karner blue butterfly?

3) What does disturbance conservation mean? Does it apply to Karner blue butterfly conservation?
4) What are some of the key factors leading to the success of the Karner Blue Butterfly HCP? What are some areas for improvement? What are some new causes for concern for the future of the Karner blue butterfly?

In 1973, the Endangered Species Act (ESA) was signed into federal law. In the first decade of the ESA's existence there was growing dissatisfaction with how the law instituted environmental protection of endangered species in the United States from the business community, especially property developers. The original version of the ESA was called the "pit bull" of environmental law at that time. Under the original ESA, any "take" (killing) of an individual member of a species was a violation of the law that basically shut down any further development of land where the kill took place. This was a perennial aggravation to developers to the point where it led to unintended perverse incentives. Given the law, there was a motivation for landowners not to disclose the presence of endangered species on their land for fear of regulations on land use. There were even reports that developers would intentionally destroy habitat where there was any likelihood for endangered species to be found in the future. This was with the hope that members of endangered species would not occupy the landowners' properties so that land could be developed without the threat of "taking" any members and the consequent possibility of prosecution.

The US Fish and Wildlife Service (USFWS) became aware of these issues and were increasingly concerned that reactions to the law were creating more harm than good for the long-term sustainability of endangered species, as habitat preservation is critically important to species survival. In 1982, provisions were made for a new tool for the conservation of habitat of endangered species as Section 10(a) was added to the ESA, relaxing some requirements of the law.[60] More specifically, Section 10(a) formally introduced incidental take permits as well as habitat conservation plans (HCPs) into federal environmental law. These measures presented a compromise in that they allowed for development of property at the same time as they served to conserve critical habitat for listed species. The agreement goes like this; developers or a state environmental agency can apply for an incidental take permit allowing landowners to develop land even if these activities result in "take" or killing of individual members of an endangered species as long as such taking is unintentional.

60 Karen Donovan, "HCPs—Important Tools for Conserving Habitat and Species," in Donald C. Baur and William R. Irvin, eds. *Endangered Species Act: Law, Policy, and Perspectives* (Chicago: American Bar Association Publishing, 2001), p. 319.

In exchange, the developer would be responsible for submitting, implementing, and financing a plan (HCP) to set aside critical habitat from development to serve as a sanctuary for endangered species and/or engage in development practices conducive to the survival of the species.

The HCP proposal is examined by the USFWS and the applicant works with their state's Department of Natural Resources (DNR or equivalent agency) to revise the plan as needed to meet USFWS guidelines. Thus, ordinarily the HCP approval process goes through the USFWS and its coordination is facilitated through state DNRs. Another crucial component of the updated ESA was the additional of a "no surprises" clause. That is, during the period of the permit, no new species or conditions can be added on or changed from the original agreement. The time period for the incidental take permit can range over any number of years (the first HCP ever implemented had an initial duration of 15 years, but some have one-hundred year agreements set).

After a public comment period, an incidental take permit can be granted if the applicant has met a number of requirements. First, there must be assurance that an HCP for the species in question will be implemented. Second, the take of said members of species will be incidental (again, no intentional culling of the species is covered under the permit). Third, the applicant must agree to minimize and mitigate the impacts of take to the greatest extent practicable. Fourth, the take will not appreciably reduce the likelihood that the species will survive and recover in the wild. Finally, there need to be assurances that the HCP will be adequately funded.[61]

But oddly, few land managers and developers took advantage of the Section 10 provisions and through the 1980's several of the problems the new provisions were designed to solve persisted. In fact, before 1994 HCPs were exceedingly rare for roughly the first decade after the addition of Section 10, with only eleven approved.

There are a number of factors that may have contributed to why there were so few HCPs established in the years shortly after their inception. One reason may simply have been lack of familiarity with the HCP process on the part of the private sector and even governmental agencies outside of USFWS. There also was some confusion about whether any habitat degradation still violated Section 9 no matter what Section 10 stipulated. Also, in a limited number of cases there are provisions under Section 7 allowing for incidental take permits without appeal to Section 10. Given uncertainty about this new provision in the law, Section 7 provisions for permits may have been more attractive. Finally, some landowners just may not have trusted that Section 10 incidental

61 Donovan, "HCPs—Important Tools," p. 132.

take permits assured that further restrictions on land use would not arise in the future.[62]

In this context, we can see why by the early 90's under the Clinton Administration there was a concerted effort by USFWS to encourage use of HCPs. USFWS developed a handbook providing a detailed outline of how to apply for incidental take permits. The strategy to promote HCPs worked, with some 429 of them approved by 2003.[63] This was mainly due to administrative changes that greatly accelerated the processing times of HCPs and of "no surprises" assurances.

The reaction to Section 10 has been mixed. To some, the initiation of incidental take permits and HCPs began a "deal with the devil" that ultimately places endangered species in greater peril. For others, turning to incidental take permits was the only viable way the ESA could be fixed to secure the long-term sustainability of endangered species protection.

Just as this shift in the law was coming into place, the USFWS was planning to add the Karner blue butterfly to the endangered species list. Originally, the range of this insect was across a continuous band from eastern Minnesota to the eastern seaboard. But by the early '90s that range was becoming much more constricted, and in some areas the population levels were quickly shrinking. Today, Karner blues are found in parts of Minnesota, Indiana, Wisconsin, Ohio, Michigan, New York, and New Hampshire.[64] No more Karners exist in Massachusetts, Pennsylvania, Illinois, or in the province of Ontario in Canada. Records of the species indicate that originally Karner blues occurred in clusters of populations (metapopulations) that shifted over a substantial area of fire-swept landscape covering thousands of acres.[65] While part of the population was lost along with critical habitat in blazes, the fires also prepared landscape for re-colonization in early post-fire vegetation, with large and rapid population build-up. However, over the past century, it is estimated that the population has declined by 99% or more over its historic range, with most of this fall occurring in the 80's and 90's. It is interesting (and relevant as we shall see) to note that though the Karner blue butterfly has low populations

62 Stanford Environmental Law Society, *The Endangered Species Act: A Stanford Environmental Law Society Handbook* (Stanford, CA: Stanford University Press, 2001), p. 133.

63 Stanford Environmental Law Society, *The Endangered Species Act*, p. 133.

64 See David R. Lentz and Jimmy S. Christianson, "The Karner Blue Butterfly: Wisconsin's Statewide Habitat Conservation Plan," in Kaush Arha and Barton H. Thompson, Jr., eds, *The Endangered Species Act and Federalism: Effective Conservation through Greater State Commitment* (New York: RFF Press, 2011), p. 147.

65 Lentz and Christianson, "The Karner Blue Butterfly," p. 148.

in other states, it was not thought to be endangered in Wisconsin when the species was listed and hasn't been since.[66] In fact, the species is still not even listed in the Wisconsin Endangered and Threatened Species Act.[67] Michigan supports significant and sustainable populations of Karner blues as well.

In January of 1992, in 57 F.R. 2241, the USFWS published a proposed rule listing the Karner blue butterfly as an endangered species. The primary causes of the species decline are not surprising; rapid land development, habitat fragmentation, silviculture, and forest succession in the absence of disturbance. The Wisconsin DNR (WDNR), given the stronger population of Karners in the state, suggested the species be listed as "threatened" instead of "endangered." The WDNR argued that given the wide distribution of a relatively healthy population of Karner blues, a more flexible conservation approach under a "threatened" species classification was appropriate and most effective for meeting conservation goals.[68] Yet, the USFWS remained worried that for the long-term viability of the species listing the Karner as endangered was warranted. The federal agency argued that the species was under a high threat of extinction throughout all or a significant portion of its current range. Consequently, the species was listed as endangered despite USFWS's recognition that viable populations existed in Wisconsin. Likely, this was also a strategy used by the agency to avoid the controversies that came with the listing of the northern spotted owl in the Pacific Northwest.

In 1993 and 1994, with its addition to the endangered species list, landowners in the state contacted the WDNR to look into applying for incidental take permits and establishing HCPs. USFWS had also contacted the WDNR as they were concerned about how the latter would respond to the species listing with so many private landowners distributed over a significant portion of the Karner blue range in the state. The intent of the WDNR from the beginning was to try as best it could to balance protection of the state's Karner blue habitat with land use for recreational, forestry, wildlife management, and right-of-way maintenance activities. This was essential as Wisconsin is the nation's largest supplier of paper. Timber has been harvested from private, state, and county forests for decades. Once timber is cleared, lupine can grow in fresh cut areas providing excellent Karner blue habitat. The same result can emerge from right-of-way maintenance performed by utility companies. There was little doubt among many stakeholders that the only way to ensure long-term protection of Karner blue habitat and the species itself in Wisconsin was through some sort of land

---

66 Lentz and Christianson, "The Karner Blue Butterfly," p. 148.

67 Lentz and Christianson, "The Karner Blue Butterfly," p. 149.

68 Lentz and Christianson, "The Karner Blue Butterfly," p. 149.

management arrangement. Regulatory relief from take was, in the opinion of the WDNR, critical to the success of the project.[69] Additionally, it wasn't merely private businesses that would need regulatory relief from the ESA. WDNR itself is the largest public land manager and forest steward in the state, and to meet this stewardship goal it would need to have the flexibility to engage in activities that may kill individual Karner blue butterflies for the betterment of the species as a whole.

The decision was made early in the process that an HCP would be the best tool going forward for protection of Karner blues. Local representatives of the USFWS had had little experience with other possible ESA instruments, such as "safe harbor" agreements. Also, given the disturbance conservation approach vital to the protection of Karner blues, the strategy had to accommodate management of a working landscape, for which the HCP model would fit well. The activities of HCP partners in land management, would, if done well, mimic the natural disturbance conditions necessary for Karner blue survival.[70]

The vision for the Karner HCP from the WDNR perspective was to target critical habitat. Also, the aim was to design the HCP so as to provide protection to all stakeholders (including small landowners) from take prosecution. While this was designed to be a Wisconsin-wide initiative, the central and northwestern parts of the state, according to species audits, was the main focus of the HCP and this area was dubbed as "The Karner Blue High Potential Range."[71]

Many state landowners were more comfortable working directly with the WDNR than with USFWS. Locals perceived a strong population of Karner blues within the state and didn't understand why they were listed. International Paper, Consolidated Papers, Georgia Pacific, and others were the first to contact the WDNR about the development of an HCP and preparation of the application for an incidental take permit. Utility companies, county forest agencies, other state agencies, and the Nature Conservancy followed this first wave. Other organizations joined later, such as the Sierra Club, the Wisconsin Woodland Owners Association, and the Audubon Society. Twenty-six landowners committing some 250,000 acres to the HCP were among the coalition of partners that signed on to the conservation effort following a public comment period.[72]

A two-tiered conservation effort for Karner blues was established. On one tier, large landowners were the focus. On this tier specific land management strategies were required of those who wanted regulatory relief within

69 Lentz and Christianson, "The Karner Blue Butterfly," p. 149.

70 Lentz and Christianson, "The Karner Blue Butterfly," pp. 150–151.

71 Lentz and Christianson, "The Karner Blue Butterfly," p. 150.

72 Lentz and Christianson, "The Karner Blue Butterfly," p. 151.

the Karner high probability zone. Those larger landowners were identified as "HCP partners" and were non-voluntary participants. The second tier consisted of relatively small landowners (usually owning 1,000 or fewer acres) who would still be covered under the incidental take permit, but without having to join the HCP proceedings. Instead of being required to join, these participants would be encouraged via outreach and education to voluntarily take conservation measures for Karner blues.

In 1994, after initial discussions occurred, the application was submitted. The WDNR and USFWS sent out information to some 3,000 recipients throughout the Karner Blue High Potential Range to explain the listing of the species and the proposed HCP. There were several public meetings to discuss the initiative. The early discussions also served as a way of identifying large landowners as possible HCP partners. The HCP was finally approved in September of 1999. Today, there are forty-one private and public partners participating in the Karner HCP, mainly from the logging industry, utility companies, and road management authorities.

As far as the roles it played in HCP development, the WDNR basically held a leadership position in implementing the plan and was the official applicant for the permit from a constellation of HCP partners. More specifically, the WNDR's representative for the plan is also the HCP Implementation Coordinator and works in consultation with the Implementation and Oversight Committee (IOC), which consists of representatives of the HCP partners to administer the plan. A further responsibility of the HCP Implementation Coordinator is to oversee partner compliance. Twice a year the USFWS Green Bay Field Office and WDNR staff meet to assess data from the HCP, consider its performance, and make any possible revisions necessary. The role of the USFWS in the Karner Blue HCP is to lend technical assistance and to help the WDNR and HCP partners adhere to the guidelines and procedures stipulated in the HCP document.[73] In addition, the Services offer opinions on the conservation approach taken to Karner blues and raise any concerns about how the HCP is functioning. Part and parcel with this, the USFWS shares appropriate biological opinions on best practices for Karner conservation. The Karner HCP is in a three-year review cycle in which the WDNR reports out on the overall status of the HCP. The WDNR also submits smaller annual reports to the USFWS.

With regard to funding, a panoply of sources are involved, many of them offering in kind services. During the implementation stage, this funding amounted to $1.5 million. HCP partners supplied the lion's share of these

73 Lentz and Christianson, "The Karner Blue Butterfly," p. 153.

services, though the WDNR kicked in about a third of the total. Landowners and participants contributed many hours of their time and expertise to developing and executing the HCP. Conducting surveys, effectiveness monitoring, and planning and enhancing habitat, required a great deal of time and effort from many hands. The WDNR received some of its funding from its regular state budget and through grants. The USFWS dispersed funds related directly to species recovery, funding research, surveys, and land acquisition. The Services also picked up the publication cost for the HCP. Finally, the HCP partners provided some monies to cover administrative expenses.[74]

The Karner Blue Butterfly HCP is unique in a couple of ways. First of all, it was the first statewide HCP established in the US. Typically HCPs (even those covering large areas) are established regionally. In many cases, they are relatively small and atomized, with individual landowners applying for exemptions to the Section 9 take prohibition. Part of the intent of developing a statewide HCP was to resist the need for each individual landowner to have to establish his/her/its own HCP. Having many individual landowners filing for incidental take permits is impracticable, as well as ecologically unsound, as it is better to have contiguous land areas under the HCP. Second, the plan is also open to a bevy of small private landowners (again generally those with less than a 1,000 acres), allowing them to participate voluntarily.

The larger HCP partners have educational and outreach responsibilities as part of the agreement. As Peter Moreno and David Lentz write, "HCP partners have pledged to encourage Karner Blue Conservation among all landowners in the Karner Blue range; including voluntary, non–partner participants."[75] This acknowledges that many landowners with smaller parcels and not required to be members of the HCP are still important to its proper function. Part of the reason for this provision is that there are so many private land managers in the Karner blue range. Additionally, one of the goals of this strategy is to build a sense of trust between state resource agencies and a coalition of smaller landowners by freeing them from regulatory requirements. From the high number of small landowners who are participating, the effort seems to be a success. Finally, as environmental legal scholar Barton Thompson points out,

---

74 Lentz and Christianson, "The Karner Blue Butterfly," p. 154.

75 Wisconsin Department of Natural Resources, "Affirming the Effectiveness of Outreach, Education and Voluntary Participation in the Wisconsin Karner Blue Butterfly Habitat Conservation Plan," Peter Moreno and David Lentz, report, (Madison, WI: Wisconsin Department of Natural Resources, 2003), p. 52.

the Karner Blue HCP is one of the few that meets the ESA Section 10 requirements by managing the working landscape.[76]

Nationally, many HCPs have been limited in their effectiveness due to frictions between landowners who develop them and regulators. However, the Karner HCP has been held up as a paradigm case of the strategy working well. Some key factors contribute to these positive results. First, as mentioned before, the Karner blue butterfly has not been an endangered species in Wisconsin. At first glance, this doesn't seem to show the HCP is really aiding with species conservation. But concentrations of Karners have not only remained high in the state, but also have increased in size and frequency over the lifespan of the plan. Also, habitat hospitable to Karners has increased in Wisconsin. These are at least two signs of the effectiveness of the HCP.

The second key factor of success with the Karner HCP is that the species is conducive to "disturbance" conservation. Karner blues need sandy pine barrens (found mainly in northern and central Wisconsin) or oak savannahs (found more in the west-central and southern parts of the state) as habitat where lupine can grow (i.e., lupine is a forb of the pea family and the only food Karner larvae eat).[77] Karner blue butterfly habitat does not completely overlap with lupine, but it tends to lie along the northern tier of the plant's range. Yet over time shrubs and trees crowd out the lupine. This is where disturbance conservation comes into play. The logging industry cuts timber in a mosaic pattern with cleared areas creating re-established habitat for Karner blues. Then, as habitat becomes cluttered with tree canopy and shrubs less hospitable for them, an adjacent area can be cleared of overgrowth allowing for lupines to reappear and serve as an area to which Karners can migrate. Additionally, periodic disturbance allows for the clearing of over-story so that myriad understory plants can thrive, providing nectar sources for adult butterflies.[78]

With respect to the goal of not losing net habitat, there are some encouraging signs that the Karner HCP is (at least roughly) meeting it. It is hard to obtain a precise measure of how much acreage is being conserved given that the range of Karner blues is always in flux. However, there are some indications that there is no net loss of habitat. An approximation from species audits,

---

76 "Managing the Working Landscape," in Daniel D. Goble et al., eds. *The Endangered Species Act at Thirty: Renewing the Conservation Promise* (Vol. 1), (Washington, D.C: Island Press, 2006), pp. 102–103.

77 United States Fish and Wildlife Service, "Karner Blue Butterfly (*Lycaeides Melissa Samuelis*)," (website), June 16, 2016. Available at http://www.fws.gov/midwest/endangered/insects/kbb/index.html (accessed July 12, 2016).

78 Lentz and Christianson, "The Karner Blue Butterfly," p. 148.

reviews of land management practices, and an examination of habitats in large land holdings in the HCP suggest the maintenance of Karner blue habitat. Also, previously unidentified habitat is continuously being found and large-sized samples of their habitat have not declined.[79]

As mentioned earlier, another objective in establishing the Karner HCP was to provide protection for small landowners from take regulation, all the while encouraging them to participate in Karner blue conservation. This was to be achieved via education and outreach, often provided by both HCP partners and the WDNR. Part of this outreach comes in the forms of events, such as the Karner Blue Butterfly Festival, which is now an annual celebration in Black River Falls. How these efforts have turned out is captured in a recent WDNR Participation Plan Three-Year Review and Assessment Report:

> The Karner Blue Butterfly HCP has done more than help the Karner blue butterfly; it has bolstered statewide enthusiasm for rare species conservation and ecosystem restoration. The plan has also rebuilt trust between landowners and natural resources agencies, at least state agencies. This is a significant accomplishment given the storied history of traditional Endangered Species Act implementation. The results of these achievements are beginning to show. The WDNR is confident that the relationships developed with landowners will encourage even greater conservation collaboration in the future.[80]

Apparently, the USFWS was impressed as well, determining that parties to the HCP were meeting the requirements of the permit and acknowledging that these efforts demonstrated the strong work of the WDNR and HCP partners.

Other more measurable successes of the Karner HCP can be documented. For example, there have been increases in the known occurrences of Karners from a baseline of 103 locations to 500 locations by 2007. There have also been increases in lupine habitat by 300% from 1998–2007. The number of voluntary participants in the HCP has increased, as has the frequency of their conservation efforts in favor of Karner blues. Timber harvests have been conducted in a way that maximizes the ability of Karners to migrate from one patch of habitat to another.

Though the process of negotiating the HCP was by no means painless, the Karner HCP scores relatively well with respect to democratic participation and somewhat well with regard to monitoring. Lumber, utilities, and road

79 Lentz and Christianson, "The Karner Blue Butterfly," p. 154.

80 Lentz and Christianson, "The Karner Blue Butterfly," p. 155.

management executives willingly came to the table and set aside critical habitat for Karners, recognizing the value of land use management conservation. They spent substantial time and treasure to engage in mitigation strategies. Eventually, though the five-year process of developing the HCP surely caused frustration at times for all parties, a compromise was made and the basic arrangement has remained in place for over fifteen years since its approval.

WDNR wanted to ensure that whatever the final format of the HCP looked like, it would be shaped in an open, participatory structure. One important reason for this full participatory approach was to alleviate the fears of landowners of being prosecuted under the auspices of Section 9 of the ESA's prohibition on take. A transparent and participatory negotiating environment would allow the space necessary for landowners to openly admit instances of Karner blues on their land and encourage them to become good stewards of Karner habitat. It appears the WDNR was even surprised by the HCP partners during the implementation of the participatory plan when in some instances the partners engaged in new voluntary conservation efforts. The WDNR's Division of Forestry also played a large role in making contributions to the HCP. It was able to build on relationships the division had already forged with paper mills and other private landowners for Karner blue conservation and educating the public on these efforts.[81]

The WDNR also seems to think that the emphasis on a participatory approach allowed other state agencies not directly involved in nature resource management to take a more constructive role in Karner blue conservation. For example, the Wisconsin Department of Transportation (WDOT) has modified its mowing regimen to accommodate the species. As alluded to earlier, over eight to twelve years, canopy cover comes to dominate an area and habitat for Karners is lost. Therefore, forestry managers, to accommodate Karner blues use a "shifting mosaic" strategy to ensure a constant supply of new habitat. Highway agencies and some utility company partners also committed to the conservation efforts by managing rights-of-way with frequent surveys and engaged in mowing practices conducive to disturbance conservation. Still other partners have employed controlled burns to not only aid Karner blues, but other species as well.

Concerning democratic participation, when the HCP proposal was floated as part of the public notification process, no concerns or objections arose. The WDNR took this as a good sign that the HCP, with its accompanying environmental impact statement, explained the participation strategy and risk

81 Lentz and Christianson, "The Karner Blue Butterfly," p. 157.

analysis in such clear detail that no contention emerged.[82] Moreover, after the statewide HCP strategy was agreed upon in principle, all parties met to consent to more specific commitments and responsibilities for the HCP partners. This needed to be done for each partner in what were called Species and Habitat Conservation agreements. These documents, when completed, were bundled with the HCP and the incidental take permit applications. Not only was species conservation listed as a desired goal, but also included were the landowner's economic objectives. From 1995–1998, WDNR, the HCP partners, and the USFWS met to continue work on the plan. Throughout the process the WDNR reported that there was strong public participation and a high degree of transparency in the development process. This included meetings with groups that would not be directly participating in the HCP, but nonetheless had a stake in the plan. The National Wildlife Federation and Defenders of Wildlife participated in some of the discussions. While the HCP was in development, it took some time to build sufficient trust for a workable plan. An unintended (and ironic) consequence of the lengthy negotiation period was that land management that could have benefited Karner blues could not be practiced because such activities could have resulted in incidental take.[83]

The monitoring of whether Karner Blue HCP partners are meeting the conditions of the incidental take permit is the responsibility of both the WDNR and the USFWS. As mentioned earlier, friction has impeded the success of other HCPs and there have been worries that the WDNR and the Services do not have adequate resources to properly monitor HCP sites. While the Karner Blue HCP has not been immune from those issues, the WDNR noted improvement over the second half of the first decade of the HCP's existence. During its first five years, it was clear that the monitoring scheme set out in the original agreement was not providing adequate feedback for good, adaptive, wildlife planning.

Shortly thereafter, the WDNR set about a new strategy to strengthen the monitoring program. This included the implementation of a formal method of cause-effect monitoring studies and the discontinuation of random sampling to identify new sites for lupine inventory. This would allow the HCP partners to simply seek out habitat where they think the greatest likelihood for increasing species number and/or areas where they would need to do a pre-management survey.

Recovery monitoring was conducted for the second year in 2009 and ten different properties located in four separate designated zones were tested.

82 Lentz and Christianson, "The Karner Blue Butterfly," p. 152.

83 Lentz and Christianson, "The Karner Blue Butterfly," p. 153.

The method implemented was distance sampling as recommended by the US Geological Service and fifty people underwent Distance Sampling Training.[84] Volunteers, including members of the DNR Wisconservation Corps (Americore), monitored the Karner HCP. On larger properties with limited staff for monitoring or with more extensive recovery requirements, student interns were provided. In other cases, property staff did the monitoring. Habitat assessments were conducted on all recovery sites, examining 16 key nectar plant species, lupine presence, and invasive species to help facilitate recommendations for habitat management needs. In the 2009 report, WDNR stated it was satisfied with the new reduced Distance Sampling method.[85] With respect to the Wisconservation volunteers (who were provided to help reduce some of the monitoring burdens from the landowners), they tended to be interested in the process, motivated to complete the tasks, and were efficient in doing so. While this specific strategy worked well, it required shifting staff over very long distances, leaving other state conservation teams short-staffed and adding greatly to travel costs. Yet, these monitoring efforts are by and large more successful than what has been achieved in other HCPs nationwide. Property owners were very helpful with monitoring, supplying relatively high numbers of field staff for the surveys and helping the Distance Sampling teams when necessary. This cooperation is crucial to any successful monitoring regime.

Beyond monitoring, there are many other activities essential to protecting Karners under the HCP, even if it means incidental take of some members. Full HCP partners are responsible for mowing, mechanical site preparation, invasive plant control, timber harvest and improvement, brush clearing, tree planting, and some prescribed burnings. According to the 2009 WDNR report, limited partner activities included mowing and removing brush.[86] These partners made small, but non-negligible contributions and were comprised of representatives from ten townships and county highway departments, working to add Karner blue habitat along 33 acres of roadsides. HCP partners helped with other conservation activities, including outreach and education about Karner blues and their habitat. Twenty partners reported outreach and education activities in 2009.

---

84 Wisconsin Department of Natural Resources, "The Karner Blue Butterfly Habitat Conservation Plan Annual Report of Activities for the Calendar Year 2009," (website), David Lentz, March 24, 2011. Available at http://dnr.wi.gov/topic/ForestPlanning/documents/2008AnnualReport.pdf (accessed July 12, 2016).

85 Wisconsin Department of Natural Resources, "The Karner Blue Butterfly," p. 9.

86 Wisconsin Department of Natural Resources, "The Karner Blue Butterfly," p. 12.

Despite the success of the Karner Blue HCP, maintenance of the program has not been completely seamless. On the one hand, one would expect to hear about more problems between landowners and regulatory bodies. However, what has happened has been more complex than that. In the Karner HCP, the main issues have arisen between the state natural resource managing agency (the WDNR) and the federal regulatory body (the USFWS). The relations are strained in this case in part because of different views on the best conservation strategy for Karner blue butterflies. The WDNR seems frustrated that they and landowners in the state do not have more autonomy over land use. The WDNR never thought Karner blues should be listed anyway (or at least not in Wisconsin). Recall that the USFWS tried to justify listing the species on the basis that its long-term viability had to be assured and since the species' population had declined over several states, the recovery program had to be categorical (including all populations of the species). Even since the implementation of the statewide HCP in Wisconsin, populations of Karner blues have not increased in other states. Hence, again for the long-term sustainability of the population, the USFWS insisted that Karner blue butterflies needed to stay listed in all locations where they currently reside. The WDNR disagreed, pointing out that for vertebrates there is some flexibility in the ESA to delist populations of endangered species in certain geographical locations where their numbers are sufficiently strong. It is only because of the language in the law that this flexibility doesn't extend to invertebrates like the Karner blue and not due to some ecologically sound reason. The WDNR's argument here is not just about effective wildlife managements, but also about justice, as they contend that Wisconsin resource agencies and landowners are being unfairly treated. For all of the good work that has been done to support Karner blue conservancy in the state, landowners do not receive substantial reward in the form of regulatory relief.

The USFWS strategy for Karner blue conservation, claims the WDNR, is too focused on the numbers, sizes, and spatial configurations of individual Karner blue populations. Instead, the WDNR has argued in favor of recovery criteria emphasizing the amount, quality, and distribution of habitat disturbance over time. One reason for this is that it difficult to keep track of a population constantly in flux, as the Karner blues population is, given their need for habitat disturbance. Another problem is that it wouldn't be economically or ecologically feasible to preserve every Karner blue population in the state. Some populations are too small and/or have shown significant population decline to the point where they will not likely be sustained despite the best conservation efforts. Instead, the WDNR argues the emphasis should be on statewide butterfly populations that have disturbed habitats readily available to them. Thus, they argue the focus should be on the distribution of disturbed populations over time. The WDNR concludes that a habitat-based strategy would better account for normal population fluctuations and perhaps even benefit other species sharing the butterflies' habitat.[87]

Because Wisconsin's metapopulation of Karner blues is seen to be strong by most groups in the state, there is little incentive to be concerned with every population of the species. This view extends into the future with a strong probability that Karner blues will remain in Wisconsin for a long time; especially since continued timber industry practices are a good engine for maintenance of disturbed habitat. But with the current law not allowing for invertebrates to be delisted in states with healthy populations, and with little evidence that conservation efforts will improve in the eastern states, it doesn't look as though Wisconsin landowners will receive regulatory relief any time soon. The WDNR worries this is a waste of resources that could be used to support conservation of other species. It also doesn't help with public perceptions that conservation is not worth it, since even when it is plain that a local listed population is doing well it remains listed indefinitely. This political point is important, as it is likely true not just with Karner blues, but also with other species that are imperiled. Ultimately, the public is going to need to see their conservation efforts as worthwhile.

There is also the concern for how to continue to fund the Karner HCP. Wisconsin's conflict of interest laws do not allow for the WDNR to solicit funding from HCP partners. What this has meant is that while time and expertise can be used from HCP partners, an important conduit for funds is eliminated. The WDNR Division of Forestry has been a significant financial contributor for covering compliance and administrative costs. But with recent budget cuts, it is hard to see how a consistent level of resources will continue to flow for maintaining the HCP. Funding to the WDNR in general has already been tight and may face further decline. Ironically, federal money that would otherwise be available for Karner blue conservation through Section 6 of the ESA is unavailable since the species is not considered to be in peril in Wisconsin. So in the eyes of the WDNR, there are a linked set of perverse implications of the current ESA. Landowners and the WDNR face increasing regulatory restrictions on land use despite doing a good job of maintaining species conservation, and yet that success doesn't allow them access to resources that could aid in such future conservation.

---

87 Wisconsin Department of Natural Resources, "The Karner Blue Butterfly," p. 158.

Beyond the concerns about money for the program, many of the initial players of the HCP have since retired or moved on to other pursuits.[88] Companies participating in the plan have changed hands and it is hard to maintain enthusiasm and commitment if they see no net direct benefit of Karner conservation to them.

Despite the successes of this unique HCP, there are additional worries that a new booming industry may be changing the landscape for future management of Karner blue habitat—frac sand mining.[89] As of this writing, only one frac sand mining company (Unimin) has even contacted the WDNR HCP Coordinator to inquire about having a Karner species audit. The audit is necessary to see if one's land has high potential for Karner Blues and if the threshold is high enough to enter the HCP to benefit from the incidental take permit.

If one closely examines the zones where there are high probabilities of Karner blue butterfly populations and where there are sandstone formations which hold the highest likelihood for finding the silica sand used for frac sand mining, the overlap is extensive. For example, high probabilities for both Karner blues and frac sand mines are contiguous in Monroe, Juneau, Wood, Adams, Clark, and Jackson counties. It is unclear how frac sand mining will fit into the disturbance conservation paradigm of Karner blue recovery. While there is disturbance of the landscape from this activity, it is highly unlikely it will be conducive to Karner blues. Mining operations for frac sand will likely last for long periods of time, taking net habitat away from the species.

There are several moral issues that arise in this case and the following isn't by any means meant to be an exhaustive list. There is the issue of whether we think land management should be part of federal laws meant to protect species. It may just be a lucky coincidence in the case of Karner blue butterfly that it happens to be conducive to disturbance from some industries. But this is not the case for most endangered species. This may be the problem for other HCPs when mitigation strategies fail (for example, what if the additional land that is set aside for species does not end up being quality habitat?). Sometimes even with having critical habitat available for endangered species, if they are forced to move they do not adjust well to migration.

One might also challenge the issue of fairness that the WDNR raised with respect to how Wisconsin's landowners are treated. While perhaps other states should pull their weight with respect to Karner conservation, is it *unfair* that Wisconsin landowners don't received regulatory relief? After all, species listing is a complex matter and does it make sense to say that because other states

88 Wisconsin Department of Natural Resources, "The Karner Blue Butterfly," p. 158.

89 See Case 8—"There's Silica in Them Thar' Hills!".

have fallen down on their conservation efforts that Wisconsin can pull away its support? Should an effective body for species protection at the federal level use regulatory authority to treat different states differently due to their level of conservation effort?

The fairness question comes to the fore again with respect to traditional, longstanding HCP partners versus companies from new or suddenly booming industries. Former Karner HCP Coordinator David Lentz and his co-author Jimmy Christianson note that Wisconsin needs an infrastructure to handle adaptive management in the face of new information and economic realities.[90] While it is hard to know what this infrastructure would be, the sentiment was prescient. The frac sand industry would definitely seem to constitute such a challenge on the economic front. When we consider the question of fairness, how fair is it to the logging, utility, land management companies, and agencies that have spent time and money establishing and maintaining an HCP to have a new industry enter the state whose companies are not playing by the same rules?

Though we likely don't want to make too much of this, there is the question of whether companies in new industries such as frac sand mining will have the same conservation ethos that has been exhibited by logging and utility companies, and state forestry divisions. In political philosophy, many communitarians argue that longstanding traditional ties to a place not only constitute political reality, but also underwrite the values that attach to that place. These "ties that bind" would presumably make citizens more likely to care deeply about the political, cultural, economic, and environmental conditions of the place in which they live. The logging industry has been deeply embedded in the state of Wisconsin for decades and has several local ties, with corporate offices in the area. In contrast, it is interesting that a large number of frac sand mining outfits are not based in the state, as their gas and oil industry headquarters are located far away. One large holder of frac sand mines and processing centers in Wisconsin is EOG Industries (formally Enron), which has its headquarters in Houston, Texas. It is possible that companies which have their decision-makers and shareholders located far away from Karner blue country will not have the same kind of stake in the conservation of land and species in Wisconsin as we have seen from the traditional in-state stakeholders in the Karner HCP. This may be part of the reason why frac sand mining companies have been so slow to contact the WDNR to schedule Karner audits, even when mines have been proposed for areas in the Karner high probability zone. It is not wild speculation to imagine that for many energy company

90 "The Karner Blue Butterfly," pp. 161–162.

executives, there isn't much awareness at all of Karner Blue habitat or the species it harbors.

Review Questions

1) Consider the financing of the Karner Blue Butterfly HCP. If it comes from USFWS, then the source is most likely taxpayers throughout the country. If it is the WDNR, then the source is likely the taxpayers of Wisconsin. If the financial sources are the HCP partners, then it is from their profits. Part of the reason why HCP partners cannot be solicited for funds now is because of conflict of interest laws. Who do you think should pay the costs for land acquisition, maintenance, and administration of the plan? Should the conflict of interest laws be changed so as not to apply to HCP funding? Is there a way to get around conflict of interest concerns for funding of these types of measures?
2) Do you think there is anything to the claim that those who live within a community may have a stronger local conservation commitment to that place (that is, to think that preserving local species and habitat is valuable) than those who don't?
3) Note the rift between state and federal regulatory agencies in how to best manage species conservation in this case study. Do you think the WDNR (which favors looser requirements and greater landowner autonomy) or the USFWS (which supports more stringent requirements on landowner activities vis-à-vis Karner conservation) has the better approach and why? Should wildlife conservation be determined on a state-by-state basis or should it be based on a larger regional or national approach?
4) What do you make of the claims of unfairness that could be made by HCP partners (such as logging companies, utilities, and the WDNR) who have "played by the rules" in committing to Karner blue butterfly conservation versus new industries that have largely ignored this commitment? Is there some way to justify frac sand mining companies not committing to the Karner Blue Butterfly HCP?

Case 38 – Animal Hearts as Human Spare Parts? –

Study Questions

1) What is xenotransplantation and how is done? Why do some think xenotransplantation is necessary?
2) What are some of the moral questions that arise with xenotransplantation?

One of more controversial recent developments in biomedical technology is the ability to use the vital organs of non-human animals to save and enhance the lives of humans. One factor leading to an intensive effort by the medical community to research *xenotransplantation* of non-human animal organs into humans is the dearth of human organs available. This situation stems from a lack of human donors. Popular media have pounced upon stories of possible primate-to-human and porcine-to-human heart transfer with great zeal in headlines such as "'Million-dollar Pigs' are Medical Marvels,"[91] and "Making Manimals."[92] Documentarians have also used catchy titles such as "Organ Farm: A History of Xenotransplantation Experiments," to draw interest to this issue.[93]

Xenotransplantation can be put under an umbrella of many new innovations in genetic engineering that have engendered controversy. There are some terms to consider in any discussion of xenotransplantation. Xenotransplantation is also sometimes called *cross-species transplantation.* Also, xenotransplantation is an alternative for what is more generally called *iso-grafting*; for example, human-to-human transplants. According to the United States Health Resources and Services Administration (HRSA), a new person is added to the American transplant waiting list every ten minutes. Furthermore, they estimate that on average, 21 people die each day while waiting for a transplant.[94] As of mid-2015, there were 122,978 total waiting list candidates needing a lifesaving organ transplant, with 79,130 of them on the active waiting list. For the first half of 2015, only 7,282 transplants had been performed. There were only 3,521 donors from January to March 2015.[95] All of these figures point to a scarcity of available organs in the United States and these data reflect what is occurring worldwide—if anything, there is even greater scarcity in many countries outside the US.

91 Luis Fabregas, "'Million-dollar Pigs' are Medical Marvels," *Pittsburgh Tribune Review*, April 9, 2006. Available at http://triblive.com/x/pittsburghtrib/news/regional/s_441762.html (accessed July 12, 2016).

92 William Saletan, "Making Manimals," *The Washington Post*, June 27, 2007. Available at http://www.washingtonpost.com/wp-dyn/content/article/2007/06/22/AR2007062201643.html (accessed July 12, 2016).

93 *Organ Farm: A History of Xenotransplantation Experiments*, Dir. Frank Simmonds, film, Frontline Films, 2001.

94 United States Department of Health and Human Services, Health Resources and Services Administration, "Organ Procurement and Transplantation Network," (website), June 26, 2015. Available at http://optn.transplant.hrsa.gov (accessed June 29, 2015).

95 United States Department of Health and Human Services, Health Resources and Services Administration, "Organ Procurement."

With this paucity of organs, biotechnology companies have examined other options beyond human iso-grafting. While the notion of using nonhuman organs in humans is still very much in its infancy, people with ailments such as kidney failure, and diseases of the heart, lungs, and liver may well be treated in the future with the use of animal parts.

Controversial transplants have occurred throughout history even between humans, with objections sometimes emerging along racial or religious lines. For example, in the 16th century, Cosmos and Damian transplanted the leg of a "black" Ethiopian onto the body of a "white" Justinian.[96]

But xenotransplantation has had a relatively short history, as researchers only began thinking seriously about the possibility in the 1960s. This interest came as a result of strides made in human transplantation, but also due to some animal-to-human transplants in the early to mid 1960's. The first experiments of transplanting chimpanzee kidneys to humans came in 1963 and 1964, with one recipient living for nine months. According to Jonathan Okeke and Chris Akpan, between 1963 and 1993, there were 31 clinical procedures involving placing whole animal organs into human recipients performed in the United States and South Africa. In these cases, the animal organs were not intended to stay in the human recipients long-term, but only to serve as a stopgap method of keeping them alive until replacement human organs could be procured. The reason for this was that xenotransplantation was still experimental.

Likely the two most well-known cases of xenotransplantation came with Baby Fae, an infant who received a baboon heart in 1984 and Jeff Getty, a patient with AIDS who received a bone marrow transplant, again from a baboon, in 1995.[97] Baby Faye was able to survive for 20 days with the assistance of the heart. Getty's body accepted the transplanted bone marrow almost immediately and remained free of any baboon-transmitted viruses—in fact, by October 1998 he no longer showed any signs of baboon marrow in his system.[98]

More recently, researchers have turned to domesticated pigs and cows as sources of other tissues for humans. They have become preferred sources over primates for a number of reasons dealing with health, logistics, and safety. Since pigs mature very quickly and have organs of comparable size and function to those of humans, this makes them better organ transfer candidates. It is also easier to raise pigs in microbiologically-controlled environments to

---

96 Jonathan C. Okeke and Chris O. Akpan, "An Inquiry into the Moral Question of Xeno-Transplantation," *Online Journal of Health Ethics* (2012) 8(1): 1–13 at 3. Available at http://aquila.usm.edu/cgi/viewcontent.cgi?article=1103&context=ojhe (accessed July 12, 2016).

97 Okeke and Akpan, "An Inquiry," p. 3.

98 Okeke and Akpan, "An Inquiry," p. 3.

greater ensure high quality organs for transplant. Monkeys are harder to raise in captivity according to these health standards and their organs are often rejected by human recipients. Scientists are more comfortable working with pigs because of little fear of new viruses being introduced into organ recipients. As pig transplant researcher John Atkinson has opined, "A new virus is a theoretical possibility, but I can't imagine that with all our contact with pigs over thousands of years, that it would be a major problem."[99] This latest research trend in xenotransplantation points up another interesting moral issue, as recent efforts are geared toward specifically breeding certain animals for the purpose of harvesting their organs. Should this be permitted?

Some research designed to ultimately aid humans involving xenotransplantation has nothing to do with transplanting whole organs into human transplant recipients. For instance, work on early human development involves securing tissue from fetuses that have been aborted and is then introduced into rodent hosts. This research is focused on examining the effects of environmental pollutants on fetal cells through exposure to a number of different hazardous materials. Tissues are prepared for transplantation into immunodeficient rodent hosts, which are then exposed to common environmental stressors and pollutants (such as arsenic, lead, and phthalates) to monitor human fetal tissue responses.[100] Other research possibilities include removing the kidneys of aborted human fetuses and implanting them in rats where the kidneys can be grown to larger sizes. The goal of such research is to grow larger human kidneys in bigger animals that can be transplanted back into other humans.

Obviously, those who have argued in favor of xenotransplantation cite the suffering and mortality of humans in need of transplants who are left without organs. If there were no gap between needy recipients and willing donors, the need for xenotransplantation likely wouldn't exist. The rapid development of potent and effective immunosuppressant drugs in the 1980s paved the way for successful transplant of many types of organs between humans who are unrelated. Despite these breakthroughs, there have not been enough donors to even come close to fulfilling the need for organs.

Those in favor of xenotransplantation also point to the likelihood that the need for organs will only become direr in the future. On the one hand, enhanced nutrition and developments in medical care are increasing the number of years humans are living. But, this also increases the likelihood of

---

99 Okeke and Akpan, "An Inquiry," p. 4.

100 Bindu Panikkar, Natasha Smith, and Phil Brown, "Reflexive Research Ethics in Fetal Tissue Xenotransplantation Research," *Accountability in Research* (2012) 19: 344–369.

more people living longer lives who would have otherwise perished due to fatal organ failures. This will correspondingly lead to increased demand for compatible organs for transplant. On the other hand, it seems unlikely that other social arrangements for organ donation will resolve this issue. In the US, even if the current policy where one must explicitly consent to being an organ donor were changed to a "presumed consent" approach (such as Spain's) where one is presumed to be an organ donor unless she expressly opts out, it doesn't seem this would adequately meet demand. Stem cell therapies may alleviate the need for organ donors in the future, if therapeutic cloning and the development of artificial organs continue at its present pace. Yet, while new research is making these options more viable, their ready commercial availability may be far in the future. Of course, all of these technologies present their own deep ethical controversies.

As mentioned earlier, there has been an emphasis on developing genetically modified pigs to render their organs more histocompatible with humans.[101] Perhaps the most vexing biological challenge that genetic engineers have encountered is rejection of donor organs by recipients' immune systems. The surface molecules of recipients detect transplanted tissue as "foreign" and thus their immune systems lodge an attack on the implanted tissue. Eventually, the goal is to overcome problems of tissue rejection following xenotransplantation through genetic engineering. One way of doing so is by coaxing donor cells to express surface proteins that deceive the immune system, leaving a less hostile environment for the donor tissue. Researchers have been encouraged that this technique might work as it has (to some extent) been successful in experiments with pig-to-(non-human)-primate transplantation.

As opposed to whole-organ xenotransplants, smaller tissue implantation from animals to humans has become common. Many people have received replacement heart valves from pigs and cows. Another area of promise in xenotransplantation has been with respect to the treatment of type 1 diabetes (T1D). In this type of diabetes, usually found in young children, patients are unable to produce insulin from islet beta cells in the pancreas. While the entire pancreas is not necessary for transplant, more than one to two pancreases worth of beta islet cell tissue is needed to treat patients with T1D. Technology has advanced to the point where harvesting cells from a pig pancreas may well be more cost effective than current treatments, especially in young diabetics. Implanting these cells would allow T1D patients to live without constant insulin injections.

---

101 Robert Sparrow, "Transplantation, Consent, and International Justice," *Developing World Bioethics* (2009) 9(3): 119–127.

As already mentioned, in the nascent research of xenotransplantation, the primary goal was not for the donor organ to remain in the recipient for the remainder of her life, but to extend her life until a viable human organ became available or to allow a patient more time at home to prepare for death, as opposed to dying in the hospital. Yet, this strategy may change due to recent research developments. It may soon be possible install whole hearts of pigs into human recipients. Recently, a pig heart was successfully transplanted into a baboon and functioned for over a year. Since the genetic difference between humans and baboons is no greater than that between a pig and a baboon, the hope is that pig-to-human transplants soon allow for long-term donor organ use.

Those opposed to xenotransplantation have presented many reasons, only some of which space will permit to be described here. Perhaps the most significant worry is the possibility of viruses being transmitted via the donor's tissue to the recipient. While this is certainly a concern, despite the number of heart valve replacements into humans culled from porcine and bovine sources, no infections have been detected. Even with that said, xenotransplantation of whole animal organs raises the issue of how infectious agents will respond to a transplanted organ in humans. Presumably, the human recipient's health would be dependent in part on the health of the new organ and it may be susceptible to pathogens that normally do not adversely affect humans.[102]

Another worry is that a pathogen introduced via an animal organ may adapt to its new environment and infect human tissue as well. Evidence suggests that pathogens (particularly viruses) can come from agents moving between hosts of different species. These concerns can be boiled down to what is called *xenozoonosis*. Some scientists think via human agricultural practices there are more opportunities for pathogens to adapt and spread from species to species.[103] In fact, the US Centers for Disease Control believe most emerging diseases around the globe are zoonotic; including 11 of the past 12 causing serious health consequences.

Several moral questions are raised by xenotransplantation. Is it ethical to use animals as spare parts for humans? Does this practice use them merely as a means to human ends and is this permissible? Is it ethical to produce genetically modified pigs with human genes? How should subjects in initial medical trials be protected from adverse medical events and should they be allowed to accept such risks? Who decides whether organs should be transplanted? With possible risks of animal diseases being passed on to humans, should such research be done? And what of third party effects—what if a pathogen

102 Sparrow, "Transplantation," pp. 119–127.

103 Sparrow, "Transplantation," pp. 119–127.

is passed on to a recipient (even if he has agreed) but spreads to those who have not consented to xenotransplantation? Should governmental agencies either directly engage in xenotransplantation research or fund other researchers to do so? How should animal suffering be dealt with in xenotransplantation research?

Notice here that it isn't only the fate of the human recipient that is at question, but also those of non-human animals and human communities. In fact, some scholars claim this is one of the characteristics that separates xenotransplantation from other experimental surgical procedures. One might go further to say it is not only *some* communities that are inherently implicated in xenotransplantation, but that the *global* community is as well. How so? Scientists involved in xenotransplantation research consider the theoretical risk that introduction of living non-human tissue into humans may result in the transmission or evolution of novel pathogens, which might then spread beyond just transplant recipients.[104] The worst outcome imagined by researchers is that xenotransplantation could lead to new viral pandemics on the scale of AIDS.[105] On the heels of these concerns, some ethicists have argued that since these possible outcomes present exceptional risks to the community, some sort of community consent is required in addition to the consent of the recipient for any kind of xenotransplantation.[106]

Thus, there is a broader moral issue at stake here. And since the concern about xenozoonosis appears to be significant, we have to get at the heart of it. What are our obligations in balancing the potential health benefits of some against the potential third party risks of others? After all, those who never consent to xenotransplantation may suffer its effects through contracting a disease offered a pathway through other humans who may contract it via xenotransplantation clinical procedures.[107]

Finally, other concerns are most properly considered metaphysical questions about personal identity. Could part of the anxiety about xenotransplantation be compared to worries about cyborgs? Cyborgs are part human and part machine, causing discomfort that they are not, in some literal sense,

---

104 Sparrow, "Transplantation," pp. 119–127.

105 Sparrow, "Transplantation," pp. 119–127.

106 Abdullah S. Daar, "Animal-to-Human Transplants—A Solution or a New Problem," *Bull World Health Organization* (1999) 77: 54–61 and Martine Rothblatt, *Your Life or Mine: How Geoethics Can Resolve the Conflict Between Public and Private Interests in Xenotransplantation* (Aldershot, U.K.: Ashgate, 2004), p. 143.

107 Jonathan Hughes, "Justice and Third Party Risk: The Ethics of Xenotransplantation," *Journal of Applied Philosophy* (2007) 24(2): 151–167.

fully human when possessing mechanical and/or electronic devices that compose part of our body. Analogously, this raises the question of whether we are fully human if we are composed of (at least some) animal parts—in a way beyond just the worry of being susceptible to new diseases. Personal identity is thought of as an important part of human life—that who we really are is of upmost importance in attempting to live a meaningful life. This concern manifests itself in different ways. For some, it is a search for their origins in wanting to know who their biological parents are if they were orphaned or adopted. For others, it may be the desire to return to some ancestral homeland from which their generation has been separated. Even without such separation, the popularity of genealogical services helping people research their family trees underscores the attraction they have to discover their origins and "who they *really* are." But does having animal parts really endanger this sense of identity? It appears the way that biotechnology is progressing, we will have the opportunity to find out.

Review Questions

1) What do you think is the strongest argument in favor of xenotransplantation? What do you think is the strongest argument against this sort of medical experimentation? Ultimately, do you think xenotransplantation is morally acceptable? Why or why not?
2) One could imagine an animal right's advocate posing the following challenge. One might assume it is morally permissible to harvest vital organs from non-human animals for use in humans because human life is more valuable than non-human life. But why assume this is true? Let's imagine that someday it is medically feasible for human hearts to be transplanted into baboons. If non-human life is at least as valuable as human life, shouldn't it at least in some cases be morally permissible for human hearts to be transplanted into needy, individual members of non-human animal species? What would be the moral case against such xenotransplantation, if any? We might even make the case stronger by imaging that medically speaking it might someday be more viable to, on average, transplant human organs into other species than the converse. Would this increase the obligation to do such transplants?
3) Besides the well-known cases of xenotransplantation from pigs and baboons to humans, there have also been transplants of vital organs from some species of non-human animals to other non-human animals. The motivation for most of these transplants has been experimental—to pave

the way for human experimentation. Are there are any moral problems in conducting this sort of research? What might they be? Would there be a problem even if such transplants were not geared toward benefit to humans (i.e., perhaps benefit for some non-human species from the use of others)?

4) Note that early clinical xenotransplantation procedures were only meant to be temporary, with human replacements anticipated as they became available. Let's say this same practice remains in contemporary xenotransplantation where there are roughly equal risks and anticipated medical outcomes from either type of transplantation. Even if you agree with the moral permissibility of the use of xenotransplantation, are there any moral worries of replacing the animal organs with human ones as they become available—such as medical waste, or placing a recipient at additional risk from enduring an extra procedure?

Case 39 – Mustangs with No Home on the Range –

Study Questions

1) Why have some called for the removal of wild mustangs from public lands in America's West?
2) Consider Paula Morin's quote in the case study below. What approach to environmental ethics does it seem to reflect? How does it apply to questions of how best to manage wild horses?
3) Are horses native to North America? What makes this a complex question to answer?

There was a time in the American West when large herds of wild horses roamed vast expanses of land. But the 1900s were punctuated with a public outcry as their numbers were diminishing. Whether they are being shot at or herded by rustlers selling them for horsemeat, mustangs have endured several forms of harassment. Those lamenting the challenges facing mustangs believe these wild horses represent a beauty and untamed sensibility that deeply reflects the freedom of the land west of the Mississippi. Mustangs are, they claim, a touchstone to the Wild West, with herds extending from Montana to California.

However, after decades of declining populations, mustangs are making a comeback, to the point where some are calling for thinning their herds. Complaints about the increased number of mustangs have come from the usual quarter—ranchers who prefer cattle on the range as opposed to wild horses.[108] But even a growing number of environmentalists are concerned that wild horses have a detrimental impact on other species of plants and animals further down the food chain. For instance, mustangs are seen as disruptive to local habitats of endangered tortoises and desert birds. As Paula Morin, the author of the book *Honest Horses*, puts it, "There's not just horses out there, there's other critters, from the desert turtle in the south to the big horn sheep in the north. We've come a long way in our awareness of the web of life maintaining the whole ecology. We do the horses a disservice when we set them apart."[109]

The recent rhetoric surrounding mustangs has changed dramatically with some calling them "feral equids." Beyond this, some environmentalists advocate euthanizing mustangs if the alternatives are boarding them at taxpayer expense, leaving them to populate and damage rangeland, or simply allowing them to die of hunger or thirst. The United States Bureau of Land Management (BLM), serving as legal custodian of wild horses on public lands, also recently proposed the use of euthanasia to curb mustang herds, after years of maintaining an Adopt-a-Horse program. This program sells wild horses to those who will care for them, and was intended to be a protective measure for mustangs, which have been endangered by starvation caused by droughts.

In 2008, there were 30,000 mustangs in holding pens at BLM facilities and with rising feed and fuel prices there are fewer adoptions than ever.[110] At that time, $21 million of the agency's $37 million Wild Horse and Burro Act budget was spent on the program with significant boarding cost increases anticipated. Also, the BLM says that for ecological reasons the wild herd needed to be cut by 6,000.[111] Since that time the numbers of wild horses on the range and held in agency facilities has only *increased* to about 40,000 horses and 10,000 burros combined in pastures and corrals.[112] The BLM believes the open rangelands it manages can only sustain 26,677 mustangs and burros combined, while its land holds 40,605. According to the National Research Council, the BLM may

108 Felicity Barringer, "On the Mustang Range, a Battle Rages on Thinning the Herd," *The New York Times*, July 20, 2008, A14.

109 Barringer, "On the Mustang Range," p. A1.

110 Barringer, "On the Mustang Range," p, A14.

111 Barringer, "On the Mustang Range," p. A14.

112 Lenny Bernstein and Brady Dennis, "U.S. Looking For Ideas to Help Manage Wild-Horse Population," *The Washington Post*, January 26, 2014. Available at http://www.washingtonpost.com/national/health-science/us-looking-for-ideas-to-help-manage-wild-horse-overpopulation/2014/01/26/8cae7c96-84f2-11e3-9dd4-e7278db80d86_story.html (accessed July 13, 2016).

have even undercounted the number of mustangs by 10–50%, with horse populations increasing by 15–20% each year. Mustangs have no natural predators with the exception of mountain lions and bears, which are scarce where wild horses roam.[113]

The BLM has 179 "herd management areas" spread across 31.6 million acres of federal land. The agency believes it needs to take action by removing mustangs because they will otherwise overwhelm the available food and water resources available, as well as destroy the surrounding habitat. The BLM currently either relocates them to private pastures when space is available or keeps the remainder in corrals. But this solution leads to other problems, such as cost—with 33,608 horses to feed in pastures at $1.30 per head each day, along with 16,160 horses in other "short-term" corrals (for roughly 18 months) at four times that cost per head, this results in a big bill for US taxpayers.

The battle lines have been drawn with advocates of wild horses maintaining that mustangs have a right to stay on the range without intervention as is codified in federal law. They consider the proposition that mustangs ought to be euthanized completely unacceptable. As Virginia Purant, the former director of the American Wild Horse Preservation Campaign says, "It's not acceptable to the American public. [The mustang] is part of the American myth. People want to know that they can come to the American West and know they can see herds of wild horses roaming. It's part of the imagery."[114]

However, those concerned about leaving mustangs roaming with no management note it isn't merely the interest of wild horses at stake. As Steven Davis, emeritus professor of animal science at Oregon State University, adds, "Many of the wild horse supporters claim that the horses have a right to be there. I reject that argument. They damage the water holes. They damage the grasses, the shrubs, the bushes, causing negative consequences for all the other plants and critters that live out there."[115]

Some describe Velma Johnston, also known as "Wild Horse Annie," as being the one who almost single-handedly pushed Congress to pass the Wild Free-Roaming Horses and Burros Act (WF-RHB Act) of 1971.[116] Her determination came in the face of stiff odds for mustangs at the time, which were being poisoned, abused, and slaughtered. There was a real concern that mustangs could be wiped out. But no one concerned about wild horses then could believe that the situation could be so different in just under a half-century later. Authorities

113 Bernstein and Dennis, "U.S. Looking For Ideas."

114 Bernstein and Dennis, "U.S. Looking For Ideas."

115 Bernstein and Dennis, "U.S. Looking For Ideas."

116 Bernstein and Dennis, "U.S. Looking For Ideas."

are now not sure what to do with so many mustangs roaming over such a large area of the western US.

According to the WF-RHB Act, the federal government is responsible for more than 40,000 mustangs over 10 states.[117] The wild horses compete with cattle and wildlife for scarce water resources due to long spells without rain and for forage that has been burned away due to increased spread and severity of wildfires. While the Adopt-a-Horse program seemed to be a good idea initially, the public has been tepid about taking on the care and responsibilities of mustang ownership. Contraceptive efforts have been limited in their effectiveness. Mustangs can't be slaughtered under federal law for any purpose including human consumption, nor can they be sold to outside interests who might do so. Of course, though wild horse advocates simply want them to be left on the land, this leads to higher numbers of mustangs and more animosity towards the animals from ranchers as well as greater worries about degraded ecosystems from wildlife ecologists. The specific case for the aggressive management of wild horses comes due to their grazing habits. They have been accused of destructive denuding of the landscape, drawing comparisons to invasive species, outcompeting deer and other foragers for food and water.

But mustang advocates claim the BLM is falling down on its duty to protect the species not only by removing its members from their native environment, but by doing so with cruel means. They cite cases of low-flying helicopters chasing panicked mustangs; sometimes injuring them. Wild horse advocates also charge that mustangs have been treated inhumanely during roundups.[118] Bruce Wegmen, an attorney who represents a wide range of animal interest groups, says the BLM's approach to mustang management violates the intent of the WF-RHB Act, "They've been doing the wrong thing since day one. Instead of protecting and preserving them, they're doing the opposite."[119]

While mustangs engage the interest of many people and represent American values to many, their origin is somewhat mysterious. While it is said the early Spanish conquistadors brought them to North America, the story is much more complex. More than 50 million years ago, a small animal about the size of a fox fed on fruits and leaves. Its body was only about a foot high at the shoulder, had a long tail, as well as a short-snouted head that likely made it look more like a dog than a horse. Its feet most likely possessed pads, but with toes that ended with a hoof instead of a claw. In modern horses, one of the toes evolved into a larger hoof and the remainder of the toes have evolved as

117 Bernstein and Dennis, "U.S. Looking For Ideas."

118 Bernstein and Dennis, "U.S. Looking For Ideas."

119 Bernstein and Dennis, "U.S. Looking For Ideas."

vestigial bumps further up the leg. Fossil hunters first found evidence of this creature about a century ago, dubbing it *eohippus* or "the dawn horse." They thought this animal was the first link in an evolutionary chain leading directly to the horses we see now.

But while researchers believe the modern horse evolved from the eohippus, it wasn't by a direct route. Paleontologists have discovered fossils demonstrating that horse ancestors varied in size. Some were originally larger species that gave way to smaller ones. The other interesting discovery was that some lines of horse-like animals had different numbers of toes over time. One early horse-like species once thought to be the direct ancestor of the modern horse was later found to be a distant cousin. Another branch of what could be thought of as a species in the horse family kept growing and roughly one million years ago produced an array of small pony-sized animals found on ancient plains worldwide in large herds. These ancestors very likely looked and acted much like mustangs do on the range today.

However, less than 10,000 years ago, many of these horse-like species went extinct in the same major era of extinction that removed many other browsing animal species, including the mammoth. A combination of climate change and hunting are possible causes for its extinction. The only surviving browsing species were the horse-like lines of Asia, as well as zebras, with horses wiped out in North America. Historians believe the more recent ancestors of wild horses currently on the American plains are descended from those brought by the Spaniards in their New World voyages of the 1500s.[120] They, of course, multiplied quickly, sometimes escaping captivity and living in the rangelands of the US. The word "mustang" is derived from the Spanish word "mustengo" meaning "ownerless beast" or "stray horse."

This foray into the genealogical history of mustangs raises interesting questions. Is it correct to say that wild horses are "native" to North America? Does this provide some reason for thinking mustangs have a right to stay on the range without intervention—by saying they are "natural" as some of their advocates suggest? What does this say about re-introduced species? Does it matter whether they were reintroduced and does this give a good reason for either advocating that they "run wild" or for killing or removing them to control their populations? These questions are not just theoretically interesting. Note that laws designed to protect wild horses have appealed to the natural

120 J. Edward de Steiguer, *Wild Horses of the West: History and Politics of America's Mustangs* (Tucson, AZ: University of Arizona Press, 2011).

standing of mustangs to justify such protection. For example, the WF-RHB Act recognizes the wild horse as an "integral component of the natural system."[121]

In a final, interesting twist to this debate is a possible argument advocating leaving wild horses on the plains, but from an anthropocentric position. Some wild horse advocates make their case *without* claiming that wild horses are indigenous to the American West or even by privileging non-human animal concerns over human interests. Instead, they note that mustangs are here now and have to be taken into account. These advocates also disagree with the conventional notion that mustangs are destructive to the environment, using the counter that wild horses are *part* of the environment, not separated from it.[122] While it may be true that mustangs alter the landscape, their removal also alters it. With respect to the argument that reducing the number of mustangs enhances the pleasure of the species as a whole, these advocates might say it is unethical to protect them (especially as they are not listed as endangered) if the only reason for doing so is because we like them and hold them up for preferential treatment. The policy should be consistent for *all* animals, whether they are horses or coyotes. *None* of these animals should be removed unless they are causing an imminent decrease in human well-being. Since this doesn't seem to be the case, they should be left alone.[123]

Review Questions

1) As noted, some environmentalists believe euthanasia is permissible if keeping mustangs is of great expense to the public, since they destroy the habitats of other species of plants and animals or will overpopulate and perhaps die of hunger and thirst. What is the approach to environmental ethics accepted here?
2) Is it morally right to euthanize members of one species of animal for the sake of others (or many others)? What is the justification for your position?
3) One method of mustang population control would be by opening a hunting season on them. Traditionally, there has not been such a season,

---

121 American Wild Horse Preservation, "Wild Horses and the Ecosystem," (website), 2015. Available at http://wildhorsepreservation.org/wild-horses-and-ecosystem (accessed July 13, 2016).

122 Chuck Klosterman, "Should We Protect Wild Horses?" *The New York Times Magazine*, March 21, 2014. Available at http://www.nytimes.com/2014/03/23/magazine/should-we-protect-wild-horses.html?_r=0 (accessed June 30, 2015).

123 Klosterman, "Should We Protect Wild Horses?".

however wild horses have been killed illegally for sport. Is there anything wrong with using hunting (whether for food or sport) as a way of controlling animal populations? If so, why and should there be a hunting season on mustangs? If not, why not, and what objections could be raised to your position?

4) Do you think the BLM should turn to euthanizing mustangs or should they continue the Adopt-a-Horse program? Do you think the BLM should use both strategies? Or should they have no direct role in managing mustangs?
5) What do you think about the argument that since wild horses are natural to the American West and imbue the traits of freedom, beauty, and grace that they should be left alone? Is one part of the argument—the notion that mustangs carry certain desirable characteristics or that they are a natural part of the West—a stronger reason for advocating leaving them on the land?

CHAPTER 10

Think Globally, Act Locally

Case 40 – Community Supported Agriculture –

Study Questions

1) What is the definition of Community Supported Agriculture (CSA)? What is the business model it uses and what are the motivations for farmers to create CSA farms and for consumers to invest in them?
2) What are supposed to be some of the benefits of CSAs? What are some of their drawbacks?
3) What are some of the moral values proponents of CSAs hold? Are these values different from those who advocate for more conventional forms of farming? Explain the similarities and differences in the moral values held by CSA and conventional farming advocates.

Advocates define Community Supported Agriculture (CSA) as simply "a community-based organization of growers and consumers."[1] Yet, since their introduction into the US from Europe and Japan in the mid-1980's, CSAs have become considered pragmatic "saviors" of the ideal of the smaller, rural farm. As a system of mutual support, groups of consumers agree to buy shares of produce from local farmers and pay lump sums at the beginning of each growing season. Farmers reciprocate by supplying agricultural goods to these consumers. Transparency and sustainability are taken to be two basic virtues of CSA schemes. Consumers know directly the origin of the produce (often grown organically or biodynamically)[2] and are promised it is fresh, safe, and naturally grown. In return, farmers are guaranteed the financial support necessary to

1 Trauger Groh and Steven McFadden, *Farms of Tomorrow Revisited: Community Supported Farms, Farm Supported Communities* (Kimberton, PA: Biodynamic Farming and Gardening Association, 1997). Community Supported Agriculture farms are referred to as "CSAs."

2 By *biodynamic agriculture*, I mean a farming method that strives to build and maintain soil fertility through the use of manures and composts from the farm. See Dorothy Suput, *Community Supported Agriculture in Massachusetts: Status, Benefits and Barriers*, Master's Thesis. (Medford, MA: Tufts University, 1992), p. 18. Or as stated in the Brookfield Farm CSA brochure, "Through a balanced crop rotation and application of bio-dynamic herbal preparations the farmer regulates the composting process and stimulates the life activity of the soils and plant[s]."

 | DOI 10.1163/9789004339996_012

operate since they receive financing for the year's crop production "up front," enabling them to produce a wide variety of crops otherwise not economically viable to grow.

CSAs are part of both the "sustainable agriculture" and "locavore" movements. Defining what locavore means is relatively easy—eating food grown close by. But what does *sustainability* mean? Devising a precise definition is more problematic, since there are several competing candidates both in content and scope for what sustainability means in the agricultural context.[3] One helpful way of defining it is in an economic and ecological sense. Economically speaking, one important variable is whether CSA farms can last over time, which requires some minimal level of economic viability. Ecologically speaking, we need to know whether CSA farms have a dramatic impact on maintaining and improving the environmental conditions of the land. This latter factor is significant, since agriculture is inherently *disruptive* to the natural environment.[4] It is naïve to say cultivation of any sort does not change the natural order of the ecosystem to which it is attached. The question is how to create an agriculture that even in its alteration of the environment does so in a way optimally "in sync" with the natural surroundings.

To fully flesh out this notion of sustainability in these twin economic and ecological senses, we might put together two different sets of criteria. The ecological efficiency strand is borrowed from environmental ethicist William Aiken's work on eco-humanistic agriculture.[5] First, we examine how the land is being used, particularly since in contemporary agriculture there is a finite and rapidly dwindling supply of arable land, even in America, which often serves as a paradigm case for agrarian abundance.[6] Secondly, the sustainability of agriculture is based on *local adaptation*.[7] This encompasses what is grown, how it is grown, how much of it is grown, as well as how these factors relate to the peculiarities of the local ecosystem. Third, any analysis of ecological efficiency

3 For a look at conflicting definitions of sustainability in agriculture, see William Aiken, "Ethical Issues in Agriculture," in Tom Regan, ed. *Earthbound* (New York: Random House, 1984), p. 277 and also Carl D. Esbjornson, "Once and Future Farming: Some Meditations on the Historical and Cultural Roots of Sustainable Agriculture in the United States," *Agriculture and Human Values* (1992) 9(3): 20–30 at 20.

4 Aiken, "Ethical Issues," p. 277.

5 Aiken, "Ethical Issues," p. 277.

6 Donald Worster notes the tradition of this vision of bounty, for even the US Bureau of Soils proclaimed (in 1909) that soil "is the one resource that cannot be exhausted; that cannot be used up." See his *Nature's Economy: A History of Ecological Ideas* (Cambridge, U.K.: Cambridge University Press, 1985), pp. 28–29.

7 Aiken, "Ethical Issues," p. 278.

(just as in the case of economic efficiency) must pay close attention to whether the agriculture in question is viable over a long period of time.[8] This means that whichever agricultural method is used has a regenerative effect on the environment. The agricultural method meets this specification if it replaces extracted resources in an ecologically safe way and replenishes the land. For example, fallowing, a common regenerative agricultural technique, can be supplemented by adding input if it is readily available and effective over a long period of time.[9]

In regard to sustainability in the economic sense, the issue is trickier. The CSA concept cannot match the massive scale of industrial agriculture in its ability to produce total output or overall efficiency.[10] Inherently, CSAs will maintain their character only by keeping a manageable size and they cannot serve a huge population of consumers as can corporate farming. Also, one goal of CSAs is to retain high quality of land and food, which means they are compelled to use methods of crop rotation that (sometimes) severely lessen the total yield of their harvests. Therefore, it would not be a valid comparison to examine the total yields of these two sorts of agriculture. Even if we were to examine the ratio of inputs to outputs and found industrial agriculture to be superior in efficiency,[11] it would not be clear CSAs are not preferable. Several outputs of CSA farming are not concrete but still important (i.e., such as quality of life and increased leisure time) and we would be asked to measure some outputs (satisfaction or quality of life) that are incommensurable to normal output indicators (such as pounds of produce).

However, CSAs are "internally efficient." They are (by and large) able to meet their budgets and produce yields greater than their inputs of hours worked. Nonetheless, in the short term, CSAs will provide little competition to corporate farming in relation to overall efficiency in serving large numbers of consumers, even if thousands of small to mid-sized CSAs were to spring up overnight.

---

8 Aiken, "Ethical Issues," p. 279.

9 An example of an input that would have a low degree of acceptability would be nonrenewable fossil fuels.

10 As agricultural ethicist Richard Conviser observes, "In terms of one of the goals of that mode of organization [large-scale corporate farming], maximizing short-term productivity, modern agriculture is indeed nonpareil." See his "Toward Agricultures of Context," *Environmental Ethics* (1984) 6: 71–85 at 71.

11 Given the capital-intensive nature of large-scale industrial farming, large-scale corporate farms will often score better on the efficiency scale if it is a technical, straightforward measure of the ratio of inputs to outputs. Corporate farms can simply buy products (such as seed and fertilizer), land, and machinery in large-scale quantities accounting for this advantage in efficiency.

The CSA concept is spreading. In December 2004, there were over 1,300 CSAs in the US and Canada. This was a slight increase from January 1999, when that number was estimated to be just over 1,000.[12] CSAs in the US alone involved 100,000 households in 2004, almost all of which buy shares of produce in CSAs. However, some households counted in this figure, instead of purchasing shares, supply part-time farm work or technical support (i.e., designing web pages and brochures; organizing community events and workshops, etc.) to their associated farms.[13] The number of CSA farms rose to 12,549 in the US alone by 2007 and leveled off by 2012 with 12,617, with all states represented.[14] While the scale of yield and the number of stakeholders fed by individual farms fluctuate according to each individual CSA's size and the length of growing season, the typical farm (5–10 acres) is able to produce enough food for 250–500 persons for an average of nine months of the year. A particularly good case study due to the detail of its budget and efficiency analysis, the Indian Line Farm in Massachusetts, feeds 300 people for 43 weeks of the year.[15]

Of course, CSAs are not replicants from some institutional design genetics lab. While the basic financing structure is similar in all CSAs, modes of payment for shares of produce can differ. As mentioned, instead of paying for the share in full at the beginning of the season, some shareholders offer compensation in non-monetary ways by working the farm or aiding in the distribution of shares.[16] While almost all CSAs utilize alternative methods of farming, such as biodynamic and organic approaches, nothing inherent in the CSA farming concept precludes conventional farming techniques.[17] Also, one farm may simply be associated with a few families (sort of an independent CSA) without any

12 Cathy Roth, "The Concept of Community Supported Agriculture," (website), 1999. Available at www.umass.edu/umext/csa/about.html (accessed December 22, 2004).

13 Groh and McFadden, *Farms of Tomorrow Revisited*, p. xvi.

14 United States Department of Agriculture, "Community Supported Agriculture," (website), 2014. Available at http://www.nal.usda.gov/afsic/pubs/csa/csa.shtml (accessed April 27, 2015).

15 Robyn Van En, *Basic Formula to Create Community Supported Agriculture* (Great Barrington, MA: Indian Line Farm, 1992), p. 1.

16 By "aiding in the distribution of the shares," I mean that while most CSAs have a drop-off point just outside of the community it serves, or merely have shareholders pick up their shares of produce at the farm, some CSAs deliver the shares directly to member households and organic grocers. This direct drop-off option can be a cumbersome task and some shareholders can "pay for" part of their shares by delivering weekly shares to other members. Interns and apprentices with appropriate transportation also may provide these services.

17 Suput, *Community Based Agriculture in Massachusetts*, p. 60.

substantive association with any other CSAs and some CSAs cluster as a small group not formally linked into a larger organization.[18] CSA farmers exercise this second option for the purposes of covering costs not only for the basic supplies of agriculture, but also for their newsletters updating farm news, the status of crops, workshops on new farming techniques, and social and cultural events such as potlucks and fairs.[19] These seemingly minor details are actually quite significant (as we will discover later). Special problems can arise depending upon the size of the CSA and whether it is involved in associative networks.

CSAs may vary greatly with respect to acreage planted and the amount of produce their harvests yield. CSAs in the US are, in the main, modest in size. Yet, the scale of yield varies depending upon region. For instance, larger-scale CSA operations predominate in the western US where traditionally larger tracts of land have been cultivated—at least more so than in the Midwest, and particularly the Northeast.[20] By 1996 in California, the typical CSA involved 500–700 members, yet according to some local CSA farmers and researchers, including Jered Lawson, some sought to expand to 2,000 shares. This 200% plus increase in shares per individual farm leads to results that Lawson admits are unnerving and counter-intuitive to the CSA ideal.

> I became a bit disillusioned by the reality that most of the California farms initiating CSA were not doing it in the same way with the same intention [of the original CSA concept]. They weren't looking to cover the farm's whole budget, or even a little percentage of it through CSA. CSA was arising as more of an add-on concept for forty to eighty to 100-acre organic farms. They already had a large part of their production and distribution geared toward the wholesale market, toward farmer's markets, or toward other forms of direct-to-consumer marketing.[21]

However, smaller-sized CSAs are the prevalent form in most of the US and involve comparatively lesser parcels of land (5–10 acres) and a significantly smaller number (20–100) of member households own shares in the harvest.[22]

18 Marcia Ostrom, "Community Farm Coalitions," in Trauger Groh and Steven McFadden, eds. *Farms of Tomorrow Revisited: Community Supported Farms—Farm Supported Communities* (Kimberton, PA: Biodynamic and Gardening Association, 1997).

19 Suput, *Community Based Agriculture in Massachusetts*, p. 25.

20 See Steven McFadden, "Large Scale Community Farms," in Trauger Groh and Steven McFadden, eds. *Farms of Tomorrow Revisited: Community Supported Farms—Farm Supported Communities* (Kimberton, PA: Biodynamic and Gardening Association, 1997), p. 82.

21 McFadden, "Large Scale Community Farms." p. 83.

22 McFadden, "Large Scale Community Farms." p. 82.

There is little evidence to support the claim that these smaller-scale farms in the non-western sections of the US have actually increased, or intend to increase, significantly in size with respect to the above indices. Most CSAs that have survived within the past few decades have only slightly increased their number of shares, having accurately calibrated the number they are able to supply at current production levels.[23]

A uniquely American phenomenon in the evolution of CSAs as opposed to their European and Asian counterparts is that many have (as alluded to earlier) formed associations of farms to perpetuate the CSA concept and ease administrative burdens. Two prime examples of these associative activities hail from the Upper Midwest including the Madison Area Community Supported Agricultural Coalition (MACSAC) and the Minnesota-Western Wisconsin Community Farm Association (MWCFA).[24] Together in an umbrella group called FairShare, they have 59 CSAs (there are 65 CSA farms total) in Wisconsin serving over 3,000 households. In Madison alone, it is estimated that CSAs serve over 25,000 residents.

These multi-farm associations are popular as tools to pool resources to effectively handle communication and educational capacities that are beyond the scope of individual CSA farms. The motivation for joining these associations is illustrated rather succinctly by one local MACSAC farmer who quipped, "... having been in a position where I was the only one who was doing it, I know for a fact that it is a lot easier to promote the concept of CSA as part of an organization than as an individual."[25] While individual CSA farms still supply crops to their own shareholder households, the associations named here have provided a sense of security in numbers; farmers can and have supplemented each others' produce in times of economic and/or climatic distress.[26]

---

23 It is also interesting to note that several anecdotal instances suggest CSA farmers prefer smaller burdens of yields and hence, only produce the number of shares that afford them a basic salary and sufficient funds to cover expenses. Instead of engaging in farming for profit, quality of life is an important element for CSA farmers. They tend to believe that "full-time farming" impinges upon quality of life. See Bob and Bonnie Gregson, *Rebirth of the Small Family Farm: A Handbook for Starting a Successful Organic Farm Based on the Community Supported Agriculture Concept* (Vashon Island, WA: IMF Associates, 1996), pgs. 4 & 17.

24 The descriptive account of MACSAC and MWCFA is borrowed heavily from the work of environmental studies researcher Marcia Ostrom.

25 Ostrom, *Community Farm Coalitions*, p. 91.

26 One instance indicative of this point comes from the first year of MACSAC's existence (1993), when a costly flood destroyed several crops and tested the commitment of the group. Farms able to do so supplied fellow member farms harder hit by the floods.

The socio-economic profile of the "typical" CSA farmer is that, generally speaking, he/she is well-educated and from a middle-class background. Often the impetus to choose farming is a lifestyle change to a more pastoral ideal of living the "good life" with nature. Most CSA farmers wish to sustain themselves in an environmentally responsible way. They believe they can weave skills acquired in the realm of academe or business into a more simplistic existence. The technical skills of some CSA farmers shatter the stereotype of the technophobic planter who eschews modern ways—many CSAs have websites, e-mail, and use social media to promote their farms and educate their subscribers. Many of the originators of the CSA concept in the US earned degrees from university programs specializing in alternative agriculture (one common one being Emerson College in the UK). Blending theoretical understanding of science, ecology, plant biology, and marketing can help supplement the practical knowledge of viable agricultural techniques acquired through "hands-on" experimentation on the farm. If nothing else, alternative agriculture is knowledge intensive and requires a firm grasp of both agricultural theory and practice.

The profiles of shareholders in CSA's often echo those of its farmers. Generally speaking, they have attended college, are environmentally conscious, and live within mid-sized to large metropolitan areas.[27] As former director of Community Supported Agriculture of North America, Robyn Van En describes, "Sharers tend to be focused on the shortcomings of our contemporary food system. They are people curious about new, different, and organic vegetables. Many are home gardeners who are aware of the work and responsibility and choose to relieve themselves of growing for their own families while keeping productive farms or gardens in their community."[28] Like their farmer counterparts, they also tend to be from middle class economic backgrounds.

The catalyst for consumers and farmers organizing into CSAs, however, is deeper than merely securing healthy produce and guaranteed markets respectively.[29] CSAs are an attempt to exert greater control over food production, distribution, and safety. It may initially seem odd that food production should be a locus of democratic participation. Unlike a positive right mandated

---

27 Large concentrations of CSAs have sprung up around San Francisco, Boston, Minneapolis, Madison, and Seattle. These also tend to be locations with several academic institutions.

28 Van En, *Basic Formula*, p. 57.

29 Even the promotion of CSAs to the general public by one of its largest and most powerful organizations (Consumer Supported Agriculture of North America—CSANA) is featured as involving more than fresh produce. For example, one CSA start-up video is entitled *It's Not Just about the Vegetables*. On this point, see Van En, *Basic Formula*, p. 2.

by the US Constitution that is meant to uphold equality of opportunity such as the right to a basic education regardless of economic status, there seems to be no substantive right to sustainable living demanding equity of voice in how our food is produced and distributed. Likewise, there is no corresponding "right to farm" in the US, demonstrating the agrarian ideal of the citizen farmer is not as thoroughgoing in the framework of US constitutional law as one might expect.[30]

While there may not be official public forums facilitating discussion of food production in the US, there is an increasingly vociferous call for direct participation. For consumers, the increasingly heated debate about the desirability of genetically modified fruits and vegetables, the distance of food production from its final destination, and particularly, the environmental damage of corporate farming, have led concerned citizens to demand greater voice and effective control over the foods they consume. Direct democratic participation in food production decision-making is imperative for CSA farmers. Smaller-scale and family farmers believe they have legitimate fear their control over food production is continuously eroding. This worry is aggravated by aggressive acquisition of smaller agribusiness companies by agribusiness behemoths (such as Monsanto) that are gaining monopolistic holds over the seed, herbicide, and pesticide markets.

Most CSAs have mission statements or charters ranging from vague goals and guidelines to a fairly complex set of rules governing the proper conduct of CSAs. For instance, MACSAC established criteria for legitimate CSAs to enter their association. Unlike most CSA associations, MACSAC has required proof that entering farms are organic or biodynamic, offer shareholders a wide variety of produce over the growing season,[31] demonstrate that its farmers have organic farming experience, and possess a proper lease or sufficient capital to invest in land.[32]

30 Paul Thompson, "Agrarianism and the American Philosophical Tradition," *Agriculture and Human Values* (Winter 1990): 3–6 at 6.

31 As an example, herb farms cannot, by themselves, enter MACSAC because they do not meet the sustainability requirement set by this organization that stipulates CSAs must supply a wide variety of produce. However, herb farmers are able to join with other "vegetable-intensive" farms in a collaborative effort to offer a wider range of produce and join the association in this capacity.

32 This final requirement is not as exclusive as it may seem, since the pledge of support by subscribers to the applying farm can be used as documentation that it has adequate financing to gain entry into MACSAC. MWCFA does not have a set list of criteria for admission to their organization. See Ostrom, *Community Farm Coalitions*, pp. 93–94.

To set these limits, however, associations of CSAs have to have a political structure and be directly participatory. While in some cases individual farmers have drawn up mission statements as a conceptual basis upon which to build their CSAs, often their "constitutions" are a deliberative effort, with all willing members developing them. Many CSAs do not keep careful records of "votes" on particular farm issues. The process usually involves low enough numbers of members in small CSAs to establish "consensus based" decision-making processes.

In general, these deliberative meetings occur weekly during growing season, at least among the members of the core group. It is usual practice for at least one of the gardeners to attend as well as the administrators and distribution members.[33] While members are welcome to attend all meetings, they are not required to do so. At the weekly meetings, as Van En explains from her own experience at the Indian Line Farm: "The status of the budget, existing and outstanding, is discussed ... and is of the utmost importance of the entire group."[34] Any budgetary changes or other major issues brought on by low harvest yield are announced in the community newsletter. If the newsletter cannot be distributed in time, all members are e-mailed or phoned to inform them about meetings with significant happenings. Meetings decrease in frequency during the winter, with the core members gathering only once a month.[35]

Weekly gatherings during growing season do not involve more important business unless there is a dire shortness of farm help, the need to make time-sensitive amendments to the budget, or if consistent inclement weather or infestation has adversely affected the crop. The yearly meeting, however, is of greater significance since this is the time when the yearly budget and the share prices are set. Other duties are decided at the annual meeting, such as the bookkeeping for the year (usually this chore is divided and rotated, depending upon the amount of work to be done).

Some positions in CSAs can require substantial time commitments. For example, significant time is spent on the telephone by those in the distribution head of the CSA, explaining recent plans and taking phone requests from potential future members (including those inquiring about internships) for information on the CSA group. Since the distribution head is part of the harvest crew and core group, they are the most aware of the concerns and can best educate the sharers about the most recent harvests.

---

33 Van En, *Basic Formula*, p. 14.

34 Van En, *Basic Formula*, p. 15.

35 Van En, *Basic Formula*, p. 7.

We have a rather detailed account of how CSAs function. Yet, description without normative justification for any major decision-making group alleged to provide moral benefit is not especially helpful. Is there an argument for why we should endorse sustainable agriculture generally (for example, is it inherently valuable or so instrumentally valuable that it is worth doing)? Are CSAs good devices for sustainable agriculture? To answer these questions, we need to reflect on two key principles; a sense of community in agriculture and the principle of direct, participatory democracy.

It *may be exceedingly difficult to defend the intrinsic value* of community and direct participatory democracy here, but CSA advocates *can likely show the instrumental value of each.* CSA proponents endorse the inherent value of community and several arguments may justify this position. The difficulty with the main argument used to establish the intrinsic value of community in the agricultural context is that it ultimately depends on a more instrumental Jeffersonian agrarianism as its foundation.

However, the intrinsic value of *sustainability* might be shown and other instrumental values such as community and direct, democratic participation might uphold it in the agricultural context. To see why sustainability is such an important normative goal, we have to see what the criticisms are of a competing approach to justifying small-scale and alternative agriculture—Jeffersonian agrarianism. While Jeffersonian agrarianism itself may not be strong enough to justify sustainability as a moral value, it still points to an argument for it defense.

But first, we have to delve a bit deeper into environmental ethics. In the American philosophical tradition, there are two contradictory ethical approaches to the environment and particularly agriculture's relation to it. One view, that I'll call the *anthropocentric view*, is a bit confusing since there are two different types of it, a utilitarian and a libertarian version. The competing approach I'll call the *Jeffersonian view*.

Many contemporary environmental ethicists regard the anthropocentric view as a pernicious exponent of unchecked property rights supported by libertarian and utilitarian principles. The general approach is anthropocentric in that humans have a basic right to appropriate and dispense with the environment and its bounty as they like. In the utilitarian thread of this standpoint, if the consequences of land use for development are more beneficial to the greatest number of members of a society than not, this value trumps all competing interests. With the cost-benefit analysis complete, society is obliged to extract resources to the point where it maximizes the good of the moral community.

From the libertarian anthropocentric view, even the principle of the "greatest good for the greatest number" is too rigid as a constraint on land use. Utilitarians might forgo greater short-term utility gains if they risk greater *long-term* costs, and thus place development restrictions on land use. But libertarians think that *any* restriction of property rights (since they regard them as absolute) is morally impermissible. Hence, the general libertarian stance would allow for short-term profiteering with high yields even if they cause irrevocably disastrous ecological devastation.

Critics of the anthropocentric view who want to show its failures simply point to American history, citing Manifest Destiny, the great deforestation of America, and the unabated game hunting that led to the establishment of America's National Park System. They add that libertarian and utilitarian land-use policies have continued largely ceaselessly in the form of present day corporate agriculture (with the exception of some environmental regulation such as the Clean Air and Clear Water Acts). Chuck Thesig, editor of *The Farm Journal* exemplifies the anthropocentric view when he sarcastically derides sustainable agriculture as a term that "makes my teeth hurt" and adds that the only sustainable agriculture is one that is profitable.[36]

For an opposing view of the relations between the environment and agriculture within the American philosophical tradition, we need look no further than to Thomas Jefferson.[37] His brand of agrarianism is often invoked by those longing for the idyllic, bucolic lifestyle of the "citizen-farmer" peacefully tending to the land. Metaphorical language of the American transcendentalists Ralph Waldo Emerson and Henry David Thoreau create vivid depictions of the inherent value of land preservation and the pastoral life. In the case of Emerson, philosopher and agricultural economist Paul Thompson tells us that the poet/philosopher was so taken by this ideal that he "built American transcendentalism on an agrarian foundation."[38] Yet, according to Jefferson's own admission, agrarianism, while an extremely valuable vision, is not inherently valuable.

Despite what Jefferson said, many contemporary environmentalists (including CSA advocates) argue Jefferson offers a defense of the ideal of actively engaged agrarianism as a first-order, intrinsic good. Environmental thinkers and writers such as Wendell Berry and Wes Jackson cite Jefferson's "ideal"

36 "What is Sustainable Agriculture?" *The Land Stewardship Newsletter* (1989) 7(2): 9.

37 Certainly Jefferson did not always compare favorably to his ideals. Jefferson's own plantation reflected practices closer to those found in expansive corporate agriculture. The description here of Jeffersonian agrarianism focuses on his theoretical conception of agriculture rather than his actual behavior.

38 "Agrarianism," p. 3.

as a "best bet" for a democratic, small-scale, community-oriented, sustainable agriculture.[39] Other modern environmentalists (most notably James Montmarquet) forcefully use what they think is Jefferson's belief in the inherent goodness of the pastoral lifestyle in their arguments against large-scale agribusiness. The Jeffersonian view on the intrinsic quality of agrarian community is likely attractive since it offers a strong tie between democracy and communitarian ideals, and simultaneously provides a principled defense of the environment.

While powerful bonds exist between Jefferson's veneration of agrarianism and democracy, it may be fruitless and disingenuous to say that this is because of a *Jeffersonian* belief that *sustainability* or even the agrarian lifestyle is intrinsically valuable. Thompson supplies three reasons why Jeffersonian agrarianism does not support sustainability as an *intrinsic* value. First of all, Jefferson's writing provides no indication there obtains an intrinsic right to farm. In a 1785 letter to James Madison, Jefferson describes land as "a common stock for man to live and labor on." Yet, even Montmarquet must admit that this definition, if it can be called one, is as he says, "Jefferson's reaction to the inequalities in France at the time."[40] The value of farming is not absolute or inherent, but contingent; located in the events that would aid in promoting a revolutionary agenda. Also, there is no right to farm in the Bill of Rights and little evidence to suggest Jefferson was so adamant about including it that he attempted to block adoption of the Bill without such a right.

Secondly, Jefferson's commitment to farming *originated* from his finding it to be an *instrumental* value.[41] The value of farming derives from its ability to serve democracy. Farmers are, in the eyes of Jefferson, virtuous because their "means of making a living [is one] that aligns private and public interest." This comes in contrast to the usual view imposed on Jefferson's writings that farming is somehow intrinsically virtuous or that farm families imbue moral perfection. The thought is more that farming promotes democratic virtues including independence and self-reliance, while still attending to the needs of the common good. In sum, farming supplies some of the most crucial moral values underpinning democracy, though only instrumentally.

---

39 See Berry, *The Unsettling of America: Culture and Agriculture* (San Francisco: North Point Press, 1977), pgs. 143–169 and Jackson, "Making Sustainable Agriculture Work," in Robert Clark, ed. *Our Sustainable Table* (San Francisco: North Point Press, 1990), pp. 132–141 at 134.

40 See his *The Idea of Agrarianism: From Hunter-Gatherer to Agrarian Radical in Western Culture* (Moscow, ID, University of Idaho Press, 1989), pp. 88–89.

41 Thompson, "Agrarianism," p. 5.

The final reason that we cannot read Jefferson as an exponent of agrarianism as an inherent good is that he places too much emphasis on the formation of a *viable* democratic state. Jefferson's *philosophical* convictions reveal an unflinching devotion to democracy and the common good. Theoretical issues of both statecraft and constitutional choice dominated his thought. Small farms fit into Jefferson's overarching view of democracy in an "emerging nation because they promised a temporary respite from the age old dilemmas of democracy."[42]

But just as we have good reason to amend the Jeffersonian "ideal" in the face of evidence demonstrating the instrumental desirability of smaller farms and the agrarian lifestyle, we may need to rethink the domain of democratic participation in which he believed. Jefferson's claims that "cultivators of the earth" are inclined to hold privileged places in democratic forums or that their lifestyle effectively serves the revolutionary virtue of democratic citizenship may seem too grandiose. It isn't clear that bucolic lifestyles serve as better "schools for democracy" than other sorts of lifestyles. On the contrary, skills and virtues of farm life may very well steer its practitioners towards isolation and parochial attitudes that atomize rather than unify citizens in civic arenas.

Even so, Jeffersonian agrarianism is useful in building the case for linking the values of sustainability, small-scale agriculture, community, and democracy. The spirit of Jefferson's thought has force if we consider that even the *instrumental value* of small-farms expands democratic participation *within* certain areas of life (such as food production and consumption) if properly organized. In turn, direct democratic deliberation is important to sustainability because it allows for persons to have control, or at least a vote, on matters of vital interest to them. With respect to control over their own food, and their relations to it, participants can determine an agriculture that will significantly serve their private interests.

Yet the sense of community is also important to sustainability (instrumentally at least) since it is not enough to have the power to make important decisions concerning food, but there must also be "ties that bind." Sustainability requires reflection on the interests of others, determining the common ideals members of these smaller face-to-face associations tend to have. Deliberations in these associations demand consideration for others and the public interest. If the goal of CSAs is to bring farmers and consumers together in ways that empower them to gain greater voice and control over food production they

---

42 Again this involves the conflict between private and public interests. See Thompson, "Agrarianism," p. 5.

desire, then it may serve as a promissory note for the democratic ideal in this area of their lives. We only need to ask if CSAs indeed deliver on this promise.

For all of this, the hope is that both ecological and human interests can be served, but human interests will ultimately win out when the two clash even on this view. However, ecological disruption would still only be justified in promoting the *vital interests* of humans, serving only essential human needs. These needs can be understood as those ensuring survival. As James Heffernan tells us:

> The survival interests of human beings ought to outweigh those of the rest of biotic community and the survival interests of the rest of the biotic community ought to outweigh the non-survival interests of human beings.[43]

This blends a sort of anthropocentric humanism with environmental integrity. Merely because human interests take precedence in this approach to the value of sustainability does not necessarily mean ecological interests are completely disregarded. Strong reasons are necessary to justify merely *some* destruction of the environment. "[E]nvironmental preservation and resource sustainability are goals that should be sought for human benefit" combines seamlessly with the eco-centric belief that "ecosystem integrity ... contributes to the long term interests of humans."[44]

One might still wonder what it is about human nature that privileges it over the countervailing interests of non-human animals? This is not easy to answer. Indeed, some non-human animals may carry an attribute that is supposed to be possessed only by humans, and some humans may, in turn, lack this essential characteristic. Yet, as Aiken stresses, one trait that may distinguish human from non-human animals is *self-conscious creativity*, which he calls the "attribute which grounds special significance to vital human interests and ultimately grounds rights claims in our case."[45]

On this view, while sentience[46] certainly is of significant value and is possessed by animals, since it is in a creature's best interests to avoid suffering, the status of creatures having self-conscious creativity (humans) would still be more significant. Indeed, self-conscious creativity entails an awareness of

---

43 "The Land Ethic: A Critical Appraisal," *Environmental Ethics* (1982) 4(3): 240–257 at 246.

44 Ian G. Barbour, *Technology, Environment and Human Values* (New York: Praeger Publishers, 1980), pp. 96–97.

45 "Ethical Issues in Agriculture," p. 272.

46 Sentience is the ability to feel pleasure and pain.

the future in ways sentience does not. Humans with self-conscious creativity are aware of their own interests and the satisfaction of life plans. Yet, executing life plans depends on long-term health (at least at a minimal level) including not only healthy food, but also a sustainable environment over the course of a lifetime. These interests are not merely contained within a single generation, but are extended to humans over time, into future generations. Yet, present and future sustainability of humans needn't project so far into the future. It is a current right of individuals to sustainable living, and remains for one's lifetime.

One may ask how the individual worth of human beings and obligations not to impinge on the sustainability of other humans in any way defends ecological interests? Won't human interests always override ecological interests? The problem is that this question assumes a choice committing an either/or fallacy. It is not necessarily true that human interests and ecological interests *are not both important ones*, even if those two categories of interests are *unequal*. We cannot assume human worth and ecological sustainability exist in a vacuum. It is only through humans placed *harmoniously* in the biotic community that sustainability remains a live option.

Notice this argument places severe restrictions on permissible use of the environment. Since human interests only override environmental ones when they are vital, a corporate agriculture only serving convenient, affluent lifestyles is not morally permissible. This has revolutionary implications on the ethical acceptability of specific agricultural practices since the use of harmful pesticides, furrowing, and monocultural farming could not be justified. Pesticides contaminate aquifers in ways severely limiting the ability of other humans to acquire clean water and destroy the ecosystems of animals that prey on agricultural pests. Monocultural farming practices reduce biodiversity, eroding sustainable ecosystems long-term. Deep level furrowing is not at all regenerative, exhausts the soil, and requires inputs of nitrates to fertilize crops annually. All of these factors demonstrate the ethical unsuitability of most contemporary agriculture as it is practiced in the US. Since increased production in this mode of agriculture destroys ecosystems, it is prima facie wrong and should be avoided. An agriculture averting the need for increased production (such as sustainable agriculture of the ilk of CSA) is thus morally preferable and may aid in reducing environmentally damaging production.

However, let us imagine that defenders of modern, corporate agriculture could prove their claim that despite some ecological damage, technology coupled with some revision of current agricultural techniques can correct for environmental resource losses (and thus meet any obligations we might have to current and future generations of humans). Would they have license to exploit the environment in the face of the Jeffersonian argument for sustainable agriculture?

It is doubtful this would be permissible, even with such corrective measures. It is hard to imagine how this could be sustainable into the long term, and as such, would remain prima facie wrong. As Aiken adds, "Even if this damage would not affect the welfare of future persons, it should, nonetheless, be avoided whenever possible. So whether science can provide substitutes is irrelevant to the question: Should we deplete biospheric resources?"[47]

With this partial defense of sustainability in place, we can test whether CSAs promote sustainability both ecologically and economically, as well as support the values of community and direct democratic participation. Recall that CSAs are internally efficient. So, one indicator we can examine is the ability of CSAs to meet their budgets. Since each CSA farm must submit a budget to its shareholders, the anticipated cost and revenue of these farms is fairly transparent. In analyses of CSAs in New England,[48] Wisconsin, and Minnesota, we find that most CSAs indeed meet their budgets.[49]

However, many CSA farmers express that even if their method of farming is not the most economically efficient form of agriculture, they see other benefits that outweigh these considerations. Many CSA farmers see themselves as establishing an alternative normative order and their farms' outputs have more to do with accomplishing important environmental goals. Measurement of these outputs is problematic and it is difficult to see how they may be quantified. On the average CSA, the workweek for the head gardener is 15–20 hours during the greenhouse period, increasing to 60 hours during the harvest seasons, and reducing back down to 20–30 following the harvest.[50] This allows them time for tending to life as a farmer, while during some less labor intensive times of the year also affords them a large degree of leisure time to pursue other interests.

But the connections that CSAs allow for both their farmer and non-farmer members also bring a high degree of satisfaction. Andrew Lorand, a member of the CSA Garden of Great Barrington, Massachusetts, underscores the importance of community sentiment as an instrumental value to garnering sustainability:

> We have a responsibility towards the way that we treat the earth and only by treating the earth in as comprehensively good a way as we know how, can we expect it to properly nourish us. If you take that as a starting point, then that's something people can connect with. It's a practical ideal. For

---

47 "Ethical Issues in Agriculture," p. 276.

48 Suput, *Community Supported Agriculture in Massachusetts*.

49 Ostrom, "Community Farm Coalitions."

50 Van En, *Basic Formula*, pp. 5–6.

> that to be possible, though, there has to be a community that says, 'we want it'...So CSA basically says, if you want to take control of your own food situation, then join us and support the gardeners.[51]

Direct democratic participation within CSAs has also been an important source of power to its members and aided tremendously in the ecological and political sustainability of CSAs. Consistently, CSAs have turned down temptations to adopt conventional farming tactics in order to grow in membership, increase output, or even decrease the cost of shares. Shareholders have, by both majoritarian and consensus decision-making processes, elected to retain sustainable agricultural practices. CSA farmer Trauger Groh observes from his own experience organizing CSA efforts across Europe and the United States the range of choices members make:

> The decision about which day of the week should be the pick-up day when shareholders come to receive their share of the harvest, is in this rights sphere...Here the agreement of equal partners has to prevail. The rights of all who are partners must be considered and taken into account. Who can justly share in these decisions? Only those in the community who fully share in the financial risk of those decisions down to their persons assets [of shares]...in the planning of the farm, landscaping, crop rotation, breeding, composting, and so forth, down to the planning of their daily work....[52]

One aspect of the democratic process within CSAs that shareholders are particularly pleased with is the transparency of the decision-making process and the clear expectations that are jointly drawn. Just one case emblematic of this sense of transparency on several CSAs comes from the CSA Garden. The prospectus presented each spring gives the expected shareholder price, the content of the shares, the number of shares, and other expected revenues, which directly equal the amount of expenditures. The prospectus also clearly states, "our gardeners plant to provide a balanced harvest each week...since there are so many variables in nature, surpluses and deficits of certain of particular crops are likely."[53]

Philosophical differences have arisen on CSAs as to whether to use biodynamic versus standard organic farming techniques, determining who has

51 McFadden, "Large Scale Community Farms," p. 124.

52 Groh and McFadden, *Farms of Tomorrow Revisited*, p. 39.

53 Groh and McFadden, *Farms of Tomorrow Revisited*, p. 125.

how much decision-making authority, and defining the roles of shareholders. However, these are seen as conflicts that would be expected within any democratic institution. There has been no mass exodus from any particular CSA on grounds of underrepresentation in decision-making. In fact, overall satisfaction is likely higher in CSAs than it would be by merely entering the marketplace or attempting to exert voice concerning food production and consumption in conventional political arenas.

The economic sustainability of CSAs weighs more heavily on the minds of all CSA proponents, especially with respect to land acquisition. Creating more CSAs is obviously directly dependent on accumulating land. The urgency of this problem is captured by CSA farmer Jack Root:

> The single biggest issue in the development of CSAs is acquiring land. So far either all CSA's have a donor who has given them the land, or someone who is very sympathetic who is practically giving them the land or they have a very tenuous agreement about the land ... In the long term, we need to create a consciousness that land for agricultural needs to be taken out of the normal economic equation. We need a CSA land trust—angels who will buy land for the CSA. But one way or the other, the agricultural land must be taken out of the marketplace.

Land acquisition will continue to be an onerous difficulty to which CSAs must attend.

We also need to consider what sorts of inputs go into CSA farms. Environmentalists routinely complain that high dependence on irrigation, petroleum, and chemicals not only contaminate the land, but also consume scarce non-renewable resources from outside the farm. This is an important concern and many CSA farms are not exempt from using petroleum for tractors. However, most CSA farmers attempt to limit usage to only when it is necessary. Unlike factory farms, the organic methods of CSA farmers do not require burrowing far into the soil for planting, and hence a tractor is not usually required.[54] Regardless, the limited use of petroleum is probably where CSAs score the worst with respect to costly inputs.

Almost all CSA farms eschew the use of all chemical pesticides and herbicides. While some use organic pesticides and herbicides, these generally decompose with other plant material and don't add toxins to the farm's soil or water table. Prospective CSA farmers are careful to look for not only good soil

54 Generally, if they are used, it is during harvest time for breaking the soil to aid in the acquisition of tuber crops.

when they seek land on which to establish their projects, but also tend to seek natural aquifers. Irrigation of crops on most CSAs is fairly limited and usually doesn't involve drawing water from off the farms' premises. Yet, if CSAs expand to over 1,000 shares in some arid areas, wasteful irrigation could occur.

This brings us to local adaptation. This is a harder measure to take since, as we have seen, CSAs exist in several different regions of the US with varying climates, soils, and degrees of water availability. Organic and biodynamic approaches are generally accommodating to *indigenous* food plants, and since most CSAs use these methods, they adapt very well to local conditions.[55] In addition to growing a wide variety of crops, CSAs also develop hedgerows, retain local trees, and use other natural boundaries on the land. Other techniques used by the majority of CSA farms include contour plowing and small plot plowing. The interplanting of species also dovetails nicely into local adaptation to the land and space available for cultivation.[56] Many CSAs even sacrifice yields by growing "useless weeds" to retain the natural variety of indigenous plants. As an added benefit, some of these plants have been discovered to act as natural deterrents to other weeds or pests destructive to crops.

Ecological efficiency is also gauged by how long farms are able to last on the land. Accordingly sustainable agriculture must be regenerative. "Self-augmenting" gardens are pervasive in the CSA format, especially on biodynamic farms. CSA farmers take composting very seriously and for many it is the centerpiece of the fertilization process. Inputs may be added if they are available and effective over a long period of time. Given this, a marginal, but increasing number of CSA farms add inputs to their own fertilizer from off-the-farm sources while simultaneously providing a fringe benefit for local communities. By volunteering to serve as a collection site for leaf vegetable composts from private homes or from city parks they keep organic matter out of landfills in a useful way.[57] More CSAs are exploring Community Supported Composting (CSC) programs and finding them to be valuable sources of soil-building material for the farm. As William Brinton of Woods End Agricultural Institute in Mt. Vernon, Maine, observes, "Implicit in the CSA movement is the building of the soil as a vitally important component... Intensive vegetable culture takes

---

55 Suput, *Community Supported Agriculture in Massachusetts*, p. 60.

56 Aiken, "Ethical Issues in Agriculture," pp. 278–279.

57 There may be some concern that plastics and other items not natural to the environment (including pesticides and herbicides) may make group composting a risky option and some CSA farmers warn against it. However, there are communities that do not spray and request that private homeowners only contribute unsprayed organic material. For more on composting, see Case 41—"What To Do with Food Waste?".

a great deal out of the soil and farmers need to put it back."[58] As a response to this need, Steven McFadden reports that "Through CSC, the community can provide the necessary organic material [often table scraps, leaves and lawn clippings], thereby easing pressure on local landfills."[59]

CSAs score high on long-term sustainability not only due to vigilance in composting, but also due to their combination of regenerative farming techniques. For example, many use methods of crop rotation to temporarily relieve areas of farmland under heavy use in recent growing seasons. This, coupled with heavy composting, allow for vital nutrients to rebuild the soil. Collaborative agricultural efforts help ensure the land can be used for generations, as it was by early agricultural societies.

By 2050, it is anticipated that there will be around 9 billion people to feed. In some areas of the world, people already suffer from malnutrition. But some argue the situation would be much worse without industrial-scale agriculture. For all the controversy surrounding corporate agriculture, the shift to industrial farms produces tremendous yields that can't be matched by family farms. Conventional farms armed with technology have revolutionized how many people can be fed. In this context, conventional farming techniques should be applauded, not vilified. Despite heavy tolls industrial agriculture has placed on the land, the answer is not to support fringe movements such as organic and biodynamic farming. Instead, the focus should be on how to make conventionally farming even more efficient and to use technology to further develop drought and disease resistant crops in ways that cause less stress on the land. The bottom line, from this perspective, is that to realistically feed an increasing population, only large-scale agribusiness can do it—we just need to make it better.

This points to the conclusion that while economic sustainability is achieved with some difficulty (but at the end of the day is secured by the majority of CSAs), ecological viability is attained to a high degree. The question is whether this conclusion is robust enough to imply that CSA is a morally preferred agriculture and hence should be supported.

Review Questions

1) Do you think that agriculture in the US should be shifted to organic and locally grown modes as much as possible? On what grounds?

58 McFadden, "Large Scale Community Farms," p. 207.

59 McFadden, "Large Scale Community Farms," p. 207.

2) Do you think we should endorse the CSA model or do you think it distracts us from other agricultural strategies that are better? Why or why not?
3) Do you think democratic participation has anything to do with agriculture? Should it? How much emphasis should be put on sustainability in agriculture (if any at all)? Explain your answer.
4) What do you think should be the most important values in agriculture? After you list some, see if any come into conflict? Are there ways to resolve these conflicts? Furthermore, can you make a case for some of these values to be ranked higher than others? If some do, why do they?

Case 41 – What to Do with Food Waste? –

Study Questions

1) How much food is wasted in the United States each year? Where does it go and why do some think it is both an environmental and ethical issue?
2) What is compost? What materials can be composted and how are they composted? Explain the steps in detail.
3) What is a curbside composting program? What are some of the possible models used in these programs?
4) In what ways are curbside composting programs successful at handling food waste? What are some of their limitations?

In the United States, over 250 million tons of garbage are produced each year. Of that, about thirty-five million tons of food waste are sent to landfills and incinerators. This is more than any other single material in municipal solid waste (MSW) that heads to the dump. In 2012 alone, thirty-six million tons of food waste were generated in the US, with only five percent of it being siphoned off for composting from the usual disposal endpoints of incinerators and landfills.[60] Advocates of large-scale composting argue that much more can and should be done to keep food waste out of landfills as much as possible for a bevy of ecological reasons. This effort will likely be led by large population centers where implementation and expansion of citywide composting programs will maximize the possible economic and environmental benefits.

---

60 United States Environmental Protection Agency, "Reducing Food Wastes for Business," (website), July 8, 2016. Available at http://www.epa.gov/foodrecovery/ (accessed July 16, 2016).

Compost is decomposed organic matter that can be used for soil amendment or as material for growing plants. It comes from food matter, mixed with yard wastes and manures in appropriate ratios, and then is placed in rows, piles, or containers. Other materials are added to "bulk up" the compost (wood chips can be ideal for this purpose) and expedite the decomposition process. Compost needs to have time to cure, thus stabilizing and maturing it. Mature compost contains humus that is dark brown or black and has an "earthy" odor.

There are basically five variables necessary for successful composting, some of which overlap. First, there needs to be a balance of nutrients and feedstock. That is, proper composting needs to achieve a delicate balance between having suitable "green" organic materials (which include manure, yard clippings, and food waste) and "brown" organic materials (such as branches, dry leaves, and wood chips). Green organic materials supply copious amounts of nitrogen, but little carbon, to the compost. Brown organic materials do the reverse, providing a great deal of carbon, but offering little nitrogen. Obtaining the correct balance comes from experimentation and lends some credence to the idea that composting is a bit of an art as well as a science.

The second variable in composting is the size of the particles used in the process. These particles need to be small, but not too small. Whether by shredding, grinding, or slicing organic matter, its surface area needs to be increased so that microorganisms can feed and grow. More homogenous compost can be achieved through the use of smaller particles. Smaller particles also insure enhanced pile insulation, allowing for a higher and more consistent temperature. However, again balance matters. Successful composting requires particles small enough to maximize surface area for microorganism feeding, but not so small as to inhibit oxygen flow.

Oxygen flow, in fact, is the third variable in composting. Using shredded newspaper and wood chips as compost additives can help with appropriate aeration of the pile. Also, placing an array of pipes in the pile or occasionally turning it helps to introduce oxygen flow into the compost. Increasing the aeration of piles accelerates their decomposition. Yet again, balance is necessary when trying to account for this variable. Allowing too much aeration into the pile is liable to dry it out too much and thus impede decomposition.

The oxygen flow variable leads us seamlessly to a fourth variable in composting—moisture content. Moisture is an essential ingredient for the survival of microorganisms in compost. Water is also the means of transporting substances within the compost and makes nutrients available to the microorganisms in the pile.

The final variable, which has been alluded to above, is temperature. To reach their level of optimal activity, microorganisms must be in an appropriate temperature range. Temperatures in the pile can rise to 140 degrees F as microorganisms increase their activity. If the first four variables are balanced well, the appropriate temperature in the core of the pile should be maintained. The high temperature also destroys pathogens and weed seeds that natural decomposition cannot.

Human-engineered composting mimics natural biological decomposition, which has occurred since the first plants existed on earth. Vegetation slowly decays after falling to the ground, offering the nutrients and minerals necessary for the flourishing of microorganisms, plants, and animals.

Those who advocate steering food waste away from landfills present a number of reasons for doing so. First of all, when food rots, methane is released in the decomposition process. Methane is a greenhouse gas with 21 times the global warming potential of carbon dioxide. Why landfills matter ecologically is that they are a large contributor to human-related methane in the US, accounting for roughly 20% percent of all methane emissions.[61] Coupled with this, municipal landfill use time can be extended if food waste is steered out of them.

Second, there are sanitation and public health concerns with simply throwing food in the trash. Food wastes disposed of in this way can attract insects, vermin, and other pests as well as generate bad odors. Durable, sealed, leakproof, reusable composting containers can greatly reduce these problems and possibly even eliminate them.

Third, there is likely an overall environmental benefit to composting. One benefit is what the United States Environmental Protection Agency calls "valuable soil amendment." Recycling food waste is a relatively cheap and effective way of improving soil health and structure. This is important because throughout the developed world there is a rapid decline in available arable land. Soil amendment is imperative to maintaining the amount of available arable land and perhaps to even increase it to reverse the trend. Compost can also aid with increasing drought resistance in soil. It also reduces the amount of water, herbicides, and pesticides used in agriculture. Likewise, composting can promote higher crop yields, while reducing or even eliminating the need for chemical fertilizers. The soil amendment abilities of composting shouldn't be given short shrift. By amending compacted, low-quality, or even contaminated soils, restoration of land can occur for generating reforestation, wetlands restoration, and habitat restoration projects. Composting can also help in removing grease, oil, solids, and heavy metal from storm-water runoff. Again, this is

61 United States Environmental Protection Agency, "Reducing Food Wastes for Business."

important in urban areas where there are increasing worries about the amount of phosphate infiltrating local aquifers.[62]

As an economic consideration, food waste turned into compost represents the building of a commodity as opposed to what is normally an expensive burden. For every ton of food waste kept out of the landfill (which is a sunk cost), the amount of useable compost is multiplied, generating the benefits mentioned. Moreover, composting is a relatively cheap alternative to landfills and chemical soil additives. Thus, the US Environmental Protection Agency concludes that composting works as an extremely efficient environmental remediation technology, perhaps 50% less in cost than conventional soil, air, and water pollution soil amendments. Composting can capture and destroy 99.6% of industrial volatile organic compounds (VOCs) in contaminated air.

Municipal composting programs feed food waste to anaerobic digesters, converting the material into renewable energy as well as valuable soil amendment. The prospect of saving space in landfills and possibly producing energy captured from methane otherwise contributing to greenhouse gases has attracted some cities to implement, or at least investigate, composting programs. Beginning in 1996, San Francisco was the first city in the US to implement a citywide composting program. By 2012, the city diverted 1 million tons of food waste from landfills. The amount of food waste collected accelerated beginning in 2008 once the city shifted from a voluntary to a mandated program, with some 600 tons being diverted from landfills. The goal is to achieve zero waste in the city by 2020. Over US ninety cities have launched curbside composting plans of their own. Portland, Oregon, began its program in October of 2011. Among other cities that have initiated composting programs include Seattle, Boulder, Denver, and Salem, Oregon. Before its program started, it was estimated that about a third of all garbage going into San Francisco dumps was compostable. When combining composting and recycling programs, the city now diverts 78% of its waste from landfills. In addition to city composting programs, twenty-four states have banned yard clippings from landfills.[63]

In 2011, the city of Madison, Wisconsin, began a composting pilot program. By 2014, about 550 households and five businesses were involved. Though the pilot program was running out of funding in the same year, there was an eleventh hour approval of the city council to supply an extra $150,000 to keep it in operation for another two years. This infusion of capital was needed mainly for

62 See Case 10—"The Lakes and Phosphorus Twins: Monona and Mendota."

63 Kate Shepard, "Why Doesn't Your City Have Curbside Composting?" *Mother Jones*, September 10, 2012. Available at http://www.motherjones.com/environment/2012/09/why-doesnt-your-city-have-curbside-composting (accessed September 21, 2014).

refurbishing an anaerobic digester. The Streets Division of Madison now plans to expand the program to include a total of 1,600 households and several area businesses. While the city ships its organic waste from its curbside composting program to a processing facility at the University of Wisconsin-Oshkosh, it is anticipated that there will be a digester built in or near Madison city limits within three years. Madison's mayor, Paul Soglin, was on board with expanding curbside collection saying, "We think the program has a high likelihood of success. The preliminary work tells us that the value of the energy we create as well as the composting will exceed the cost of the program."[64]

America's largest metropolis, New York City, has joined the ranks of those with curbside composting programs. Former mayor Michael Bloomberg announced in 2013 that the city should increase composting of food scraps. At the time there was talk of a mandate that every household in the city must separate compost out from other trash as part of the waste program. Since then, the city's Department of Sanitation has developed a pilot program called "Zero Waste" that collects food scraps and yard waste.[65]

While there appears to be promise in composting on a wide scale in large municipalities, there are reasons for hesitating to sign on. Just as it has been in the case of other recycling programs for plastic, glass, metals, aluminum, and paper, there will be little public buy-in for composting without the process being convenient and relatively nuisance-free. One of the main complaints about food waste recycling programs is the smell. This concern gained traction in New York City where some claimed the density of the population renders curbside composting impracticable. But Bloomberg seems to be convinced of the usefulness of composting programs. As he mentioned in his 2013 State of the City address, food waste is "New York City's final recycling frontier. We bury 1.2 million tons of food waste in landfills every year at a cost of nearly $80 dollars per ton. That waste can be used as fertilizer or converted to energy at a much lower price. That's good for the environment and for taxpayers."[66]

---

64 Xig Wang, "City's Organic Collection Program to Expand: Turn Organic Waste into Energy," *Madison Commons*, (website), Feb. 23, 2014, Available at http://madisoncommons.org/?q=content/city's-organics-collection-pilot-program-to-expand-turn-organic-waste-into-energy (accessed September 19, 2014).

65 New York City Department of Sanitation, "Food Scraps + Food Waste," (website), 2015, Available at http://www1.nyc.gov/assets/dsny/zerowaste/residents/food-scraps-and-yard-waste.shtml (accessed July 16, 2016).

66 Brian Clark Howard, "How Cities Compost Mountains of Food Waste," *National Geographic*, June 18, 2013. Available at http://news.nationalgeographic.com/news/2013/06/130618-food-waste-composting-nyc-san-francisco/ (accessed August 21, 2016).

While facing some challenges, many stewards of municipal composting programs carefully plan to make the practice as convenient as possible. Measures to ease citizens into city-wide composting programs have included smaller-scale voluntary pilot programs, curbside pickup, supply of indoor and outdoor bins, supply of compostable bags, and schedule coordination for same day pick-up of trash, recycling, and composting. Also, some local waste management departments are shooting videos and uploading them online to their respective websites to describe the composting program in an attempt to keep citizens informed. Other cities use social media to offer updates on their programs.[67]

Towns and cities promoting curbside compost collection still need to be aware of several issues when developing viable programs. One regards who will collect the waste. A city's relationship with a local vender who will haul it away can make or break a composting program. Of those cities with effective residential organic waste programs, 70% contract with a single private vendor to haul trash, recycling, and compost. Only 10% of cities have waste services run by the municipality and only 6% have multiple vendors for waste management.[68]

Part of the reason for enlisting single vendor service for all waste disposals is to maximize route density to create more efficient and profitable programs. In San Francisco's case, the city's longstanding relationship with their waste management hauler has allowed the stability necessary for significant long-term investment in the program, especially in future capacity. This has helped make San Francisco's composting program arguably the most effective in the country by several measures. However, this is not to say other formats can't necessarily work. Seattle divides waste management responsibilities between two separate vendors whose hauling responsibilities are split geographically. This allows not only for market competition, but also sufficient route density and efficient collection of waste. Other cities allow for more than one vendor for collection responsibilities, with little to no geographic overlap.

What exactly will be collected is another key question that cities and towns should ask themselves when considering whether to establish curbside

---

67 See the City of Cambridge, Massachusetts, Department of Public Works, "Composting Questions," (website), 2014. Available at http://www.cambridgema.gov/TheWorks/OurServices/RecyclingAndTrash/FAQRecyclingAndRubbish/CompostingQuestions/CompostPilot.aspx (accessed Oct. 30, 2014).

68 Juri Freeman and Lisa A. Skumatz, "Best Management Practices in Food Scrap Programs," *Ecoconservation Institute*, (website), 2010. Available at http://www.foodscrapsrecovery.com/EPA_FoodWasteReport_EI_Region5_v11_Final.pdf (accessed July 16, 2016).

composting programs. There are different options as to what sorts of organic material can be collected. Many city programs began with collecting yard waste primarily. However, adding food waste is the quickest and cheapest way to implement a new program. Over 90% of programs accept meat and dairy. In San Francisco's pilot program, it was found that many more participants signed on when meat and dairy products were added to the composting program than when only vegetable matter was allowed.[69]

Another important consideration is whether paper will be included in the composting program. In some communities, about 50% of all waste by weight is composed of paper material. An effective way of increasing participation rates is to allow paper soiled with food waste (think pizza boxes!) to be included. Of course, what can be accepted is contingent on what a town or city is able to process effectively. This issue comes to the fore when thinking about whether to accept compostable serving ware. Processors who deal in large volumes of compostables generally accept compostable service ware, yet this is not true of processors who handle less volume.

How compost is going to be collected is also significant for communities to think about when beginning a composting program. There is the issue of how often compost will be collected and the general collection schedule (i.e., will compost be collected at the same time as trash and recycling?). The majority of curbside programs offer weekly collection. Some reasons for this rate of frequency are to address sanitation concerns as well to stem participant complaints of bad odors. However, collection schedules can be contingent on climate as well. During the winter months, particularly in cold weather regions, collection may slow to once every two weeks. In cities where curbside composting is going well, a weekly composting schedule can be accompanied by a bi-weekly trash collection schedule. This can serve as an incentive to compost even more.[70]

With respect to bins, most composting programs supply wheeled carts with lids that can hold 32 to 96 gallons of organic waste. This variance in sizes allows for programs to charge differential amounts based on the amount of waste produced. Most programs provide compostable bags that line kitchen bins in which scraps of food can conveniently stowed until they are ready to be taken

---

69 Jack Macy, "San Francisco Takes Residential Organics Collection Full-Scale," *BioCycle Magazine*, January 1, 2000. Available at https://www.environmental-expert.com/articles/san-francisco-takes-residential-organics-collection-full-scale-1068 (accessed July 18, 2016).

70 Massachusetts Institute of Technology, Sloan School of Business, "Getting Started: 10 Questions for Cities and Towns Considering Residential Curbside Composting," (website), 2014. Available at http://mitsloan.mit.edu/actionlearning/media/documents/s-lab-projects/Guide-to-Composting.pdf (accessed September 22, 2014).

to larger outdoor bins. However, the trend has been to discourage the use of these bags for a couple of reasons. The first is that compostable bags are expensive for consumers. But also, processors find that they do not break down easily during the composting process. Alternatively, many composting programs encourage the use of paper bags to line the composting bin. Of course, some participants may balk at this solution in cases of wet waste.

What also sparks debate between administrators of composting programs is what type of kitchen bin should be recommended. Many composting programs offer participants 1–2 gallon bins for temporary storage of food waste inside the residence. Some programs, however, have found a significant number of participants do not use the supplied kitchen bins and thus they are an unnecessary expense. As an alternative, some municipalities have resorted to educating their participants about using other containers already available in the house.

Another important issue to think about for towns and cities considering composting programs is where food wastes will be taken after pick-up. A significant consideration is what a municipality's capacity is for processing organic waste. For a period of time the city of Cambridge could not expand its program from a modest drop-off pilot to curbside composting because of a lack of compost processing capacity in eastern Massachusetts. In contrast, more successful composting programs such as those in San Francisco and Seattle have large volume capacity processing facilities nearby, with the capability of handling 500 tons of waste a day. In contrast, most of the processors in eastern Massachusetts have been small farms that can only process 15 tons per day. Cities also need to consider how far away a processing facility is and can learn from Denver's experience. The city had to halt a very successful and popular pilot curbside composting program in 2010 when it became too expensive to ship waste the long distance to its closest processing facility.

Economic factors play a significant role in the viability of composting programs and the decision on the funding mechanism shouldn't be taken lightly. Municipalities generally fund waste collection in one of three ways. The first way is through revenues garnered via property taxes. The second is by using fixed fees from households. The third way is by a Pay-As-You-Throw (PAYT) model. Overall, PAYT programs are used in 26% of US communities to finance waste disposal. However, studies have shown that in cities with successful curbside composting programs, 80% utilize a PAYT system.[71] Part of this difference may be attributable to a rate structure whereby communities that

---

71 United States Environmental Protection Agency, "2006 PAYT Programs, " (website), 2006. Available at http://www.epa.gov/epawaste/conserve/tools/payt/states/06comm.htm (accessed September 22, 2014).

use PAYT charge less for organic waste, thus offering participants a major financial incentive to join the program. A prime example of this is Boulder, Colorado's curbside composting program. The city's PAYT program creates economic incentives as citizens are offered 32, 64, or 96-gallon trash containers with rates increasing with size. In 2006, when the initial pilot began, participants were also given a 32-gallon compost bin. As a consequence of this strategy, there was a 12% drop in the number of households using the 64 and 96-gallon bins.[72] For some cities, PAYT programs are not viable and an alternative pay structure is to have a minimal composting service fee automatically placed in the cost of all citizens' waste management bills. Boulder mandated this strategy and it did raise the monthly rates for costumers by about $2.50 a month on average.

But not only do benefits of the program have to accrue to environmentally-minded people, they have to be realized by citizens generally to build and sustain a successful composting program. Part of this means the municipality must also experience cost recovery. The amount of savings per household will depend on the region. Much depends on the frequency of collection and the amount of waste different locales collect. In many situations participants will be required to pay a bit more for collection services. In cases where citizens are asked to pay more for their waste collection services due to the implementation of a composting program, encouraging these households to divert organic waste into composting bins may lessen these worries. In the long run, their municipality might decrease fees since it is generally cheaper to process composting than to pay the cost for maintaining (and building more) landfills. Also, an emphasis on the environmental benefits may encourage some citizens that all things considered paying a bit more for composting services is worth it.

Local organizations can go a long way toward helping the implementation and maintenance of composting programs. Some communities have found it useful to hold a conference on organics to catalyze public interest. This can serve as a forum for interested organizations to discuss both the promises and perils of beginning different waste collection strategies. The sooner that local government, ordinary citizens, waste haulers, and organics processors can meet and deliberate together, the better. This way they might come to some consensus more quickly about how to handle compost. This is likely to also allow these groups to develop the willpower that will aid them with respect to sometimes difficult negotiations. Community groups such as neighborhood associations and local environmental interest organizations can be essential

72 Lisa Skumatz, "Boulder Colorado Pilot Recycling Program Report," Skumatz Economic Research Associates, report, September 2007.

to providing some of the grassroots support needed to implement new programs. An example of this can be found in Arlington, Massachusetts, where Sustainable Arlington played a pivotal role in a push for a composting program in the town.

State environmental agencies can also lend support to composting programs if asked by local municipalities, as they can provide valuable technical assistance. They can also sometimes offer financial assistance, as was the case for programs in Denver and Cambridge through state grants contributing to covering start-up costs. In more established programs, such as San Francisco's, greater longstanding contributions are possible as the city's Department of the Environment offers $600,000 each year to non-profits to provide education, marketing, innovative strategies, and other composting and recycling initiatives. Colleges and universities have in some cases played a major role in initiating and implementing new composting programs. They may have access to grants and other sorts of funding to begin new programs. At times, they even provide (via their students) the labor and enthusiasm to begin new composting initiatives.[73]

It can't be overstated how much of a role an effective public education campaign can have for a successful curbside composting program. Public perception is extremely important and having some strong motivation via a clear message for the importance of the program is essential to its viability. Seattle is a case in point. In 2009, the city launched a widespread curbside solid waste removal program, including a mandate for residential composting. But preceding full implementation of the plan was a successful ad campaign informing citizens of why it was being instituted. 90% of Seattle residents surveyed stated they were aware of the program before its full-scale implementation and the same percentage were "very satisfied" with it. This was despite the fact that the program included a 42% waste removal rate increase and affected 530,000 customers. Citizens were informed in multiple ways, including by newspapers, city meetings, newsletters, city websites, and instructions on the new trash receptacles.

San Francisco takes this strategy a step further by employing a full-time team of "zero-waste experts." They provide technical assistance to households and businesses, conduct waste audits, and inform citizens door-to-door when changes in the plan are instituted. The city also leverages its strong relationships with local media to emphasize positive aspects of the program to drive up participation rates.

---

73 MIT, Sloan School of Business, "Getting Started."

This brings us to the need of possible legislative or other governmental support for composting programs. On the local level, this sometimes means changes in city codes to facilitate a new program, even though 90% of participation in composting programs nationwide is voluntary. But as mentioned earlier, San Francisco and Seattle's plans are mandatory. In Boulder, independent waste management companies are required to haul away compost and charge an embedded service for it in trash collection bills. The Department of Environmental Protection in Massachusetts put a plan in place to ban scraps from commercial waste in 2014. Local governments can also play an active role by speeding up the permitting of organics processing facilities in an attempt to encourage composting programs. They can also lower permitting fees, to make processing more economically feasible.

Among the lessons learned by cities in various stages of development of composting programs is that implementation is not seamless. A variety of factors can overwhelm the efforts of cities to begin and then maintain composting programs. One of the chief complaints about composting boils down to the *"yuck factor."* That is, being opposed to an idea or situation solely on the basis that it is deemed disgusting, gross, or unnatural. As mentioned earlier, residents sometimes resist composting due to fears of bad odors, vermin, and even disease. Education may help alleviate these concerns. Composting is a learning experience and once people are used to methods such as wrapping food scraps in paper, or freezing them, or layering them with yard waste, advocates believe composting fears will dissipate. Advocates will also point out that with the exception of people who use garbage disposals, there is nothing really new about composting besides the use of a separate bin. After all, where was food waste going *before* composting programs if not in a garbage disposer? Likely, it was going in the trash anyhow with all of the same issues one would have with composting.

There are also worries about animals being attracted to organic waste, especially in rural areas or places otherwise close to wilderness. For example, Boulder has had to face this challenge. In the city 25% of residents have reported some instance of animal disturbance in their composting bins. However, proper education can mitigate this issue. Obtaining the correct bin is part of the solution. Also, leaving bins out at night when animals are active should be prevented when possible.

Likely the most pressing complaints concerning curbside composting programs are about their cost to consumers and the lack of processing capacity for certain communities. Pilot composting programs in Denver and Burnsville, MN were suspended due to these concerns.[74] And while local officials in both

74 MIT, Sloan School of Business, "Getting Started."

communities remain hopeful that future programs will be more successful, these cases serve as cautionary tales to towns and cities as they try out their own new programs. This capacity problem can be alleviated if larger processing facilities are made more readily available and this should be established well before any significant new composting program is put into place.

While composting programs might seem to be relatively uncontroversial, a few issues beyond costs and aesthetics may arise. Some may be opposed to composting programs because they see relatively little need for them given their costs and/or denial that they will really do much to stem climate change. Of course, some don't believe humans make much of a contribution to climate change or even deny that climate change is occurring. They bristle at the notion that the government (even merely local government) plays a role to impose what they see as "nanny state" provisions.

Joined with this is the concern about governmental mandates in general and that these impositions entail undue constraints on individual liberty. For those endorsing libertarian political views, there is deep suspicion about the ability and desirability of what is seen as state coercion of individuals. For libertarians, the only legitimate role of the state is to protect private property rights (derived from humans having property rights in themselves). So beyond protecting people from harm from others and keeping their property safe from those who might steal or destroy it, the state has no right to tax individuals for other measures, such as environmental protection. Nor does the state have the right to "protect people from themselves."

What this reduces to is also the idea that the state should not impose any particular values on its citizens. That is, democratic states should be neutral with respect to the good life, providing no particular advantage or encouragement to one set of values over others.[75] Libertarians who take this neutrality provision seriously will argue that, for example, the state doesn't have the right to force citizens to enroll into a program even (and perhaps especially) with an important social value as its goal. While some may care deeply (and be willing to be heavily taxed) for provisions to health care, others may not share that primary goal. With that said, people should not be taxed for this provision in the eyes of libertarians and they certainly should not be required to participate (say by being forced to buy insurance). According to the logic of this argument, "saving the environment" though it may be a value to some people, is not to others, and the state has no right to champion that particular value and force citizens to do so as well.

75 Anthony Mason, "Autonomy, Liberalism, and State Neutrality," *The Philosophical Quarterly* (1990) 40 (160): 433–452.

With this said, any mandate to join a composting program, though not credibly as significant a worry perhaps as being forced to join a health insurance plan, will very likely be anathema to those with libertarian inclinations. Even short of being required to participate in such a program, libertarians will balk at the idea of paying taxes for a service they see as unnecessary. Ultimately, libertarians will take a position close to that of the philosopher Robert Nozick, who suggested that those who care so much about health care or the environment could join with others who care to voluntarily raise and use their own money to ensure such provision.[76]

Review Questions

1) What do you think are the ethical duties local governments have in reducing food waste going into landfills? Are there any obligations? If so, what would they be? If not, why not?
2) Consider the argument from the libertarian perspective in this case study. Suppose one could convince a libertarian who is not a climate change skeptic that curbside composting is convenient, cost effective for the city and consumers, and efficaciously keeps food waste out of landfills, thus reducing GHG emissions and the need for new land for dumps. In reply, the libertarian still insists that mandatory composting is an unjust policy because it violates state neutrality and forces citizens to do something they don't want to do (and pay for). Do you agree with this argument? Why or why not? If mandatory curbside composting is not legitimate, are there any local mandatory public programs that are?

Case 42 – Beware the Invaders! –

Study Questions

1) What is an invasive species?
2) What is intrinsic value and what is extrinsic value? What is anthropocentric value? What are biocentric and ecocentric values? What kind of value is most likely being appealed to for the value of native species?
3) What are some examples of invasive species noted in this case study and why are they classified as invasive species? Can you think of other species in your state or region that would qualify as invasive species? On what grounds?

---

76 Robert Nozick, *Anarchy, State and Utopia* (New York: Basic Books, 1974).

For the past two decades, an increasing number of invasive species have been detected in ponds, lakes, and rivers around the United States. While there are other conduits for the introduction of such species, one of the most common may also be one of the most overlooked—the use of boats on multiple lakes. The instances of fish being infected with parasites and lakes being inundated with choking weeds from other ecosystems in what were formally pristine bodies of water have been growing dramatically since the late 1980's. This problem is compounded by the fact that invasive species are extremely adept at exploding in population size and once introduced into a new area can spread at astonishing rates.

The negative impact of invasive species on ecosystems has been significant. Globally, it is estimated that $1.4 trillion has been spent to deal with invasive species. The cost to the US economy from invasive species is likely in the billions of dollars.[77] The most cited paper on the economic cost of invasive species in the US reports that they cost the nation some $120 billion dollars per year.[78] In 2011 alone, the Department of the Interior spent $100 million dollars on invasive species prevention, early detection and rapid response, control and management, research, outreach, international cooperation, and habitat restoration. In the Great Lakes region, it costs $200 million annually merely to control invasive species usually brought into waterways by the ballast water discharge or on hulls of ships. Zebra mussels alone, an invasive species in Wisconsin, have cost citizens into the millions of dollars, caused nightmares for public utilities officials, and proven to be a significant nuisance to private lakeshore landowners.

Invasive species infestations clearly cause negative impacts on property values. In Wisconsin, Eurasian watermilfoil (an aquatic plant) has reduced lakefront property values by 13% and in Vermont by 16%. Non-native species of flora and fauna can also impede agricultural productivity and add significant service costs to the maintenance of public utilities. They can destroy native fisheries and thus decrease tourist dollars from game fishers. Outdoor recreation can also be adversely affected by invasive species. Beyond all of this, the overall health of ecosystems can be imperiled by substantial infestations of non-native species of plants and animals.

77 United States Fish and Wildlife Service, "The Cost of Invasive Species," (website), January 2012. Available at http://www.fws.gov/verobeach/PythonPDF/CostofInvasivesFactSheet.pdf (accessed November 14, 2014).

78 David Pimentel, Rudolpo Zuniga, and Doug Morrison, "Update on the Economic and Environmental Costs Associated with Alien-Invasive Species in the United States," *Ecological Economics* (2005) 52: 273–288.

In Executive Order 13112, an invasive species is defined as "an alien species whose introduction does or is likely to cause economic or environmental harm or harm to human health."[79] One of the typical ways invasive species harm native species is through predation. But non-native species can also do so through destroying or eroding quality habitat or by outcompeting native species for scarce food resources. What is usually not discussed about the extinction of fauna is that one of the leading causes of their elimination is by the successful invasion of non-native species. For instance, more than 400 of the 1,300 species on the Endangered Species List are there at least partly due to displacement by, predation by, or competition with invasive species. This is also true for another 180 candidate species that may be protected under the Endangered Species Act in the future. In freshwater lakes, one of the leading causes of fish extinction and endangerment is invasive species. Sometimes an individual invasive species can imperil a number of other species simultaneously. For example, on the island of Guam the brown tree snake is thought to be the main culprit in the extremely sharp decline of native forest birds and the modern extinction of at least 10 species of them.

With the extent of the damage on a regional and national level caused in ecosystems by invasive species understood, we can now turn to a few particular cases of such damage in a local environment. Wisconsin's lakes and rivers have been historically known as being relatively pristine and free of invasive species when compared with other states in the Midwest. However, that status is changing. The problem of invasive species has come at such a rapid pace that while there were some controls against them already in place, it was only relatively recently that a formal law was enacted by the Wisconsin Department of Natural Resources (WDNR) directly aimed at the curbing or elimination of non-native species in the state. Wisconsin Statute 23.22 defines an invasive species much like the USFWS does, declaring that they are "nonindigenous species whose introduction causes or is likely to cause economic harm or environmental harm or harm to human health."[80] Under Wisconsin natural resources codes, in Chapter NR 40, invasive species are listed in two categories, as either "prohibited" or "restricted" with different regulations and control requirements attached to each designation.[81] *Prohibited invasive species* are not yet within the state's borders (or if so are only in a few places), are likely to cause economic and/or environmental harm, and it is still possible to eradicate them and prevent their spread statewide. *Restricted invasive*

79 Exec. Order No. 13112, 64 C.F.R. 6183 (1999).

80 *Wis. Stat. § 23.22.*

81 This law went into effect in January 2009.

species, on the other hand, are already found widely within the state, have already demonstrated environmental and/or economic costs, and their eradication is unlikely.

Two invasive species of animals that tend to be on the top ten lists for "most destructive" nationally are both found in Wisconsin's waterways—zebra mussels and Asian carp. Both species are under the "restricted" classification of invasive species as they already have been well established in the state's waterways. However, in the case of Asian carp the current and pressing battle is to keep them from spreading into the Great Lakes, where they could have a devastating impact on fisheries in several states as well as Canada. Asian carp are known to be voracious eaters, feasting on the same food resources as those used by native fish, and now threaten the $7 billion dollar game fishing industry of the Great Lakes.[82] 46% of endangered species in the Great Lakes are threatened by invasive species already. With widespread introduction of Asian carp into this large water system, it is estimated that a great deal more harm will be done to the fishery.

While the carp themselves have not been found in Great Lakes waters yet, their DNA has been recently detected. Carp come in a number of varieties, including bighead, black, silver, large-scale silver, and grass. The term "Asian carp" has been applied by the media to describe this group of invasive fish species native to Asia that pose immediate and severe danger to fisheries throughout the Midwest. While collectively all of these fish pose problems, state environmental regulatory agencies throughout the Great Lakes region have particular worries about bighead and silver carp. These varieties swim upstream to breed, with their eggs and larvae drifting downstream to develop. Both types are very fast growing as they can mature to 100 pounds. They have high fecundity rates as individual females can lay up to one million eggs.

Though it might seem odd to say, Asian carp can also be a danger to fishers and other boaters. Large silver carp are able to leap high out of the water (especially if disturbed by boat engines) and injure people on the water. These fish can turn themselves into projectiles and at 40 pounds can pack a surprising wallop to the unsuspecting boater, sometimes causing injury.

Bighead and silver carp are filter feeders; they strain phytoplankton and zooplankton, making them food resource competitors with other game fish and native mussels throughout the region. Their presence has caused population decline in other fish species, including game fish such as lake whitefish, walleye, and perch. Because bighead and silver carp are so good at filter feeding, fish farmers producing catfish originally imported these species to the

---

82 The Great Lakes comprise the largest freshwater system on Earth.

southeastern US to clean algae and suspended matter out of holding ponds. They were also used to clean up wastewater treatment lagoons.

While it is not known precisely how they managed to escape aquacultural facilities in the Southeast, some speculate that large floods in the mid-1990's caused catfish farm ponds to spill over their banks into small local waterways connected to the Mississippi River Basin. From there, all varieties of Asian carp have spread rapidly throughout the US, including northward toward the Great Lakes. One important conduit to the Great Lakes comes via the Illinois River, which is linked to Lake Michigan by the Chicago Area Waterway System (CAWS)—a series of ship and city sanitation department canals.

An array of devices has been deployed in an attempt to keep carp out of the Great Lakes. For example, electronic river barriers in portions of the CAWS are used to establish a field through which carp cannot swim. This is part of larger and more comprehensive plans for dealing with Asian carp, including the Great Lakes Restoration Initiative and the Asian Carp Control Strategy Framework. The United States Geological Survey, the Illinois Department of Natural Resources (IDNR), and Southern Illinois University have joined forces to fight the carp infestation, with the effort being lead by the President's Council on Environmental Quality. This consortium assembled along the Illinois River in the summer of 2013 to test the viability of using multiple tools to stop the progress of Asian carp toward the Great Lakes. In the history of aquatic pest control, regulators have found that battling invasive species on multiple fronts while using a variety of tools has been the most effective control method. This approach reduced the population of the sea lamprey, an invasive fish in Illinois waterways, by 90%. Among the tools used against Asian carp have been food attractants, sound barriers, sonar fish tracking, and commercial fishing.[83]

The demonstration project site was set just outside of Morris, Illinois, not too far from the far western suburbs of Chicago. There were four explicit goals of the project devised by the research team. The first was to evaluate the effectiveness of using multiple forms of technology to monitor and control Asian carp populations. The second was to demonstrate and identify Asian carp control and monitoring measures for resource managers. In this effort, five tools were evaluated; population assessments, habitat assessments, food attractants, seismic water gun control barriers, and commercial fishing.

To become clear on the pervasiveness of an invasive species population, researchers have to know the total population and how it is distributed in a

83 Grass carp is an example of an invasive species proven to be a desired commercial commodity, usually exported to Asian countries, but sometimes sold in US markets.

body of water. Using a state-of-the-art, high-definition, sonar detection system (including side beam sonar and split-beam sonar), researchers surveyed the carp population before, during, and after the full integration of all of the control tools. The advantage of using sonar for fish monitoring is that detection can be made through dark and cloudy water. The boat's transducer sends out sonar pulses and the echoes are then detected to create a digital image of what is in the water. More specifically, the split beam sonar uses two beams along the side of the boat—these provide information on fish size, swimming patterns, and direction of movement in three dimensions. The side beam sonar system uses a high-definition towing device suspended from the side of the boat. This system provides high image clarity of fish and river bottom structures. The resulting echograms (or sound pictures) of carp concentrations allowed researchers to estimate the total population size.

Along with getting a handle on the size of the carp infestation and their movement behavior, it is essential for scientists to also evaluate environmental factors conducive to carp populations. Sound waves from an acoustic Doppler current profiler coupled with a GPS unit to determine position were used to map the bathymetry (i.e., the depth of a body of water) at the demonstration site. This gave researchers valuable data about the volume of water in the river. Sonar equipment also measured depth, velocity, and flow into and out of the backwater at the demonstration site. This allowed the scientists on the team to calculate fish densities and provided a population assessment. A superstation (or high tech stream gauge) measured water height and water quality data. All of these data were used to show how water quality and flow could affect concentrations of Asian carp. This would allow researchers to know how to most effectively and efficiently distribute the food attractant. The information on water quality and river flow provided a meta-analysis of how carp select habitat to predict their future movements.

Researchers found that the carp progression had stalled at the Dresden pool just northeast of the demonstration site. The key was to find out why the vanguard of the Asian carp surge has stopped in this area, as this might give a clue for preventing their future spread. With respect to the food attractant, researchers discovered that silver and bighead carp were most attracted to an algal food mixture. These food attractants lure carp to be trapped. The researchers utilized two dry food quality algae in the mixture—chlorella and spirulina mixed with water. Underwater video showed a feeding frenzy when the algae mixture was presented to schools of carp. Research teams delivered coolers with the wet mixture to the riverbanks near the feeding sites. Liquid food attractant was then pumped into the water. Six feeding stations were set

up to attract carp into the deeper backwater, which drew them behind a water gun containment barrier. Fish were observed in real time using sonar technology during the demonstration.

With a reliable lure method tested, it was time to use it in concert with other methods for carp containment. Many strategies are used to halt carp progression toward the Great Lakes, including seismic containment water guns. These have also been deployed to trap carp in holding areas where they can be more easily harvested. Other strategies have included electrical, chemical, and sound barriers. With seismic containment water guns, sound pressure waves are shot into the water. Originally, this technology was developed for the purpose of oil exploration, but was also found to impact fish. The water gun is metal and works by firing a piston in multiple chambers, creating a high-pressure jet. A compressor is connected to the gun to build the pressure necessary for its operation. Before water guns were used for fish repulsion at the demonstration site, they were tested for possible ecological effects by using a pressure map to gauge the strength and dispersal of waves emanating from them. After this initial experimental phase, researchers tested them on populations of carp and other fish in laboratory ponds using underwater cameras, telemetry, and sonar to monitor fish movement.

Then, once the laboratory pond studies were completed, the water guns were deployed in the Illinois River Side Channel. Here sonar was used to see if fish crossed the temporary sound barrier. Upon successful completion of the pond and field experiments, a water gun containment barrier was established at the demonstration site at Morris. Two water guns were used to contain the carp in the backwater lake. They were fired continuously for three days and feeding stations lured carp to one side of the barrier and then commercial fishing was allowed in the backwater pond. High-resolution sonars were placed on both sides of the barrier to monitor the carp. By examining fish movement patterns, scientists could learn vital information to maximize the positioning of water guns in the future.

The effectiveness of commercial fishing in combination with the other tools already described was also assessed in the demonstration project. The hope is that intensive commercial fishing can reduce the size of the carp population in the area. The research team already had some evidence to suggest this strategy might work, as the IDNR began contracting with commercial fishers for Asian carp removal in 2010 to try to alleviate the number of carp surging against the electronic barriers placed in the CAWS. Between 2010 and 2013, commercial fisherman placed 1,000 miles of nets, removing some 100,000 carp weighing over 900 tons. Beginning in 2013, the IDNR has also experimented with using heavy Great Lakes pound nets in the backwaters of the Illinois River. These

are designed for extended use and serve as surface to bottom fish traps large enough to cover the mouths of backwater lakes. Once the pilot project proved successful, this technique was combined with the feeding stations and a containment barrier to allow for carp harvest. Then, for three days, the site was divided into zones with commercial fishers allowed to fish for four-hour periods each day, and then randomly switched to fish other zones to reduce capture bias. At the end of the demonstration period, the fish were counted and then processed into liquid fertilizer. Some of the fish were weighed, measured, and retained for future analysis. In the three-day integrated demonstration trial, 1,000 carp weighing 15,000 pounds were removed from the Illinois River backwater site.

Overall, the demonstration project proved successful. The joint effort by the multiagency team gleaned a large catalogue of information as the methods used for monitoring the fish populations worked well. The control measures were also quite effective in containing and reducing the number of carp at the demonstration site through intensive fishing. Consequently, the group will continue to expand their model, developing more field test sites on other rivers and tributaries.

But carp are not the only threat to the Great Lakes. Zebra mussels have also wrecked havoc in the region. When massing in great concentrations they cause heavy algae growth due to their feces and vomit. This in turn creates anaerobic conditions that choke out other plants and animals. Zebra mussels are extremely small animals, with adults measuring just one-fourth to one-and-a-half-inches long—about the size of an adult human fingernail. They can be identified by their D-shaped shells, which have brownish and yellow alternating stripes. Female zebra mussels have a high fecundity rate as each can produce 10,000–50,000 eggs per year. The eggs develop into free-living larvae that begin to form shells. The larvae, within two-to-three weeks, begin attaching to objects using what are called *byssal threads*, with a preference for firm surfaces. It is the only freshwater mussel that can attach to objects. Zebra mussels originated in Eastern Europe and most likely were introduced to the Great Lakes through the discharge of ballast water from European ships. They were first discovered in the Great Lakes in 1988.

Zebra mussels are more than a nuisance to lakeshore property owners and recreational users as this species is particularly good at clogging water pipes. They are attracted to areas with strong water flow and hence are often found attached to intake hoses. Lakefront property owners who water their lawns using pipes drawing directly from lakes commonly report issues with plugged lines caused by clusters of mussels. Sometimes the mussels attach to boat motors and over time can clog parts of the engine where water-cooling occurs.

This can lead to the overheating of boat motors and the collection of mussels on boat hulls create drag and can cause steering issues.

Though more of an annoyance than a significant issue, the shells of such mussels may cause cuts and scrapes to swimmers (think of a lower-degree barnacle problem on ocean waterfronts) and anglers can lose lures as sharp shells may cut their lines. Infestations can become so pervasive that swimmers and beachgoers need to wear foot protection to use the shoreline for recreational purposes. While they are not predators of other native mussels, the smaller zebra mussels sometimes attach to the larger native mussels en masse and eventually kill them.

Zebra mussels cause major problems for public utilities. The same clogging issues that occur for private lakefront landowners do as well for municipal water treatment facilities. Power plants using exchange pumps to cool plant infrastructure have high maintenance costs due to zebra mussels infiltrating pipes.

But large infestations of an invasive species can prove to be complex in their impacts as they may have some positive consequences for lakes. For example, zebra mussels are very good at filtering water and extracting contaminants that cause water to be opaque. Thus, increased water clarity may be attributed to the introduction of zebra mussels into a body of water, which seems like an unintended positive effect. However, while the water is clearer, it is not necessarily cleaner. Light sensitive fish need some vegetation for shade, but with effective filter feeding by zebra mussels, these fish can lose habitat as they have to dive deeper to avoid excessive light. Also, heavy light penetration allows for overly dense vegetation growth further down the water column. This makes it too easy for smaller fish to hide and in turn can stunt the population of the larger fish who prey on them.

Considering the damage done by the spread of zebra mussles, what can be done to control them? First of all, commercial vessels are not the only causes of infestations. While the introduction to invasive flora and fauna in larger bodies of water can often be attributed to large-scale industrial shipping and with those species hitching a ride on their hulls, the more pervasive spread of invasive species to smaller river tributaries, freshwater lakes, and ponds comes from recreational boating. Environmental regulatory agencies in states throughout the Great Lakes region have tried to battle the zebra mussel menace via an expansive education and outreach campaign targeted at recreational water users. The Wisconsin and Minnesota DNRs have collaborated extensively to utilize YouTube and other social media to spread the word about identifying and reporting invasive species, as well as sharing tips on safe boating and fishing practices in hopes of stopping their spread.

Among the practices for preventing the spread of zebra mussels include inspecting your boat and all boating equipment, scrubbing propellers and anchors upon each exit from a lake or pond, and being sure to remove all aquatic life from these items. It is also recommended to remove all mud from the boat and engine as zebra mussel larvae may be mixed in with it. (Perhaps not surprisingly, this advice doesn't merely apply to zebra mussels. Some invasive plants can reproduce from a single strand barely two inches long.) Other instructions include draining all water from the boat, motor, live wells, bait wells, and bilge, as well as disposing of live bait in a trash receptacle—and not simply throwing it out on the ground or in the water. In addition, rinsing your boat and all fishing equipment in hot tap water—104 degrees F or higher—is vital for killing adult zebra mussels and their larvae. If hot water is not accessible, another possibility is to completely dry your boat and fishing equipment by leaving them out in the sun for five days before entering any new body of water.[84]

There are some issues of philosophical relevance here and some lessons from the cases of invasive species above. For one, the line between a *native* and *non-native species* is not as clear as it seems when reading regulatory documents on the subject. An historical account of what we call "native species" suggests that what we currently call "native species" are not always indigenous to the areas they currently inhabit. Consider rabbits in England. As recounted by the historian Lynn White, Aldous Huxley was chagrinned about what he called "man's unnatural treatment of nature and its sad results."[85] Nearing his death he found that the hills and dales once adorned by grasses he had loved as a child were later overgrown with unsightly brush. Rabbits had kept the brush in check, but also destroyed farmer's crops. As a result, farmers introduced a disease (myxomatosis) that eliminated the rabbits. What White points out is the irony that back in 1176, domesticated rabbits had actually been *introduced* to England, presumably to provide an abundant protein source to peasants.

This tale challenges us to ask the question of what exactly does it mean for a species to be "natural" or "native" to a region? Also, the terms "native" or "natural" here for certain species do not seem to be value-neutral. There appears to be a presumption that "native" things are good and "non-native" or "alien" species are bad. But this assumption could be challenged. Consider that if dinosaurs still existed in the state of Connecticut (and assuming that we are

84 See the University of Wisconsin Sea Grant Institute Website, "Tips for Preventing the Spread of Invasive Species," (website), 2013. Available at http://seagrant.wisc.edu/Home/Topics/InvasiveSpecies/Details.aspx?PostID=1078 (accessed November 14, 2014).

85 "The Historical Roots of Our Ecological Crisis," *Science* (1967) 155(3767): 1203–1207 at 1203.

also somehow around), we would likely have to admit that they are "native" species though this would cause us some pause about how "good" they are. It is indeed unlikely that we would believe we should protect them despite them being species "natural" or "native" to the area. We would most likely think, assuming the Tyrannosaurus Rex were a great menace to human health and welfare, that they would have little value and hence we might think them good candidates for eradication, or at least should be treated as a dangerous threat to be controlled. The point is that whether a species is "natural," "native," or "indigenous" to an area really isn't the issue with respect to its anthropocentric value. The issue is more whether a species is valued. Would anyone have cared if Asian carp "infested" Wisconsin waters if it were prized as a game fish, were not so bony, and didn't outcompete other species that we enjoy having around?

The equivocation of "native" and "valuable" is easier to see when we reflect on the definition of an invasive species according to state and federal law. While it seems to be a *necessary condition* that invasive species are not "indigenous" to the area, this is not a *sufficient condition*. It also must be demonstrated that such a species either is or is likely to cause substantial environmental or economic harm, or to endanger human health. What isn't always made explicit is that a species being "native" to a location is often relative to time period and the value judgments of humans. The issue seems to come down to what humans in a particular location tend to value, that is whether something is a nuisance or harmful to them or to other things (other species or ecosystems) they care about. The problem with invasive species is not that are "from outside the area," but that they are in some sense undesirable. Thus, the value of native and non-native species alike is not based on something *intrinsic* to the species as a whole or its individual members. This value is grounded in *human* interests, and thus policies against invasive species are (in the legal context anyhow) justified by some *anthropocentric* theory of value.

Another idea often discussed among scholars and policymakers applicable to cases of invasive species are the *unintended and unanticipated adverse effects* of a practice. Note that the introduction of Asian carp into the US was supposed to be for the benefit of the aquaculture and waste treatment industries (with carp contained in storage ponds), but eventually led to undesirable consequences when they escaped. The initial reason for bringing in Asian carp was for a desirable goal (to take aggressive action to clean up ponds and reverse the adverse effects of fish farming and sewage lagoons).

However, one thing to be taken under consideration with any introduction (or reintroduction) of a species of plant or animal is "what is plan b?" just in case the species is released into the wild whether with intent or by accident.

Unfortunately, there was no "plan b" with Asian carp—no contingency plan in place to insure against their adverse impacts in case they escaped confinement. But of course, part of the reason there was no plan in place was because no one knew what would happen if Asian carp somehow escaped into the wild. One might wonder what the application of the *precautionary principle* might be here. What preventative measures would the precautionary principle possibly require? What has been called the "*strong precautionary principle*" (SPP) would suggest that since there might always be uncertainty about whether a new species initially meant to deal one with one problem could cause another, there should at least be strict regulations on its transport. But at most, would the SPP require a ban on the transport of *all* species of plants and animals? Would this sort of prohibition be too draconian and impractical? After all, one of the chief complaints about the precautionary principle taken in its strictest sense is that it would likely militate against *all* actions and unduly stunt economic and scientific progress.

However, there may still be problems even if a more lenient version of the precautionary principle were in place. A less strict version called the "*weak precautionary principle*" (WPP) might go like this: lack of scientific evidence does not preclude action if damage would otherwise be serious or irreversible. The WPP allows for preventative measures in the face of uncertainty, but does not require them. The usual applications of the WPP include safety practices such as avoiding certain dangerous neighborhoods at night or buckling your seatbelt. Likely the problem with WPP, however, is that it is not clear how such a principle would be helpful in trying to prevent the possible spread of invasive species because it is too weak. As we have seen, there is sufficient scientific evidence to demonstrate that zebra mussels and Asian carp cause harm to other species of non-human animals and costly economic woes for humans. Furthermore, this evidence would certainly be grounds for appeal to the WPP. Yet of course, all the WPP would be strong enough to endorse are the actions we already employ—while it is recommended to do so, it is not currently required that boaters wash down their boats to prevent introduction of zebra mussels into additional waterways. This is likely one of the reasons why zebra mussels continue to spread and why a rather robust educational campaign has been launched to try to stop it.

The issues of blame and culpability arise in the case of Asian carp where there has been a great deal of feuding between state governments. For example, the northern Great Lakes states of Wisconsin, Michigan, and Minnesota have called on states further south such as Illinois to take more aggressive actions against carp. The more northerly Great Lakes states argue that the electronic barrier in the CAWS is not a "fail-safe" measure of preventing carp surge. Instead, they believe the only permanent solution will be for the CAWS system to be disconnected from the Great Lakes. Illinois has remained steadfast in keeping the canals open, citing deep negative economic consequences not only for the city of Chicago, but also for the rest of the state if the system were cut off from Great Lakes access. Yet, the northern Great Lakes states argue that their economic case is just as viable—exposure of the Great Lakes to Asian carp would spell ruin to the tourist, recreational fishing, and boating sectors vital to their economies.

One might argue there is no evidence the state of Illinois intentionally or unintentionally released Asian carp into the wild. That is, in the case of intent, Illinois can't be shown to have caused an act of *commission*—actually doing something that caused harm to others. In this sense, one might question in what way Illinois is culpable for any possible future damages done to the Great Lakes region by the migration of Asian carp. Moreover, why should Illinois be held fully responsible for the substantial burdens of Asian carp control (as were outlined in extensive detail earlier in this case study)? Should the states of Wisconsin, Michigan, and Minnesota send financial aid and offer technical assistance to Illinois to combat Asian carp?

One might still insist the state of Illinois has a moral obligation to combat Asian carp because if people are in close proximity to help prevent a great wrong from occurring, this is sufficient ground for requiring them to help. Not doing so would be wrongful because it would be a *moral omission*. That is, one can be held morally culpable for *failing* to do something that could prevent harms to others. The paradigm case that is supposed to support this line of argument comes from the work of Peter Singer, a contemporary utilitarian philosopher. He gives the example of a child drowning in a pond and of a passerby seeing the child struggle. Singer claims that if it wouldn't be too risky for the passerby to act (that is, he can swim) and causes him little more than inconvenience by doing so, he is morally required to save the child.[86] Analogously, one might argue that officials in a position to help (state and federal environmental regulatory agencies) in the state of Illinois, by virtue of their proximity to a highly destructive invasive species approaching a major conduit to the Great Lakes, have a similar obligation. One might add that since Illinois built

86 Do note that in the US, there is no legal requirement to save someone whose life is imperiled unless the person in a position to save is an emergency responder. That is, there are no Good Samaritan laws in the US.

the last section of the conduit (the CAWS) to the Great Lakes, they may have an additional responsibility to keep harmful species out.

Finding moral culpability in the case of zebra mussels might be more difficult. While there have never been formal accusations or calls for reparations, might there be an argument that the original sources of the infestations might be held accountable? The ships that brought zebra mussels on their hulls and in their ballast waters look to be the first link in the chain of causes of the infestation of US waterways. Should the shipping companies be held liable or perhaps their government(s) for not requiring protocols for hull cleaning before embarkation? But deriving responsibility from this fact is obviously tricky. How would the exact culprits be identified? Also, all signs point to zebra mussel introduction to American waters as being accidental. Likely little thought was put into whether to release ballast water into the Great Lakes by transoceanic freighters. Even if one might argue that zebra mussels in American and Canadian waters came from Europe, should the European Union (EU) or the shipping companies be held responsible for introducing the invasive species to the Great Lakes? Furthermore, how would compensation be handled? Should the EU or the shipping companies pay for the costs of containing the problem of zebra mussels and repairing public utilities facilities damaged by them in the US and Canada? These are extremely complex moral and policy issues that won't likely be settled any time soon.

Review Questions

1) What kinds of regulations (if any at all) should states and the federal government have at their disposal to cease or lessen the damage done by invasive species?
2) What are the individual responsibilities of boaters using lakes to help keep them cleared of invasive species? Do you have a moral obligation to clean your boat? Why or why not?
3) Let's assume for the moment that you believe all non-human animals have intrinsic value. In this case, could you *consistently* call for controls of zebra mussels and Asian carp on the grounds that they cause economic and perhaps personal harms to *humans* through their destruction of habitats for recreational fishers? How might that argument go?
4) Does this case study show that the precautionary principle is useless as a principle for environmental protection? Is there a version of the principle you can think of that may be of greater use? Why or why not?

Case 43 – A Different Approach to Snow Removal –

Study Questions

1) What are some of the issues caused by salt and other materials applied to roadways in the deicing process?
2) What are some strategies used to reduce contaminants originating from roadway deicing?

Snow can have a significant impact of local ecosystems since it accumulates many contaminants from the atmosphere. But its removal can cause even more environmental worries to those in charge of lessening water pollution. Snow plowed off from roadways can contain salt, salt additives, asbestos, heavy metals, oil, grease, and nutrients. All of these contaminants can enter nearby soils in heavy concentrations and seep into local water systems.

To reduce this adverse environmental impact and attempt to lower costs of its municipal snow removal service, the city of Beloit, Wisconsin, has taken an unorthodox approach. Its crews use a solution of salt water and beet juice proven to be extremely effective in deicing streets. It also saves money by reducing some of the rock salt that would otherwise be used and is actually better at melting snow and ice. Part of this effectiveness arises because beet juice freezes at a lower temperature than salt, allowing the solution to work at lower temperatures than salt brine can alone. The beet juice solution is derived from the juice left over from the process of removing sugar from sugar beets. Sometimes crews spray roads before snow and ice fall to help salt stick to the surface. The mixture reduces the amount of salt needed on roadways and mitigates its corrosive properties. Also, it is attractive because it doesn't have harmful chloride effects on local water sources.[87] Beloit has now used beet juice for over a decade.

But what are the specific worries about rock salt use on roads? To understand this, we have to examine the characteristics of rock salt more carefully. All salt comes from seas one way or another. Solar salt and sea salt are harvested from existing seas, while "halite" salt originates from ancient seabeds and is mined. Rock salt is the more common name for "halite" and has the

87 Katie DeLong, "Experimenting with Beet Juice on Frozen Roads," *The Milwaukee Journal*, February 11, 2008. Available at http://www.todaystmj4.com/news/local/15508127.html (accessed July 15, 2015).

chemical formula (NaCl or sodium chloride). Seawater contains dissolved Na+ and Cl- ions and when it evaporates salt is left behind. The snow removal system in place is important to the salt industry as about two-thirds of all salt produced in the US ends up on transportation conduits. Only about 6% is refined into table salt. Rock salt is also sometimes used in the roadbeds of highways in northern climates to help reduce the freezing point of water on roads, thereby not allowing it to freeze at 32 degrees F on road surfaces.

Rock salt causes the most effective melt of snow and ice when the temperature is at 20–25 degrees F. This means that for many winter storms, its efficacy is actually quite high for clearing roads. But rock salt can do damage not only to the local environment, but also to the road itself. In the 20–25 degree F range, rock salt does minimal to moderate damage to concrete. This level of damage, while accounted for, does not weigh heavily against using rock salt on state and municipal roadways. With this said, for most local departments in charge of clearing highways, roads, and streets, rock salt is the most common applicant to roads during winter storms because it is fairly effective over a wide range of temperatures and is often the most cost effective solution.

However, rock salt has its limitations in its use as a deicer. First of all, even salt water freezes at 0 degrees F. This means that when temperatures are very cold (as they can be for significant time periods in the northern parts of the United States and Canada), rock salt is mainly ineffective in producing snowmelt. It is also very corrosive. It can eat away the steel of automobiles and if you ask anyone who has lived in the northern US or Canada for some period of time, they can likely regale you with stories of rusted out vehicles. Rock salt can also corrode the rebar in concrete bridges, which makes its use a significant contributor to the sorry state of America's bridge network. More recently developed deicers use different sorts of salts, including calcium chloride ($CaCl_2$) and magnesium chloride ($MgCl_2$). These compounds further reduce the freezing point of water and produce exothermic reactions that enhance the melting capacity of the deicer.

There are several effects of rock salt as it spills out into the local ecosystem after it either bounces off from the road when it is applied and falls into gutters, or is washed into storm water drains via snowmelt. It can collect and raise the level of salinity to the point where grasses, plants, and trees along roadsides can be harmed. Water quality can also obviously be negatively affected. After all, it is not as if rock salt simply disappears after it is applied. A study conducted in the Twin Cities area of Minnesota revealed that roughly 70% of salts applied to roads stay within the local watershed. This was the first study to analyze just how much salt is retained in the watershed of an entire metro area.

In areas near oceans a great proportion of road salt may likely enter the sea[88] raising the salinity levels of marine estuaries and adversely affecting shoreline ecosystems. Saline infiltration into fresh groundwater is a substantial problem since once this sort of contamination occurs, it is extremely difficult to remove and only at great expense.

Migratory birds can also suffer adverse health effects from rock salt since it is sometimes difficult for them to distinguish between corns of rock salt and seeds that are part of their diet. The salt crystals they ingest can be severely hazardous to these birds. There are also effects on humans, though not as severe, but costly in their own way, as rock salt can damage carpets as it is easily tracked into homes and businesses. This can result in not only destroying carpets, but also in damaging floors (particularly hardwood ones). For these reasons, some states limit the amount of salt that can be used by transportation department crews in dealing with snow and ice removal.

There are also broader effects of accumulations of rock salt that cause more widespread, systemic, and significant damage over time. For example, it can get into the drinking water supply, which renders the water too saline and toxic to both plants and animals. This is also a health issue for people on sodium-restricted diets. Even if the problem is not severe enough to cause harm to health, there still might be other negative effects such as making the taste of water less palatable from small amounts of saline intrusion. Beyond infiltrating the drinking water supply, rock salt can disperse into lakes via storm drains during heavy rains and snow pack melts. Heavy doses of both sodium and chloride ions that derive from diluted roadway salt can disrupt the ability of freshwater animals to regulate how fluid passes in and out of their bodies.[89] Amphibians can be negatively affected in acute ways as their skin is permeable and thus high levels of saline can produce toxic results. Additionally, altered salinity levels of individual ponds or lakes can impact how water columns mix during changes in the seasons—leading to high salinity pockets near the bottom and biological dead zones.[90]

Each winter, roughly more than 20 million tons of sodium chloride are applied to US roadways. This is about 13 times the amount of salt used by the entire food industry.[91] Surface water salt pollution is determined by chloride

88 Nina Rastogi, "Salting the Earth: Does Road Salt Harm the Environment," *Slate*, February 16, 2010. Available at http://www.slate.com/articles/health_and_science/the_green_lantern/2010/02/salting_the_earth.html (accessed July 1, 2015).

89 Rastogi, "Salting the Earth."

90 Rastogi, "Salting the Earth."

91 Rastogi, "Salting the Earth."

concentration. The US Environmental Protection Agency (USEPA) recommends that chloride levels be kept under 230 milligrams per liter as measured over the course of four days. This is equivalent to one teaspoon of salt per five gallons of water. Yet, in a 2009 study released by the US Geological Survey, about 40% of the urban and suburban streams tested in the northern US contained chloride levels at or above the level recommended by the USEPA at some point during the sampling period from 1991–2004. The deicing of roads was cited as the most significant contributor to these high levels of chloride.[92]

In response to these ecological concerns, as well as to recent shortages in road salt (a case in point was the winter of 2013–2014 with its heavy snowfall in many parts of the country along with deep freezes from polar vortex events), many highway departments have had to become innovative in their snow removal strategies. Beet juice has been used for many reasons. We have already mentioned that it has a lower freezing point than salt brine and is also sticky enough to hold rock salt onto the road to allow it to melt snow and ice more effectively. This is important as it is estimated that 30% of rock salt corns simply bounce off from the road when they are applied, meaning that not only do they contribute to the chloride problem, but they do so in a wasteful way without having any positive melting effects on roadways. By comparison, there is only a 3% loss of rock salt when it is applied with beet juice. Usually, beet juice is mixed with rock salt with most road managers fixing on the "magic formula" of 80% rock salt to 20% beet juice. However, beet juice can also be applied by itself as a treatment for roads, especially when temperatures are below 20 degree F.

Admittedly, in examining the multiple news reports about the shift to beet juice and other organic materials to fight snow and ice, the primary motivations involved have been more economic than environmental. Salt is relatively inexpensive until there are shortages. Beet juice is much more expensive than rock salt, but it greatly reduces the amount of rock salt needed on roadways. When rock salt is in short supply and the inventory needs to be stretched to cover a long winter season, beet juice can help. Whatever salt is used needs to stay on the road longer and beet juice allows for that to happen. Some think salt/beet juice mixtures result in a win-win-win scenario (in terms of cost, effectiveness, and environmental stewardship). So even though the initial financial hit is greater when highway departments purchase beet juice, it allows for so much less salt use that it tends to recover its cost.

Public works officials in Beloit believe that it is actually more effective using its beet juice/salt mixture in clearing roads than using rock salt alone, and thus

92 Rastogi, "Salting the Earth."

increases public safety. As the city's public works supervisor Bruce Slagoski put it, "It's making it easier for residents to travel through town, it cuts down on some of our labor time . . . and it will help us to sleep." When he was asked what makes sugar beet juice effective in melting ice and snow he added, "The sugar inside of the beet just releases. It then works as a heater, causing ice to melt faster. It works better on the snow and ice and it's actually better for the environment because you're using those smaller quantit[ies] and being organic—it's kind of natural. Economically it helps—it makes it a little better for us as a community so we can afford to do more."[93]

But even the use of beet juice may be no panacea for solving the environmental damage caused by road salt since it could cause its own problems. Beet juice may seep into nearby riverways and the nitrates of which it is composed may contribute to oxygen depletion in water sources. Beet juice is messy and municipalities using it sometimes receive calls from concerned motorists about how to remove the brown sticky film from their vehicles. It can also congeal, though it has to be around -10 to -15 degrees F for this to occur.

However, some of these potential downsides of using beet juice can be diminished easily. Highway officials note it can be removed from vehicles with ordinary soap and water, causing no permanent staining. Beet juice issues can also be lessened if drivers heed warnings to stay back from the spray and avoid tailgating public works trucks. Beet juice is less corrosive than salt and since it contributes to using less salt this also mitigates worries that deicers are rusting out vehicles. Some even claim the reaction *between* beet juice and salt renders the salt less corrosive.

Even still, some worry that beet juice is not as eco-friendly as originally thought. Beet juice is actually a by-product extract created when making beet molasses and is combined with other beet waste products.[94] This means it is an organic material and has a rotten smell. But it isn't the odor that has caught the attention of the Madison and Dane County Department of Health (DOH) in Wisconsin. The agency has examined the use of beet juice, questioning its safety for local ecosystems. As an organic material just like others such as yard waste that enter lakes in runoff, it requires oxygen to break down its carbon (see Case 10—"The Lakes and Phosphorus Twins: Monona and Mendota").

93 "Vegetable Juice Helps Beloit to Beet Out Snow and Ice," *WTVO and WQRF News* (Television series episode), January 30, 2014, Available at http://www.mystateline.com/fulltext-news/d/story/vegetable-juice-helps-beloit-to-beet-out-snow-and/12370/s8O1—rjnEy-ah-z8b2mBQ (accessed July 18, 2016).

94 Dani Molnar, "Beat It, Beet Juice," *The Municipal*, (website), January 13, 2013. Available at http://www.themunicipal.com/2013/01/beat-it-beet-juice/ (accessed July 1, 2015).

While beet juice is biodegradable, if it's consuming too much oxygen as it decomposes the DOH worries it could contribute to hypoxia, which reduces the amount of oxygen fish and other aquatic organisms need for survival.[95]

But beet juice supporters see ways around these concerns. Christine Walsh, the streets supervisor for the city of Beloit and Denver Preston, the general sales manager of winter products for KTech Specialty of Ashley, Indiana, both insist these problems are avoidable. Using smaller amounts of beet juice will eliminate the problem of runoff as Walsh notes, "When we're spraying rock salt [with beet juice], it's maybe seven gallons over one mile, just enough to get it wet."[96] She also observes that Madison's streets division already doesn't use much salt brine on roads, keeping runoff to a minimum. Walsh thinks that the 80% salt, 10% organic beet juice and 10% calcium chloride ratio is ideal, but she recognizes street workers who mix the concoction before the trucks go out use a figure closer to 15% beet juice, and adjust the amount according to the weather.

Preston notes that while Beloit has had positive experiences using beet juice, he recommends that when a streets department shifts in its approach to using organic alternatives mixed with rock salt, it shouldn't use beet juice, but instead beet molasses extract—for reasons other than environmental worries. He asserts that beet juice is inconsistent in its application and quite messy. "It's waste stream. Waste stream products are just that. They're nothing but waste from the process of making molasses and table sugar. As they're processing the beet, they're removing all of the waste, all the components they don't want such as beet pulp, sand, and dirt."[97]

And these wastes are just the things that are sold to municipal streets divisions and can clog the lines of sprayers. Some divisions don't have filters to remove these impurities—sometimes even finding whole sugar beets at the bottoms of tanks that made it through processing. While diluting beet juice with water could lessen the problem, this in turn causes another issue as obviously, water freezes. However, this can also be resolved, according to both Preston and Walsh, by using chlorides.

Ultimately, Preston thinks it is time to move on from using beet juice to using beet molasses products instead. Unlike beet juice, which has very little sugar and can be inconsistent in how much it contains, beet molasses product contains consistent sugar content—as it must since it is used in baking.

---

95 Molnar, "Beat It, Beet Juice."

96 Molnar, "Beat It, Beet Juice."

97 Molnar, "Beat It, Beet Juice."

In the future, some roadways may not need any chemical applicants to remain snow free, at least if cutting edge technology used for airport runways becomes economically feasible enough to penetrate the transportation market. The British company ICAX constructs runways for airports using recycled solar energy to clear ice by heating runways.[98] One of the greatest hazards in winter air travel is (unsurprisingly) a runway slippery from ice and snow buildup. This means in effect the runway may not be long enough for some planes if they cannot use their brakes effectively. A second danger is that corrosion is accelerated by the use of potassium-based deicers that can damage runways over time. The third problem, as with normal roadway deicing, is the adverse environmental impact of using toxic chemicals as they are washed off the runways into adjacent grounds and streams, leeching into soils and contaminating local habits of flora and fauna.

But all three of these problems can be resolved using a recycled solar energy system. ICAX has developed an interseasonal heat transfer system that accumulates and stores solar energy from the summer season for use in the winter. The energy is sequestered in thermal banks lodged into the ground until it is needed during cold months to de-ice runways by raising their surface temperature above the freezing point. First, an asphalt solar collector gathers energy into the system. It is then transferred to the thermal banks that can heat the ground to over 77 degrees F. Black road surfaces have the tendency to absorb heat to a point where they radiate it back as quickly as they absorb it; that is, the surface temperature of roads in direct sunlight can often reach 59 degree F higher than the ambient air temperature. However, the asphalt developed by ICAX collects this heat via a fluid circulating through an array of pipes embedded in the road and deposits the energy into the thermal bank. The temperature across a large thermal bank infrastructure can be increased from its natural temperature of 50 degrees F to 77 degrees F over the course of the summer months. In a trial in Toddington, England, the Transit Research Laboratory reported on a stretch of road where the ICAX system was tested:

> On 7 February 2007 extreme surface temperatures of –6 C (21 F) and –3 C (26 F) [occurred] for a few hours on the unheated and heated sections respectively. The following morning snow fell and briefly settled although the road surface was being maintained between 0°C (32 F) and +0.5°C (33 F) at the time. Apart from this one incident, no other issues

98 ICAX, "ICAX Solar Runways Systems Clear Ice from Airplane Parking Stands Using Under Runway Heating," (website), 2015. Available at http://www.icax.co.uk/Solar_Runways.html (accessed July 19, 2016).

> were encountered in keeping the road surface above freezing. Heating was activated automatically on 16 more days than the 28 days on which salting took place on the nearby motorway. Further tests would help to refine the parameters used for triggering surface heating and to conserve heat for more extreme temperature conditions. Numerical modeling suggests that a significant improvement in winter maintenance performance is likely if the heat pipes below the road surface were at shallower depths of cover than the 120mm (~4 ½ inches) used in this study.[99]

While it is clear that more research will need to be conducted to make this technology widespread and cost-effective, there is the possibility that chemical deicers may be phased out in the future (or at least reduced). But as it stands now, public works departments and streets divisions will have to choose which method or methods of road treatment are most appropriate, and that decision may not simply depend on economics.

Review Questions

1) Local transportation departments sometimes have competing obligations with respect to snow removal. On the one hand, there is the directive to keep public roadways as safe as possible for the public. However, given recent hard winters and depleted salt supplies, the public safety concern does not "win out at any price" as these agencies also have to consider the cost of rock salt and other deicers in determining how much to apply to roadways. Also, conventional deicers can carry environmental hazards, leading streets divisions to use less salt and/or consider alternatives such as beet juice, cheese brine, pickle juice, molasses, and even vodka! How do you think these competing principles of safety versus cost should be weighed against each other? Should any decrease in the level of public safety be permitted if there are economic concerns? Should there be decreases in the level of public safety due to countervailing environmental concerns? What if there is no unit of measure that can be used to make this calculation? How is it possible to measure these concerns against each other in a utilitarian calculation to come to a policy decision?

---

99 ICAX, "TRL Reports on IHT Ice Clearing at Toddington Using Under Road Heating," (website), April 2008. Available at http://www.icax.co.uk/toddington_results.html (accessed July 19, 2016).

2) Do you think there should be *public* investment in research and development of new technologies for deicing such as those being experimented with by ICAX using solar energy? Why or why not?

Case 44 – The Deicing Dilemma –

Study Questions

1) What are some of the local effects of deicing driveways and sidewalks? How do these compare with the effects of deicers on streets, roads, and highways?
2) What happens to the pH of soil when it is exposed to rock salt?
3) How do the use of organic materials compare with the use of chemical deicers with respect to effectiveness in clearing sidewalks and driveways?

Imagine that you live on a residential street in a town or city, or along a rural route. It is two days after a storm that dumped a foot of snow. There is a significant warm up in the temperature during the daylight hours melting the snow, topping out at 45F by late in the afternoon. Yet, the evening forecast is for a cold front to arrive. Overnight temperatures are slated to drop to 27F, which will be cold enough to refreeze any water remaining on the sidewalk from the day's melt and leave it slippery underfoot. The ground is frozen, so there is little heat to rely on there. Let's suppose further that your driveway is on an incline, perhaps with storm water drains at the end of it that run to a nearby lake, or has gutters and a culvert meant to divert water to a local marsh. What will you do? Will you simply do nothing, thinking that not many people pass by (or not caring if they do)? Or will you treat it with a deicer to try to ensure safe passage for yourself and others—perhaps trying to avert a nasty fall, lawsuit, or both?

This is a situation often faced by those who live in snowy and icy climates during the wintertime. The coming and going of ice and snow on sidewalks and driveways near homes and businesses can highly fluctuate over the winter and into the spring as well. We have already seen (Case 43—"A Different Approach to Snow Removal") that with deicing streets, roads, highways, and runways there can be a significant widespread effect on local ecosystems with the use of conventional deicers.

Yet there are local effects of snow removal by public works divisions on private residents and businesses. Salt applied to roads can wash off and not only raise soil salinity, but also damage trees and shrubs some 650 feet away. While it would likely take overapplication of salt on rural and suburban driveways of residents, it has happened, effectively turning road shoulders into salt licks attracting elk, moose, deer, and birds. Roadside salt buildup thus contributes to a common problem; vehicles hitting these animals, causing safety issues for human and non-human animals alike, along with extensive damage to motor vehicles.[100]

Salt application will corrode your sidewalks, driveway, and vehicle. Heavy treatments over time can even harm your domesticated pets, shrubs, trees, and other plants. After all, salt is water-soluble and water is the great transportation system in nature—salt may have landed on the driveway, but it's not going to stay there. Your driveway is likely impermeable, thus that sodium and chloride-laced water is going to become part of the sediment somewhere else. Also, when salt leeches into the soil, this can cause heavy metals to get drawn out and be carried by flowing water into local waterways. It's likely going to go into your adjacent lawn, flowerbeds, or into the storm water system. This is why in some towns and cities there are decals placed on storm drains reminding residents that what goes in them ends up in lakes and rivers. Some storm drains and pipes lead to treatment plants, but most don't. And even if storm water is treated, sodium and chloride are not filtered out, meaning that those elements simply end up in the watershed.[101] Salty water lowers soil pH, leaving it more acidic, and inhibits water and nutrient uptake into plants. In the presence of excessive salt, plants will seek more water to balance out the pH. Yet, the water necessary for dilution may not be available.[102]

If you happen to have wild animals that eat grass near your driveway, heavy salt use may leave them with less food. Even if grasses are able to grow, they may hold high concentrations of salt, poisoning animals ingesting them. Sodium chloride is also an irritant and in especially high concentrations can cause superficial burns to animals coming into contact with it—usually on the pads of paws.

Thus, residents may have to think about ethical questions regarding how they should care for their walks and driveways in the wintertime. We are faced with the question of whether private citizens, on property they own or oversee, should use certain types of deicers on passageways for which they are responsible. In some municipalities, while the city owns the sidewalks that run along

100 Rastogi, "Salting the Earth."

101 See Case 24—"Foul Ball: Ballpark's Effluent out of Play?"

102 Zach Mertz, "Why You Shouldn't Use Salt to Melt Ice," *Boston Globe*, December 24, 2015. Available at http://www.boston.com/yourtown/news/weymouth/2012/12/salt_in_the_wound_why_you_shou.html (accessed July 19, 2016).

the property lines of private residents' homes, those property owners still shoulder the burden of maintenance.

The bottom line is that a very small amount of sodium chloride can have a very large negative effect on freshwater quality. This includes degrading water quality beyond the threshold for safe consumption. And of course, while one individual property owner salting her driveway will not cause serious effects downstream, whole neighborhoods or communities doing so can make a broad and serious negative cumulative impact.

Public works departments have used a number of alternatives to simply applying rock salt or the stronger calcium chloride on roadways, including beet juice, cheese brine, pickle juice, molasses and even vodka ... yes, vodka.... But some homeowners have also tried these out on their own sidewalks and driveways. Unfortunately, the information on what is the most effective (in terms of melting) is only anecdotal. One streets division worker for the City of Edmonton, Alberta, Canada (a city that knows something about cold weather!) has compared rock salt with several other substances for deicing. He notes that a variety of melting products sold commercially are only really different with respect to how much calcium chloride they contain. However, this worker concludes that rock salt is best at clearing sidewalks—better than pickle juice or beet juice. Other solutions are alleged to work on melting ice almost as well as rock salt, including a mixture of alfalfa hay, beetroot juice (which has very high sugar content), a bit of gravel, and a tiny portion of table salt. This mixture has little to no environmental impact in comparison to rock salt.[103]

While rock salt by itself may be better at melting snow and ice than other organic alternatives, assuming it is difficult for most residents to get their hands on BeetHeat[104] and other items, we need to know how these deicers work. Beet juice, beet molasses, pickle juice, and other organic substances do not melt ice by themselves. They are usually combined with a bit of salt and form a sticky substance allowing more salt to stick to targeted surfaces longer, hence enhancing their melting capacity. Sugars in these substances also interact with the ice, lowering its melting point and allowing for more effective melting at colder temperatures than the usual 15 degrees F threshold for saline.

Of course, other chemical deicers will melt ice and snow at colder temperatures than rock salt, including calcium chloride. The latter will do so at a faster

103 See Zach Mertz, "Rock Salt," *YouTube* (online video), December 15, 2012. Available at https://www.youtube.com/watch?v=1202ahpjeCw (accessed July 2, 2015).

104 This is beet molasses mixture from KTech, a company referred to in Case 43—"A Different Approach to Snow Removal."

rate than rock salt and is not as harmful to plants and animals. But calcium chloride has its own downsides. Its melt rate is so quick that it can leave a residue of moisture behind that may refreeze. Calcium chloride also eats into both metal and concrete and can damage some trees. Chloride also can't be separated from water and is left in the environment forever.

Ultimately, one might object that issues such as how we should deice sidewalks and driveways don't really mean that much environmentally and at most is a matter of personal ethics. The complaint is that life can't be so micromanaged and besides, how could it *really* matter if I put a bit of rock salt outside my home? After all, risks surround us. Am I really going to trouble myself much about the possible effects of my deicing when there are bigger and more important things to worry about such as life, my job, picking up kids from school, and all the other minutia of a given day? Sure, I would rather not fall on ice and hurt myself. I may even prefer that neighbors don't fall either (or at least I don't want to hear from them when they do). So, I make a small attempt to keep that from happening and toss out the cheapest and/or most convenient deicer—enough said!

The problem is that the way we take care of our property *does* have some impact on the environment, even if very small. And *collectively*, our contributions can combine with others to make a rather significant difference. It may then boil down to a question of character—what type of person do you want to be? Do you want to be in the set of people who contribute to the problem (even with the knowledge that you could do otherwise)? After all, perhaps lying to people for a significant part of your life is not all that bad—maybe it won't have a terrible consequence for your life or those of your family and friends. Even if others find out that you're lying and don't trust you as much (or take what you say with a grain of salt), this surely isn't the worst thing that will befall you, or others in your life. But my guess is that you would not find that to be a good reason to tell lies and you definitely wouldn't want to make a habit of it. Likewise, you might still have to confront the question of sorting through all of the other little things over which you have control (realizing what it is you do) and see if conducting yourself that way is "honorable" or "true to yourself."

Review Questions

1) Do you think private owners of houses or apartment buildings have any obligation to avoid using salt or calcium chloride on driveways and sidewalks to protect themselves, family members, and or pets from possible harms of these applicants? Why or why not? Do you think you have an

obligation not to do so for others? How about neighbors you don't know well or people in your region? Non-human animals? The ecosystem? Why or why not?

2) What do you think about the idea that even when your actions may have a negligible effect, you still should act in a certain way with a certain type of disposition because that is what is involved in a morally good character? Explain your answer. How would you apply your answer to whether or how you should go about deicing the driveway or sidewalk?

Case 45 – First 'Soccer Moms,' Now 'EcoMoms' –

Study Questions

1) What is an "EcoMom"? How did the EcoMom Alliance begin?
2) What are some of the goals of the EcoMom Alliance? How is the group structured?
3) What are the core principles of the EcoMom Alliance? What are some of the organization's initiatives? To what degree have they been successful?

In a setting that may look much like a women's book club meeting, we find a passionate discussion about whether to use anti-bacterial hand sanitizers or endorse waste-free school lunches. This is a local branch of the EcoMom Alliance, which as of 2008 had some 9,000 members across the United States[105] and a year later had grown to 11,000 strong.[106] As of 2015, according to information obtained from its website, the organization boasts 30,000 followers, fans, and subscribers in their social media network.

The EcoMom Alliance is part of a broader movement dedicated to the "green mom," which has been spawned by websites and blogs with titles such as cleanandgreenmom.blogspot.com and eco-chick.com.[107] The EcoMom Alliance itself has grown in size and prominence, as there are now not only local groups in the United States, but also in Mexico, Brazil, New Zealand, and other countries. Local groups discuss such subjects as how to reduce energy

---

105 Patricia Leigh Brown, "For New Wave of 'EcoMoms,' Saving Earth Begins at Home," *The New York Times*, February 16, 2008, p. A1.

106 Lindsay Palmer, "Earthshaking Moms: Mothers Across the Country are Jumping for Eco Joy," *Redbook*, March 19, 2009. Available at http://www.redbookmag.com/life/mom-kids/advice/a4888/eco-mom-alliance/ (accessed July 3, 2015).

107 Patricia Leigh Brown, "For New Wave of 'EcoMoms'," p. A1.

usage by washing clothes during non-primetime periods (such as after 7:00 p.m.) while using only cold water and biodegradable laundry detergents, as well as the virtues of using "smart" power strips that shut off power to household appliances when they are not in use.

Part of this need to talk about household environmental issues, admit members of the EcoMom Alliance, comes in response to their role as homemakers and feeling anxiety about how their daily activities (e.g., nightly baths, etc.) affect the environment. The membership of the EcoMom Alliance tends to be affluent, with particularly strong followings in Northern California. It is estimated that each citizen of Northern California's Marin County (a place which has its own local group) has a carbon footprint of 27 acres, which means that it takes that many acres to support the lifestyle of that person. The national average carbon footprint in the United States is 24 acres per person.

But another reason for joining may be ethical, as Tim Kasser, a psychologist at Knox College suggests. He thinks the members of EcoMom Alliance "are fighting a values battle. They are surrounded by materialism trying to figure out how to create a life more oriented toward intrinsic values".[108]

The mission of the organization is geared toward helping mothers concerned about environmental degradation plan and execute practical sustainability actions for both their families and communities:

> EcoMom® Alliance is a non-profit 501-c-3 organization nurturing, connecting, and empowering mothers to create a healthy and sustainable world. Our programs reach over a million families annually, informing and inspiring healthier living in communities like yours. By sharing connection, information, and resources, we support mothers in making healthier choices in their homes and communities, the effects of which ripple out and protect our children and environment for generations to come.[109]

The range of EcoMom Alliance activities has grown over the years, from members devising conservation strategies, to raising awareness about new products that might pose safety hazards, to organizing farm tours so that members can better understand where their food comes from.

The EcoMom Alliance began with Kimberly Danek Pinkson, who is concerned about the environment generally, but also with how ecological problems

---

108 Brown, "For New Wave of 'EcoMoms'," p. A10.

109 EcoMom Alliance, "Our Mission," (website), 2016, Available at http://www.ecomomalliance.org (accessed July 19, 2016).

relate to her family's health. In 2006, she co-produced an event for World Environment Day. Afterward, she was sitting with friends at a picnic and heard that she'd inspired one of them to install fluorescent light bulbs (as she adds, "you know, the curly-cue ones") in place of incandescent bulbs. Pinkson was touched that she had such an effect on her friend. But others were stimulated as well. As Pinkson described it, eco-friendly ideas started spreading throughout the picnic like wildfire, "... everyone started sharing planet-saving ideas. I suddenly realized the power moms have to effect [sic] global change."[110]

When asked in an interview about how she motivates moms to engage in green activities in their own communities, Pinkson noted that the EcoMom Alliance website has several resources to train EcoMom leaders. She observed that as of 2009 there were 300 EcoMom leaders, who after training host parties to present the EcoMom Challenge—the group's 10 steps for a sustainable future. The list of eco-friendly activities can include planting a tree or buying carbon offsets. The groups then meet monthly to determine how the basic principles of the EcoMom Alliance can be best implemented by their local groups to serve their communities. Pinkson reports that some of these meetings have been broadcast on the Internet as webinars on the Alliance's website so that groups can learn from each other. However, Pinkson explains that local areas can be very different and some ideas in one place won't necessarily work in others. She stresses that the EcoMom Alliance is meant to be (and remain) a grassroots organization.

Even still, there is something common to all of the local groups of the EcoMom Alliance—they serve as a united front presenting a very powerful voice.[111] Its members have participated in a number of initiatives since its inception. Some local groups petition local supermarkets, lobbying for more organic produce to be available and to impose plastic bag bans (as San Francisco now has, due at least in part to public outcry). Other local groups advocate for recycling and waste-free lunches in their childrens' local schools. Some groups have organized and executed "One Night Off" campaigns to turn off lights, televisions, and other appliances for one night per week to save electricity and draw attention to energy conservation. For moms who have trouble finding eco-friendly products for the home, the online EcoMom Marketplace provides a forum for doing so.

The organization also has a theory of social change, building on the premise that throughout history women have initiated major social and political

110 Palmer, "Earthshaking Moms."

111 Palmer, "Earthshaking Moms."

reform movements.[112] The group stresses that moms have a proud tradition ranging from fighting fevers to battling social injustices. The EcoMom Alliance calls the ways that change can be brought about "momfluencials"—methods of using buying power to change the market to promote both an environmentally and economically healthy future. This market clout is seen as a pragmatic way of initiating positive change for present and future generations.

The EcoMom Alliance taps into a desire of its members to return to some sense of balance, both in their own lives and in their relationships to the environment. EcoMom Alliance cites a 2006 Centers for Disease Control (CDC) study reporting that females in the US feel overextended, isolated, and depressed. EcoMom Alliance interprets this phenomenon as a sign that women have in a sense lost touch with themselves and the natural world.[113] The point of the organization is to help women sustain themselves and their families and reclaim some of that balance by providing useful resources, tools, education, and products.

The sustainability that the EcoMom Alliance promotes is one part environmental, but another part about the lives of women and their families. In this sense, the organization has a global and holistic platform for a sustainable lifestyle: 1) To provide an opportunity for women to gather and share inspiration and tools for healthy and balanced living. 2) To provide web-based, market-driven solutions to reach a conscious, healthy, and sustainable consumer-driven economy. 3) To provide an eco-media channel informing people about green products and news. 4) To provide a lifestyle brand representing the highest quality in sustainable business, living, and communication.[114]

As with any organization, one would expect it to change over time, evolving in certain ways to face new social and political realities and concerns. Pinkson has observed this as well. Even though the organization's core principles have remained the same, it has also advanced over time. From the beginning, Pinkson stressed that the EcoMom Alliance was more than just about the external environment, with the mantra "Sustain Yourself, Sustain Your Family, Sustain Our World." But she says she has had a deeper understanding of how much work is needed in aligning interests to raise women's voices. As she puts it, "I've realized though, even more deeply, how much we need each other. How much we need to support each other in raising women's voices. We are part of the environment and our wombs are every child's first environment. There

---

112 EcoMom Alliance, "Program Philosophy," (website), 2015. Available at http://www.ecomomalliance.org/page/theory-of-change (accessed August 21, 2016).

113 EcoMom Alliance, "Program Philosophy."

114 EcoMom Alliance, "Program Philosophy."

is no 'other' or 'out there'."[115] This more holistic conclusion came in part, says Pinkson, due to a more robust understanding that while there are great challenges, there are great opportunities since the environment is threatened in many different ways.

Along with this evolution of the group's approach to environmental issues, there is a misconception to clear up. The organization's name might suggest it only welcomes female membership. However, while the EcoMom Alliance was founded by a mom and consists mostly of mothers, it is not exclusive to women, as anyone can join simply by signing up, and getting involved in the organization in some way. One wonders if there will be an "EcoDad" Alliance anytime soon.

Review Questions

1) One might claim that organizations such as EcoMom Alliance are merely ploys by affluent people (who consume the most resources) to curb their "eco-guilt" by joining groups that seem to be conscious of environmental problems, but fail to truly address how their consumerist lifestyles cause such issues in the first place. That is, their efforts to find carbon offsets for air travel and other "luxury emissions" show their commitment to sustaining the environment is not sincere. How do you react to this type of criticism of an environmental group? Do the intentions of the members of these organizations matter or is any kind of greater awareness about environmental issues a good thing generally?
2) Why do you think there isn't an "EcoDad" Alliance? Does this say anything about the role of men in environmental movements? Does this say anything about the role of men in purchasing everyday items for the home kitchen or bathroom?

Case 46 – A Garden for Rain –

Study Questions

1) What is a rain garden and what sort of ecological purpose does it serve? Where are rain gardens often used?

115 Quoted in Amy Hook-Sopko, "The More You Know Eco: EcoMom Alliance," *Green Child Magazine*, (website), August 1, 2010. Available at http://www.greenchildmagazine.com/ecomom-alliance/ (accessed July 3, 2015).

2) How does the soil in rain gardens filter out pollutants? What are the physical and chemical filtering mechanisms used?

Rain gardens (also known as *bioretention systems*) have been lauded for several reasons, including that they provide good habitat for local species of plants and animals. But their central function is to serve as a filtration system that helps absorb water that would otherwise pass over the land surface and carry debris and contaminants down streets, sometimes (but rarely) into sewer and water treatment systems, and eventually into lakes, streams, and aquifers—some of which are used for public drinking water.[116] Considering what some see as extremely damaging effects of collective runoff from urban, suburban, and town areas, these gardens offer a possible solution to watershed environmental degradation. According to the United States Environmental Protection Agency, about 50% of rainfall will run off immediately when it lands on a city block when 75% of its surface is impermeable.

Instead of water simply flowing over impermeable surfaces such as asphalt and concrete leading quickly and directly into local bodies of water, storm water captured in rain gardens has the opportunity to sink into the soil, filtering out impurities. These bioretention mechanisms are often located in parking lot islands or in small residential land areas where there is a buildup of storm water outflows (such as subdivisions). However, they can also be placed outside private residences. Rain gardens are often fitted with hardy grasses, shrubs, trees, and other plants whose dense roots help keep soil in place. There is also a thin layer of mulch lying in the plant beds to slow the movement of the water and act as a sponge soaking it in. This allows nitrogen and phosphorus compounds to be filtered out of the water.

Rain gardens look like small depressions in the land and are fairly shallow, allowing for surface runoff to enter them. The idea is for them to replicate several of the pollutant removal capacities of forested areas. During a storm, surface water collects on the garden's mulch. Runoff from larger storms is generally diverted into storm water systems and rain gardens are usually ineffective for large downpours. But water from smaller rains or slower spring snow and ice melts can enter a rain garden and filter through the mulch and prepared soils. The water, after soaking in and having time for filtering, can then enter perforated pipes to return to the storm water drainage system.

Rain gardens are generally used in small-site, heavily urbanized areas, but with modifications can be utilized in many different settings. Bioretention systems can be applied to myriad climatic conditions throughout the US. Rain

116 See Case 24—"Foul Ball: Ballpark's Effluent out of Play?"

gardens are ideally suited to dense urban areas with nearby infrastructure such as parking lots with large impermeable surfaces. They require an appreciable amount of space (about 5% of the area that drains to them), but can be incorporated into existing parking lot islands.

Rain gardens are also very useful in what could best be called "storm water runoff hotspots"—areas where there is a large flow of runoff carrying high concentrations of pollutants. These concentrations tend to be higher than what are found in normal runoff water, such as areas beside the parking lots of gas stations or convenience stores. Rain gardens can be placed in high concentration runoff areas to even treat storm water if there is an impermeable liner used at the bottom of the filtration bed.

Local storm water systems are sometimes already in place and there is the question of how to redesign developed areas to reduce runoff carrying pollutants into local bodies of water. This requires a *storm water system retrofit* designed to not only improve water quality and protect channels downstream, but also to reduce local flood risks. In these retrofits, landscaped areas are altered to employ bioretention design principles in an existing area or if a new parking surface is being added. Bioretention retrofits are usually best applied in smaller land areas, as they are extremely expensive.

Yet, there is an impetus to do this work, especially in highly developed communities struggling to meet regulatory standards, such as pollution reduction targets under recently issued municipal storm water permits or to meet local TMDLs (*Total Maximum Daily Load*). *TMDL* is a regulatory term of the United States Clean Water Act that is a calculation of the concentration of pollutants local water bodies can have short of violating federal water quality standards.[117] For example, communities along the Chesapeake Bay have had to adjust their storm water systems to accommodate EPA TMDL requirements using an "upstream" approach. Accordingly, they have constructed rather extensive urban storm water retrofits. Additionally, each of the six Chesapeake Bay states (including Virginia, Maryland, Delaware, New York, Pennsylvania, and West Virginia) plus the District of Columbia, have considered more extensive use of storm water retrofits as part of their strategies in meeting nutrient and sediment reduction targets for existing urban development under the Chesapeake Bay TMDL.[118]

117 United States Environmental Protection Agency, "Impaired Waters and Total Maximum Daily Loads," (website), March 12, 2015. Available at http://water.epa.gov/lawsregs/lawsguidance/cwa/tmdl/ (accessed July 6, 2015).

118 Chesapeake Stormwater Network, "Urban Stormwater Retrofits," (website), 2015. Available at http://chesapeakestormwater.net/bay-stormwater/baywide-stormwater-policy/urban-stormwater-workgroup/retrofits/ (accessed July 6, 2015).

Specific habitats harboring species of fish are highly sensitive to changes in temperature. Trout, for example, do not tolerate even small upticks in water temperature very well. In these areas, the options for storm water management are quite limited. Here, rain gardens can be a very useful storm water treatment option near coldwater trout streams. Water only pools in rain gardens for a short period of time, thus decreasing the amount of time it can be warmed by the ground. Some rain gardens adjacent to parking lots have been shown to actually decrease the temperature of water.

But how does the soil clean water and which hydrological features help this process? There are two main ways; one is physical and the other is chemical. The proportions of silt, clay, sand, and organic matter are the main determinants in the movement of water through soils. The first principle of cleaning water is that the longer it has contact with soil, the cleaner it will be. Water tends to move slowly through soils that are more "clayey." This is mainly because the passages through clay are very small since clay is dense and its particles are small and tightly packed. On the other hand, water moves relatively quickly through sandy soils because of large spaces between the grains. This lack of time for water interaction with sand grains combined with smaller surface area means water running through sandy soils will not be as clean as water passing through clay and silt. Chemically, both clay particles and organic matter have positive charges that attract some chemicals and keep them moving through the soil.

The effectiveness of rain gardens as a tool for storm water management has been noted.[119] It has also been demonstrated that with some minor retrofitting, such bioretention systems can greatly enhance the treatment of polluted rainwater running off from roofs. Soils, even when they are not stripping pollutants from water, can at least interact with chemicals and neutralize them into a less toxic form.

Some earlier studies had shown that nitrates are poorly retained, and there is still some skepticism that they are all that effectively removed from contaminated waters by bioretention systems. But Michael Dietz and John Clausen's findings suggest that with some alternations in rain garden design specific types of contaminants may be targeted for sequestration. As Allan Davis of the University of Maryland notes, "Two interesting notes in [Dietz and Clausen's] paper. One, they are trying to focus on nutrients and nitrogen. A corollary to

119 Liz Thrall, "Rain Gardens Stem Urban Stormwater Flows," *Environmental Science and Technology* (2006) 40: 1093–1094. For the study, see Michael Dietz and John Clausen, "A Field Evaluation of Rain Garden Flow and Pollutant Treatment," *Environmental Science and Technology* (2006) 40: 1335–1340.

that is that working with saturation gives us an engineering tool that we can play with."[120]

Rain gardens became popular as a tool for storm water management beginning in the 1980's and have risen in profile since then. 1987 was a watershed year for bioretention systems as an amendment to the Clean Water Act allowed the EPA to impose a bevy of new requirements on storm water discharge from construction sites, industries, and municipal sewer systems. In 1990, Prince Georges County, Maryland, was the first municipality to use rain gardens to reduce the concentration of contaminants from reaching local waterways. Several counties, states, and municipalities soon followed Prince Georges' lead.

Roger Bannerman of the Wisconsin Department of Natural Resources observes just how much of a boon the use of bioretention systems can be to the general strategy for storm water treatment, "We are depending more and more on rain gardens to help us achieve our storm-water management goals."[121]

But champions of rain gardens cite another potential benefit of their use—they are aesthetically pleasing. Native plants can hold up to heavy water flow and pooling, produce beautiful blossoms, and create a lush landscape. Also with extra water available to them, these plants tend to thrive as opposed to being located in areas where rapid runoff does not provide sufficient time for the roots to effectively absorb water. But beyond the beauty of rain gardens, there is the question of whether there are moral obligations to install these sorts of landscaping features. It is likely that most people do not know that storm water is not treated before it enters into aquatic systems. Perhaps this educational process will inspire the development of more rain gardens and provide a valuable tool for cleaning water before it enters our local waterways.

Review Questions

1) Some environmentalists have virtually given up on the possibility of effective, large-scale, international accords for attacking climate change and other environmental problems. Instead, they have focused their energies on local, state, and regional environmental initiatives. On the other hand, there is skepticism that pervasive and difficult environmental problems are really going to be solved in this piecemeal way and thus national and multi-national cooperation is necessary. Do you think it is worth taking local action on environmental problems even if there is the

120 Thrall, "Rain Gardens," p. 1094.

121 Thrall, "Rain Gardens," p. 1094.

chance someday that this work might be undone by large-scale environmental catastrophe?

2) Do you think the reasons for creating rain gardens could appeal to any or all of the approaches to environmental ethics described in this book's introduction (e.g. biocentrism, Leopold's land ethic, ecofeminism)? Which one(s)?

CHAPTER 11

Are These Really Environmental Problems?

Case 47 – Noise Pollution? –

Study Questions

1) What is noise pollution and what are purported to be some of its negative effects on human and non-human animals?
2) What is the "Bensenville Pause" and what is its cause?

The Noise Pollution Clearinghouse introduces information on its website with the following quote: "Good neighbors keep their noise to themselves."[1] Noise pollution includes human-devised ambient disturbances, more specifically by transportation (air travel, motor vehicle travel, rail travel, water travel, etc.), industrial or construction projects, and security systems. City councils, neighborhood associations, and other concerned citizens have called for noise ordinances, noise barriers, and sometimes, outright moratoriums on building or industrial activity that annoys citizens.

The traditional definition of noise pollution is simply unwanted or disturbing noise. Those who have lodged noise complaints generally have done so due to sound that they claim disrupts their normal activities such as sleeping and carrying on conversations. Sounds can be regarded as a nuisance when they interrupt the activities that allow residents to enjoy life under "everyday" conditions.

Unsurprising, since noise can't be seen, smelled, or tasted it has gone under the radar as a type of pollution. Sounds surround us, but we have the ability to tune out some of them and sometimes just focus on a voice (e.g., having a conversation at a crowded party). Yet, for some people, sounds can reach the level of annoyance (sometimes by escalation in volume or simply by its persistence) where they can no longer be ignored or tolerated. Some argue this "annoyance" is not trivial; it impacts them deeply with serious health consequences. According to some studies, noise pollution adversely affects the lives of millions of people, degrading their fitness. Among the types of ailments afflicting people who have exposure to high-level and persistent noise include stress-related illnesses, hearing loss, sleep disruption, high blood pressure,

1 "Mission," (website), 2014. Available at http://www.nonoise.org (accessed July 20, 2016).

 | DOI 10.1163/9789004339996_013

speech interference, and productivity loss. Of these adverse medical effects attributed to noise pollution, perhaps the most prevalent and well-known is Noise Induced Hearing Loss (or NIHL).

Some of the most common cases of noise pollution arise from the transportation sector. Bensenville, Illinois, lies near O'Hare International Airport, one of the busiest airports in the US. As Charles W. Schmidt puts it, "Anyone who lives in Bensenville, Illinois, knows about the 'Bensenville pause.' According to long-time resident Pat Johnson, it goes like this: as the roar of a jetliner departing from nearby O'Hare International Airport becomes a blasting shriek, the residents of this small town stop talking and wait. Conversations pick up as the plane goes by, but they soon pause again; planes fly over Bensenville every three to four minutes. 'Sometimes, it's hard to fall asleep,' Johnson says, 'You do, but then you wake up again. The noise interrupts churches and classrooms. There are times you can't even talk on the phone.'"[2]

Schmidt acknowledges that while the Bensenville case is likely extreme, it is much more common than people think. He argues that millions of Americans have to endure noise pollution from various sources such as road traffic, car alarms, amplified music, and of course, planes. Companies promote everyday items to be heard not only by the primary listener, but also by the rest of the neighborhood. Consider the (now defunct) ad campaign from Sony marketing amplifiers and speakers that would "Disturb the Peace." The company proposed "new ways to offend" those around the block.[3]

While it might seem sounds from industrial sites or near airports are the most serious noise pollution cases, perhaps the most prevalent and severe problems come from automotive traffic. Two researchers, Tania Mihaiescu and Radu Mihaiescu, conducted a study of the Romanian town of Bistrita in 2009, examining the effects of persistent traffic noise on its 70,000 or so residents.[4] They made measurements of the ambient noise in major traffic corridors throughout the town using a 2250 Controller Integrating Sound Level Meter. Equivalent noise levels were measured at nine different locations (each at 30 minute intervals). Unsurprisingly, the highest value of continuous equivalent acoustic levels (or Leq) were found on 1st and 2nd category roads where the authors describe the most intensive traffic activity. The most heavily noise polluted points were found on Auto Bridge and Han Ran crossings. The highest recorded Leq values in the study were between 83 and 84 dB(A). It is important

---

2 Charles W. Schmidt, "Noise That Annoys: Regulating Unwanted Sound," *Environmental Health Perspectives* (2005) 113(1): A42–A44.

3 Schmidt, "Noise That Annoys," pp. A42–A44.

4 "Traffic Noise Pollution in Urban Areas: Bistrita Town," *ProEnvironment* (2009) 2: 222–225.

to note that sound is measured on a logarithmic scale, so 80 dB(A) is two times as loud as 70 dB(A). To provide some context for the these figures, 84 dB(A) is the level of sound roughly equivalent to standing by a diesel truck going 40 mph from 50 feet away.[5] Overall, the range of Leq was from 61–84 for all roads in Bistrita. While that may seem to be annoying, but not particularly harmful, humans exposed to this noise level for 8 hours or longer can experience possible permanent hearing damage. It is also worth noting the highest acoustic levels occurred between 7:00–7:30 am during rush hour traffic, while many residents may still be trying to sleep. And of course, we have to consider the effects when living near one of these exposure areas over a long period of time.

These findings were not anomalies occurring at one or two high traffic areas. Local regulations governing noise pollution in Romania (STAT 1009–88—"Urban Acoustics") establish limits on permissible noise levels in urban areas throughout the country. In Romanian cities, there are different zones where differing levels of noise are permitted depending on the type of area it is (i.e., business districts allow higher decibel levels than residential areas). Studied sites were selected as representative samples of each roadway classification, such as intense traffic, medium traffic, and low traffic streets in residential areas.

Of the nine monitored locations, seven had noise levels exceeding established limits. Even on the busiest street, the absolute highest admissible acoustic level allowed is 75 dB(A). However, at many of the data collection locations with the 75 dB(A) limit, the noise level was 8–9 dB(A) higher than the legal standard. This was not just true of high traffic areas, as even at a residential site where the limit is 45 dB(A), the measured noise level was at 62 dB(A);[6] though this is not thought to be an ear damaging level of sound, it is still likely "annoying" to those who live on a residential street.

Mihaiescu and Mihaiescu note that the volume of traffic noise has been rising in Bistrita in recent years and only projects to increase. They also report that a corresponding grassroots movement is calling for officials to find ways to reduce noise. Mihaiescu and Mihaiescu recommend that a combination of noise barriers, lower vehicle noise levels, and better urban planning should be used to lower noise pollution in Bistrita.[7] More specifically, this would involve decreasing noise at its sources (e.g., vehicles) by attaching silencing devices. Additionally, vegetation buffer zones could block noise from traffic in

---

5 See Industrial Noise Control, "Comparative Examples of Noise Levels: Comparative Examples of Noise Sources, Decibels, and Their Effects," (website), 2010. Available at http://www.industrialnoisecontrol.com/comparative-noise-examples.htm (accessed July 7, 2015).

6 Mihaiescu and Mihaiescu, "Traffic Noise Pollution," p. 224.

7 Mihaiescu and Mihaiescu, "Traffic Noise Pollution," p. 224.

adjacent neighborhoods. It may even be necessary to divert traffic or improve the traffic light cycle at some intersections.[8]

There is, of course, the question of whether this is really a moral issue. After all, these cases tend to be in urban and/or suburban environments, and one might argue that noise is simply a cost of living in these places. But, Les Blomberg, Executive Director of the Noise Pollution Clearinghouse disagrees, calling noise pollution "aural litter" or "audible trash." He puts the point rather dramatically, "That is how people experience community noise: as someone else's garbage thrown into their space."[9]

The issue has been regarded with some degree of seriousness in the legal realm. Over the years, some legal regulations have been placed on noise, especially in the European Union. Generally, these regulations are aimed at noise reductions in the construction and transportation sectors. For instance, Switzerland prohibits all airline departures from 11:30 PM to 5:00 AM. Some countries have laws protecting residents from noise, but have had difficulties enforcing them, as is the case with Japan's 1960 and 1967 national laws against noise pollution. These laws have not really done much to stem general increases in ambient noise or noise from the transportation sector, such as from airplanes and motor vehicles. They have also done little to provide coherent directives to local governments for effectively regulating noise.

There has been a relative regulatory void on noise in the US for several years. The Environmental Protection Agency once had regulatory control over noise pollution issues. The agency used the Office of Noise Abatement and Control (ONAC) to design policies to regulate noise. EPA staff wrote model noise control measures that could be fitted to local municipalities upon request. The Noise Control Act of 1972 also allowed the EPA regulatory authority as this federal law was designed to protect Americans from "noise that jeopardizes their health and welfare."

But funding cuts impaired ONAC from proposing new standards on noise from transportation and construction machinery. This meant that there was no comprehensive national noise policy as these sectors are common and major causes of noise. Without federal dollars to help fund noise reduction measures, states left to try to implement any new measures and noise control often lost fights for state dollars. Transportation noise, of course, did not go away despite the budget constraints. Curiously, governmental agencies with missions that are sometimes at odds with noise reduction have been left to manage it. A prime example of this is the Federal Aviation Administration

8 Mihaiescu and Mihaiescu, "Traffic Noise Pollution," p. 224.

9 Quoted in Schmidt, "Noise that Annoys," p. A42.

(FAA), which has in its purview the authority to control airline noise. Of course, a core part of the mission of the FAA is to promote airline industry growth. In a similar way, the Federal Highway Administration has jurisdiction over traffic noise. However, again, the FHA has at its core a commitment to upgrade and expand highways to promote the efficient movement of traffic. This means that even when local communities have pressed for noise pollution control, the FAA and the FHA have stymied the efforts. More interestingly, even in cases where individual airports have proposed measures to reduce noise levels for communities in their service areas, FAA officials have blocked them if they include limiting flight connections that could impede commerce.[10]

Some US cities have established noise control regulations. Portland, Oregon, with funding from the EPA and the Housing and Urban Development Department at the time, drafted a noise ordinance in 1975. This was a landmark piece of legislation as it has served as the template for subsequent noise control measures in other US and Canadian metropolitan areas. Most places with noise ordinances restrict sound from surmounting a certain threshold from roughly 10pm to 6am. During the daytime, these thresholds are set at a higher decibel level.

Yet, laws without enforcement are not really laws and enforcement of noise ordinances is decidedly uneven. It is not uncommon for some local authorities to simply ignore noise complaints, partly because law enforcement personnel will claim to have more important things to do than get neighbors to quiet down. Warnings may be issued, but since taking offenders to court is expensive, the warning may be the most aggressive action taken by law enforcement in many municipalities. A notable exception to this is in Portland, where fines can be as much as $5,000 per infraction. It is also possible there to issue multiple citations to a violator in the same day.

If we look at the etymology of the word "noise," it possesses a negative connotation as it comes from "nausea," meaning seasickness. An interesting feature of noise pollution is that sound enters air, which we generally think of as a commons or a public good. Public goods are goods with special features, with an especially relevant one being *non-excludability*. This means that it is not possible to provide the good without it being possible for others to enjoy it. For example, I could listen to a delightful tune from a neighbor's radio, but I can enjoy the ditty even without my neighbor's knowledge and with an open window and my ears as the only necessary devices. Conversely, it is not possible to exclude those who abuse the good—my enjoyment of blaring heavy metal music from my new high output speakers will spill over into my neighbor's "airspace." If there is a ruckus she doesn't care to hear, and if it is loud enough,

10 Schmidt, "Noise that Annoys," p. A42.

she will have a hard time blocking it out—and there will likely be some cost to her in trying to do so.

Thus my actions will have negative externalities—spillover effects to which you didn't consent. If these constitute harm, there are grounds for calling on noise emitters to cease and desist. We can recall utilitarian philosopher John Stuart Mills' harm principle, as we did earlier.[11] In effect, it says that people have the right to do what they wish just so long as they don't cause harm to others. In cases of industrial operations where there are loud sounds (e.g., areas around airports or with heavy machinery or automotive traffic), when decibel levels cause hearing deterioration and loss, it seems the *harm principle* is violated. It will likely be harder to the show that psychological issues or causing stress will violate the harm principle, since lots of other conditions in which people live cause these issues and they may not be straightforwardly seen as harms.

We can also imagine the following challenge. What if I contested that it really isn't that big a deal that I like to listen to my music loud, even if certain neighbors complain? Lowering the volume would lessen my pleasure from listening to it, and being forced by the police (by them serving me a citation) to quiet my music is an undue constraint on my liberty. Furthermore, what if I argued that I am not really *harming* you by playing my music loudly? You may not like my music; you may even claim that it is "driving you nuts," but it isn't really causing you deep psychological trauma. After all, a psychologist could examine you afterward and find that you are agitated, but likely not "insane." Thus, in effect, you may be *offended*, but surely your being "offended" is not the same as causing you harm. Perhaps harms should be avoided even if preventing them constrains my liberty, but this can't mean liberty should be constrained because people are offended. Offense is subjective—one person's art may be another person's garbage; pornography is desired by some and causes disgust in others. Liberty is important and shouldn't be curtailed unnecessarily. What do you say to this challenge? We are left with the questions of what noise pollution brings about, how pervasive is it, and what steps are legitimate to take against it.

Review Questions

1) Do you think noise pollution is an ethical issue? Why or why not? Do you think it is an *environmental* issue, why or why not?
2) As almost anyone who has or currently lives near a busy airport can attest to, living in such conditions takes some getting used to at best, and is

11 See Case Study 8—"There's Silica in Them Thar' Hills!".

barely tolerable at worst. Yet, cheaper land and houses are often found near airports, making such areas attractive to homebuyers with modest means and constrained buying options. Would you live in Bensenville, Illinois? If you did live there, what would you do about the noise? Should airlines be expected to reduce sound? Should airports petition airlines to schedule departures and landings when they are least disturbing to residents on flight paths nearby (perhaps by setting a "no fly" curfew)?

Case 48 – Light Pollution? –

Study Questions

1) How is light pollution defined in this case study? Why is light pollution seen as being problematic?
2) What are some of the steps the city of Calgary, Alberta, Canada took to reduce light pollution?

In 2008, Calgary, Alberta, Canada began the process of installing streetlamps designed to reduce light emissions dramatically throughout the city as part of its EnviroSmart Streetlights Retrofit Project. This was a primary component of the city's effort to combat *light pollution*—which is most straightforwardly defined as unwanted, artificial light obscuring the stars.[12] In many large metropolitan areas viewers will, at most, be able to make out only a couple dozen stars above.

Initially, the rational for changing the city's lighting system had little to do with seeing the night sky. One of the chief reasons why Calgary commenced this project was because the streetlight system was the single largest consumer of electricity in the city. Officials were aware that energy prices can fluctuate wildly and didn't want to get stuck in another budgetary bind as they had been in 2001 when energy prices spiked to record levels. Hence, Calgary's roads division sought ways to greatly reduce the cost of lighting local streets. Their logic was that lower wattage lighting fixtures would consume less electricity and correspondingly operation costs would fall. From 2002 to 2010, Calgary saved 107,000 MWh of energy—a cost savings of $11 million from reduced

12 David Henderson, "Valuing the Stars: On the Economics of Light Pollution," *Environmental Philosophy* (2010) 7(1): 17–26.

energy consumption. Their transformation also reduced carbon emissions by 124,000 tons.[13]

But there are other reasons to reduce public lighting. Wildlife conservationists cite widespread dispersed light as a problem for various species of animals. For about a billion years, non-human animals have evolved with a fairly regular and reliable day-night schedule of illumination levels in their environment. This expectation is imbedded in the DNA of all organisms. Regular light intervals "entrains" or regularizes basic anatomical activities and functions of flora and fauna, including humans. Therefore, increased levels of ambient light during "irregular" hours have a negative affect on all animals.[14] This issue is gaining more attention.

There are some well-known cases of animal species negatively impacted by more recent widespread disperse lighting. Nocturnal bats possess only rods in their retinas and are blinded by bright lights. As larger portions of forested areas, such as those in Amazonia, are cleared and human development enters into previously underdeveloped areas, land bats are negatively impacted. Roads in recently cleared areas bring streetlights and new farms that have access to towns and cities to sell their meat and produce. Yet, it is not uncommon for many farms to fail as the cleared rainforest has soil poorly suited for crops. Once these farms are unproductive, they are abandoned, but the streetlights and roads remain.[15]

Among the risks nocturnal animals face from light pollution include greater exposure to predators. This reduces the time they have to find shelter, food, and mates. This is important, as half of all animals are either nocturnal or crepuscular (i.e., active at dawn and dusk). This list includes skunks, deer, foxes, and pets such as cats, rabbits, hamsters, and dogs. Our artificial lights intrude on their normal activities.

US roadways contribute to lots of wasted light, with much of it emitted by upward-pointing 300W halogen bulbs. Conventional street lighting misses

13 City of Calgary, "International Spotlight on City's EnviroSmart Streetlights: BBC Team to Look at How Streetlights are Saving The Environment and Money in Calgary," (website), August 20, 2010. Available at http://newsroom.calgary.ca/news/international-spotlight-on-city-168299 (accessed July 8, 2015).

14 Florida Atlantic University, Department of Physics, "Light Pollution Harms the Environment," (website), 2015. Available at http://physics.fau.edu/observatory/lightpol-environ.html (accessed July 20, 2016). Also see Case 29—"Not Seeing the Forests for the Deserts."

15 Daniel Lewanzik and Christian Voigt, "Artificial Light Puts Ecosystem Service of Frugivorous Bats at Risk," *Journal of Applied Ecology* (2014) 51(2): 388–394.

about 50% of its intended target.[16] Use of LED lights pointing downward could much more efficiently, effectively, and safely guide travellers.

Calgary has received praise and/or awards for its "green" streetlight retrofit project from numerous organizations, such as the Royal Astronomical Society of Canada (RASC), the International Dark Sky Association (IDSA), the Calgary Awards, the Alberta Emerald Foundation for Environmental Excellence, Natural Resources Canada, The American Public Works Association, and the prestigious Dubai International Best Practices Awards. The city's efforts at light pollution reduction prompted the Chairman of the Light Pollution Abatement Committee of the RASC (Calgary Center) to say, "The EnviroSmart Streetlight Retrofit Program is a cornerstone for light pollution reduction efforts in Calgary. It has reduced energy consumption and associated carbon dioxide emissions, improved visibility, lessened the impact on the nocturnal environment and reduced sky glow."[17] The city's light pollution reduction efforts have not gone unnoticed internationally in countries such as the United Kingdom, where light pollution is having a major impact. A team of journalists and a film crew came to learn from Calgary's example in hopes of translating what the city has done to Britain. As BBC producer and director Keir MacKenzie put it, "Calgary has decided to invest time and money tackling the problem. We're here to see how Calgary's efforts have helped reduce energy waste and again reveal the night sky to its residents."[18]

And many people certainly find value from a clear night sky and the ability to see the "wonders of the heavens." Those moving from rural areas to cities often lament that whole constellations of stars cannot be seen at night. But how much value should be placed on the ability to see lots of stars? Some would hazard there is no price that could be placed on the pleasure of stargazing. For others who have barely glanced at the sky, there is nothing "out there" worth all that much and balk at the idea that human development has much to apologize for in providing the modern convenience of light.

Philosopher David Henderson argues that unobstructed views of the night sky possess an underappreciated value, even if the market fails to value it.[19] He points to the plurality of values things in the world can have and how they can be incommensurable—there is no way to measure these values against each other. After all, how much is a sunset or a first kiss from a desired suitor worth? As with the case of noise pollution (see Case 47—"Noise Pollution"?), there are "public goods" which are free, have no connection to labor, and are

16 April Reese, "Take Back the Night," *E* (2000) 11(3): 22–24.

17 Reese, "Take Back the Night," pp. 22–24.

18 Reese, "Take Back the Night," pp. 22–24.

19 Henderson, "Valuing the Stars."

non-rival (i.e., inspire no competition). These goods lead to what the economists call *market externalities*, which are consequences of economic exchanges that are experienced by third parties. While they can be positive or negative, Henderson argues that reduction of light pollution would result in more positive externalities than negative ones.

Immanuel Kant once called the starry sky "the noblest spectacle that can ever be presented to the human senses."[20] While Henderson seems to agree with Kant, he doesn't think putting a price tag on such phenomena should be given up so quickly, especially since what *seems* to be immeasurable may well be measurable. This price may actually show just how great a value the night sky can have. We also have to acknowledge there are benefits to outdoor lighting at night. It can lend a huge hand to security. It is can also do a great deal to advertise goods and services we need. But certainly, there are downsides to this lighting, including wasted energy, degradation of wildlife habitat, hazards related to glare, hindrance to astronomy, and of course, diminished grandeur of the night sky (in a nod to Kant).[21]

As mentioned earlier, outdoor fixtures in the US are woefully inefficient to the tune of an estimated $2.2 billion lost in energy costs to the nation each year.[22] Even when lights are properly directed to illuminate desired targets, their wattage is unnecessarily high, not to mention that lighting is on at times and in places when and where it serves no purpose. Fugitive light worldwide accounts for about two-thirds of people on Earth today being practically unable to see the night sky. Astronomers bear special costs when conducting research. Light interference has moved them further away from most developed areas hosting universities and research centers to their make observations. This adds greatly to travel costs to conduct astronomical research, often relegated to remote locations such as Hawaii or the mountains of Chile. Amateur stargazers greatly outnumber the pros. Even conservative estimates suggest they buy $250 million worth of astronomical equipment.[23]

Estimates of the cost of wasting energy are fairly easy to calculate, with perhaps the exception of hidden costs, such as contributions to climate change by producing electricity that is unnecessarily wasted. Of course, some costs to astronomy can be assessed since they involve concrete tools for investigation, such as expensive telescopes for professionals that also come at a surprisingly

20 Immanuel Kant, *Critique of Practical Reason*, in Mary J. Gregor, trans. and ed. *Practical Philosophy* (Cambridge, UK: Cambridge University Press, 1996).

21 Henderson, "Valuing the Stars."

22 International Dark Sky Association, "IDA Practical Guide 2: Introduction to Light Pollution," (website), 2009. Available at http://darksky.org (accessed July 20, 2016).

23 Henderson, "Valuing the Stars," p. 20.

high cost for amateurs. While it is virtually impossible to put a value on all the wildlife adversely impacted by light pollution, this doesn't mean that *no value* can, at least if we add up the costs of studying and recovering wildlife adversely affected by light pollution such as bats, moths, deer, sea turtles, birds, many insects, and, of course, all nocturnal animals. These costs constitute what Henderson calls "a lower bound" and the upshot is that even when we use very conservative figures for those goods we *can* put a price on, cost-benefit analysis reveals that we should retrofit our lighting systems.[24]

However, does cost-benefit analysis really get to the moral question? To be fair to Henderson, he tries to make the economic case not because he rejects the moral case for reducing light pollution. Instead, he wants to strengthen a possible moral case against light pollution by showing it isn't simply *impossible* to add up at least some of its more tangible costs. Henderson adds that if on top of the benefits that would accrue to us by reforming out lighting systems, we consider the aesthetic and other pleasures of the night sky and the moral values we're offered, we should combat light pollution.

Review Questions

1) Do you think light pollution is an environmental issue? Is it an *ethical* issue?
2) Do you think cities should avoid light pollution? If so, why? If not, why not?
3) What if installing new streetlights reduces light emissions, decreases energy waste, and saves money, but poses safety issues for humans in some neighborhoods? Would this be an acceptable tradeoff? Or must advocates of new lighting systems have to show convincingly that their plan provides at least roughly the same amount of illumination as would conventional lighting to help ensure safety?

Case 49 – An Intergalactic Threat –

Study Questions

1) What are NEOs? Which ones are the most worrying to astronomers and why?
2) What is terraforming? What is the relationship between terraforming and possible threats to the Earth from comets and meteors?

24 Henderson, "Valuing the Stars," p. 20.

3) What are some arguments in favor of terraforming? What are some objections to it?

Some scientists predict the odds of a potentially devastating space rock hitting the Earth this century could be as high as one in ten. NASA estimates there may be 20,000 potentially hazardous objects (mainly comets and asteroids) in the general vicinity of Earth.[25] The Earth is dented by both ancient and relatively new collisions, such as what is now known as the Tunguska event (1908), in which what was most likely a meteoroid exploded at an estimated 28,000 feet over central Siberia with the force of 185 Hiroshima bombs. This explosion destroyed approximately 800 square miles of forest. All things considered, this was still a relatively small object; simulators indicate that the cause of the Tunguska event may have only been 90 feet across.[26]

In response to the possibility of future larger collisions, NASA has been trying to figure out if and when potential "civilization killers"—space objects 1 kilometer or more in diameter with orbits within 30 million miles of earth—will possibly hit. Among the effects of larger space rocks hitting the Earth would include global suffering from widespread destruction, dust storms that would leave skies dark for decades, and severe acid rain. (The confluence of extreme shock waves and intense heat at the point of impact creates nitric and nitrous acids that can produce rain as corrosive as battery acid.)[27]

Of course, the Earth has been hit many times by asteroids and comets over the past 4.5 billion years.[28] Every day our planet is pelleted with more than 100 tons of dust and sand-grain-sized particles. The vast majority of these collisions go undetected by us, with many objects simply burning up in the atmosphere before reaching our surface. With that said, the orbits of a number of large interplanetary materials can come close enough to overlap with that of our planet. Collectively, these materials are called *Near Earth Objects* (*NEOs*) and pose a risk to us. Collisions can cause catastrophic effects locally or on a global scale depending on the size of the NEO. There isn't so much a question of whether such a collision will take place, but more of a matter of when.

25 Gregg Easterbrook, "The Sky is Falling," *The Atlantic* (2008) 301(5): 74–84 at 77.

26 Andrew C. Revkin, "Maybe Chicken Little wasn't Paranoid After All," *The New York Times*, July 6, 2008, p. B2.

27 Easterbrook, "The Sky is Falling," p. 78.

28 University of Hawaii, Institute for Astronomy, "The Threat to Earth from Asteroids and Comets," (website), 2005. Available at http://pan-starrs.ifa.hawaii.edu/public/asteroid-threat/asteroid_threat.html (accessed July 9, 2015).

Strong scientific evidence suggests that mass extinctions were caused by NEOs smashing into the Earth, as documented in our fossil records.[29] A graphic reminder of the damage large objects can have when crashing into a planet came as recently as 1994, when the Shoemaker-Levy 9 comet split apart into 21 separate fragments and collided with Jupiter's atmosphere. Some fragments were 2 km in diameter and if they had hit the Earth, the impacts would have resulted in a global disaster.[30]

Looking holistically at Earth's history there is an intriguing dualistic relationship between life on our planet and NEOs. While the beginnings of life here are mysterious, we do know there are some necessary elements for its genesis and propagation. Carbon, water, and oxygen are necessary for life. High levels of carbon and water are contained in some space rocks. If a planet doesn't already have these elements, they have to come from elsewhere, such as other objects floating (or more likely whizzing by) in space. So, it stands to reason that despite the potential for great destruction of life, NEOs also hold the keys to future regeneration.

The first billion years of the Earth's existence didn't allow for any life to exist due to the continuous impacts of comets and asteroids, which in turn made the Earth's surface too hot for the collection of enough water and carbon-based molecules sufficient for organic activity. Just at the end of a period of heavy bombardment of the Earth by asteroids and meteors, it is hypothesized that our first life emerged on the planet (at least for the 3.5 billion years for which we have fossil records). When life did begin on Earth, it took off rather rapidly. The problem is finding an explanation for how this process occurred so quickly with little water and carbon. Yet, as was mentioned, carbon and water molecules are also found in copious amounts in comets. It could be that once the bombardment of Earth calmed down a bit, subsequent collisions of comets with our planet supplied it with the building blocks necessary for life—occurring sometime from 3.5 to 3.8 million years ago.[31]

But again it may be that what gives life may also someday take it away in the very distant future. What follows is a description of one of the most disastrous collisions of a comet with the Earth:

29 University of Hawaii, Institute for Astronomy, "The Threat to Earth."

30 University of Hawaii, Institute for Astronomy, "The Threat to Earth."

31 National Aeronautics and Space Administration. Jet Propulsion Laboratory. "Near-Earth Objects and Life on Earth," (website), 2015. Available at http://neo.jpl.nasa.gov/neo/life.html (accessed July 21, 2016).

> Comets have this peculiar duality whereby they first brought the building blocks of life to Earth some 3.8 billion years ago and subsequent cometary collisions may have wiped out many of the developing life forms, allowing only the most adaptable species to evolve further. It now seems likely that a comet or asteroid struck near the Yucatan peninsular in Mexico some 65 million years ago and caused a massive extinction of more than 75% of the Earth's living organisms, including the dinosaurs. At the time, the mammals were small burrowing creatures that seemed to survive the catastrophic impact without much difficulty. Because many of their larger competitors were destroyed, these mammals flourished. Since we humans evolved from these primitive mammals, we may owe our current preeminence atop the Earth's food chain to collisions of comets and asteroids with the Earth.[32]

The logical question that follows from these thoughts on the dangers of NEOs and how they have life-extinguishing potential is: just how much of a chance is there that a large NEOs could collide with Earth and cause extensive damage? The estimate of the number of NEOs has risen rapidly over the years. In 1980, there were thought to be 86 near-Earth asteroids and comets. That number increased to 170 by 1990 and by 2000 the total had greatly increased to 921. In 2008, there were reportedly 5,388, and as of July 2016, there were a total of 14,577 NEOs discovered. Of these NEOs, some 872 are asteroids that are 1 km or greater and some 1,714 have been classified as *Potentially Hazardous Asteroids* (*PHAs*). PHAs are asteroids that can get closer than roughly 4,650,000 miles to Earth and are larger than 500 ft. in diameter. NASA emphasizes that PHAs will not necessarily make impact with Earth, only that there is the possible threat of collision. However, they do need to be closely monitored so that their orbits can be updated, allowing for easier predictions of whether they will hit.[33]

There are also some anecdotal cases of near misses. For example, on March 18, 2004, a meteor almost 100 feet in diameter called 2004 FH passed the Earth not far above the orbit of telecommunications satellites. At the speed it was travelling and given its size it could have destroyed a city.[34] In 2015, a giant asteroid over a half mile wide dubbed 2014 YB35, soared past Earth coming

32 National Aeronautics and Space Administration. Jet Propulsion Laboratory. "Near-Earth Objects."

33 National Aeronautics and Space Administration, Jet Propulsion Laboratory, "What is A PHA?" (website), 2015. Available at http://neo.jpl.nasa.gov/neo/groups.html (accessed July 9, 2015).

34 Easterbrook, "The Sky is Falling."

within 2.8 million miles from our planet. The pass was about 11.7 times the distance of the moon from the Earth, but in astronomical terms this was a near miss. If this asteroid had hit our planet, the impact would have been equivalent to a thousand of the largest nuclear weapons combined, resulting in devastation for hundreds of miles around. It could have caused tsunamis, earthquakes, or firestorms depending on whether it crashed on land or in the sea. This asteroid is scheduled to pass by us again, in 2023, but this time only 9 moon distances away.[35] Yet perhaps most worryingly was the January 2015 near miss by 2004 BL86, one of the largest asteroids to pass by our planet in many years.[36] It came within 745,000 thousand miles during its closest approach or about 3.1 times the distance of the Earth to the moon.

In response to these threats, the first stage is obviously tracking them. But what further measures could be taken to protect Earth from space materials? NASA has collaborated with the National Nuclear Safety Administration (NNSA), which is currently in charge of dismantling nuclear weapons. In addition to NNSA's chief mission, some nuclear warhead components are being set aside for possible deployment against potential asteroid strikes.[37] The plan is not so much to destroy PHAs as it is to deflect them off course from colliding with Earth.

There is an interesting connection between the issue of light pollution[38] and global security in this case study. Some claim light pollution is a public safety issue, as part of the national protection strategy initially involves observations of potentially harmful space rocks. Scientists need to be alert to all space materials 1 km in size or greater that can cause catastrophic damage. By some estimates the impact of a 1 km meteor on Earth would be equivalent to the energy of 100,000 megatons of TNT and was about the size of the one thought to have wiped out the dinosaurs. But detection efforts are increasingly inhibited by light interference impeding astronomers from observing deeper into space for potential threats. If public policies for reducing light pollution fail, there will be an increased need to launch telescopes capable of escaping our atmosphere to keep sentry for PHAs. But this would contribute to another

---

35 CBS San Francisco, "Giant 'Potentially Hazardous' Asteroid Hurtling Toward Earth," (website), March 27, 2015. Available at http://sanfrancisco.cbslocal.com/2015/03/27/giant-asteroid-earth-2014-yb35/ (accessed July 21, 2016).

36 CBS San Francisco, "Giant 'Potentially Hazardous' Asteroid."

37 Jeremy Bender, "The US is Setting Aside Nuclear Weapons for Potential Defense Against Asteroids," *Business Insider*, October 1, 2014. Available at http://www.businessinsider.com/us-is-saving-nuclear-weapons-for-astroids-2014-10 (accessed July 9, 2015).

38 See Case 48—"Light Pollution"?

daunting issue; the accumulation of space junk from abandoned satellites and parts of old rockets that endangers spacecraft.[39]

As a moral issue, James S.J. Schwartz has argued that space exploration is necessary because it serves to protect the Earth and its inhabitants from extraterrestrial threats such as meteorites.[40] One might ask what practical effect there is to knowing that a planet-killing asteroid or meteorite will crash into Earth if there is little to be done about it? But defenders of space exploration believe we must take a long-term perspective in making these kinds of assessments. For example, could a nuclear defense system protect us by potentially blowing up or deflecting NEOs as they come closer to Earth? Could we learn how to live on other planets to escape if we were warned soon enough about a highly probable strike from a PHA? Part of preparing for this future would entail support for space exploration now to figure out if colonizing other planets is feasible.

There are objections to these arguments. One is on pragmatic and prudential grounds; supporting space exploration is simply too costly and the hurdles of technological advance necessary to either deflect PHAs, or to fly away to escape them, are too great to overcome. Another objection is more principled; the complaint is that the projects of promoting space exploration and promoting environmentalism are fundamentally opposed to each other. A final objection to space exploration comes from the philosopher Robert Sparrow, who has argued against *terraforming*[41] (colonizing and manipulating other planets to render them more Earth-like for the sake of preserving and enhancing human and non-human animal interests). Sparrow argues that terraforming is done for purposes that lead to two blameworthy vices—blindness to aesthetic value and the sin of hubris.[42]

Although having the capacity to escape a catastrophic incident on Earth (and the impending impact of a 1 km in diameter or larger meteor into the Earth's atmosphere would seem to quality) might meet Sparrow's objection even on his terms. This is because he thinks there is an exception to the *prima*

---

39 See Case 50—"Why Not Just Toss It Into Space"?

40 James S.J. Schwartz, "Our Moral Obligation to Support Space Exploration," *Environmental Ethics* (2011) 33(1): 67–88.

41 Robert Sparrow, "The Ethics of Terraforming," *Environmental Ethics* (1999) 21: 227–245.

42 Sparrow, "The Ethics of Terraforming," pp. 227–245. A good description of Sparrow's view is also found in Robert Heath French, *Environmental Philosophy and the Ethics of Terraforming Mars: Adding the Voices of Environmental Justice and Ecofeminism to the Ongoing Debate*. Unpublished Master's Thesis (Denton, TX: University of North Texas, 2013), pp. 9–13.

facie rule that terraforming a planet is wrongful. For he says the sole way of morally justifying a terraforming project would be if "no alternative existed for the survival of the human race."[43]

This case is interesting because it offers intertwined issues that seem like something out of a sci-fi movie, but on the other hand, reflect perplexing problems that might be all too real. Space objects have caused astronomically high damage in the past, but the chance of that happening again (and within what timeframe) is uncertain. Our abilities at detecting such intergalactic threats are decreasing due to our flourishing and development as a species; how do we balance these effects? If we choose the status quo with respect to inefficient lighting (or even increase it), the need for early NEO detection still doesn't go away. We then face the prospect of deploying more devices in the atmosphere that remain when they become non-operational and pose further threats to other functional equipment upon which we depend. What should we do?

Review Questions

1) When considering the competing claims we have on governmental resources, to what degree should NASA fund protective measures from potentially hazardous space materials?
2) Do you think there are some morally justifiable reasons for or against terraforming? Why or why not?

Case 50 – Why Not Just Toss It into Space? –

Study Questions

1) How would you define "the environment"? Would you consider space to be part of the environment? Why or why not?
2) What is space junk? Roughly how much space junk is in orbit around the Earth?
3) What are some of the hazards of space junk? Are the main worries about people on Earth getting hit by it or about something else?

The United States Space Surveillance Network reports there are roughly 13,000 man-made objects larger than 4 inches in diameter orbiting the Earth. This doesn't even include thousands of smaller objects in space, including bits of

43 Quoted in French, *Environmental Philosophy*, p 11.

plastic, metal, and paint of which there are closer to 500,000 larger than a marble according to NASA. Nicholas Johnson, the chief scientist and program manager for NASA notes, "Of those 13,000 objects, 40% came from breakups of both spacecraft and rocket bodies. This is a growing environmental problem."[44]

Space junk is standardly defined as "any uncontrolled object in orbit that is greater that 10 cm (~4 inches)." This is basically because an object the size of a small box or larger can be tracked.[45] Beginning with Sputnik, launched by the then Soviet Union in 1957, humankind not only produced the first satellite, but also the first space junk. Larger space debris is a major worry. Even during launch bits fall away from rockets due to the extreme vibrations of take off, such as ice from liquid oxygen used as a propellant. The lower booster rockets themselves detach when explosive bolts are detonated (because you don't want any part of the lower booster staying attached!) to allow the spacecraft to continue its journey.

There are several reasons to worry about space junk. While the media sensationalizes the slim possibility of humans being hit by space debris, the real concern is of it colliding with working satellites and space stations. There are roughly one thousand operational satellites currently orbiting the Earth. Satellites have a plethora of purposes, including telecommunications, GPS navigation, meteorology, resource management, search and rescue, national security and troop support, treaty monitoring, and of course, scientific discovery.[46] These days, satellites even assist with farming. Increasingly, satellite technology is used to provide farmers with LANDSAT data to assess crop health, and to know how much fertilizer or water to place on specific land areas. Space junk collisions can thus cause both minor and major disruption to service. Satellites are also very expensive. In 2009, an inoperable Cosmos satellite collided with a working Iridium satellite, completely destroying it—a $55 million dollar craft was quickly obliterated. Smaller collisions occur on a regular basis. Because of these close encounters, most spacecraft are now designed to at least withstand incidental impacts from smaller bits of space junk.[47]

---

44 Stefan Lovgren, "Space Junk Cleanup Needed, NASA Experts Warn," *National Geographic*, January 19, 2006. Available at http://news.nationalgeographic.com/news/2006/01/0119_060119_space_junk.html (accessed July 21, 2016).

45 See John Henderson's presentation as part of Lawrence Livermore National Laboratories "Science on Saturday" Lecture Series entitled, "Space Junk: Traffic Cops in Space," *YouTube* (online video), December 20, 2012. Available at https://www.youtube.com/watch?v=UoIcXxbyX1s (accessed July 10, 2015).

46 John Henderson, "Space Junk."

47 John Henderson, "Space Junk."

But even these smaller bits of debris can do some oversized damage. For example, a mere fleck of paint hit the first Space Shuttle flight causing a 1 mm across radiating crack in the windshield. On a later Space Shuttle voyage, a slightly larger one actually penetrated the windshield halfway through. Also, parts of functioning satellites with large surface areas, such as solar arrays, are vulnerable to small bits of space debris that destroy photovoltaic cells on contact and thereby reduce their power-generating capacity. Over time these smaller level collisions can terminate the operation of satellites or shorten their service time.

Of course on Earth, thrown flecks of paint don't penetrate automobile windshields. But in space, we have some counterintuitive results of collisions between objects. A 1 cm bite-sized candy hurled at a cement wall at 22,000 mph would deeply penetrate the wall. In a collision at that speed with the same amount of mass, the kinetic energy released at the point of impact would be equivalent to the same mass of a high explosive. In this case then, a candy bar would have higher explosive capability than a hand grenade. While satellites and space stations can be materially and structurally reinforced to combat the reality of crashes into space debris, this will likely not be enough to keep them functioning. This is why redundancy in spacecraft systems is key—there is two of everything on board, so that if the primary system goes down, the backup one is ready to replace it.[48]

One of the obvious problems with satellites is that since they have to be launched high into orbit, the primary goal is to make them as light as possible. They are usually designed to withstand hits from debris up to 1 cm in size, but overall this leaves them vulnerable to dings from even relatively small pieces of space debris. The popular orbits for most satellites are about 400 miles above us. A few hundred satellites are further out and have geosynchronous orbits, about a tenth of the way to the moon. Telecommunications and weather satellites tend to orbit at the same rate as the Earth's rotation, so that they can remain directly over a single spot on the Equator. Consequently, while there seems to be a lot of space, many orbiting objects are actually confined to a relatively small area, thus increasing their chances of colliding over time as more traffic and detritus of old vehicles is added to the orbits. Some of this debris is cleaned out of space from the lowest orbits by falling back to Earth. Our upper atmosphere tends to slow objects down and they re-enter the atmosphere where they can be burned off within a few months or years. However,

48 John Henderson, "Space Junk."

just a few hundred miles above the Earth's atmosphere is so thin that anything reaching that level will basically stay there indefinitely.[49]

A growing consensus believes there is simply too much debris orbiting Earth, as was noted at the 6th European Conference on Space Debris in Darmstadt, Germany, held in April 2013. The head of the European Space Agency's Space Debris Office, Henrik Klinkrad, warned in his address, "There is a wide and strong expert consensus on the pressing need to act now to begin debris removal activities. Our understanding of the growing space debris problem can be compared with our understanding of the need to address Earth's changing climate some 20 years ago."[50]

A NASA computer model predicts that even without any more launches of satellites, the amount of debris in low orbit will remain the same until 2055. Presumably, it will then begin to increase as larger pieces break up from collisions. One example of those crashes was cited earlier. It wasn't just that a $55 million dollar satellite was lost in the 2009 collision, it also created two more clouds of space debris in orbit covering a larger area where they might do additional damage. Along with current efforts at curtailing the number of space objects being launched, larger pieces of space junk will need to be removed.

Free-floating, uncontrolled objects pose a threat to space activities, including robotic missions and human space flight. Additionally, clouds of superheated gas can cause electrical shorts and thus disable satellites. A case in point is the 2005 collision between a 31 year-old US rocket and a fragment from the third stage of a Chinese launch vehicle (which itself had exploded in 2000). The US Air Force runs an advertisement showing control center operators detecting a threat to a space station and quickly responding to it. While the ad makes it look (rather misleadingly) like the space station maneuvers out of the way of an oncoming piece of a dead satellite just in the nick of time before impact, the true capabilities of moving space stations and satellites away from threats are impressive nevertheless. Some satellites have the ability to be repositioned in a matter of hours to avoid potentially hazardous encounters with space debris. As for the massive International Space Station, it can be moved out of harm's way within 24 hours. Early detection of close encounters

---

49 Steve Olson, "The Danger of Space Junk," *The Atlantic*, July 1998. Available at http://www.theatlantic.com/magazine/archive/1998/07/the-danger-of-space-junk/306691/ (accessed July 22, 2016).

50 Mike Wall, "Space Junk Threat Demands Immediate Action, Experts Say." *Space.com*, (website), April 25, 2013. Available at http://www.space.com/20832-space-junk-cleanup-needed.html (accessed July 10, 2015).

is key to mitigating damage. However, only about a quarter of all operating space vehicles currently orbiting the Earth have this capacity.[51]

Nations involved in space exploration are very aware of the problem of orbital debris, with an organization (the Inter-Agency Space Debris Coordination Committee) devoted to exploring the problem of space junk and what to do about it. The committee has representatives from the space agencies of ten countries as well as the European Space Agency. Unfortunately, cleaning up space junk will be technologically difficult. Two common ideas floated are to develop a craft that can grab debris or to develop a laser to slow large fragments, causing them to fall to Earth more frequently. Both suggestions were technically unfeasible and cost prohibitive over a decade ago.[52] However, with engineering advances they are coming closer to reality now.

Nonetheless, the space junk problem is not going away. As Johnson states, "It's like any environmental problem. It's growing. If you don't tackle it now, it will only become worse, and the remedies in the future are going to be even more costly than if you tackle it today."

A novel approach to tracking space debris uses "space cops."[53] The program is called Space-Based Telescopes for Actionable Refinement of Ephemeris (STARE) and can predict the trajectory of a satellite to within less than 160 feet.[54] STARE uses a system of mini-satellites to detect and then refine a targeted satellite's path. STARE is best equipped to monitor low-orbit satellites to try to prevent collisions. Yet, keeping track of all satellites is a daunting task and current technology can only detect a satellite's location down to within roughly half-a-mile. This level of uncertainty results in about 10,000 false alarms a year. In response, satellite operators usually fail to follow up on warnings of possible collisions, at least until one occurs.[55] The hope is that STARE can greatly reduce the level of location uncertainty, perhaps down to within 330 feet (a distance from about home plate to the outfield fences of a major league baseball park). Researchers from the Lawrence Livermore National Laboratory are also working on these monitoring issues, as they hope to reduce location uncertainty even further, down to 160 feet. They plan to test a system of low-orbiting nanosatellites to better monitor space traffic.

---

51 Henderson, "Space Junk."

52 Lovgren, "Space Junk Cleanup Needed."

53 Philip Ross, "Mini-Satellite 'Space Cops' Could Control Satellite Traffic and Help Prevent Collisions," *International Business Times*, January 25, 2014. Available at http://www.ibtimes.com/mini-satellite-space-cops-could-control-satellite-traffic-help-prevent-collisions-1548024 (accessed July 22, 2016).

54 Ross, "Mini-Satellite."

55 Ross, "Mini-Satellite."

Other solutions to the problem reveal an eye to the future of design. One strategy is to ensure that all future satellites are sustainable. This requires a cradle-to-grave plan to minimize the impact of spacecraft on the Earth's orbit, including proper disposal of satellites at the ends of their lives. Some experts think this strategy, along with removing old pieces of space junk, called *deorbiting*, needs to be adopted without too great of a concern for cost. According to Klinkrad, removing space junk is now a global issue demanding international cooperation, "While measures against further debris creation and actively deorbiting defunct satellites are technically demanding and potentially costly, there is no alternative to protect space as a valuable resource for our critical satellite infrastructure. Their direct costs and the costs of losing them far exceed the cost of remedial activities."[56]

Interesting legal questions arise from policing the removal of space junk or adjudicating cases of spacecraft liability. Since this is an emerging area of jurisprudence, these are also moral questions as we try to figure out what ought to be done with space junk. Questions of property rights emerge as we consider what should happen when apparatus are removed from space. What should happen when someone complains their company's device is removed without their consent? Does this violate a property right? Has someone just stolen another's property from space? Other questions of responsibility come to the fore. Who is responsible for collisions between satellites? Is there an obligation for parties to monitor their own spacecraft, even after they are beyond their functional lifespans? Other questions of global justice emerge. With deorbiting of satellites, there is the issue of equitable distribution of orbits around the Earth—what sort of limits should there be on how many satellites countries can deploy? Who is responsible for removing spent satellites? This issue is heightened by the fact that one country may launch other country's satellites. For example, Russian and other European rockets have for some time now launched several US commercial satellites.

Finally, there is an interesting parallel between despoiling the Earth and space. In just over a half-century, we have, as Steve Olson describes it, "seriously polluted the final frontier."[57] For many years, those in the space industry didn't really acknowledge there could be a problem with space junk. After all, it was the last, great frontier, with an immense area that couldn't possibly be filled. But now it seems that orbital arteries are rapidly clogging, challenging the notion that space is akin to the Wild West.

---

56 Wall, "Space Junk Threat."

57 Olson, "The Danger of Space Junk."

Review Questions

1) Do you agree with Johnson's assessment that space junk is an environmental problem? If so, why? If not, why not?
2) While one might agree that space junk is indeed an environmental problem, he might reject that we should make cleaning it up a priority. Despite some danger to the lives of astronauts and property from space junk, there are certainly many more humans, animals, and plants on Earth threatened by many other environmental problems closer to us. Hence, there is greater urgency for dealing with environmental issues on Earth before handling those in space. How would you respond to this argument?
3) Some scientists project that even if nations were willing to refrain from launching new spacecraft for the next 200 years, eventually the amount of space junk will exceed the current amount due to further collisions between existing large fragments. Also, we learned about the difficulties of cleaning up debris already in orbit. With these factors in mind, should there be a ban or some other limitation on space launches? Is so, why? If not, why not?
4) Increasingly, space missions are international endeavors. Who should be responsible for the clean up of space junk? Should only the most affluent nations bear the cost, or should other countries share in it (especially if they are benefitting from the device that is launched into space)? What is the justification for whatever distribution of costs you develop?
5) How does this case study relate to the problem of future generations, if at all?
6) Recall that for ecocentric environmental ethicists, the wrongfulness of environmental degradation is not to individual members of species or to species themselves, but to the ecosystem, the environment, or to the Earth itself. Assume you agree with ecocentrists on this point, do you think the same argument can be made with respect to space? Can it be wrongful *to space* for cluttering it with garbage in the same way as it is wrongful to an ecosystem when we pollute rivers and streams?

Case 51 – An Unconventional Mr. Clean (and Crime Fighter Too?) –

Study Questions

1) What were some public services not available in New Orleans for some time after Hurricane Katrina caused substantial damage to the city?

2) How did Sidney Torres IV get started in the waste management business? What set his approach apart from the usual delivery of basic public services in the city?
3) Why did Torres criticize New Orleans mayor Mitch Landrieu?

For many months after Hurricane Katrina hit New Orleans in 2005, basic utility services were spotty throughout the city. The French Quarter always had a reputation for being dirty before the storm, but after Hurricane Katrina, garbage really began to pile up. As trash spilled out on to the streets, private business owners became increasingly irritated by the lack of dependable waste management services. One businessman attempting to find solutions to these problems was a young hotel owner, Sidney Torres IV, who researched private waste management firms to take care of his own hotels' refuse.[58] In an attempt to get the city back on its feet, Torres had allowed FEMA officials and secret service agents to stay at his hotels in the French Quarter.[59] But, in the absence of dependable publically funded local disposal service, he soon ran into a problem—who could handle trash from his properties since service was so sporadic? When Torres priced out private companies to deal with it instead, he found their services prohibitively expensive. So, he hired drivers and rented a few trucks to haul trash away on his own, adding his name and number to the vehicles.

Other business owners in the area saw what Torres was doing and asked if he could remove their garbage for a fee. He soon found himself adding trucks. Recognizing that others were looking for cheap and efficient waste management, Torres established a company called SDT Waste and Debris Services. With his "movie star" good looks and only prior business experience as an hotelier, Torres became the unlikely local "garbage king," winning bids to collect waste first in St. Bernard Parish, and then in New Orleans. City officials stipulated they wanted more than basic services from any waste management company winning the contract. They required attention to detail, recognizing a cleaner French Quarter was essential to luring tourist dollars back into the city. Waste Management, the company which had previously held the contract, didn't even bid as they said they couldn't meet the city's criteria. But SDT was able to jump in, win the contract, and fill a waste management void.

Within a day of winning the contract, French Quarter residents and tourists were astounded at how clean the streets were, now relieved of putrid

58 Nicole Alper, "The Rembrandt of Refuse," *American Way*, March 15, 2008, p. 34.

59 See the interview with Wendy Bounds in "Cleaning Up in Trash Collection," *CNBC—American Dream Segment* (Television series episode), February 7, 2007. Available at https://www.youtube.com/watch?v=t8MgVZM1KRU (accessed July 22, 2016).

stenches. SDT grew in popularity by not only scrubbing clean the city's famous entertainment district, but also by doing so with pizzazz. Company employees sprayed down the streets with lemon and eucalyptus scented solutions. Overflowing trashcans and loose litter were gone.[60] SDT deployed crews for sixteen hours a day, flushing the streets and sidewalks with pressurized water, with its workers dressed in "hip" black and white uniforms. SDT crews were constantly sweeping, prodding one journalist to compare it to a scene of street sweepers at Disneyworld.[61]

Torres was known to check up on his employees on a golf cart to ensure they were providing the best quality service. He used energy efficient street sweepers and had an aroma specialist design a special lemon scent (called "Superfresh") to use for cleaning the French Quarter's streets. SDT used a $500,000, state-of-the-art surveillance system to track discarded trash. Torres ramped up the business, employing over 150 staff members with 55 trucks as part of the fleet by early 2007 and expanding into other parishes (counties). And these impressive results came just in time for the biggest annual event in the city, Mardi Gras.

Torres earned the nickname "Trashanova" as he brought the garbage industry into the limelight, recruiting both local and national celebrities to endorse his business through print and digital media. What really set his company apart were its glitzy ad campaigns and shiny, polished trucks. The likes of rock musicians Lenny Kravitz (for whom Torres once worked) and Kid Rock shot ad spots. SDT also ran advertisements featuring Trombone Shorty, a locally well-known and respected musician. Torres willingly invested major cash into these efforts, spending over $60,000 from December 2006-February 2007. And for all of that, he was making money—with revenues of $25–30 million in 2007. SDT held over 4,000 commercial accounts in the New Orleans area. The venture was seen throughout the city as a "roll-up-your-sleeves" approach to a basic services problem utilizing innovative methods, resonating positively with many residents. As Torres put it, "From the beginning, starting with one truck right after Katrina, really getting into the business, you know, I saw a need and I jumped in to help."[62]

60 "SDT Waste and Debris," *WWL TV News—New Orleans* (Television series episode), February 26, 2007, Available at https://www.youtube.com/watch?v=mQRoeDf9hqo (accessed June 26, 2015).

61 "Cleaning Up in Trash Collection."

62 Michelle Krupa and Bruce Eggler, "Trash Magnate Sidney Torres Wants to Bring Glitz to Nation's Garbage Collection," *The New Orleans Times-Picayune*, June 4, 2011. Available at http://www.nola.com/politics/index.ssf/2011/06/trash_magnet_sydney_torres_wan.html (accessed June 26, 2015).

SDT was acclaimed an unqualified success, with the areas served by the company looking pristine. Residents asked for Torres' autograph. Local shop owners were so impressed by SDT's work that Torres was selected as Grand Marshal for a Mardi Gras parade.[63] He even starred in (an albeit short-lived) reality television show called "Trashmen" on The Learning Channel featuring his garbage collection service.

In 2011, Torres sold the business to IESI Corp., using the proceeds to launch a new investment firm (STDIV), which invests in small and mid-sized family businesses. But Torres' affection for the company he spawned led him to work as a consultant for IESI (locally called Progressive Waste Solutions)—a national player in the trash business with a strong regional presence in southern Louisiana. He helped with the transition and focused on bringing the same glitz to IESI's marketing that he had given to SDT. Torres also strove to take the unconventional image of the trash industry he brought to New Orleans to the rest of the country, "Trash is sexy in New Orleans. We've got to make trash sexy across the United States."[64] Ultimately, Torres was lauded for taking a stinky and dirty industry and making it seem, well ... cool.[65]

Progressive's contract didn't last long, as a new firm (Empire Janitorial Services) began collecting garbage and scrubbing the streets in both the French Quarter and the Central Business District. The bidding process for the contract was based purely on price and EJS won the $3.9 million annual contract by submitting the lowest bid.[66] But in July 2016, Torres announced via an Instagram post that he was diving once again into the trash business through his new venture IV Trash, beginning to serve New Orleans in August 2016.[67] But at first, the company will only do private trash removal, contracting with individual hotels, restaurants, and other local businesses. IV Trash may bid selectively on municipal waste disposal contracts in the future.[68]

Since entering, exiting, and re-entering the trash business, Torres has taken on other causes in New Orleans, including crime prevention in the French

63 Alpert, "The Rembrandt of Refuse," p. 36.

64 Krupa and Eggler, "Trash Magnate."

65 Krupa and Eggler, "Trash Magnate."

66 Jeff Adelson, "New Company is Handling French Quarter, New Orleans CBD Trash Removal; It's Not Sidney Torres' Firm," *The New Orleans Advocate*, January 27, 2015. Available at http://theadvocate.com/news/neworleans/neworleansnews/11412323-123/new-company-is-handling-french (accessed June 26, 2015).

67 Jennifer Larino, "'Trash King' Sidney Torres IV is Back in the New Orleans Disposal Business," *New Orleans Times-Picayune*, July 18, 2016. Available at http://www.nola.com/business/index.ssf/2016/07/sidney_torres_iv_returns_trash.html (accessed September 2, 2016).

68 Larino, "'Trash King' Sidney Torres IV."

Quarter. He began a "Keep the Quarter Safe" campaign emphasizing the need for city government to develop a comprehensive plan to address a high number of robberies, burglaries, assaults, and shootings. Crime has hit close to home for Torres, whose Esplanade Avenue house was burglarized in December of 2014.[69] He released two commercial spots, both shot in black and white, featuring French Quarter residents airing their concerns and calling for an action plan to better fight crime. In announcing the spots, Torres noted, "The French Quarter is our jewel. It's the face New Orleans and all of Louisiana shows to the world. People have to be safe there and this is a problem we can tackle together as a city. Col. Edmonson and the state police are ready to help if there's a plan. Now the mayor and his administration need to come forward with the plan, and fast."[70]

One commercial spot was more pointed, as Torres called out New Orleans mayor, Mitch Landrieu, to do more to protect the French Quarter. Torres has accused Landrieu of being too passive in the face of the crime surge in the Quarter, imploring him to step up efforts to prevent crime and show leadership. Torres claims residents have had a difficult time even getting the attention of the mayor concerning crime in the city.[71] He also asserts that while the problem is serious and needs immediate attention, it can also be easily fixed since the mayor is the city's equivalent to a business's CEO. And like all CEOs, the mayor needs to be held accountable if he/she is not doing his/her job.[72]

Torres proposed using 25 off-duty officers in the French Quarter to provide protection from sundown to sunrise, positioning them at every exit and entrance of the district. He estimated it would cost $2.8 million to fund the operation each year. Torres also emphasized that the Quarter is a relatively small area (6-by-13 blocks) and thus should be manageably policed. As the "the gem of the city" and absolutely vital to the tourist industry the city relies on, Torres says it needs to be safe.

While Landrieu said he appreciated any civic input into addressing crime issues in the city, he took a swipe at Torres, stating that "He made millions

---

69 "Sidney Torres Takes Crime Reality to TV," *The NOLA Defender*, January 12, 2015. Available at http://www.noladefender.com/content/sidney-torres-takes-crime-reality-tv-videos (accessed June 26, 2015).

70 "Sidney Torres Takes Crime Reality to TV."

71 See Natasha Robbins' January 7, 2015 news segment on *Fox 8 Newscast* reporting on Torres' call for Landrieu to be more aggressive on crime. "We Can Do This Together: Demand a Plan From Our Mayor!" *YouTube* (online video), Available at https://www.youtube.com/watch?v=Mw7t9ZXjXMo (accessed July 17, 2015).

72 See Bill Capo's report, "Torres Ad Demands Mayor Act to Protect FQ," *Eyewitness News—4WWL* (television series episode), January 7, 2015. Available at https://www.youtube.com/watch?v=Mw7t9ZXjXMo (accessed June 26, 2015).

and millions and millions of dollars off of garbage contracts in the French Quarter, and maybe he should just take some of that money and do it himself if he thinks its so easy. It's just not."[73] Landrieu added that the city was already taking steps to address the uptick in the area's crime and reminded Torres of the ultimate obligations of city government. "Public safety is the number one priority in this city. I am glad that Mr. Torres, who has a lot of money, is wanting to help. But remember, we have to govern the whole city and not just the French Quarter. Everybody in this city deserves to be safe."[74] He added that the city's police force received pay raises and is redeploying assets to do everything possible to handle crime. Michael Harrison, New Orleans' chief of police, increased the size of a special task force designed to address the surge in violent crime in particular.

Some worried at the time that Torres would air a second commercial warning visitors planning on coming to New Orleans about the crime issue and thus negatively impact tourism. Torres planned to air the ads in nearby Baton Rouge, and perhaps in Texas, Mississippi, and Florida as well.[75] When he was questioned by a reporter about the concerns of scaring tourists from the city, Torres retorted by saying, "Do you think it's fair that we advertise how great we are as a city, but then we put them [tourists] out on the street to be brutally beaten, to be robbed? I mean, to me, that's not a good situation. I think let's fix the problem and get this under control because if you want to get out there and advertise about the city being great... which it is the best city in the United States, you have to make sure you have the proper protection, and make sure you have the right plans put together, so that these people don't get hurt."[76]

Eventually, after Torres persisted, the mayor and the New Orleans Police Department allowed him to use a private police force in the Quarter. Torres pumped $380,000 into the enterprise. With the New Orleans police force reduced to around 500 officers, it looked to Torres as though the city would not be able to properly protect him and his neighbors in the Quarter.[77] His outfit utilizes a number of high-tech approaches to crime prevention, with local residents able to report criminal activity using a smart phone application. This

73 "Torres Ad Demands Mayor Act to Protect FQ."

74 "Torres Ad Demands Mayor Act to Protect FQ."

75 "Torres Ad Demands Mayor Act to Protect FQ."

76 See Meg Gatto's report in "NOPD Announces Steps to Reduce Violent Crime," *Eyewitness News 4—WWL* (Television series episode), January 7, 2015. Available at https://www.youtube.com/watch?v=Mw7t9ZXjXMo (accessed June 26, 2015).

77 David Amsden, "Who Runs the Streets of New Orleans?" *The New York Times Magazine*, July 30, 2015. Available at http://www.nytimes.com/2015/08/02/magazine/who-runs-the-streets-of-new-orleans.html (accessed September 2, 2016).

will summon a private police patrol from the French Quarter Task Force to help. The Task Force patrols in black Polaris Rangers (which resemble militarized golf carts) retrofitted with surveillance equipment linked to a computerized crime-reporting network.[78] Torres' idea is not novel if we consider that in the US, private police officers outnumber public ones 3-to-1. The US Justice Department reports that there are 450 public-private law enforcement partnerships through America and actively encourages these collaborative efforts.[79]

This case study is not only a profile of an influential member of a struggling community getting involved in local public affairs, but also a glimpse into how a weather event can have myriad long-term impacts on a place years after it occurs. It also raises several other questions. It is interesting that not long ago leading environmental organizations largely ignored urban air and water quality issues. In the mid-1980's, when citizens of a South Central Los Angeles neighborhood tried to keep out a 1,600-ton-per-day solid waste incinerator, the Sierra Club and the Environmental Defense Fund withheld their support as they didn't consider the issue "adequately environmental."[80] Have these impressions changed? If so, in what ways have they? Is urban waste management an environmental problem? Also, what should the role of private businesses be with respect to what have for roughly the last century in the US been considered public services? There are surely tensions between citizens and city officials over appropriate roles with regard to addressing crime. Are there fair ways to resolve them?

Review Questions

1) As mentioned, a number of basic utilities and services (e.g., fire and police protection, water and gas services, garbage collection) have been considered public responsibilities (or at least paid for by tax dollars). Yet, for a variety of reasons, there has been a push to privatize all these basic services. Do you agree that cities should purchase the services of private companies to collect trash? Do you see any possible problems with this arrangement? Would you be more (or less) worried if all private citizens and companies had to contract out *individually* with waste management companies? What about for police protection?

---

78 David Amsden, "Who Runs the Streets . . . ?".

79 David Amsden, "Who Runs the Streets . . . ?".

80 Ronald Sandler and Phaedra C. Pezzullo, "Introduction," in Ronald Sandler and Phaedra C. Pezzullo, eds. *Environmental Justice and Environmentalism: The Social Justice Challenge to the Environmental Movement* (Cambridge, MA: MIT Press, 2007), p. 10.

2) It is clear Torres is quite affluent and has appreciable resources to marshal for issues that matter to him. Local reporters suggested that if Torres hadn't been a crime victim himself, he wouldn't be so adamant in his persistence to put a spotlight on law enforcement in the city with a particular focus on the French Quarter (where he lives). Also, while violence in the French Quarter is worrisome, Mayor Landrieu noted he doesn't have the luxury of managing crime in just one part of New Orleans—his responsibility is to the whole city. Do you think Torres' motivations matter morally in this case? Explain your answer. Do you think focusing on protection of the French Quarter is somehow morally shortsighted? Why or why not? Is there anything morally problematic with very wealthy people being able to purchase media time (or create private police forces) to promote public causes, even if they might be to the benefit of other citizens? Imagine you are a resident of New Orleans who doesn't live in the Quarter. Might you justifiably have qualms with Torres' actions?

Bibliography

Acaroglu, Layla. "Where Do Old Cellphones Go to Die?" *The New York Times*. May 4, 2013. Available at http://www.nytimes.com/2013/05/05/opinion/sunday/where-do-old-cellphones-go-to-die.html?_r=0 (accessed February 26, 2015).

Achenbach, Joel. "Short-term Fixes for Long-Term Climate Problems? Not So Fast, Experts Say." *Washington Post*. February 8, 2015. Available at http://www.washingtonpost.com/national/health-science/upcoming-report-on-technological-fixes-for-climate-change-adds-to-debate/2015/02/07/07bbe150-ac6e-11e4-9c91-e9d2f9fde644_story.html (accessed March 26, 2015).

Adelson, Jeff. "New Company is Handling French Quarter, New Orleans CBD Trash Removal; It's Not Sidney Torres' Firm." *The New Orleans Advocate*. January 27, 2015. Available at http://theadvocate.com/news/neworleans/neworleansnews/11412323-123/new-company-is-handling-french (accessed June 26, 2015).

Aiken, William. "Ethical Issues in Agriculture." In Tom Regan, ed. *Earthbound*. New York: Random House, 1984.

Alaska Department of Revenue. Permanent Fund Dividend Division. "Summary of Dividend Applications and Payments." (website), 2013. Available at https://pfd.alaska.gov/Division-Info/Summary-of-Applications-and-Payments (accessed June 6, 2016).

Allen, Myles R., David J. Frame, and Charles F. Mason. "The Case for Mandatory Sequestration." *Nature Geoscience* (2009) 2: 813–814.

Alper, Nicole. "The Rembrandt of Refuse." *American Way*. March 15, 2008, pp. 34–36.

American Heart Association. "Overweight in Children." (website), August 2014. Available at http://www.heart.org/HEARTORG/GettingHealthy/Overweight-in-Children_UCM_304054_Article.jsp (accessed June 21, 2016).

American Wild Horse Preservation. "Wild Horses and the Ecosystem." (website), 2015. Available at http://wildhorsepreservation.org/wild-horses-and-ecosystem (accessed July 13, 2016).

Amsden, David. "Who Runs the Streets of New Orleans?" *The New York Times Magazine*. July 30, 2015. Available at http://www.nytimes.com/2015/08/02/magazine/who-runs-the-streets-of-new-orleans.html (accessed September 2, 2016).

Ananda, Rady. "Solar Radiation Management, Geoengineering, and Chemtrails: IPCC Warns Policymakers Not to Stop 'Solar Radiation Management.'" *Global Research*. (website), November 5, 2013. Available at http://www.globalresearch.ca/solar-radiation-management-geoengineering-and-chemtrails/5356632 (accessed July 5, 2016).

Anderssen, Erin. "The Cod Collapse." *The UNESCO Courier* (1998) 51(7&8): 44–46.

Anscombe, G.E.M. "Modern Moral Philosophy." *Philosophy* (1958) 33: 1–19.

Anthony, Paulie. "Cell Phone Toxins and the Harmful Effects on the Human Body When Recycled Improperly." *E-Cycle*. (website), October 15, 2013. Available at http://www.e-cycle.com/cell-phone-toxins-and-the-harmful-effects-on-the-human-body-when-recycled-improperly/#sthash.NPrNRJ50.dpuf (accessed March 2, 2015).

Archer, Michael. "Second Chance for Tasmanian Tigers and Fantastic Frogs." *TEDx De-extinction/National Geographic* (online video), 2013. Available at http://longnow.org/revive/tedxdeextinction/ (accessed December 2014).

Arctic Power. "Faces of ANWR: Caribou." (website), 2015. Available at http://www.anwr.org/features/pdfs/faces-caribou.pdf (accessed June 6, 2016).

Aristotle, *Nicomachean Ethics*. Trans. W.D. Ross. New York: Oxford University Press, 1998.

Aristotle, *Politics*. Trans. Ernest Barker. Oxford, U. K.: Oxford University Press, 1948.

Attanasi, E.D. "Undiscovered Oil Resources in the Federal Portion of the 1002 Area of the Arctic National Wildlife Refuge: An Economic Update." Open-File report. Reston, VA: United States Geological Survey, 2005. Available at http://pubs.usgs.gov/of/2005/1217/pdf/2005-1217.pdf (accessed May 28, 2015).

Bader, Christine. "What Do Chief Sustainability Officers Actually Do?" *The Atlantic*. May 5, 2015. Available at http://omnifeed.com/article/www.theatlantic.com/business/archive/2015/05/what-do-chief-sustainability-officers-actually-do/392315/ (accessed June 3, 2015).

Barbour, Ian G. *Technology, Environment and Human Values*. New York: Praeger Publishers, 1980.

Barclay, Eliza. "Why Some Chefs Just Can't Quit Serving Bluefin Tuna." *National Public Radio*. (website), January 7, 2015. Available at http://www.npr.org/sections/thesalt/2015/01/07/375366742/why-some-chefs-just-cant-quit-serving-bluefin-tuna (accessed July 8, 2016).

Barcott, Bruce. "Kill the Cat that Kills the Bird?" *The New York Times*. December 2, 2007. Available at http://www.nytimes.com/2007/12/02/magazine/02cats-v--birds-t.html?pagewanted=all (accessed June 11, 2015).

Barringer, Felicity. "On the Mustang Range, a Battle Rages on Thinning the Herd." *The New York Times*. July 20, 2008, pgs. A1 and 14.

Barry, John. *Rethinking Green Politics*. London, Sage, 1999.

Barton, David N. "Payments for Ecosystem Services: Costa Rica's Recipe." *International Institute for Environment and Development* (website), November 29, 2013. Available at http://www.iied.org/payments-for-ecosystem-services-costa-rica-s-recipe (accessed July 6, 2016).

Barton, Thomas. "Frac Sand Mining Divides Bridgeport." *Dubuque Telegraph Herald*. June 11, 2014. Available at http://www.thonline.com/news/tri-state/article_30eae459-9a93-5bf5-9dc3-31e281f2c3cf.html (accessed September 10, 2016).

Bass, Rick. *Caribou Rising: Defending the Porcupine Herd, Gwich-'in Culture, and the Arctic National Wildlife Refuge*. San Francisco: Sierra Club Books, 2004.

Baxter, William. *People or Penguins: The Case for Optimal Pollution*. New York: Columbia University Press, 1974.

Behm, Don. "Honey Creek Tainted with Human Waste." *Milwaukee Journal Sentinel*. January 21, 2007. Available at http://www.jsonline.com/news/milwaukee/29416799.html (accessed May 13, 2015).

Behm, Don. "Waukesha Concedes It Can't Make Deadline for Radium-Free Water." *Milwaukee Journal Sentinal*. November 19, 2014. Available at http://www.jsonline.com/news/waukesha/waukesha-concedes-it-can-meet-deadline-for-radium-free-water-b99393845z1-283255831.html (accessed May 15, 2015).

Bender, Jeremy. "The US is Setting Aside Nuclear Weapons for Potential Defense Against Asteroids." *Business Insider*. October 1, 2014. Available at http://www.businessinsider.com/us-is-saving-nuclear-weapons-for-astroids-2014-10 (accessed July 9, 2015).

Bereano, Philip. "Patent Nonsense—Patent Pending: The Race to Own DNA." *Seattle Times*. August 27, 1995, p. B5.

Bernstein, Lenny and Brady Dennis. "U.S. Looking For Ideas to Help Manage Wild-Horse Population." *The Washington Post*. January 26, 2014. Available at http://www.washingtonpost.com/national/health-science/us-looking-for-ideas-to-help-manage-wild-horse-overpopulation/2014/01/26/8cae7c96-84f2-11e3-9dd4-e7278db80d86_story.html (accessed June 30, 2015).

Berry, Wendell. *The Unsettling of America: Culture and Agriculture*. San Francisco: North Point Press, 1977.

Berry, Wendell. *What Are People For?* New York: North Point Press, 1990.

Biegelsen, Amy. "America's Looming Burial Crisis." *The Atlantic*. October 31, 2012. Available at http://www.citylab.com/housing/2012/10/americas-looming-burial-crisis/3752/ (accessed June 5, 2015).

BirdLife International. "Spotlight on Threatened Birds." (website), 2014. Available at http://www.birdlife.org/datazone/sowb/spotthreatbirds (accessed June 11, 2015).

Blackstone, William T. "Ethics and Ecology." In William T. Blackstone, ed. *Philosophy and Environmental Crisis*. Athens, GA: University of Georgia Press, 1972, pp. 16–42.

Bookchin, Murray. *The Ecology of Freedom*. Palo Alto, CA: Cheshire Books, 1982.

Bookchin, Murray. *The Philosophy of Social Ecology*. Montreal: Black Rose Books, 1990.

Bookchin, Murray. "What is Social Ecology?" *Environmental Ethics*. Upper Saddle River, NJ: Prentice Hall, 2001, pp. 62–76.

Bourne, Joel K., Jr. "What Obama's Drilling Bans Mean for Alaska and the Arctic." *National Geographic*. February 5, 2015. Available at http://news.nationalgeographic.com/news/2015/02/150205-obama-alaska-oil-anwr-arctic-offshore-drilling/ (accessed May 26, 2015).

Brand, Stuart. "The Dawn of De-extinction: Are You Ready?" *TED Long Beach* (online video), 2013. Available at http://www.ted.com/talks/stewart_brand_the_dawn_of_de_extinction_are_you_ready.html (accessed December 2014).

Brei, Vinicius and Steffen Bohm. "Corporate Social Responsibility As Cultural Meaning Management: A Critique of the Marketing of 'Ethical' Bottled Water." *Business Ethics: A European Review* (2011) 20(3): 233–252.

Brennan, Deborah Sullivan. "Bluefin Fishery On the Hook, New Limits Looming for Popular Gamefish." *San Diego Union Tribune*. September 12, 2014. Available at http://www.utsandiego.com/news/2014/sep/12/bluefin-tuna-limits-fishing/ (accessed May 21, 2015).

Brown, Patricia Leigh. "For New Wave of 'EcoMoms,' Saving Earth Begins at Home." *The New York Times*. February 16, 2008, pp. A1 and A10.

Buchanan, Allen. *Ethics, Efficiency, and the Market*. Totowa, NJ: Rowman and Allenfeld, 1985.

Burnham, Douglas. "Immanuel Kant: Aesthetics." *Internet Encyclopedia of Philosophy*. ISSN 2161–0002. Available at http://www.iep.utm.edu (accessed April 29, 2015).

CBS San Francisco. "Giant 'Potentially Hazardous' Asteroid Hurtling Toward Earth." (website), March 27, 2015. Available at http://sanfrancisco.cbslocal.com/2015/03/27/giant-asteroid-earth-2014-yb35/ (accessed July 21, 2016).

California Energy Commission. "Ethanol as a Transportation Fuel." (website), 2016. Available at http://www.consumerenergycenter.org/transportation/afvs/ethanol.html (accessed July 6, 2016).

Carlton, Jim. "Some NFL Teams are Going Green." *Wall Street Journal*. May 18, 2014. Available at http://www.wsj.com/articles/SB100014240527023046779045795378826915550494 (accessed June 8, 2015).

Carrington, Damian. "The Maldives is the Extreme Test Case for Climate Change Action." *The Guardian*. September 26, 2013. Available at http://www.theguardian.com/environment/damian-carrington-blog/2013/sep/26/maldives-test-case-climate-change-action (accessed June 8, 2016).

Carlson, Allen. "Environmental Aesthetics." *Stanford Encyclopedia of Philosophy*. (website), January 26, 2015. Available at http://plato.stanford.edu/entries/environmental-aesthetics/ (accessed April 29, 2015).

Carlson, Allen and Arnold Berleant. *The Aesthetics of Natural Environments*. Peterborough, ON: Broadview Press, 2004.

Carson, Rachel. *Silent Spring*. Boston: Houghton Mifflin, 1962.

Census of Marine Life. "Bluefin Tuna—Past and Present." (website), 2009. Available at http://www.coml.org/comlfiles/press/CoML_tuna_release_public.pdf (Accessed July 22, 2016).

Centers for Disease Control and Prevention. "Reported Cases of Lyme Disease by State or Locality, 2004–2013." (website), March 19, 2015. Available at http://www.cdc.gov/lyme/stats/chartstables/reportedcases_statelocality.html (accessed June 9, 2015).

Centers for Disease Control and Prevention. "Strategies to Prevent Obesity and Other Chronic Diseases: The CDC Guide to Strategies to Increase the Consumption of Fruits and Vegetables." Atlanta: U.S. Department of Health and Human Services, 2011. Available at http://www.cdc.gov/obesity/resources/recommendations.html (accessed January, 7, 2015).

Chesapeake Stormwater Network. "Urban Stormwater Retrofits." (website), 2015. Available at http://chesapeakestormwater.net/bay-stormwater/baywide-stormwater-policy/urban-stormwater-workgroup/retrofits/ (accessed July 6, 2015).

Christopher, Tom. "Can Weeds Help Solve the Climate Crisis." *The New York Times Magazine*. June 29, 2008, pp. 42–47.

Church, George. "Hybridizing with Extinct Species." *TEDx De-extinction/National Geographic* (online video), 2013. Available at http://longnow.org/revive/tedxde extinction/ (accessed July 24, 2016).

Cimitile, Matthew. "Corn Ethanol Will Not Cut Greenhouse Gas Emissions." *Scientific American*. April 20, 2009. Available at http://www.scientificamerican.com/article/ethanol-not-cut-emissions/ (accessed June 22, 2015).

City of Calgary. "International Spotlight on City's EnviroSmart Streetlights: BBC Team to Look at How Streetlights are Saving The Environment and Money in Calgary." (website), August 20, 2010. Available at http://newsroom.calgary.ca/news/international-spotlight-on-city-168299 (accessed July 8, 2015).

City of Cambridge, Massachusetts. Department of Public Works. "Composting Questions." (website), 2014. Available at http://www.cambridgema.gov/TheWorksOurServices/RecyclingAndTrash/FAQRecyclingAndRubbish/CompostingQuestions/CompostPilot.aspx (accessed Oct. 30, 2014).

City of Waukesha. "Application Summary: City of Waukesha Application for a Lake Michigan Diversion with Return Flow." (website), October 2013. Available at http://www.ci.waukesha.wi.us/c/document_library/get_file?uuid=a972a2e4-d45b-4748-9948-17c0ce17b692&groupId=10113 (accessed May 15, 2015).

"Cleaning Up in Trash Collection." *CNBC—American Dream Segment* (Television series episode), February 7, 2007. Available at https://www.youtube.com/watch?v=t8MgVZM1KRU (accessed July 22, 2016).

Colbert, Elizabeth. *The Sixth Extinction: An Unnatural History*. New York: Henry Holt and Company, 2014.

Collings, David. *Stolen Future, Broken Present: The Human Significance of Climate Change*. Ann Arbor, MI: Open Humanities Press, 2014.

Conviser, Richard. "Toward Agricultures of Context." *Environmental Ethics* (1984) 6: 71–85.

Corley, Cheryl. "Burials and Cemeteries Go Green." *National Public Radio* (Radio transcript), December, 16, 2007. Available at http://www.npr.org/templates/story/story.php?storyId=17232879 (accessed July 22, 2016).

Corn, M. Lynne, Bernard A. Gelb, and Pamela Baldwin. "Arctic National Wildlife Refuge: Controversies for the 109th Congress." *Congressional Research Service Issue Brief for Congress*. IB10136. 2015. Available at http://fpc.state.gov/documents/organization/50817.pdf (accessed May 26, 2015).

Crutzen, Paul. "Albedo Enhancement by Stratospheric Sulphur Injections: A Contribution to Resolve a Policy Dilemma?" *Climatic Change* (2006) 77: 211–219.

D'Arms, Justin, "Envy." *The Stanford Encyclopedia of Philosophy*. January 22, 2009. Available at http://plato.stanford.edu/archives/win2013/entries/envy/ (accessed June 3, 2016).

D'Eaubonne, Francoise. *Le Feminisme ou la Mort*. Paris: Pierre Horay, 1974.

Daar, Abdullah S. "Animal-to-Human Transplants—A Solution or a New Problem." *Bull World Health Organization* (1999) 77: 54–61.

Damon, Arwa. "Ruthless Drought in West Timor Puts Children in Crisis." *CNN*. (website), July 6, 2008. Available at http://www.cnn.com/2008/WORLD/asiapcf/07/06/westtimor.children/ (Accessed June 17, 2016).

Davenport, Carol. "Rising Seas." *The New York Times*. March 27, 2014. Available at http://www.nytimes.com/interactive/2014/03/27/world/climate-rising-seas.html?_r=0 (accessed February 18, 2015).

Davey, Monica. "Wisconsin Bill Would Allowing Hunting of Once-Rare Crane." *The New York Times*. February 23, 2012. Available at http://www.nytimes.com/2012/02/24/us/wisconsin-consider-hunting-of-sandhill-crane.html (accessed February 10, 2015).

Davison, Issac. "No Need for Fracking Ban in New Zealand." *New Zealand Herald*. November 27, 2012. Available at http://www.nzherald.co.nz/business/news/article.cfm?c_id=3&objectid=10850290 (accessed April 9, 2015).

Dean, Cornelia. "New Orleans Journal: From Gator Fodder to Haute Cuisine." *The New York Times*. July 27, 1993. Available at http://www.nytimes.com/1993/07/27/us/new-orleans-journal-from-gator-fodder-to-haute-cuisine.html (accessed July 15, 2015).

DeLong, Katie. "Experimenting with Beet Juice on Frozen Roads." *The Milwaukee Journal*. February 11, 2008. Available at http://www.todaystmj4.com/news/local/15508127.html (accessed July 15, 2015).

Dembek, Cathryn, Jennifer L. Harris, and Marlene Schwartz. "Trends in Television Food Advertising to Young People: 2013 Update." White paper. Yale Rudd Center. May 2014. Available at http://www.yaleruddcenter.org/resources/upload/docs/what/reports/RuddReport_TVFoodAdvertising_6.14.pdf (accessed January 7, 2015).

Dembosky, April. "Biodegradable Coffins." *The New York Times Magazine*. December 9, 2007, p. 57.

DePillis, Lidia. "The Boom-Proof Economy: How to Handle a Fracking Bust." *Washington Post*. January 15, 2015. Available at http://www.washingtonpost.com/news/storyline/wp/2015/01/15/the-boom-proof-economy-how-to-handle-a-fracking-bust/ (accessed July 13, 2016).

de Steiguer, J. Edward. *Wild Horses of the West: History and Politics of America's Mustangs*. Tucson, AZ: University of Arizona Press, 2011.

Devall, Bill and George Sessions. *Deep Ecology: Living as if Nature Mattered*. Layton, UT: Gibbs & Sons, 1985.

"De-Vine Biofuel: Ethanol from Kudzu." *NBC Nightly News* (Television series episode), December 16, 2008. Available at https://www.youtube.com/watch?v=MhYux3QIJgo (accessed May 1, 2015).

de Vos, Asha. "Ending Overfishing." *TED Ed* (online video), May 21, 2012. Available at https://www.youtube.com/watch?v=F6nwZUkBeas (accessed July 8, 2016).

Dietz, Michael and John Clausen. "A Field Evaluation of Rain Garden Flow and Pollutant Treatment." *Environmental Science and Technology* (2006) 40: 1335–1340.

Donovan, Karen. "HCPs—Important Tools for Conserving Habitat and Species." In Donald C. Baur and William R. Irvin, eds. *Endangered Species Act: Law, Policy, and Perspectives*. Chicago: American Bar Association Publishing, 2001.

Easterbrook, Gregg. "The Sky is Falling." *The Atlantic* (2008) 301(5): 74–84.

EcoMom Alliance. "Our Mission." (website), 2016. Available at http://www.ecomomalliance.org (accessed July 19, 2016).

EcoMom Alliance. "Program Philosophy." (website), 2015. Available at http://www.ecomomalliance.org/page/theory-of-change (accessed July 3, 2015).

Ehrenfeld, David. "Transgenics and Vertebrate Cloning as Tools for Species Conservation." *Conservation Biology* (2006) 20: 723–732.

El Bassam, Nasir. *Handbook of Bioenergy Crops: A Complete Reference to Species, Development and Applications*. Washington, D.C.: Earthscan LLC, 2010.

Ellick, Adam B. "A Ban on Cockfighting, but Tradition Lives On." *The New York Times*. July 6, 2008. Available at http://www.nytimes.com/2008/07/06/us/06fight.html?pagewanted=all&_r=0 (accessed August 21, 2016).

Enriquez, Darryl. "Mayor Optimistic on Lake Water." *Milwaukee Journal Sentinel*. January 31, 2008, p. 1.

Esbjornson, Carl D. "Once and Future Farming: Some Meditations on the Historical and Cultural Roots of Sustainable Agriculture in the United States." *Agriculture and Human Values* (1992) 9(3): 20–30.

European Commission. "Waste Electrical and Electronic Equipment." (website), March 5, 2015. Available at http://ec.europa.eu/environment/waste/weee/index_en.htm (accessed March 6, 2015).

Exec. Order No. 13112, 64 C.F.R. 6183 (1999).

"Exit Strategies." *Economist* (2010) 396(8700): 74–5. Available at http://ezproxy.lvc.edu:2048/login?url=http://search.ebscohost.com/login.aspx?direct=true&db=aph&AN=53783708&site=ehost-live (accessed June 7, 2015).

Fabregas, Luis. "'Million-dollar Pigs' are Medical Marvels." *Pittsburgh Tribune Review*. April 9, 2006. Available at http://triblive.com/x/pittsburghtrib/news/regional/s_441762.html (accessed July 12, 2016).

Fahey, John. "A New Century of Exploration." *TEDx De-extinction/National Geographic* (online video), 2013. Available from http://longnow.org/revive/tedxdeextinction/ (accessed December 15, 2014).

Feinberg, Joel. *Rights, Justice and the Bounds of Liberty: Essays in Social Philosophy*. Princeton, NJ: Princeton University Press, 1980.

Florida Atlantic University. Department of Physics. "Light Pollution Harms the Environment." (website), 2015. Available at http://physics.fau.edu/observatory/lightpol-environ.html (accessed July 20, 2016).

Folch J., M.J Cocero, P. Chesne, J.L. Alabart, V. Dominguez, Y. Cognie, A. Roche, A. Fernadez-Arias, J.L. Marti, P. Sanchez, E. Echegoyen, J.F. Beckers, A.S. Bonastre, X. Vignon. "First Birth of an Animal from an Extinct Sub-species (*Capra Pyrenaica Pyrenaica*) by Cloning." *Theriogenology* (2009) 71: 1026–1034.

Foley, Jonathan. "It's Time to Rethink America's Corn System." *Scientific America*. March 5, 2013. Available at http://www.scientificamerican.com/article/time-to-rethink-corn/ (accessed July 24, 2015).

Food and Agriculture Organization of the United Nations. "Deforestation and Net Forest Area Change: Facts and Figures." (website), November 4, 2011. Available at http://www.fao.org/forestry/30515/en/ (accessed April 10, 2015).

Food and Agriculture Organization of the United Nations. "Global Forest Resources Assessment." (website), 2010. Available at http://www.fao.org/docrep/013/i1757e/i1757e.pdf (accessed April 15, 2015).

Francione, Gary L. and Anna Charlton. *Eat Like You Care: An Examination of the Morality of Eating Animals*. Louisville, KY: Exempla Press, 2015.

Franke, J.M. *The Invasive Species Cookbook: Conservation through Gastronomy*. Wauwatosa, WI: Bradford Street Press, 2007.

Freeman, Juri and Lisa A. Skumatz. "Best Management Practices in Food Scrap Programs." *Ecoconservation Institute*. (website), 2010. Available at http://www.food-scrapsrecovery.com/EPA_FoodWasteReport_EI_Region5_v11_Final.pdf (accessed July 16, 2016).

French, Robert Heath. *Environmental Philosophy and the Ethics of Terraforming Mars: Adding the Voices of Environmental Justice and Ecofeminism to the Ongoing Debate*. Unpublished Master's Thesis. Denton, TX: University of North Texas, 2013.

"Frequently Asked Questions About Lionfish." *SafeSpear.com.* (website), 2015. Available at http://www.safespear.com/v.php?pg=59 (accessed July 7, 2016).

Fry Funeral Home. "Green & Natural Burials." (website), 2011. Available at http://www.fryfuneralhome.com/services/green-burial/ (accessed June 4, 2015).

Fuller, Thomas and Arthur Revkin, "Climate Plan Looks Beyond Bush's Tenure." *The New York Times.* December 16, 2007. Available at http://www.rainforestcoalition.org/documents/nytimes_12-16-071.pdf (accessed June 8, 2016).

Funk, MacKenzie. *Windfall: The Global Business of Global Warming.* New York: Penguin Group USA, 2014.

Gardiner, Stephen. "Is 'Arming the Future' with Geogineering Really the Lesser Evil?" In Stephen Gardiner, Simon Caney, Dale Jamieson, and Henry Shue, eds. *Climate Ethics: Essential Readings.* Oxford, U.K.: Oxford University Press, 2010.

Gardiner, Stephen. "Some Early Ethics of Geoengineering the Climate: A Commentary on the Values of the Royal Society Report." *Environmental Values* (2011) 20: 163–181.

Garvey, James. *The Ethics of Climate Change: Right and Wrong in a Warming World.* London: Continuum Press, 2008.

Gehrke, Robert. "Utah May Toughen Cockfighting Laws After Years of Trying." *The Salt Lake Tribune.* March 10, 2015. Available at http://www.sltrib.com/news/2276551-155/utah-may-toughen-cockfighting-laws-after (accessed June 10, 2015).

Gerdeman, Dina. "What Do Chief Sustainability Officers Do?" *Forbes.* October 8, 2014. Available at http://www.forbes.com/sites/hbsworkingknowledge/2014/10/08/what-do-chief-sustainabilty-officers-do/ (accessed June 2, 2015).

Gerland, Patrick, Adrian E. Raftery, Hana Sevcikova, Nan Li, Danan Gu, Thomas Spoorenberg, Leontine Alkema, Bailey K. Fosdick, Jennifer Chunn, Nevena Lalic, Guiomar Bay, Thomas Buettner, Gerhard K. Heilig, and John Wilmoth. "World Population Stabilization Unlikely This Century." *Science* (2014) 346(6206): 234–237.

Gewirth, Alan. "Human Rights and Future Generations." In Michael Boylan, ed. *Environmental Ethics.* Upper Saddle River, NJ: Prentice Hall, 2001, pp. 207–211.

Giry, Stephanie "Climate Conflict." *The New York Times Magazine.* December 9, 2007, pp. 60–62.

Global Environment Facility. "Conserving Timor Leste's Rich Forest Land." (website), 2015. Available at https://www.thegef.org/gef/node/11123 (accessed August 23, 2016).

Godlovitch, Sam. "Valuing Nature and the Autonomy of Natural Aesthetics." *British Journal of Aesthetics* (1998) 38(2): 180–197.

Golden, Kate. "Yahara Beach Closures Highlight Algae, Bacteria Threats Statewide." *Wisconsin Watch.* April 23, 2014. Available at http://wisconsinwatch.org/2014/04/yahara-beach-closures-highlight-algae-bacteria-threats/ (accessed January 27, 2015).

Greenpeace. "Canadian Atlantic Fisheries Collapse." (website), 2007. Available at http://archive.greenpeace.org/comms/cbio/cancod.html (accessed May 21, 2015).

Gregson, Bob and Bonnie Gregson. *Rebirth of the Small Family Farm: A Handbook for Starting a Successful Organic Farm Based on the Community Supported Agriculture Concept.* Vashon Island, WA: IMF Associates, 1996.

Groh, Trauger and Steven McFadden. *Farms of Tomorrow Revisited: Community Supported Farms, Farm Supported Communities.* Kimberton, PA: Biodynamic Farming and Gardening Association, 1997.

Gross, Terry. "After Dump, What Happens to Electronic Waste?" *National Public Radio* (Radio transcript), December 21, 2010. Available at http://www.npr.org/templates/transcript/transcript.php?storyId=132204954 (accessed February 27, 2015).

Grunwald, Michael and Susan Glasser. "Experts Say Faulty Levees Caused Much of Flooding." *The Washington Post.* September 21, 2005. Available at http://www.washingtonpost.com/wp-dyn/content/article/2005/09/20/AR2005092001894_pf.html (accessed June 25, 2015).

Hammack, Richard W., William Harbert, Shikha Sharma, Brian W. Stewart, Rosemary C. Capo, Andy J. Wall, Arthur Wells, Rodney Diehl, David Blaushild, James Sams, Garret Veloski. *An Evaluation of Fracture Growth and Gas/Fluid Migration as Horizontal Marcellus Shale Gas Wells are Hydraulically Fractured in Greene County, Pennsylvania.* NETL-TRS-3–2014; EPAct Technical Report Series; U.S. Department of Energy, National Energy Technology Laboratory: Pittsburgh, PA, 2014.

Hardin, Garrett. "The Tragedy of the Commons." *Science* (1968) 162(3895): 1243–1248.

Harker, Alexandra. "Landscapes of the Dead: An Argument for Conservation Burial." *Berkeley Planning Journal* (2012) 25(1): 150–159.

Harris, Emily. "Toxic E-Waste." *NOW* (Television news transcript), July 19, 2002. Available at http://www.pbs.org/now/printable/transcript_ecycling_print.html (accessed March 6, 2015).

Harvey, Fiona. "Overfishing Causes Pacific Bluefin Tuna Numbers to Drop 96%." *The Guardian.* January 9, 2013. Available at http://www.theguardian.com/environment/2013/jan/09/overfishing-pacific-bluefin-tuna (accessed July 8, 2016).

Hayden, Erika Check. "'Patent Trolls' Target Biotechnology Firms." *Nature* (2011) 477: 521. Available at http://www.nature.com/news/2011/110928/full/477521a.html (accessed April 28, 2015).

Hayward, Tim. "Anthropocentrism: A Misunderstood Problem." *Environmental Values* (1997) 6(1): 49–63.

Heffernan, James P. "The Land Ethic: A Critical Appraisal." *Environmental Ethics* (1982) 4(3): 240–257.

Henderson, David. "Valuing the Stars: On the Economics of Light Pollution." *Environmental Philosophy* (2010) 7(1): 17–26.

Henderson, John. "Space Junk: Traffic Cops in Space." *YouTube* (online video), December 20, 2012. Available at https://www.youtube.com/watch?v=UoIcXxbyX1s (accessed July 10, 2015).

Hepburn, Ronald. "Contemporary Aesthetics and the Neglect of Natural Beauty." In Bernard Williams and Alan Montefiore, eds. *British Analytical Philosophy*. London: Routledge and Keegan Paul, 1966.

Hertsgaard, Mark. *Hot: Living Through the Next Fifty Years on Earth*. Boston: Houghton Mifflin Harcourt, 2011.

Hewett, Stephen and Timothy Simonton, "Walleye Management Plan: Moving Management into the 21st Century," Administrative Report #43. Madison, WI: Wisconsin Department of Natural Resources, January 1998.

Hobbes, Thomas. *Leviathan*. New York: Penguin Classics, 1981.

Holyoke, John. "Biologists Says Maine Deer Herd Rebounding Nicely," *Bangor Daily News*. March 29, 2013. Available at http://bangordailynews.com/2013/03/29/outdoors/biologist-says-maine-deer-herd-rebounding-nicely/ (accessed June 9, 2015).

Hook-Sopko, Amy. "The More You Know Eco: EcoMom Alliance." *Green Child Magazine*. (website), August 1, 2010. Available at http://www.greenchildmagazine.com/ecomom-alliance/ (accessed July 3, 2015).

Horne, Jed. "Five Myths about Hurricane Katrina." *The Washington Post*. August 31, 2012. Available at http://www.washingtonpost.com/opinions/five-myths-about-hurricane-katrina/2012/08/31/003f4064-f147-11e1-a612-3cfc842a6d89_story.html (accessed August 22, 2016).

Horton, Benjamin P., Stefan Rahmstorf, Simon E. Englehart, and Andrew C. Kemp. "Expert Assessment of Sea-level Rise by AD 2100 and AD 2300." *Quaternary Science Reviews* (2014) 84: 1–6.

Howard, Brian Clark. "How Cities Compost Mountains of Food Waste." *National Geographic*. June 18, 2013. Available at http://news.nationalgeographic.com/news/2013/06/130618-food-waste-composting-nyc-san-francisco/ (accessed July 16, 2016).

Hoxie, Neil J., J.P. Davis, J.M. Vergerant, R.D. Nashold, and K.A. Blair. "Cryptosporidiosis-Associated Mortality Following a Massive Waterborne Outbreak in Milwaukee, Wisconsin." *American Journal of Public Health* (1997) 87(12): 2032–2035.

Hughes, Jonathan. "Justice and Third Party Risk: The Ethics of Xenotranplantation." *Journal of Applied Philosophy* (2007) 24(2): 151–167.

"Hunting Sandhill Cranes Isn't for Everyone, but It Isn't Unethical." *Chicago Sun-Times*. April 6, 2014. Available at http://chicago.suntimes.com/uncategorized/7/71/195127/hunting-sandhill-cranes-isnt-for-everybody-but-it-isnt-unethical/ (accessed February 10, 2015).

I Am An Animal: The Story of Ingrid Newkirk and PETA. Dir. Matthew Galkin. HBO documentary, 2007.

ICAX, "ICAX Solar Runways Systems Clear Ice from Airplane Parking Stands Using Under Runway Heating." (website), 2015. Available at http://www.icax.co.uk/Solar_Runways.html (accessed July 19, 2016).

ICAX, "TRL Reports on IHT Ice Clearing at Toddington Using Under Road Heating," (website), April 2008. Available at http://www.icax.co.uk/toddington_results.html (accessed July 19, 2016).

Industrial Noise Control. "Comparative Examples of Noise Levels: Comparative Examples of Noise Sources, Decibels, and Their Effects." (website), 2010. Available at http://www.industrialnoisecontrol.com/comparative-noise-examples.htm (accessed July 7, 2015).

Intergovernmental Panel on Climate Change. "Assessment Report." 2013, 19.3.4.1 Available at http://www.ipcc.ch/ipccreports/tar/wg2/index.php?idp=671 (accessed February 16, 2015).

Intergovernmental Panel on Climate Change. "Summary for Policymakers." In T.F. Stocker, D. Qin, G.-K. Plattner, M. Tignor, S.K. Allen, J. Boschung, A. Nauels, Y. Xia, V. Bex and P.M. Midgley, eds. *Climate Change 2013: The Physical Science Basis. Contribution of Working Group I to the Fifth Assessment Report of the Intergovernmental Panel on Climate Change*. Cambridge, UK and New York, NY, USA: Cambridge University Press, 2013.

Intergovernmental Panel on Climate Change. "Summary for Policymakers." In C.B. Field, V.R. Barros, D.J. Dokken, K.J. Mach, M.D. Mastrandrea, T.E. Bilir, M. Chatterjee, K.L. Ebi, Y.O. Estrada, R.C. Genova, B. Girma, E.S. Kissel, A.N. Levy, S. MacCracken, P.R. Mastrandrea, and L.L. White, eds. *Climate Change 2014: Impacts, Adaptation, and Vulnerability. Part A: Global and Sectoral Aspects. Contribution of Working Group II to the Fifth Assessment Report of the Intergovernmental Panel on Climate Change*. Cambridge, UK and New York, NY, USA: Cambridge University Press, 2014, pp. 1–32.

Intergovernmental Panel on Climate Change. Working Group I. "Climate Change 2013: The Physical Science Basis." *Fifth Assessment Report of the Intergovernmental Panel on Climate Change*. Cambridge, UK and New York, NY, USA: Cambridge University Press, 2013.

Intergovernmental Panel on Climate Change. Working Group II. "Impacts, Adaptation and Vulnerability." *Fifth Assessment Report of the Intergovernmental Panel on Climate Change*. Cambridge, UK and New York, NY, USA: Cambridge University Press, 2014.

International Coral Reef Initiative. "What are Corals?" (website), 2015. Available at http://www.icriforum.org/about-coral-reefs/what-are-corals (accessed March 19, 2015).

International Crane Foundation. "Endangered Whooping Crane Shot in Louisiana." (website), January 22, 2015. Available at http://www.savingcranes.org/whats-new/2015/01/endangered-whooping-crane-shot-in-louisiana/ (accessed February 15, 2015).

International Crane Foundation. "George and Tex." *YouTube* (online video), October 28, 2011. Available at https://www.youtube.com/watch?v=fsDYhdaJL98 (accessed July 9, 2016).

International Crane Foundation. "Historic Whooping Crane Numbers." (website), 2011. Available at https://www.savingcranes.org/images/stories/site_images/conservation/whooping_crane/pdfs/historic_wc_numbers.pdf (accessed February 3, 2015).

International Crane Foundation. "Suspected Shooting of a Whooping Crane in Aransas Bay, Texas." (website), 2015. Available at http://www.savingcranes.org/whats-new/2015/02/5307/ (accessed February 13, 2015).

International Dark Sky Association. "IDA Practical Guide 2: Introduction to Light Pollution." (website), 2009. Available at http://darksky.org (accessed July 20, 2016).

International Scientific Committee for Tuna and Tuna-like species in the Northern Pacific. "Pacific Bluefin Tuna Working Group." (website), July 1, 2014. Available at http://isc.ac.affrc.go.jp/working_groups/pacific_bluefin_tuna.html (accessed May 19, 2015).

"Interview with Brandy Gallagher." *YouTube* (online video). 2014. Available at https://www.youtube.com/watch?v=HcyIWpAhv6s (accessed June 21, 2016).

Jabr, Ferris. "Will Cloning Ever Save Endangered Animals?" *Scientific American*. March 11, 2013. Available at http://www.scientificamerican.com/article.cfm?id=cloning-endangered-animals (accessed December 10, 2014).

Jackson, Wes. "Making Sustainable Agriculture Work." In Robert Clark, ed. *Our Sustainable Table*. San Francisco: North Point Press, 1990, pp. 132–141.

Johnsgard, Paul A. *Sandhill and Whooping Cranes: Ancient Voices Over America's Wetlands*. Lincoln, NE: University of Nebraska Press, 2011.

Johnson, Baylor L. "Ethical Obligations in a Tragedy of the Commons." *Environmental Values* (2003) 12(3): 271–287.

Jonassen, Wendi. "7 Years After Katrina, New Orleans is Overrun by Wild Dogs." *The Atlantic*. August 24, 2014. Available at http://www.theatlantic.com/national/archive/2012/08/7-years-after-katrina-new-orleans-is-overrun-by-wild-dogs/261530/ (accessed July 17, 2015).

Kahn, Elizabeth. "The Tragedy of the Commons as an Essentially Aggregative Harm." *Journal of Applied Philosophy* (2014) 31(3): 223–236.

Kant, Immanuel. *Critique of Practical Reason*. In Mary J. Gregor, Trans. and Ed. *Practical Philosophy*. Cambridge, UK: Cambridge University Press, 1996.

Kant, Immanuel. "Duties to Animals and Spirits." Trans. Louis Infield. *Lectures on Ethics*. New York: Harper and Row, 1963.

Kant, Immanuel. *Grounding for the Metaphysics of Morals*. Trans. James Ellington. Indianapolis: Hackett, 1981.

Kennedy, Tony. "Local Officials Dealing Themselves a Piece of Frac Sand Boom." *Star Tribune*. December 26, 2012. Available at http://www.startribune.com/local-officials-dealing-themselves-a-piece-of-frac-sand-boom/184771921/ (accessed June 3, 2016).

Klosterman, Chuck. "Should We Protect Wild Horses?" *The New York Times Magazine*. March 21, 2014. Available at http://www.nytimes.com/2014/03/23/magazine/should-we-protect-wild-horses.html?_r=0 (accessed June 30, 2015).

Knight, Joe. "DNR: Bore Holes Incorrectly Abandoned." *Eau Claire Leader-Telegram*. December 1, 2014. Available at http://www.leadertelegram.com/News/Front-Page/2014/08/08/DNR-Bore-holes-incorrectly-abandoned.html (accessed June 3, 2016).

Kramer, Virginia. "Conservation Groups, Native Organizations Celebrate Wilderness Recommendation for the Arctic National Wildlife Refuge." *Sierra Club*. (website), January 25, 2015. Available at http://content.sierraclub.org/press-releases/2015/01/conservation-groups-native-organizations-celebrate-wilderness-recommendation (accessed May 27, 2015).

Krueger, Jennifer. "Open Water Spearing in Northern Wisconsin by Chippewa Indians During 2009." Odanah, WI, GLIFWC, January 2010.

Krupa, Michelle and Bruce Eggler. "Trash Magnate Sidney Torres Wants to Bring Glitz to Nation's Grabage Collection." *The New Orleans Times-Picayune*. June 4, 2011. Available at http://www.nola.com/politics/index.ssf/2011/06/trash_magnet_sydney_torres_wan.html (accessed June 26, 2015).

LaFond, Kaye. "Waukesha Presses First Test of Great Lakes Water Compact." *Circle of Blue*. (website), July 9, 2014. Available at http://www.circleofblue.org/waternews/2014/world/waukesha-presses-first-test-great-lakes-water-compact/ (accessed May 15, 2015).

Lallanilla, Marc. "Green House Gas Emissions: Causes and Sources." *LiveScience.com*. (website), February 10, 2015. Available at http://www.livescience.com/37821-green house-gases.html (accessed February 25, 2015).

Lanting, Frans. *Madagascar: A World Out of Time*. New York: Aperture Foundation, Inc., 1990.

Lanza, Robert. "Cloning of an Endangered Species (*Bos gaurus*) Using Interspecies Nuclear Transfer." *Cloning* (2000) 2: 79–84.

Lanza, Robert. "The Use of Cloning and Stem Cells to Resurrect Life." *TEDx De-extinction/National Geographic* (online video), 2013. Available from http://longnow.org/revive/tedxdeextinction/ (accessed December 2014).

Larino, Jennifer. "'Trash King' Sidney Torres IV is Back in the New Orleans Disposal Business." *The New Orleans Times-Picayune*. July 18, 2016. Available at http://www.nola.com/business/index.ssf/2016/07/sidney_torres_iv_returns_trash.html (accessed September 2, 2016).

Lentz, David R., and Jimmy S. Christianson. "The Karner Blue Butterfly: Wisconsin's Statewide Habitat Conservation Plan." In Kaush Arha and Barton H. Thompson, Jr. eds. *The Endangered Species Act and Federalism: Effective Conservation through Greater State Commitment*. New York: RFF Press, 2011.

Leopold, Aldo. *A Sand County Almanac*. Oxford, U.K.: Oxford University Press, 1949.

Leopold, Aldo. *A Sand County Almanac and Other Writings on Ecology and Conservation*. Curt Meine, ed. New York: Literary Classics of the United States, 2013.

Levitt, Steven D. and Stephen J. Dubner. *Super Freakonomics: Global Cooling, Patriotic Prostitutes, and Why Suicide Bombers Should Buy Life Insurance*. New York: Harper Collins, 2009.

Lewanzik, Daniel and Christian Voigt. "Artificial Light Puts Ecosystem Service of Frugivorous Bats at Risk." *Journal of Applied Ecology* (2014) 51(2): 388–394.

Loewengart, Victoria. "Urban Mining of Rare Earth Elements in the United States: A Win-Win Proposition." Unpublished paper, 2011. Available at http://www.academia.edu/2578513/Urban_Mining_of_Rare_Earth_Elements_in_the_United_States_A_Win-Win_Proposition (accessed June 9, 2016).

Loi, Pasqualino, Grazyna Ptak, Barbara Barboni, Josef Fulka, Pietro Cappai, and Michael Clinton. "Genetic Rescue of an Endangered Mammal by Cross-species Nuclear Transfer Using Post-mortem Somatic Cells." *Nature Biotechnology* (2001) 10: 962–964.

Lovgren, Stefan. "Space Junk Cleanup Needed, NASA Experts Warn." *National Geographic*. January 19, 2006. Available at http://news.nationalgeographic.com/news/2006/01/0119_060119_space_junk.html (accessed July 21, 2016).

MacKenzie, William R., Neil J. Hoxie, Mary E. Proctor, M. Stephen Gradus, Kathleen A. Blair, Dan E. Peterson, James J. Kazmierczak, David G. Addiss, Kim R. Fox, Joan B. Rose, and Jeffrey P. Davis. "A Massive Outbreak in Milwaukee of *Cryptosporidium* Infection Transmitted Through the Public Water Supply." *New England Journal of Medicine* (1994) 331: 161–7.

MacPherson, Mark. "Frac Sand Mining Company Donates $500,000 to City of Independence." *WKBT*. (website), March 6, 2014. Available at http://www.news8000.com/news/frac-sand-mining-company-donates-500000-to-city-of-indepe-nence/24852046 (accessed July 24, 2016).

Macy, Jack. "San Francisco Takes Residential Organics Collection Full-Scale." *BioCycle Magazine*. January 1, 2000. Available at https://www.environmental-expert.com/articles/san-francisco-takes-residential-organics-collection-full-scale-1068 (accessed July 18, 2016).

Madison Metropolitan Sewage District. "Phosphorus Harvesting." (website), 2015. Available at http://www.madsewer.org/Programs-Initiatives/Phosphorus-Harvesting (accessed June 7, 2016).

Major League Baseball Advanced Media, "Chase Field History," (website), 2015. Available at http://arizona.diamondbacks.mlb.com/ari/ballpark/information/index.jsp?content=chase_history (accessed July 14, 2015).

Maine Department of Inland Fisheries and Wildlife. "Deer: Whitetails in the Maine Woods." (website), 2013. Available at http://www.maine.gov/ifw/wildlife/species/mammals/deer.html (accessed June 9, 2015).

Makower, Joel. "Ellen Weinreb Checks the Pulse of the CSO Profession." *GreenBiz.* (website), October 21, 2014. Available at http://www.greenbiz.com/blog/2014/10/21/ellen-weinreb-checks-pulse-cso-profession (accessed June 3, 2015).

"Maldives Cabinet Makes a Splash." *BBC News.* (website), October 17, 2009. Available at http://news.bbc.co.uk/2/hi/8311838.stm (accessed June 8, 2016).

"Maldives Facts." *National Geographic Atlas.* 8th Edition, 2008. Available at http://travel.nationalgeographic.com/travel/countries/maldives-facts/ (accessed June 8, 2016).

Marcotty, Josephine. "Wis. Sand Mine Spills Cause Call for Penalties against Minn. Firms." *Star Tribune.* June 12, 2012. Available at http://www.startribune.com/wis-sand-mine-spills-cause-call-for-penalties-against-minn-firms/158518655/ (accessed June 3, 2016).

Mardt, Marah and Carl Safina. "Covering Ocean Acidification." *Yale Climate Connections.* (website), June 24, 2008. Available at http://www.yaleclimateconnections.org/2008/06/covering-ocean-acidification-chemistry-and-considerations/ (accessed April 6, 2015).

Markham, Derek. "How E-waste is Becoming a Big, Global Problem." *National Public Radio* (Radio transcript), January 11, 2013. Available at http://www.npr.org/2013/01/11/169144849/how-e-waste-is-becoming-a-big-global-problem (accessed March 3, 2015).

Marvier, Michelle A, Eli Meir, and Peter M. Kareiva. "How Do the Monitoring and Control Strategies Affect the Chance of Detecting and Containing Transgenic Weeds?" In Klaus Ammann et al., *Methods of Risk Assessment of Transgenic Plants.* Vol. 3. Basel: Springer, 1999.

Mason, Anthony. "Autonomy, Liberalism, and State Neutrality." *The Philosophical Quarterly* (1990) 40 (160): 433–452.

Massachusetts Institute of Technology. Sloan School of Business. "Getting Started: 10 Questions for Cities and Towns Considering Residential Curbside Composting." (website), 2014. Available at http://mitsloan.mit.edu/actionlearning/media/documents/s-lab-projects/Guide-to-Composting.pdf (accessed September 22, 2014).

Matulka, Rebecca. "History of the Electric Car." *Energy.Gov.* (website), September 15, 2014. Available at http://energy.gov/articles/history-electric-car (accessed June 4, 2015).

McFadden, Steven. "Large Scale Community Farms." In Trauger Groh and Steven McFadden, eds. *Farms of Tomorrow Revisited: Community Supported Farms—Farm Supported Communities.* Kimberton, PA: Biodynamic and Gardening Association, 1997.

McLaren, Duncan. "Where's the Justice in Geoengineering?" *The Guardian.* March 14, 2015. Available at https://www.theguardian.com/science/political-science/2015/mar/14/wheres-the-justice-in-geoengineering (accessed July 5, 2016).

"Meet the Frackers—Part 1." *60 Minutes New Zealand* (Television series episode), March 18, 2012. Available at https://www.youtube.com/watch?v=K-aN5DatUmE (accessed August 22, 2016).

Mertz, Zach. "Rock Salt," *YouTube* (online video), December 15, 2012. Available at https://www.youtube.com/watch?v=1202ahpjeCw (accessed July 2, 2015).

Mertz, Zach. "Why You Shouldn't Use Salt to Melt Ice." *Boston Globe*. December 24, 2015. Available at http://www.boston.com/yourtown/news/weymouth/2012/12/salt_in_the_wound_why_you_shou.html (accessed July 19, 2016).

Mihaiescu, Tania and Radu Mihaiescu. "Traffic Noise Pollution in Urban Areas: Bistrita Town." *ProEnvironment* (2009) 2: 222–225.

Mill, John Stuart. *Utilitarianism*. Oskar Priest, ed. Indianapolis, IN: Bobs-Merrill, 1957.

Missouri Department of Conservation. "Kudzu Control." (website), 2015. Available at http://mdc.mo.gov/your-property/problem-plants-and-animals/invasive-plants/kudzu-control (accessed July 7. 2016).

Molnar, Dani. "Beat It, Beet Juice." *The Municipal*. (website), January 13, 2013. Available at http://www.themunicipal.com/2013/01/beat-it-beet-juice/ (accessed July 1, 2015).

Monaghan, Nicholas J. "'Drill, Baby, Drill!': The Arctic National Wildlife Refuge and America's Energy Reckoning." *Notre Dame Journal of Law, Ethics & Public Policy* (2009) 23: 649–672.

Monbiot, George. *Feral: Rewilding the Land, the Sea, and Human Life*. Chicago: University of Chicago Press, 2014.

Montmarque, James. *The Idea of Agrarianism: From Hunter-Gatherer to Agrarian Radical in Western Culture*. Moscow, ID, University of Idaho Press, 1989.

Mooalem, Jon. "The Afterlife of Cellphones." *The New York Times Magazine*. January 13, 2008, pp. 40–41.

Moyers, Bill. "Toxic E-Trash: Recycling Facts and Laws." *NOW* (Television news transcript), July 7, 2001. Available at http://www.pbs.org/now/science/ecycling.html (accessed March 6, 2015).

Murray, Dale. "A Lockean Argument against Gene Patenting." *Business and Professional Ethics Journal* (2001) 20(3&4): 129–143.

Myers, Ransom and Boris Worm. "Rapid Worldwide Depletion of Predatory Fish Communities." *Nature* (2003) 423: 280–283.

"NOPD Announces Steps to Reduce Violent Crime." *Eyewitness News 4—WWL* (Television series episode), January 7, 2015. Available at https://www.youtube.com/watch?v=Mw7t9ZXjXMo (accessed June 26, 2015).

NSF International, "Mission." (website), 2015. Available at http://www.nsf.org (accessed April 22, 2015).

Nabuurs, G.J., O. Masera, K. Andrasko, P. Benitez-Ponce, R. Boer, M. Dutschke, E. Elsiddig, J. Ford-Robertson, P. Frumhoff, T. Karjalainen, O. Krankina, W.A. Kurz, M. Matsumoto, W. Oyhantcabal, N.H. Ravindranath, M.J. Sanz Sanchez, X. Zhang.

"Forestry." In B. Metz, O.R. Davidson, P.R. Bosch, R. Dave, L.A. Meyer, eds. *Climate Change 2007: Mitigation. Contribution of Working Group III to the Fourth Assessment Report of the Intergovernmental Panel on Climate Change*. Cambridge, U.K., and New York: Cambridge University Press, 2007.

Naess, Arne. "The Shallow and the Deep: Long-Range Ecology Movements." *Inquiry* (1973) 16: 95–100.

Nash, Kate. "2 Cockfighting Bans May Die." *Albuquerque Journal*. February 23, 2005. Available at http://www.abqjournal.com/xgr/308380xgr02-23-05.htm (accessed July 14, 2015).

National Aeronautics and Space Administration. "Tropical Deforestation." *Earth Observatory*. (website), 2015. Available at http://earthobservatory.nasa.gov/Features/Deforestation/deforestation_update3.php (accessed July 6, 2016).

National Aeronautics and Space Administration. Jet Propulsion Laboratory. "Near-Earth Objects and Life on Earth." (website), 2015. Available at http://neo.jpl.nasa.gov/neo/life.html (accessed July 21, 2016).

National Aeronautics and Space Administration. Jet Propulsion Laboratory. "What Is A PHA?" (website), 2015. Available at http://neo.jpl.nasa.gov/neo/groups.html (accessed July 9, 2015).

National Geographic. "Deforestation." (website), 2015. Available at http://environment.nationalgeographic.com/environment/global-warming/deforestation-overview/ (accessed July 6, 2016).

National Geographic. *TEDx Deextinction/National Geographic* (online video), 2013. Available at http://longnow.org/revive/tedxdeextinction/ (accessed December 20, 2014).

National Institutes of Health. "What are Overweight and Obesity?" (website), July 13, 2012. Available at http://www.nhlbi.nih.gov/health/health-topics/topics/obe (accessed July 24, 2015).

National Oceanic and Atmospheric Administration. "Invasive Lionfish Threaten Coral Reefs and Fisheries." (website), December 15, 2014. Available at http://www.nmfs.noaa.gov/stories/2014/12/12_01_14impacts_of_invasive_lionfish.html (accessed June 24, 2015).

National Oceanic and Atmospheric Administration. PMEL Carbon Program. "Ocean Acidification: The Other Carbon Dioxide Problem." (website), 2015. Available at http://www.pmel.noaa.gov/co2/story/Ocean+Acidification (accessed March 11, 2015).

National Oceanic and Atmospheric Administration. "Pacific Bluefin Tuna: Overview." *Fishwatch US Seafood Facts*. (website), February 24, 2015. Available at http://www.fishwatch.gov/seafood_profiles/species/tuna/species_pages/pac_bluefin_tuna.html (accessed May 18, 2015).

National Oceanic and Atmospheric Administration. PMEL Carbon Program. "What is Ocean Acidification? The Chemistry." (website), 2015. Available at http://www.pmel.noaa.gov/co2/story/What+is+Ocean+Acidification%3F (accessed March 11, 2015).

National Public Radio. "A Shadow Economy Lurks in an Electronics Graveyard." (Radio transcript), January 4, 2015. Available at http://www.npr.org/2015/01/04/374780916/a-shadow-economy-lurks-in-an-electronics-graveyard (accessed February 28, 2015).

Natural Resources Foundation of Wisconsin. "Welcome Whoopers." Fact sheet. 2001.

Nelsen, Eleanor. "Will Recycling Phosphorus Help Stop Algae Blooms?" *KQED Science*. (website), October 7, 2014. Available at http://science.kqed.org/quest/2014/10/07/will-recycling-phosphorus-help-stop-algae-blooms/ (accessed January 23, 2015).

Nesper, Larry. *The Walleye War*. Lincoln, NE: University of Nebraska Press, 2002.

"New Details Released in 'Big Blue' Cockfighting Operation." *WSAZ News 3*. (website), October 10, 2014. Available at http://www.wsaz.com/home/headlines/Floyd-County-Ky-Couple-to-be-Sentenced-in-Cockfighting-Case-278682911.html (accessed July 8, 2016).

New York City Department of Sanitation. "Food Scraps + Food Waste." (website), 2015. Available at http://www1.nyc.gov/assets/dsny/zerowaste/residents/food-scraps-and-yard-waste.shtml (accessed July 16, 2016).

New Zealand Parliamentary Commissioner for the Environment, "Drilling for Oil and Gas in New Zealand: Environmental Oversight and Regulation." (website), June 2014. Available at http://www.pce.parliament.nz/assets/Uploads/PCE-OilGas-web.pdf (accessed April 10, 2015).

New Zealand Parliamentary Commissioner for the Environment. "Drilling For Oil and Gas in New Zealand: Oversight and Regulation." (online video), June 2014. Available at http://www.pce.parliament.nz/publications/all-publications/drilling-for-oil-and-gas-in-new-zealand-environmental-oversight-and-regulation/ (accessed April 8, 2015).

Newkirk, Ingrid. "Let's Face It: There is No Such Thing as Humane Meat." *Huffington Post*. February 26, 2013. Available at http://www.huffingtonpost.com/ingrid-newkirk/humane-meat_b_2765996.html (accessed June 24, 2015).

Newkirk, Ingrid. "PETA and Co-Founder Ingrid Newkirk at Animal Rights National Conference 2014." *YouTube* (online video), August 26, 2014. Available at https://www.youtube.com/watch?v=EusaXzYThZ4 (accessed June 24, 2015).

Nixon, Rob. "Congress Blocks New Rules on School Lunches." *The New York Times*. November 15, 2011. Available at http://www.nytimes.com/2011/11/16/us/politics/congress-blocks-new-rules-on-school-lunches.html?_r=0 (accessed January 7, 2015).

Noise Pollution Clearinghouse. "Mission." (website), 2014. Available at http://www.nonoise.org (accessed July 20, 2016).

Norton, Bryan. "Environmental Ethics and Weak Anthropocentrism." *Environmental Ethics* (1984) 6(2): 131–148.

Notaras, Mark. "Does Climate Change Cause Conflict?" *Our World.* (website), November 27, 2009. Available at http://ourworld.unu.edu/en/does-climate-change-cause-conflict (accessed June 12, 2016).

Novak, Ben. "How to Bring Passenger Pigeons All the Way Back." *TEDx De-extinction/National Geographic* (online video), 2013. Available at http://longnow.org/revive/tedxdeextinction/ (accessed December 2014).

Nozick, Robert. *Anarchy, State, and Utopia.* New York: Basic Books, 1974.

Okeke, Jonathan C. and Chris O. Akpan, "An Inquiry into the Moral Question of Xeno-Transplantation," *Online Journal of Health Ethics* (2012) 8(1): 1–13. Available at http://aquila.usm.edu/cgi/viewcontent.cgi?article=1103&context=ojhe (accessed July 12, 2016).

Olson, Steve. "The Danger of Space Junk." *The Atlantic.* July 1998. Available at http://www.theatlantic.com/magazine/archive/1998/07/the-danger-of-space-junk/306691/ (accessed July 22, 2016).

Omang, Joanne. "In the Tropics, Still Rolling Back the Forest Primeval." *Smithsonian* (1987) 17(12): 56.

O'Neill, John. *Ecology, Policy and Politics.* London: Routledge, 1993.

O'Neill, John. "The Varieties of Intrinsic Value." *Monist* (1992) 75: 119–137.

O'Neill, Onora. "Environmental Values, Anthropocentrism, and Speciesism." *Environmental Values* (1997) 6(2): 127–142.

Operation Migration. "Proposed Sandhill Crane Hunt." (website), 2013. Available at http://operationmigration.org/InTheField/2013/07/24/proposed-sandhill-crane-hunt/ (accessed February 7, 2015).

Organ Farm: A History of Xenotransplantation Experiments. Dir. Frank Simmonds. film. Frontline Films, 2001.

Organization for American Historians, "Through the Eye of Katrina: Past as Prolouge?" *The Journal of American History* (special issue), (2007) 94: 693–876. Available at http://journalofamericanhistory.org/projects/katrina/resources/levee.html) (accessed June 25, 2015).

Ostrom, Marcia. "Community Farm Coalitions." In Trauger Groh and Steven McFadden. eds. *Farms of Tommorrow Revisited: Community Supported Farms—Farm Supported Communities.* Kimberton, PA: Biodynamic and Gardening Association, 1997.

Pacelle, Wayne. "Weak Laws Bring Cockfighters, Mafia Flocking into Kentucky." *Huffington Post.* (website), May 8, 2014. Available at http://www.huffingtonpost.com/wayne-pacelle/weak-laws-bring-cockfight_b_5290848.html (accessed June 10, 2015).

Palmer, Lindsay. "Earthshaking Moms: Mothers Across the Country are Jumping for Eco Joy." *Redbook.* March 19, 2009. Available at http://www.redbookmag.com/life/mom-kids/advice/a4888/eco-mom-alliance/ (accessed July 3, 2015).

Panikkar, Bindu, Natasha Smith, and Phil Brown. "Reflexive Research Ethics in Fetal Tissue Xenotransplantation Research." *Accountability in Research* (2012) 19: 344–369.

Pasquinelli, Sara N. "To Drill or Not to Drill: The Arctic National Wildlife Refuge v. the 'Need' for U.S. Energy Independence." *Golden Gate University Law Review* (2010) 33: 554–555.

Passmore, John. *Man's Responsibility for Nature*. New York: Scribner, 1974.

Pearson, Thomas W. "Frac Sand Mining in Wisconsin: Understanding Emerging Conflicts and Community Organizing." *Culture, Agriculture, Food and Environment* (2013) 35(1): 30–40.

Penn State University Extension. Colleges of Natural Sciences. "Water Facts 9: Bottled Water." 2015. Available at http://extension.psu.edu/natural-resources/water/drinking-water/water-testing/pollutants/bottled-water (accessed April 22, 2015).

People for the Ethical Treatment of Animals. "Hunting." (website), 2015. Available at http://www.peta.org/issues/animals-in-entertainment/cruel-sports/hunting/ (accessed June 10, 2015).

People for the Ethical Treatment of Animals. "Top Ten Veg-Friendly Major League Ballparks." (website), May 18, 2011. Available at http://www.peta.org/blog/top-10-vegetarian-friendly-major-league-ballparks/ (accessed August 20, 2014).

People for the Ethical Treatment of Animals. "Wayne Parnell Strips Down to Stop Circus Cruelty." (website), May 27, 2011. Available at http://www.petaindia.com/features/wayne-parnell-new-peta-ad/ (accessed June 24, 2015).

Pimentel, David, Rudolpo Zuniga, and Doug Morrison. "Update on the Economic and Environmental Costs Associated with Alien-Invasive Species in the United States." *Ecological Economics* (2005) 52: 273–288.

Pimm, Stuart. "The Case Against Species Revival." *National Geographic*. March 12, 2013. Available from http://news.nationalgeographic.com/news/2013/03/130312--deextinction-conservation-animals-science-extinction-biodiversity-habitat-envi ronment/ (accessed December 15, 2014).

Plumwood, Val. "Ecofeminism: An Overview and Discussion of Positions." *Australian Journal of Philosophy* (1986) 64: 120–38.

Poinar, Hendrik. "Not All Mammoths Were Wooly." *TEDx De-extinction/National Geographic* (online video), 2013. Available at http://longnow.org/revive/tedxde extinction/ (accessed December 2014).

Pollan, Michael. *Omnivore's Dilemma, A Natural History of Four Meals*. New York: Penguin Books, 2006.

Porter, Eduardo. "Are the Grandkids Worth the Money?" *The New York Times*. March 14, 2008. Available at http://www.nytimes.com/2008/03/14/opinion/14fri4.html?_r=0 (accessed June 9, 2016).

Posnick, Lauren and Henry Kim. "Bottled Water Regulation and the FDA." *United States Food and Drug Administration, Center for Food Safety and Nutrition*. (website), 2002. Available at http://vm.cfsan. fda.gov/~dms/botwatr.html (accessed April 28, 2007).

Posnick, Lauren and Henry Kim. "February/March 2002 Ask the Regulators—Bottled Water Regulation and the FDA." *United States Food and Drug Administration.* (website), 2002. Available at http://www.fda.gov/Food/FoodborneIllnessContaminants/BuyStoreServeSafeFood/ucm077079.htm (accessed April 24, 2015).

Powers, Ann. "Sea-Level Rise and its Impact on Vulnerable States: Four Examples." *Louisiana Law Review* (2012) 73(1): 151–301. Available at http://digitalcommons.pace.edu/cgi/viewcontent.cgi?article=1863&context=lawfaculty (accessed February 24, 2015).

Prengamen, Kate. "Frac Sand Boom Creates Thousands of Jobs." *Wisconsin Watch.* (website), August 19, 2014. Available at http://wisconsinwatch.org/2012/08/sand-boom-creates-jobs/ (accessed Oct. 31, 2014).

Rastogi, Nina. "Salting the Earth: Does Road Salt Harm the Environment." *Slate.* (website), February 16, 2010. Available at http://www.slate.com/articles/health_and_science/the_green_lantern/2010/02/salting_the_earth.html (accessed July 1, 2015).

Rawls, John. *A Theory of Justice.* Cambridge, MA: Harvard University Press, 1971.

Reese, April. "Take Back the Night." *E* (2000) 11(3): 22–24.

Regan, Tom. *The Case for Animal Rights.* Berkeley, CA: University of California Press, 1983.

Regents of the University of Michigan. "Global Deforestation." (website), 2015. Available at http://www.globalchange.umich.edu/globalchange2/current/lectures/deforest/deforest.html (accessed April 15, 2015).

Revkin, Andrew C. "Maybe Chicken Little Wasn't Paranoid After All." *The New York Times.* July 6, 2008, p. B2.

Rochon, Kateryn. "How to Properly Remove a Tick." *YouTube* (online video), June 25, 2013. Available at https://www.youtube.com/watch?v=27McsguL2Og (accessed June 10, 2015).

Rolston, Holmes. "Duties to Endangered Species." *Encyclopedia of Environmental Biology* (1995) 1: 517–528.

Rolston, Holmes. "Is There an Ecological Ethic?" *Ethics* (1975) 85: 93–105.

Ross, Brian. "Washington Keeps Cockfighting Legal in Three States." *ABC News.* (website), March 30, 2015. Available at http://abcnews.go.com/WNT/story?id=131142 (accessed July 14, 2015).

Ross, Philip. "Mini-Satellite 'Space Cops' Could Control Satellite Traffic and Help Prevent Collisions." *International Business Times.* January 25, 2014. Available at http://www.ibtimes.com/mini-satellite-space-cops-could-control-satellite-traffic-help-prevent-collisions-1548024 (accessed July 22, 2016).

Ross, Stephanie. "Review of *The Aesthetics of the Natural Environment.*" *Aesthetics.* (website), 2006. Available at http://www.aesthetics-online.org/reviews/index.php?reviews_id=44 (accessed April 29, 2015).

Roth, Cathy. "The Concept of Community Supported Agriculture." (website), 1999. Available at www.umass.edu/umext/csa/about.html (accessed December 22, 2004).

Rothblatt, Martine. *Your Life or Mine: How Geoethics Can Resolve the Conflict Between Public and Private Interests in Xenotransplantation*. Aldershot, U.K.: Ashgate, 2004.

Routley, Richard. "Do We Need a New, an Environmental Ethics?" In *Philosophy and Science, Morality and Culture, Technology and Man: Proceedings of the XVth World Congress of Philosophy*. Sofia, Bulgaria: Varna, 1973, pp. 205–210.

Rudd Center. "Analysis of Nielsen Data." (website), 2014. Available at http://www.sugarydrinkfacts.org/resources/SugaryDrinkFACTS_Report_Results.pdf (accessed January 7, 2015).

Sainsbury, Keith and Tom Polacheck, "The Use of Biological Reference Points for Defining Recruitment Overfishing, with an Application to Southern Bluefin Tuna." In *Population Dynamics for Fisheries Management*. Perth, WA: Australian Society for Fish Biology, 1994, pp. 265–274.

Saletan, William. "Making Manimals." *The Washington Post*. June 27, 2007. Available at http://www.washingtonpost.com/wp-dyn/content/article/2007/06/22/AR2007062201643.html (accessed July 12, 2016).

Sandel, Michael. *Justice: What's the Right Thing to Do?* New York: Farrar, Straus and Giroux, 2010.

Sandler, Ronald. *Character and Environment: A Virtue-Oriented Approach to Environmental Ethics*. New York: Columbia University Press, 2007.

Sandler, Ronald. "The Ethics of Reviving Long Extinct Species." *Conservation Biology* (2013) 28(2): 354–360.

Sandler, Ronald. *The Ethics of Species*. Cambridge: Cambridge University Press, 2012.

Sandler, Ronald and Phaedra C. Pezzullo. "Introduction." In Ronald Sandler and Phaedra C. Pezzullo, eds. *Environmental Justice and Environmentalism: The Social Justice Challenge to the Environmental Movement*. Cambridge, MA: MIT Press, 2007.

Sarnacki, Aislinn. "Maine Tick Myths, Revealed." *Bangor Daily News*. November 7, 2012. Available at http://bangordailynews.com/2012/11/07/outdoors/maine-tick-myths-revealed/ (accessed June 10, 2015).

Scallan, Niahm. "Melting Permafrost Called Ticking Timebomb." *Toronto Star*. December 1, 2011. Available at http://www.thestar.com/news/canada/2011/12/01/melting_permafrost_called_ticking_time_bomb.html (accessed June 1, 2015).

Schmidt, Charles W. "Noise That Annoys: Regulating Unwanted Sound." *Environmental Health Perspectives* (2005) 113(1): A42–A44.

Schneider, Stephen. "Geoengineering: Could—or Should—We Do It?" *Climatic Change* (1996) 33: 291–302.

Schwartz, James S.J. "Our Moral Obligation to Support Space Exploration." *Environmental Ethics* (2011) 33(1): 67–88.

"SDT Waste and Debris." *WWL TV News—New Orleans* (Television series episode), February 26, 2007. Available at https://www.youtube.com/watch?v=mQRoeDf9hqo (accessed June 26, 2015).

Seetharaman, Deepa and Edwin Clark. "Apple Poaching Auto Engineers to Build Battery Division: Lawsuit." *Reuters*. February 19, 2015. Available at http://www.reuters.com/article/2015/02/19/us-apple-autos-lawsuit-idUSKBN0LN04Y20150219 (accessed June 4. 2015).

Serafeim, George and Kathleen Perkins Miller. "Chief Sustainability Officers: Who They Are and What They Do?" In Rebecca Henderson, Ranjay Gulati, and Michael Tushman, eds. *Leading Sustainable Change: An Organizational Perspective*. New York: Oxford University Press, 2014, ch. 8.

Semple, Robert B. Jr. "Remember Kyoto? Most Nations Don't." *The New York Times*. December 3, 2011. Available at http://www.nytimes.com/2011/12/04/opinion/sunday/remember-kyoto-most-nations-dont.html?_r=0 (accessed February 17, 2015).

Shapiro, Beth. "Ancient DNA: What it is and What it Could Be." *TEDx De-extinction/National Geographic* (online video), 2013. Available from http://longnow.org/revive/tedxdeextinction/ (accessed December 2014).

Shepard, Kate. "Why Doesn't Your City Have Curbside Composting?" *Mother Jones*. September 10, 2012. Available at http://www.motherjones.com/environment/2012/09/why-doesnt-your-city-have-curbside-composting (accessed September 21, 2014).

Sherkow, Jacob and Henry T. Greely. "What if Extinction is Not Forever?" *Science* (2013) 340: 32–33.

Shrader-Fredette, Kristen. "Gene Patents and Lockean Constraints." *Public Affairs Quarterly* (2006) 20(2): 135–161.

"Sidney Torres Takes Crime Reality to TV." *The NOLA Defender*. January 12, 2015. Available at http://www.noladefender.com/content/sidney-torres-takes-crime-reality-tv-videos (accessed June 26, 2015).

Simberloff, Daniel, Sara Kuebbing, Martin A. Nunez, and Romina D. DiMarco. "Why Eating Invasive Species is a Bad Idea." *Ensia*. (website), September 9, 2014. Available at http://ensia.com/voices/why-eating-invasive-species-is-a-bad-idea/ (accessed June 23, 2015).

Simms, Richard. "'Bizarre Foods" TV Host Andrew Zimmern Samples Area Sandhill Cranes." *Nooga.com*. (website), December 12, 2013. Available at http://www.nooga.com/164635/bizarre-foods-tv-host-andrew-zimmern-samples-area-sandhill-cranes/ (accessed February 10, 2015).

Singer, Peter. "All Animals are Equal." *Philosophical Exchange* (1974) 1: 103–116.

Singer, Peter. *Animal Liberation*. New York: New York Review/Random House, 1975.

Singer, Peter and Jim Mason. *The Way We Eat: Why Out Food Choices Matter*. New York: Rodale, 2006.

Sissenwine, Michael P. and J.G. Shepherd. "An Alternative Perspective on Recruitment Overfishing and Biological Reference Points." *Canadian Journal of Fisheries and Aquatic Sciences* (1987) 44: 913–918.

Skumatz, Lisa. "Boulder Colorado Pilot Recycling Program Report." Skumatz Economic Research Associates. report. September 2007.

"Small Town in Arctic is Sinking in the Slush." *The Milwaukee Journal.* August 11, 1960. Available at https://news.google.com/newspapers?nid=1499&dat=19600811&id=jBQoAAAAIBAJ&sjid=4yUEAAAAIBAJ&pg=7329,415706&hl=en (accessed June 1, 2015).

Smithsonian Institute. "What is Coral? A Coral Polyp and Zooxanthllae." (website), 2015. Available at http://ocean.si.edu/ocean-photos/what-coral-coral-polyp-and-zooxanthellae (accessed July 3, 2016).

Sonnenblick, Karen, Sherry Klosiewski, and Beth B. Kienbaum. *A Closer Look at Whooping Cranes: Whooping Crane Education in Wisconsin and Eastern North America.* Madison, WI: Wisconsin Department of Natural Resources and Whooping Crane Eastern Partnership, 2008.

Soper, Kate, Martin Ryle, and Lyn Thomas. *The Politics and Pleasures of Consuming Differently.* London: Palgrave Macmillan, 2008.

Sparrow, Robert. "The Ethics of Terraforming." *Environmental Ethics* (1999) 21: 227–245.

Sparrow, Robert. "Transplantation, Consent, and International Justice." *Developing World Bioethics* (2009) 9(3): 119–127.

Stanford Environmental Law Society, *The Endangered Species Act: A Stanford Environmental Law Society Handbook.* Stanford, CA: Stanford University Press, 2001.

Starck, Jeff. "Walleye Wars Evolve into Peace on the Lakes." *Wausau Daily Herald.* June 21, 2011. Available at http://archive.wausaudailyherald.com/article/20110522/WOF16/105220714/Walleye-Wars-evolve-into-peace-lakes (accessed August 24, 2015).

Stehm, Tom. "Species Status and Fact Sheet: Whooping Crane." US Fish and Wildlife Service. (website), 2001. Available at http://www.fws.gov/northflorida/whoopingcrane/whoopingcrane-fact-2001.htm (accessed February 2, 2015).

Stockholm Environment Institute. "Climate Overview: Timor Leste (East Timor)." (website), 2008. Available at https://weadapt.org/knowledge-base/small-islands-and-climate-change/timor-leste-east-timor (accessed May 8, 2015).

Suput, Dorothy. *Community Supported Agriculture in Massachusetts: Status, Benefits and Barriers.* Masters Thesis. Medford MA: Tufts University, 1992.

Tagging of Pelagic Predators. "Pacific Bluefin Tuna." (website), May 18, 2015. Available at http://www.topp.org (accessed July 8, 2016).

Taylor, Paul. "The Ethics of Respect for Nature." *Environmental Ethics* (1981) 3: 197–213.

Taylor, Paul. *Respect for Nature.* Princeton, NJ: Princeton University Press, 1986.

Thaler, Richard T. and Cass Sunstein. *Nudge: Improving Decisions about Health, Wealth, and Happiness*. New Haven, CT: Yale University Press, 2008.

The Canadian Press. "Prime Minister Harper Visits Inuvik." *Global News*. (website), January 8, 2014. Available at http://globalnews.ca/news/1068681/harper-visits-inuvik-to-make-announcement/ (accessed June 2, 2015).

The Humane Society of the United States. "Cockfighting Fact Sheet." (website), March 9, 2011. Available at http://www.humanesociety.org/issues/cockfighting/facts/cockfighting_fact_sheet.html?credit=web_id80597821 (accessed June 10, 2015).

The Island President. Dir. Jon Shenk. film. Actual Films, 2012.

The National Academy of Sciences. "Climate Intervention: Carbon Dioxide Removal and Reliable Sequestration—Reflecting Sunlight to Remove Earth." *Report in Brief*. (website), 2015. Available at http://dels.nas.edu/resources/static-assets/materials-based-on-reports/reports-in-brief/climate-intervention-brief-final.pdf (accessed March 26, 2015).

The Pew Charitable Trusts. "Rebuilding Pacific Bluefin Tuna: A Science-Based Blueprint." (website), July 11, 2014. Available at http://www.pewtrusts.org/en/about/news-room/news/2014/07/14/rebuilding-pacific-bluefin-tuna (accessed May 20, 2015).

The Price of Sand. Dir. Jim Tittle. film. Actual Films, 2013.

The Royal Society. "Geoengineering the Climate: Science, Governance, and Uncertainty." (policy document 10/09), September 2009. Available at https://royalsociety.org/~/media/Royal_Society_Content/policy/publications/2009/8693.pdf (accessed July 5, 2016).

Thesig, Chuck. "What is Sustainable Agriculture?" *The Land Stewardship Newsletter* (1989) 7(2): 9.

Thompson, Barton H. Jr. "Managing the Working Landscape." In Dale D. Goble, J. Michael Scott, and Frank W. Davis, eds. *The Endangered Species Act at Thirty: Renewing the Conservation Promise*. Vol. 1. Washington, D.C.: Island Press, 2006.

Thompson, Paul. "Agrarianism and the American Philosophical Tradition." *Agriculture and Human Values* (Winter 1990): 3–6.

Thrall, Liz. "Rain Gardens Stem Urban Stormwater Flows." *Environmental Science and Technology* (2006) 40:1093–1094.

Timor-Leste Ministry of Economy and Development. National Biodiversity Working Group. "Timor-Leste's Fourth Annual Report to the UN Convention on Biological Diversity." White paper. October 2011. Available at https://www.cbd.int/doc/world/tl/tl-nr-04-en.pdf (accessed June 17, 2016).

"Tioga County 'A Cautionary Tale' about Fracking Busts." *Washington Post*. January 16, 2015. Available at http://triblive.com/business/businessprojects/onthegrid/gas/7583745-74/gas-pennsylvania-county (accessed August 22, 2016).

"Torres Ad Demands Mayor Act to Protect FQ." *Eyewitness News—4WWL* (Television series episode), January 7, 2015. Available at https://www.youtube.com/watch?v=Mw7t9ZXjXMo (accessed June 26, 2015).

Tuck, Richard. *Free Riding*. Cambridge, MA: Harvard University Press, 2008.

"UN Report Says Climate Change Will Strongly Effect Timor-Leste's Development and Growth Challenges and Opportunities." *LDC News Service*. (website), September 16, 2013. Available at http://ldcnews.com/un-report-says-climate-change-will-strongly-effect-timor-lestes-development-and-growth-challenges-and-opportunities/ (accessed June 17, 2016).

"UNFCCC: Warsaw COP 'Pivotal Moment to Step Up Climate Action'." *Clean Tech Poland*. (website), November 9, 2013. Available at http://cleantech.cleantechpoland.com/?page=news&id=113&link=unfccc-warsaw-cop-pivotal-moment-to-step-up-climate-action- (accessed June 8, 2016).

United Nations Educational, Scientific, and Cultural Organization. United Nations Environmental Programme. Food and Agriculture Organization. *Tropical Forest Ecosystems*. Paris: UNESCO, 1978.

United Nations Industrial Development Organization, "What is CSR?" (website), 2015. Available at http://www.unido.org/en/what-we-do/trade/csr/what-is-csr.html (accessed June 5, 2015).

United States Army Corps of Engineers. Mississippi Valley Division. "MRC History." (website), 2015. Available at http://www.mvd.usace.army.mil/About/MississippiRiverCommission(MRC)/History.aspx (accessed July 16, 2015).

United States Congressional Budget Office. *The Impact of Ethanol Production on Food Prices and Greenhouse Gas Emissions*. (website), April 2009. Available at https://www.cbo.gov/sites/default/files/111th-congress-2009-2010/reports/04-08-ethanol.pdf (accessed July 6, 2016).

United States Department of Agriculture. "Community Supported Agriculture." (website), 2014. Available at http://www.nal.usda.gov/afsic/pubs/csa/csa.shtml (accessed April 27, 2015).

United States Department of Agriculture. Food and Nutritional Services. "Healthier School Day: Tools for Schools Serving Whole-Grain Rich." (website), 2014. Available at http://www.fns.usda.gov/healthierschoolday/tools-schools-whole-grain (accessed November 12, 2014).

United States Department of Energy. "Energy Efficiency Trends in Residential and Commercial Buildings." (website), October 2008. Available at http://apps1.eere.energy.gov/buildings/publications/pdfs/corporate/bt_stateindustry.pdf (accessed June 22, 2016).

United States Department of Energy. "Ethanol Vehicle Emissions." (website), December 16, 2014. Available at http://www.afdc.energy.gov/vehicles/flexible_fuel_emissions.html (accessed June 22, 2015).

United States Department of Health and Human Services. Health Resources and Services Administration. "Organ Procurement and Transplantation Network." (website), June 26, 2015. Available at http://optn.transplant.hrsa.gov (accessed June 29, 2015).

United States Department of the Interior. *Conserving Borderline Species: A Partnership between the United States and Canada*. Washington, DC: Government Printing Office, 2001. Available at https://www.fws.gov/endangered/esa-library/pdf/borderline_species01.pdf (accessed February 9, 2015).

United States Energy Information Administration. "Arctic Oil and Natural Gas Resources." (website), 2012. Available at http://www.eia.gov/todayinenergy/detail.cfm?id=4650 (accessed June 17, 2016).

United States Energy Information Administration. "Frequently Asked Questions." (website), 2012. Available at http://www.eia.gov/tools/faqs/faq.cfm?id=58&t=8 (accessed January 23, 2013).

United States Energy Information Administration. "Potential Oil Production from the Coastal Plain of the Arctic National Wildlife Refuge: Updated Assessment." (website), 2002. Available at http://pubs.usgs.gov/of/2005/1217/pdf/2005-1217.pdf (June 6, 2016).

United States Environmental Protection Agency. "2006 PAYT Programs." (website), 2006. Available at http://www.epa.gov/epawaste/conserve/tools/payt/states/06comm.htm (accessed September 22, 2014).

United States Environmental Protection Agency. "Cadmium Compounds (A)." (website), 2000. Available at http://www.epa.gov/airtoxics/hlthef/cadmium.html (accessed February 28, 2015).

United States Environmental Protection Agency. "Combined Sewage Overflows." (website), December 15, 2014. Available at http://water.epa.gov/polwaste/npdes/cso/ (accessed May 13, 2015).

United States Environmental Protection Agency. "Impaired Waters and Total Maximum Daily Loads." (website), March 12, 2015. Available at http://water.epa.gov/lawsregs/lawsguidance/cwa/tmdl/ (accessed July 6, 2015).

United States Environmental Protection Agency. "Major Crops Grown in the United States." (website), April 11, 2013. Available at http://www.epa.gov/oecaagct/ag101/cropmajor.html (accessed July 24, 2015).

United States Environmental Protection Agency. "Reducing Food Wastes for Business." (website), July 8, 2016. Available at http://www.epa.gov/foodrecovery/ (accessed July 16, 2016).

United States Fish and Wildlife Service. "Arctic National Wildlife Refuge: Potential Impacts of Proposed Oil and Gas Development on the Arctic Refuge's Coastal Plain, Historical Overview and Issues of Concern." (website), 2001. Available at http://library.fws.gov/Pubs7/arctic-oilandgas-impact.pdf (accessed May 28, 2015).

United States Fish and Wildlife Service. "Foster a Land Ethic that Would Make Aldo Leopold Proud: Large Water Birds, An Identification Guide." *Newsroom*. (website), June 23, 2014. Available at http://www.fws.gov/midwest/news/730.html (accessed July 12, 2016).

United States Fish and Wildlife Service. "Karner Blue Butterfly (*Lycaeides Melissa Samuelis*)." (website), June 16, 2016. Available at http://www.fws.gov/midwest/endangered/insects/kbb/index.html (accessed July 12, 2016).

United States Fish and Wildlife Service. "Overview: Arctic National Wildlife Refuge." (website), 2015. Available at http://www.fws.gov/refuges/profiles/index.cfm?id=75600 (accessed May 26, 2015).

United States Fish and Wildlife Service. "The Cost of Invasive Species." (website), January 2012. Available at http://www.fws.gov/verobeach/PythonPDF/CostofInvasivesFactSheet.pdf (accessed November 14, 2014).

United States Food and Drug Administration. "FDA Regulates the Safety of Bottled Water Beverages." *Food Facts*. Fact sheet. Washington, DC: US Government Printing Office, 2009. Available at http://www.fda.gov/downloads/Food/FoodborneIllnessContaminants/UCM239631.pdf (accessed April 24, 2015).

United States Geological Survey. "Status Survey and Conservation Action Plan: Whooping Crane." *The Cranes*. Fact sheet. 2013. Available at http://www.npwrc.usgs.gov/resource/birds/cranes/grusamer.htm (accessed February 3, 2015).

United States House of Representatives. Committee on Natural Resources. "ANWR: Producing American Energy and Creating American Jobs." (website), February 15, 2015. Available at http://naturalresources.house.gov/anwr/ (accessed June 6, 2016).

United States National Atmospheric and Oceanic Administration Coral Reef Conservation Program. "Symbiotic Algae." (website), 2015. Available at http://coral-reef.noaa.gov/aboutcorals/coral101/symbioticalgae/ (accessed March 19, 2015).

University Corporation for Atmospheric Research. "Greenhouse Effect." (website), 2011. Available at http://scied.ucar.edu/longcontent/greenhouse-effect (accessed July 18, 2016).

University of Hawaii. Institute for Astronomy. "The Threat to Earth from Asteroids and Comets." (website), 2005. Available at http://pan-starrs.ifa.hawaii.edu/public/asteroid-threat/asteroid_threat.html (accessed July 9, 2015).

University of Wisconsin-Madison. Nutrient and Pest Management Program. "Phosphorus Movement from Land to Water." *Wisconsin Crop Management*. November, 4, 2010. Available at http://ipcm.wisc.edu/download/pubsNM/P_Movement_landwater.pdf (accessed January 23, 2015).

University of Wisconsin Sea Grant Institute Website. "Tips for Preventing the Spread of Invasive Species." (website), 2013. Available at http://seagrant.wisc.edu/Home/Topics/InvasiveSpecies/Details.aspx?PostID=1078 (accessed November 14, 2014).

Van En, Robyn. *Basic Formula to Create Community Supported Agriculture*. Great Barrington, MA: Indian Line Farm, 1992.

"Vegetable Juice Helps Beloit to Beet Out Snow and Ice." *WTVO and WQRF News* (Television series episode), January 30, 2014. Available at http://www.mystateline.com/fulltext-news/d/story/vegetable-juice-helps-beloit-to-beet-out-snow-and/12370/s8O1--rjnEy-ah-z8b2mBQ (accessed July 18, 2016).

Wall, Mike. "Space Junk Threat Demands Immediate Action, Experts Say." *Space.com*. (website), April 25, 2013. Available at http://www.space.com/20832-space-junk-cleanup-needed.html (accessed July 10, 2015).

Wall, Tim. "Cell Phones Toxic to Humans and Earth." *Discovery News*. (website), October 5, 2012. Available at http://news.discovery.com/human/cell-phones-toxic-121005.htm (accessed March 2, 2015).

Wang, Michael, May Wu, and Hong Huo. "Life Cycle Energy and Greenhouse Gas Emission Impacts of Different Corn Ethanol Plant Types." *Environmental Research Letters* (2007) 2: 1–13.

Wang, Xiu. "City's Organic Collection Program to Expand: Turn Organic Waste into Energy." *Madison Commons*. (website), Feb. 23, 2014. Available at http://madison commons.org/?q=content/city's-organics-collection-pilot-program-to-expand-turn-organic-waste-into-energy (accessed September 19, 2014).

Warren, Karen. *Ecofeminist Philosophy: A Western Perspective on What It is and Why It Matters*. Oxford, U.K.: Rowman & Littlefield, 2000.

Warren, Karen. "The Power and Promise of Ecological Feminism." *Environmental Ethics* (1987) 9: 3–20.

"We Can Do This Together: Demand a Plan From Our Mayor!" *YouTube* (online video), January 7, 2015. Available at https://www.youtube.com/watch?v=Mw7t9ZXjXMo (accessed July 17, 2015).

Weight of the Nation. Dir. Dan Chaykin. HBO documentary. 2012.

Weinreb. Ellen. "CSO Back Story: The Evolution of the Chief Sustainability Officer." White paper. 2014. Available at http://weinrebgroup.com/wp-content/uploads/2014/10/CSO-Back-Story-II.pdf (accessed June 3, 2015).

Weissman, Jordan. "Two Ways We Now Know the Fracking Boom is Causing Earthquakes." *The Atlantic*. August 27, 2013. Available at http://www.theatlantic.com/business/archive/2013/08/here-are-2-ways-we-now-know-the-fracking-boom-is-causing-earthquakes/279102/ (accessed June 6, 2016).

Westervelt, Amy. "Does Climate Change Really Cause Conflict?" *The Guardian*. March 9, 2015. Available at http://www.theguardian.com/vital-signs/2015/mar/09/climate-change-conflict-syria-global-warming (accessed May 7, 2015).

Whatley, Rick and Walter Bresette. *Walleye Warriors: An Effective Alliance Against Racism and for the Earth*. Philadelphia: New Society Publishers, 1994.

White, Lynn. "The Historical Roots of Our Ecological Crisis." *Science* (1967) 155(3767): 1203–1207.

White, Mel. "North American Birds Declining as Threats Mount." *National Geographic*. June 21, 2013. Available at http://news.nationalgeographic.com/news/2013/06/130621-threats-against-birds-cats-wind-turbines-climate-change-habitat-loss-science-united-states/ (accessed June 11, 2015).

"Why Are Gene Patents Controversial." *The Economist*. April 18, 2013. Available at http://www.economist.com/blogs/economist-explains/2013/04/economist-explains-why-gene-patents-controversial (accessed April 28, 2015).

Wilkinson, Charles F. "Treaty and Fishing Rights of the Wisconsin Chippewa," The Oliver Randall Lecture, Occasional Paper from the University of Wisconsin Law School, April 19, 1990.

Williams, Michael. *Deforesting the Earth: From Prehistory to Global Crisis*. Chicago: University of Chicago Press, 2002.

Wilson Center. "About Environmental Change and Security Program." (website), 2015. Available at http://wilsoncenter.org/about-ecsp (accessed May 6, 2015).

Wisconsin Department of Natural Resources. "Affirming the Effectiveness of Outreach, Education and Voluntary Participation in the Wisconsin Karner Blue Butterfly Habitat Conservation Plan." Peter Moreno and David Lentz. report. Madison, WI: Wisconsin Department of Natural Resources, 2003.

Wisconsin Department of Natural Resources. "Blue Green Algae." (website), 2014. Available at http://dnr.wi.gov/lakes/bluegreenalgae/ (accessed January 29, 2015).

Wisconsin Department of Natural Resources. "City of Waukesha Water Diversion Application." (website), May 2015. Available at http://dnr.wi.gov/topic/wateruse/waukeshadiversionapp.html (accessed May 15, 2015).

Wisconsin Department of Natural Resources. "City of Waukesha Water Diversion Application: Background." (website), January 6, 2016. Available at http://dnr.wi.gov/topic/WaterUse/waukesha/background.html (accessed June 7, 2016).

Wisconsin Department of Natural Resources. "Silica Sand Mining in Wisconsin." (website), January 2012. Available at http://dnr.wi.gov/topic/mines/documents/silicasandminingfinal.pdf (accessed June 3, 2016).

Wisconsin Department of Natural Resources. "The Karner Blue Butterfly Habitat Conservation Plan Annual Report of Activities for the Calendar Year 2009." David Lentz. (website), March 24, 2011. Available at http://dnr.wi.gov/topic/ForestPlanning/documents/2008AnnualReport.pdf (accessed July 12, 2016).

Wis. Stat. § 23.22.

Wis. Stat. § 281.346(1)(tm).

Wis. Stat. § 281.348(3).

Worster, Donald. *Nature's Economy: A History of Ecological Ideas*. Cambridge, U.K.: Cambridge University Press, 1985.

Wright, Laura. "Thawing Permafrost Sinks Building, Hikes Costs in North." *Canadian Broadcasting Corporation News*. (website), November 18, 2011. Available at http://www.cbc.ca/news/canada/north/thawing-permafrost-sinks-buildings-hikes-costs-in-north-1.1108686 (accessed June 1, 2015).

Yeang, Ken. "What is Green Design and Planning? Five Propositions." In Andy van den Dobbelsteen, Machiel van Dorst, and Arjan van Timmeren, eds. *Smart Building in a Changing Climate*. Amsterdam: Techne Press, 2009.

Zimmer, Carl. "(Some) Extinction is (Not Necessarily) Forever." *TEDx De-extinction/ National Geographic* (online video), 2013. Available at http://longnow.org/revive/ tedxdeextinction/ (accessed December 12, 2014).

Zuckerman, Gregory. "Breakthrough: The Accidental Discovery That Revolutionized American Energy." *The Atlantic*. November 6, 2013.

Index

www.ingramcontent.com/pod-product-compliance
Lightning Source LLC
Chambersburg PA
CBHW031318230926
51848CB00001B/8
9789004339989